Lecture Notes in Computer Science

Lecture Notes in Artificial Intelligence 15875

Founding Editor

Jörg Siekmann

The series Lecture Notes in Artificial Intelligence (LNAI) was established in 1988 as a topical subseries of LNCS devoted to artificial intelligence.

The series publishes state-of-the-art research results at a high level. As with the LNCS mother series, the mission of the series is to serve the international R & D community by providing an invaluable service, mainly focused on the publication of conference and workshop proceedings and postproceedings.

Xintao Wu · Myra Spiliopoulou · Can Wang ·
Vipin Kumar · Longbing Cao · Xiangmin Zhou ·
Guansong Pang · Joao Gama
Editors

Data Science: Foundations and Applications

29th Pacific-Asia Conference
on Knowledge Discovery and Data Mining, PAKDD 2025
Sydney, NSW, Australia, June 10-13, 2025
Proceedings, Part VI

Springer

Editors
Xintao Wu
University of Arkansas
Fayetteville, AR, USA

Myra Spiliopoulou
Otto-von-Guericke-University Magdeburg
Magdeburg, Germany

Can Wang
Griffith University
Nathan, QLD, Australia

Vipin Kumar
University of Minnesota
Minneapolis, MN, USA

Longbing Cao
Macquarie University
Macquarie Park, NSW, Australia

Xiangmin Zhou
RMIT University
Melbourne, VIC, Australia

Guansong Pang
Singapore Management University
Singapore, Singapore

Joao Gama
University of Porto
Porto, Portugal

ISSN 0302-9743 ISSN 1611-3349 (electronic)
Lecture Notes in Artificial Intelligence
ISBN 978-981-96-8294-2 ISBN 978-981-96-8295-9 (eBook)
https://doi.org/10.1007/978-981-96-8295-9

LNCS Sublibrary: SL7 – Artificial Intelligence

This Springer imprint is published by the registered company Springer Nature Singapore Pte Ltd.
The registered company address is: 152 Beach Road, #21-01/04 Gateway East, Singapore 189721, Singapore

General Chairs' Preface

On behalf of the Organizing Committee, we are delighted to welcome all participants to the 29th Pacific-Asia Conference on Knowledge Discovery and Data Mining (PAKDD 2025). Since the first PAKDD conference in 1997, PAKDD has established itself as one of the leading international conferences on data mining and knowledge discovery. PAKDD provides an international forum for researchers and industry practitioners to share their new ideas, original research results, and practical development experiences across all areas of data mining. This year, after its three previous editions in Sydney (2004), Gold Coast (2013), and Melbourne (2018), PAKDD was held in Australia for the fourth time in the fascinating city of Sydney, during June 10–13, 2025.

We extend our sincere gratitude to the researchers who submitted their research work to the PAKDD 2025 conference. The main conference of PAKDD 2025 had three tracks, Main Track, Survey Track, and Special Track on LLMs for Data Science. This year we also organized a Special Session on Data Science Foundations and Applications (DSFA). The conference program was further enriched with ten high-quality tutorials, five workshops on cutting-edge topics, a doctoral consortium, an AI forum, and an AI school. We would like to express our sincere thanks to all authors for their efforts in research, as well as in preparing high-quality presentations. We also express our appreciation to all the collaborators and sponsors for their support, trust, and cooperation. We were honored to have three distinguished keynote speakers joining the conference: Qiang Yang (Hong Kong University of Science and Technology & Webank, China), Hiroshi Ishiguro (Osaka University, Japan), and Jon Whittle (Data61, Australia), each with high reputations in their respective areas. We enjoyed their participation and talks, which made the conference one of the best academic platforms for knowledge discovery and data mining.

We would like to express our sincere gratitude for the contributions of the members and reviewers in the Senior Program Committee and the Program Committee, led by program chairs Xintao Wu, Myra Spiliopoulou, and Can Wang, and track chairs including Jae-Gil Lee and Leo Zhang on the Survey Track, and Dongjin Song and Ernestina Menasalvas on the Special Track on LLMs for Data Science. It is through their untiring efforts that the conference had an excellent technical program. We are also thankful to the other Organizing Committee members: including Chairs on the DSFA Special Session: Xiangmin Zhou, Guansong Pang, Joao Gama, and Longbing Cao; Workshop Chairs Shuhan Yuan, Fragkiskos Malliaros, and Xin Zheng; Tutorial Chairs Hisashi Kashima, Xin Wang, and Chen Zhao; Publicity and Media Chairs Usman Naseem and Fei Zhou; Industrial Chairs Xiaoli Li and Lina Yao; Proceedings Chairs Yanqiu Wu and Yu Yao; Finance Chair Xiaohan Yu; and Web and Content Chairs Yanxiang Wang and Wenfeng Hou. We would like to particularly thank the Local Arrangement Team, Tongliang Liu (Chair), Xuhui Fan, Shoujin Wang, and Jiaojiao Jiang. The conference would not have been successful without their dedication, commitment, and efforts. We are also grateful to the PAKDD Steering Committee for their guidance, the hosting organizations, Macquarie University, and all our sponsors for continuously providing institutional and financial support to PAKDD 2025.

Finally, we hope that you found our program interesting and thought-provoking and will preserve a cherished memory of the conference and Sydney. Once again, our warmest welcome to you all!

June 2025

Vipin Kumar
Longbing Cao

PC Chairs' Preface

It is our great pleasure to present the 29th Pacific-Asia Conference on Knowledge Discovery and Data Mining (PAKDD 2025) as Program Committee Chairs. PAKDD is one of the longest-established and leading international conferences in the areas of data mining and knowledge discovery. It provides an international peer-reviewed forum for presenting and sharing new ideas, original research results, and practical development experiences from all data science areas, including data mining, data warehousing, machine learning, artificial intelligence, databases, statistics, knowledge engineering, visualization, decision-making systems, and emerging applications.

The PAKDD 2025 technical program featured three tracks, the Main Track, the Survey Track, and a Special Track on Large Language Models in Data Science. The Main Track continued its tradition of being the premier forum for the presentation of research results and experience reports on knowledge discovery, data science, and machine learning. The Survey Track was introduced in 2025 for the first time, to promote the dissemination of insightful survey papers: survey papers are intended to provide a structured synthesis of a particular topic in the area of data mining, including but not limited to theoretical foundations of mining, inference, and learning, big data technologies, as well as security, privacy, and integrity, for the perusal of junior researchers and of experts from other research fields. As the rapid advancements in Large Language Models (LLMs) have opened new avenues for innovation and research across various domains, particularly in the field of data science, we introduced the Special Track on LLMs, which aimed to explore the transformative potential of LLMs for data science, bringing together researchers, practitioners, and industry experts to discuss the latest developments, challenges, and opportunities in this rapidly growing area.

This year's conference was highly competitive. The three tracks received 696 submissions with 557 to the Main Track, 35 to the Survey Track, and 104 to the Special Track on LLMs. The selection process was competitive, rigorous, and thorough, involving 345 PC and 94 Senior PC (SPC) members. Each submission was double-blindly reviewed by at least 3 reviewers with many receiving 4 or more reviews, followed by a discussion, and then further evaluated by a designated SPC member, the Track chairs, and the PC program co-chairs. The final selection considered multiple factors such as novelty, technical depth, experimental validation, and relevance to the conference. We accepted 134 papers in the Main Track, 10 in the Survey Track, and 24 in the LLM Track, thus having an acceptance rate of approximately 25% for the main conference. Considering the high quality of some of the rejected submissions, we decided to introduce a new Special Session on Data Science – Foundations and Applications (DSFA), to which we transferred 68 of the original submissions. All papers accepted to the main conference program and to the DSFA Special Session were presented orally during the conference. We would like to thank the Chairs of the Survey Track and the LLM Track, the SPC members, PC members, and external reviewers for their invaluable efforts in ensuring a timely, fair, and highly effective paper review and selection procedure. Their diligence, dedication, and expertise helped produce the high-quality program for PAKDD 2025.

In addition to the exciting technical presentations, our program also featured three outstanding keynotes, invited industry talks, ten high-quality tutorials, five workshops, and a doctoral consortium. We hope our PAKDD 2025 program provided a wonderful journey for researchers, users, practitioners, and developers.

June 2025

Xintao Wu
Myra Spiliopoulou
Can Wang

Organization

Organizing Committee

General Chairs

Vipin Kumar	University of Minnesota, USA
Longbing Cao	Macquarie University, Australia

Program Chairs

Xintao Wu	University of Arkansas, USA
Myra Spiliopoulou	Otto von Guericke University Magdeburg, Germany
Can Wang	Griffith University, Australia

Local Arrangement Chairs

Tongliang Liu (Local Chair)	University of Sydney, Australia
Xuhui Fan (Local Co-chair)	Macquarie University, Australia
Shoujin Wang (Local Co-chair)	University of Technology Sydney, Australia
Jiaojiao Jiang (Local Co-chair)	University of New South Wales, Australia

Industry Track Chairs

Xiaoli Li	Agency for Science, Technology and Research, Singapore
Lina Yao	University of New South Wales, Australia

Survey/Vision Track Chairs

Jae-Gil Lee	Korea Advanced Institute of Science and Technology, South Korea
Leo Zhang	Griffith University, Australia

Special Track on Large Language Models for Data Science

Dongjin Song	University of Connecticut, USA
Ernestina Menasalvas	Universidad Politecnica de Madrid, Spain

Special Session: Data Science: Foundations and Applications (DSFA)

Emily Xiangmin Zhou	Royal Melbourne Institute of Technology University, Australia
Guansong Pang	Singapore Management University, Singapore
Joao Gama	University of Porto, Portugal
Longbing Cao	Macquarie University, Australia

Tutorial Chairs

Hisashi Kashima	Kyoto University, Japan
Xin Wang	University of Calgary, Canada
Chen Zhao	Baylor University, USA

Workshop Chairs

Shuhan Yuan	Utah State University, USA
Fragkiskos Malliaros	Paris-Saclay University, France
Xin Zheng	Griffith University, Australia

Doctoral Consortium Chairs

Jun Shen	University of Wollongong, Australia
Krishna Reddy	International Institute of Information Technology Hyderabad, India
Juhua Hu	University of Washington, USA

Publicity and Media Chairs

Usman Naseem	Macquarie University, Australia
Fei Zhou	Shenzhen University, China

Finance Chair

Xiaohan Yu	Macquarie University, Australia

Proceeding Chairs

Autumn Wu	Macquarie University, Australia
Yu Yao	University of Sydney, Australia
Zhangkai Wu	Macquarie University, Australia

Web and Content Chairs

Yanxiang Wang	University of New South Wales, Australia
Wenfeng Hou	Macquarie University, Australia

Steering Committee

Chair

Longbing Cao	Macquarie University, Australia

Vice Chair

Gill Dobbie	University of Auckland, New Zealand

Treasurer

Gill Dobbie	University of Auckland, New Zealand

Members

Ramesh Agrawal	Jawaharlal Nehru University, India
Gill Dobbie	University of Auckland, New Zealand
Joao Gama	University of Porto, Portugal
Zhiguo Gong	University of Macau, China
Hisashi Kashima	Kyoto University, Japan
Hady W. Lauw	Singapore Management University, Singapore
Jae-Gil Lee	KAIST, South Korea
Dinh Phung	Monash University, Australia
Kyuseok Shim	Seoul National University, South Korea
Geoff Webb	Monash University, Australia
Raymond Chi-Wing Wong	Hong Kong University of Science and Technology, China
De-Nian Yang	Academia Sinica, Taiwan
Min-Ling Zhang	Southeast University, China

Life Members

Longbing Cao	Macquarie University, Australia
Ming-Syan Chen	National Taiwan University, Taiwan
David Cheung	University of Hong Kong, China
Joshua Z. Huang	Shenzhen Institutes of Advanced Technology, Chinese Academy of Sciences, China
Masaru Kitsuregawa	University of Tokyo, Japan
Rao Kotagiri	University of Melbourne, Australia
Ee-Peng Lim	Singapore Management University, Singapore
Huan Liu	Arizona State University, USA
Hiroshi Motoda	Osaka University, Japan
Jian Pei	Duke University, USA
P. Krishna Reddy	International Institute of Information Technology Hyderabad, India
Jaideep Srivastava	University of Minnesota, USA
Thanaruk Theeramun Kong	Thammasat University, Thailand
Tu-Bao Ho	Japan Advanced Institute of Science and Technology, Japan
Vincent S. Tseng	National Yang Ming Chiao Tung University, Taiwan
Takashi Washio	Osaka University, Japan
Kyu-Young Whang	Korea Advanced Institute of Science & Technology, South Korea
Graham Williams	Australian National University, Australia
Chengqi Zhang	University of Technology Sydney, Australia
Ning Zhong	Maebashi Institute of Technology, Japan
Zhi-Hua Zhou	Nanjing University, China

Past Members

Arbee L. P. Chen	Asia University, Taiwan
Hongjun Lu	Hong Kong University of Science and Technology, China
Takao Terano	Tokyo Institute of Technology, Japan

Senior Program Committee

Albert Bifet	Télécom Paris, France
Allan Tucker	Brunel University London, UK
Arianna Dagliati	University of Pavia, Italy

Aris Anagnostopoulos	Sapienza University of Rome, Italy
Arlei Silva	Rice University, USA
Bin Cui	Peking University, China
Bo Han	Hong Kong Baptist University, China
Bo Liu	University of Technology Sydney, Australia
Byung Suk Lee	University of Vermont, USA
Chang-Tien Lu	Virginia Tech, USA
Chao Chen	RMIT University, Australia
Chao Lan	University of Oklahoma, USA
Chen Gong	Nanjing University of Science and Technology, China
Christan Grant	University of Florida, USA
Christopher Quinn	Iowa State University, USA
De-Nian Yang	Academia Sinica, Taiwan
Depeng Xu	University of North Carolina at Charlotte, USA
Dongsheng Luo	Florida International University, USA
Eirini Ntoutsi	University of the Bundeswehr Munich, Germany
Elio Masciari	University of Naples Federico II, Italy
Feng Chen	UT Dallas, USA
Feng Zhou	Renmin University of China, China
Fenglong Ma	Pennsylvania State University, USA
Fuzhen Zhuang	Beihang University, China
Gang Li	Deakin University, Australia
Giovanna Nicora	University of Pavia, Italy
Guangdong Bai	University of Queensland, Australia
Guansong Pang	Singapore Management University, Singapore
Guoxin Su	University of Wollongong, Australia
Han-Jia Ye	Nanjing University, China
Hanghang Tong	University of Illinois Urbana-Champaign, USA
Hao Liao	Shenzhen University, China
Hao Wang	Sichuan University, China
Hisashi Kashima	Kyoto University, Japan
Hongzhi Yin	University of Queensland, Australia
Hua Wei	Arizona State University, USA
Huaming Chen	University of Sydney, Australia
Hui Tian	Griffith University, Australia
Jeffrey Chan	Royal Melbourne Institute of Technology University, Australia
Jessica Lin	George Mason University, USA
Jiayu Zhou	University of Michigan, USA
Jingrui He	University of Illinois Urbana-Champaign, USA

Xiaowei Jia	University of Pittsburgh, USA
Xindong Wu	Hefei University of Technology, China
Xuequan Lu	University of Western Australia, Australia
Xuhui Fan	Macquarie University, Australia
Xuyun Zhang	Macquarie University, Australia
Yanchang Zhao	Commonwealth Scientific and Industrial Research Organization, Australia
Yang Cao	Institute of Science Tokyo, Japan
Yanjun Zhang	University of Technology Sydney, Australia
Yifeng Zheng	Hong Kong Polytechnic University, China
Ying Li	AwanTunai, Singapore
Yingxia Shao	Beijing University of Posts and Telecommunications, China
Yiqun Xie	University of Maryland, USA
Yongkai Wu	Clemson University, USA
Yue Xu	Queensland University of Technology, Australia
Zhe Jiang	University of Florida, USA
Zhongfei Zhang	Binghamton University, SUNY, USA

Program Committee

Aditya Malusare	Purdue University, USA
Akihiro Yamaguchi	Kyushu University, Japan
Akira Sakai	Fujitsu Laboratories Ltd., Japan
Alessandro Berti	University of Pisa, Italy
Alex Memory	Johns Hopkins University Applied Physics Laboratory, USA
Alfredo Cuzzocrea	Università della Calabria, Italy
Alycia Carey	University of Arkansas, USA
Amir Gandomi	University of Technology Sydney, Australia
Andrzej Skowron	University of Warsaw, Poland
Aneesh Komanduri	University of Arkansas, USA
Aofei Chang	Pennsylvania State University, USA
Arjun Roy	University of the Bundeswehr Munich, Germany
Arnaud Soulet	University of Tours, France
Arun Maiya	Institute for Defense Analyses, USA
Atsuhiro Takasu	National Institute of Informatics, Japan
Badrul Sarwar	CloudAEye, USA
Baodi Liu	China University of Petroleum (East China), China
Benjamin Halstead	University of Auckland, New Zealand

Bi-Ru Dai	National Taiwan University of Science and Technology, Taiwan
Bijay Prasad Jaysawal	National Cheng Kung University, Taiwan
Bijaya Adhikari	University of Iowa, USA
Bikash Chandra Singh	Islamic University, Bangladesh
Bo Li	Griffith University, Australia
Bo Peng	Ohio State University, USA
Bradley Malin	Vanderbilt University Medical Center, USA
Brendon Woodford	University of Otago, New Zealand
Bruno Veloso	University of Porto, Portugal
Buyue Qian	Xi'an Jiaotong University, China
Cam-Tu Nguyen	Nanjing University, China
Canyu Chen	Illinois Institute of Technology, USA
Carson Leung	University of Manitoba, Canada
Chandra N.	Rajiv Gandhi University of Knowledge Technologies, India
Chandra Mohan Dasari	Indian Institute of Information Technology Sri City, India
Chang Ge	University of Minnesota, USA
Chao Huang	University of Hong Kong, China
Chaoran Cui	Shandong University of Finance and Economics, China
Chaoxi Niu	University of Technology Sydney, Australia
Charini Nanayakkara	Australian National University, Australia
Chen Zhao	Baylor University, USA
Chengxi Zang	Cornell University, USA
Chengzhi Piao	Hong Kong Baptist University, China
Chenhan Zhang	Macquarie University, Australia
Chih-Chieh Hung	National Chung Hsing University, Taiwan
Chih-Hang Wang	National Chung Cheng University, Taiwan
Chih-Ya Shen	National Tsing Hua University, Taiwan
Chotirat Ratanamahatana	Chulalongkorn University, Thailand
Chowdhury Farhan Ahmed	University of Dhaka, Bangladesh
Chowdhury Mohammad Rakin Haider	Purdue University, USA
Chuan-Ju Wang	Academia Sinica, Taiwan
Chuan-Wei Kuo	National Yang Ming Chiao Tung University, Taiwan
Chuangyin Dang	City University of Hong Kong, China
Chul Ho Lee	Texas State University, USA
Chun Hao Chen	National Kaohsiung University of Science and Technology, Taiwan
Congrui Huang	Microsoft Research Asia, China

Dandan Lin	Shenzhen Institute of Computing Sciences, China
Daniele Pala	University of Pavia, Italy
Daokun Zhang	Santa Clara University, USA
Debanjan Datta	Virginia Tech, USA
Dheeraj Kodati	Apollo Computing Laboratories, USA
Dinusha Vatsalan	Macquarie University, Australia
Djellel Difallah	New York University, USA
Eduardo Ogasawara	Centro Federal de Educação Tecnológica Celso Suckow da Fonseca, Brazil
Egawati Panjei	Southern Methodist University, USA
Eliana Pastor	Politecnico di Torino, Italy
Emmanouil Panagiotou	Freie Universität Berlin, Germany
Erhu He	University of Pittsburgh, USA
Esra Akbas	Georgia State University, USA
Fan Wu	Central South University, China
Fanglan Chen	Virginia Tech, USA
Fatema Siddika	Iowa State University, USA
Fengxin Li	Renmin University of China, China
Florent Masseglia	Inria, France
Francesco Marcelloni	Università di Pisa, Italy
Fuwei Zhang	Beihang University, China
Giovanni Maria Biancofiore	Polytechnic University of Bari, Italy
Giuseppe Albi	Università degli Studi di Pavia, Italy
Giuseppe Manco	Institute for High-Performance Computing and Networking, Italy
Guimu Guo	Rowan University, USA
Guiquan Liu	University of Science and Technology of China, China
Guoliang He	Wuhan University, China
Guoyin Wang	Chongqing University of Posts and Telecommunications, China
Guozhong Li	King Abdullah University of Science & Technology, Saudi Arabia
Haibing Lu	Australian National University, Australia
Hang Yu	Shanghai University, China
Haoran Wang	University of Sheffield, UK
He Cheng	Utah State University, USA
Heitor Gomes	Victoria University of Wellington, New Zealand
Hirofumi Suzuki	Fujitsu Limited, Japan
Hong-Han Shuai	National Yang Ming Chiao Tung University, Taiwan
Hongjun Wang	Southwest Jiaotong University, China

Hongtao Wang	North China Electric Power University, China
Hongyang Chen	Zhejiang Lab, China
Hridoy Sankar Dutta	Deakin University, Australia
Huaxiang Zhang	Shandong Normal University, China
Huaxiong Li	Nanjing University, China
Hui Chen	Macquarie University, Australia
Huixin Zhan	Texas Tech University, USA
Ickjai Lee	James Cook University, Australia
Ilkay Yildiz Potter	BioSensics, USA
Ja-Hwung Su	National University of Kaohsiung, Taiwan
Jean Paul Barddal	Pontifical Catholic University of Paraná, Brazil
Jean-Paul	University of Bern, Switzerland
Jees Augustine	Microsoft Research, USA
Jeffrey Ullman	Stanford University, USA
Ji Xu	Guizhou University, China
Jia Chen	University of California Riverside, USA
Jian Kang	University of Rochester, USA
Jianwu Wang	University of Maryland, Baltimore County, USA
Jianzhong Qi	University of Melbourne, Australia
Jiaxin Jiang	National University of Singapore, Singapore
Jie Wang	Southwest Jiaotong University, China
Jin Li	University of Technology Sydney, Australia
Jin Liu	Shanghai Maritime University, China
Jin-Hui Wu	Nanjing University, China
Jingyuan Wang	Beihang University, China
Jinpeng Chen	Beijing University of Posts and Telecommunications, China
Jiuding Duan	Allianz Global Investors, Japan
Jonathan Zhu	Carnegie Mellon University, USA
Josep Domingo Ferrer	Universitat Rovira i Virgili, Spain
Jun-Ki Min	Korea University of Technology and Education, South Korea
Jungeun Kim	Inha University, South Korea
K. Selçuk Candan	Arizona State University, USA
Kai Shu	Illinois Institute of Technology, USA
Kaiqi Zhao	University of Auckland, New Zealand
Kaixin Wang	Beijing University of Technology, China
Kalidas Yeturu	Indian Institute of Technology Tirupati, India
Karuna Bhaila	University of Arkansas, USA
Ke Deng	RMIT University, Australia
Ken-ichi Fukui	Osaka University, Japan
Ki Yong Lee	Sookmyung Women's University, South Korea

Ki-Hoon Lee	Kwangwoon University, South Korea
Kohei Hatano	Kyushu University, Japan
Koji Maruhashi	Fujitsu Laboratories Ltd., Japan
Kok-Leong Ong	Royal Melbourne Institute of Technology University, Australia
Konstantin Avrachenkov	Inria, France
Kouzou Ohara	Aoyama Gakuin University, Japan
Krishnamoorthy Sivakumar	Washington State University, USA
Krisztian Buza	Corvinus University of Budapest, Hungary
Kun-Ta Chuang	National Cheng Kung University, Taiwan
Kunpeng Liu	Portland State University, USA
Lamine Diop	École pour l'Informatique et les Techniques Avancées, France
Lecheng Zheng	University of Illinois Urbana-Champaign, USA
Lei Duan	Sichuan University, China
Lei Gu	Nanjing University of Posts and Telecommunications, China
Lei Li	Hong Kong University of Science and Technology, China
Lei Zhang	Anhui University, China
Libin Wang	Hong Kong University of Science and Technology, China
Ling Yue	Rensselaer Polytechnic Institute, USA
Lingwei Chen	Wright State University, USA
Lingyang Chu	McMaster University, Canada
Liyuan Gao	Texas Tech University, USA
Longchao Da	Arizona State University, USA
Luca Cagliero	Politecnico di Torino, Italy
Maciej Grzenda	Warsaw University of Technology, Poland
Maguelonne Teisseire	French National Research Institute for Agriculture, Food and Environment, France
Majid Farhadloo	University of Minnesota, USA
Mandar Chaudhary	eBay, USA
Mario Prado Romero	Gran Sasso Science Institute, Italy
Martin Pavlovski	Samsung Electronics America, USA
Matej Mihelčič	University of Zagreb, Croatia
Matjaž Gams	Jožef Stefan Institute, Slovenia
Mengdi Huai	Iowa State University, USA
Mengjie Zhang	Victoria University of Wellington, New Zealand
Mengyao Ma	University of Queensland, Australia
Milos Radovanovic	University of Novi Sad, Serbia
Mingyang Zhou	Shenzhen University, China

Minh-Hao Van	University of Arkansas, USA
Mohamad El-Hajj	MacEwan University, Canada
Mohammad Masud	United Arab Emirates University, United Arab Emirates
Mrinal Das	Indian Institute of Technology Palakkad, India
Muhammad Daffa Robani	AwanTunai, Indonesia
Nan Zhang	Wenzhou University, China
Neil Miten Daftary	LinkedIn Corporation, USA
Nemanja Djuric	Aurora Innovation, USA
Nguyen Le Minh	Japan Advanced Institute of Science and Technology, Japan
Nhan Pham	IBM, USA
Nianjun Zhou	IBM Research, USA
Ning Liu	Shandong University, China
Olivera Kotevska	Oak Ridge National Laboratory, USA
Ou Wu	Tianjin University, China
P. Radha Krishna	National Institute of Technology Warangal, India
Panagiotis Liakos	National and Kapodistrian University of Athens, Greece
Paolo Garza	Politecnico di Torino, Italy
Paul Wollenhaupt	University of Oslo, Norway
Pedro Ferreira	University of Porto, Portugal
Peer Kröger	Christian Albrechts University Kiel, Germany
Peipei Li	Hefei University of Technology, China
Peng Cheng	Tongji University, China
Peng Wang	Southeast University, China
Pengpeng Zhao	Soochow University, China
Pengyang Wang	University of Macau, China
Ping Chen	University of Massachusetts Boston, USA
Ping Luo	Chinese Academy of Sciences, China
Prateek Verma	University of Arkansas, USA
Qi Liu	University of Science and Technology of China, China
Qian Li	Curtin University, Australia
Qiang Tang	Luxembourg Institute of Science and Technology, Luxembourg
Qiao Xue	Nanjing University of Aeronautics and Astronautics, China
Qing Liu	Zhejiang University, China
Quangui Zhang	Liaoning Technical University, China
Rahul Mishra	International Institute of Information Technology Hyderabad, India
Ranran Bian	University of Sydney, Australia

Raymond Chi-Wing Wong	Hong Kong University of Science and Technology, China
Renhe Jiang	University of Tokyo, Japan
Ricardo Cardoso Pereira	University of Coimbra, Portugal
Riccardo Guidotti	University of Pisa, Italy
Riccardo Ortale	Institute for High-Performance Computing and Networking, Italy
Robert Churchill	Georgetown University, USA
Rosalba Giugno	University of Verona, Italy
Roy Ka-Wei Lee	Singapore University of Technology and Design, Singapore
Ruliang Xiao	Fujian Normal University, China
Runlong Yu	University of Pittsburgh, USA
Runze Yang	Macquarie University, Australia
Sadok Ben Yahia	University of Southern Denmark, Denmark
Satyaki Sikdar	Indiana University, USA
Sebastian Moreno	Universidad Adolfo Ibáñez, Chile
Sen Wang	University of Queensland, Australia
Shaofu Yang	Southeast University, China
Shen-Shyang Ho	Rowan University, USA
Shengyu Chen	University of Pittsburgh, USA
Shichao Zhang	Central South University, China
Shifei Ding	China University of Mining and Technology, China
Shivani Kumar	Indraprastha Institute of Information Technology Delhi, India
Shiyu Yang	Guangzhou University, China
Shridhar Devamane	KLE Institute of Technology, India
Shuai Li	University of Cambridge, UK
Shuang Wang	Clemson University, USA
Shubham Jain	VISA Research, USA
Shuichi Miyazawa	Graduate University for Advanced Studies, Japan
Sibo Wang	Chinese University of Hong Kong, China
Simone Ludwig	North Dakota State University, USA
Sina Khanmohammadi	University of Oklahoma, USA
Snigdhansu Chatterjee	University of Maryland, Baltimore County, USA
Song Wang	University of Virginia, USA
Songcan Chen	Nanjing University of Aeronautics and Astronautics, China
Sourav Dutta	Huawei Research Centre, Ireland
Sungsu Lim	Chungnam National University, South Korea
Sunil Aryal	Deakin University, Australia

Tadashi Nomoto	National Institute of Japanese Literature, Japan
Tai Le Quy	University of Koblenz, Germany
Tao Chen	China University of Geosciences, China
Tao Wang	University of Chinese Academy of Sciences, China
Tetsuya Yoshida	Nara Women's University, Japan
Thomas Carr	University of North Carolina at Charlotte, USA
Tianrui Li	Southwest Jiaotong University, China
Tianxing Wang	Griffith University, Australia
Tim Kreuzer	Stockholm University, Sweden
Tim Oates	University of Maryland, Baltimore County, USA
Ting Guo	University of Technology Sydney, Australia
Tong Chen	University of Queensland, Australia
Tong Zeng	Virginia Tech, USA
Tuan Le	New Mexico State University, USA
Vana Kalogeraki	Athens University of Economics and Business, Greece
Vicenç Torra	Umeå University, Sweden
Viet Vo	Swinburne University of Technology, Australia
Vikram Goyal	Indraprastha Institute of Information Technology Delhi, India
Vinay Madanbhavi Shashidhar	University of Arkansas, USA
Vinoth Kumar Kolluru	Stevens Institute of Technology, USA
Wang Lizhen	Yunnan University, China
Wei Gao	Nanjing University, China
Wei Hu	Nanjing University, China
Wei Ye	Tongji University, China
Wei Emma Zhang	University of Adelaide, Australia
Weicheng Wang	Hong Kong University of Science and Technology, China
Weifeng Liu	China University of Petroleum (East China), China
Wenbin Zhang	Florida International University, USA
Wendy Hui Wang	Stevens Institute of Technology, USA
Wengen Li	Tongji University, China
Wenjie Xi	George Mason University, USA
Wenjun Miao	Beihang University, China
Wenpeng Lu	Qilu University of Technology, China
Wentao Li	University of Leicester, UK
Xiang Lian	Kent State University, USA
Xiangmin Zhou	Royal Melbourne Institute of Technology University, Australia

Xiao Cheng	University of New South Wales, Australia
Xiaoan Bao	Zhejiang Sci-Tech University, China
Xiaobo Zhang	Southwest Jiaotong University, China
Xiaocao Ouyang	Southwestern University of Finance and Economics, China
Xiaochen Wang	Pennsylvania State University, USA
Xiaodong Yue	Shanghai University, China
Xiaojuan Cheng	Deakin University, Australia
Xiaoling Yang	Southwest Jiaotong University, China
Xiaoyang Wang	University of New South Wales, Australia
Xikun Zhang	University of Sydney, Australia
Xin Geng	Southeast University, China
Xin Luo	Shandong University, China
Xin Wang	University of Calgary, Canada
Xin Zheng	Griffith University, Australia
Xinguo Feng	University of Queensland, Australia
Xingzhi Guo	Stony Brook University, USA
Xinyuan Wang	George Mason University, USA
Xuhui Fan	Macquarie University, Australia
Yan Liang	Amazon, USA
Yan Cao	University of Texas at Dallas, USA
Yang-Sae Moon	Kangwon National University, South Korea
Yanhao Wang	East China Normal University, China
Yanwei Yu	Ocean University of China, China
Yasuo Tabei	RIKEN Center for Advanced Intelligence Project, Japan
Ye Zhu	Deakin University, Australia
Yi-Hui Chen	Chang Gung University, Taiwan
Yi-Shin Chen	National Tsing Hua University, Taiwan
Yifan Wang	University of Hawaii, USA
Yifeng Gao	University of Texas Rio Grande Valley, USA
Yijun Lin	University of Minnesota, Twin Cities, USA
Yikai Zhang	Guangzhou University, China
Yimin Wen	Guilin University of Electronic Technology, China
Yiming Tang	Hefei University of Technology, China
Yiqun Zhang	Hong Kong Baptist University, China
Yixin Liu	Monash University, Australia
Yiyang Yang	Guangdong University of Technology, China
Yongcan Huang	University of Georgia, USA
Yongsheng Yu	Macquarie University, Australia
Yu Lin	University of Technology Sydney, Australia
Yu Yang	Education University of Hong Kong, China

Yu-Bin Yang	Nanjing University, China
Yucong Dai	Clemson University, USA
Yue-Shi Lee	Ming Chuan University, Taiwan
Yufei Han	Inria, France
Yujin Huang	University of Melbourne, Australia
Yunjun Gao	Zhejiang University, China
Yuqing Sun	Shandong University, China
Yushan Jiang	University of Connecticut, USA
Zeming Yao	Swinburne University of Technology, Australia
Zengchang Qin	Beihang University, China
Zeren Shui	University of Minnesota, USA
Zhangkai Wu	Macquarie University, Australia
Zhaohong Deng	Jiangnan University, China
Zheng Liu	Nanjing University of Posts and Telecommunications, China
Zhiguo Gong	University of Macau, China
Zhilin Zhao	Sun Yat-sen University, China
Zhiyi Tian	University of Technology Sydney, Australia
Zhiyuan Chen	University of Maryland Baltimore County, USA
Zhong Chen	Southern Illinois University, USA
Zhongying Zhao	Shandong University of Science and Technology, China
Zichong Li	Rensselaer Polytechnic Institute, USA
Zihan Wang	University of Queensland, Australia
Zike Zhang	Hangzhou Normal University, China
Zili Zhang	Southwest University, China
Ziqi Chen	Ohio State University, USA
Ziqi Xu	Royal Melbourne Institute of Technology, Australia
Ziyi Yin	Pennsylvania State University, USA

Sponsors

Foundational Sponsor

Business Events Sydney

Silver Sponsor

School of Computing, Macquarie University

Frontier AI Research Centre, Macquarie University

Bronze Sponsors

Office of the Chief Scientist & Engineer, NSW Government

The University of Sydney

Monash University

Supporters

University of New South Wales

The University of Technology Sydney

Publication

Springer

Contents – Part VI

Trustworthiness

Learning on Complex Data

Survey Track

Large Language Models for Cybersecurity Education: A Survey of Current Practices and Future Directions

Nan Sun[1(✉)], Yuantian Miao[2], Xiaoxing Mo[3], and Jun Zhang[4]

[1] University of New South Wales, Canberra, ACT 2612, Australia
nan.sun@unsw.edu.au
[2] University of Newcastle, Newcastle, NSW 2308, Australia
[3] Deakin University, Waurn Ponds, VIC 3216, Australia
[4] Swinburne University of Technology, Melbourne, VIC 3122, Australia

Abstract. The increasing complexity of cyber threats demands innovative approaches to cybersecurity education that overcome the limitations of traditional teaching methods. This survey paper presents the first comprehensive review of Large Language Models (LLMs) applications in cybersecurity education, examining how these advanced AI systems can address current pedagogical challenges. While existing surveys have explored LLMs in general education or specific cybersecurity applications, our work uniquely focuses on the intersection of LLMs and cybersecurity education. We analyze empirical studies demonstrating LLMs' effectiveness in delivering interactive, personalized learning experiences that adapt to individual student needs. The survey examines current implementations that leverage LLMs' capabilities to create dynamic training materials, provide real-time feedback, and simulate real-world scenarios. We particularly emphasize how LLMs can offer scalable, cost-effective solutions that make cybersecurity education more accessible while maintaining currency with evolving threats. The paper concludes by identifying promising future directions for LLM integration in cybersecurity education, providing valuable insights for educators, researchers, and curriculum developers working to enhance cybersecurity training frameworks.

Keywords: Cybersecurity · Education · Large Lanuage Model

1 Introduction

Cybersecurity education is essential as it empowers both individuals and organizations to effectively address the growing challenges in today's digital world [2]. Without proper knowledge, people are exposed to threats like phishing, malware, and data breaches, while organizations risk financial loss, damaged reputations, and potential legal consequences from cyberattacks [36]. A well-rounded cybersecurity education encourages proactive risk management, promotes safer online

X. Wu et al. (Eds.): PAKDD 2025, LNAI 15875, pp. 3–20, 2025.
https://doi.org/10.1007/978-981-96-8295-9_1

behavior, and enhances the ability to respond to incidents, all while strengthening security infrastructure. Ultimately, it plays a vital role in safeguarding personal data, critical systems, and the broader digital economy.

Traditional cybersecurity education typically centers around lectures, textbooks, and hands-on labs presented in structured classroom settings or online courses with similar formats. While these methods build foundational knowledge, they have several limitations. The passive learning environment may fail to engage all students effectively, and opportunities to practice real-world scenarios are often limited to simulations, which may not fully capture the complexity of actual cyber threats [21]. Additionally, traditional materials tend to be static, making it difficult to keep up with the rapidly evolving threat landscape, potentially leaving learners with outdated information and skills gaps [14]. Another challenge is that access to expensive equipment and specialized software for practical training can pose significant barriers for many learners [32].

The potential for leveraging large language models (LLMs) in cybersecurity education presents a unique opportunity to address the shortcomings of traditional approaches [41]. LLMs can foster more engaging and personalized learning experiences through interactive tutorials, adaptive quizzes that respond to individual performance, and real-world simulations that evolve based on learner actions. Their ability to generate human-like text enables the creation of up-to-date training materials that reflect the constantly shifting cybersecurity landscape, ensuring that learners stay informed with the most current knowledge and techniques. Additionally, LLMs can deliver on-demand explanations, break down complex concepts, and provide personalized feedback, making the learning process more dynamic and effective. Importantly, LLMs offer scalable, cost-effective training solutions, making cybersecurity education more accessible to a broader range of learners.

There are several surveys at the intersection of cybersecurity and LLMs. For instance, Li et al. [19] examined recent LLM research related to educational capabilities, such as mathematics, writing, programming, reasoning, and knowledge-based question answering, aiming to explore their potential in developing the next-generation intelligent education system. Zhang et al. [44] and Ferrag et al. [11] reviewed LLM applications in areas like hardware design security, intrusion detection, software engineering, design verification, cyber threat intelligence, malware detection, and phishing and spam detection. To the best of our knowledge, our survey is the first to explore how **LLMs** can enhance **cybersecurity education**.

In this survey, we collect current works on applying LLM for cybersecurity education and conduct a comprehensive review. As outlined in Fig. 1, firstly, we examine the effectiveness of LLMs in cybersecurity education by collecting and analyzing empirical studies. Secondly, we summarize current applications and techniques that enhance learning experiences through LLMs. Lastly, we propose future directions for utilizing LLMs in cybersecurity education, aiming to develop innovative strategies and tools to address emerging challenges. Our findings provide insights into how LLMs can be used to teach complex cybersecurity concepts effectively, benefiting educators, researchers, policymakers, and curriculum developers who seek to integrate LLMs into educational frameworks.

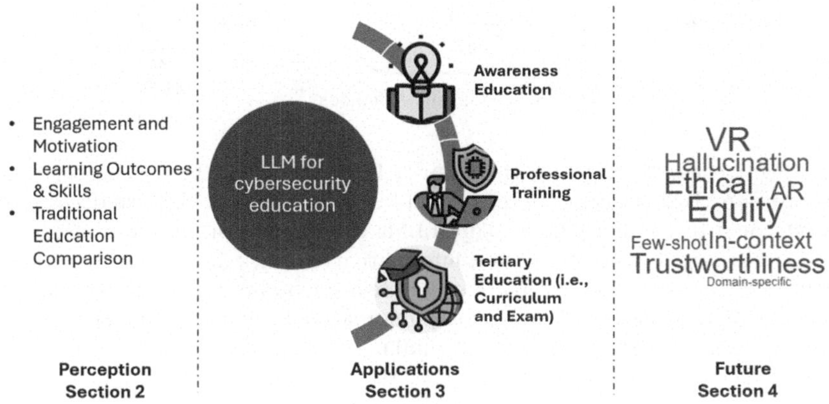

Fig. 1. LLM for Cybersecurity Education

2 Effectiveness of LLMs in Cybersecurity Education

LLMs, with their sophisticated ability to understand and generate human-like text, present a promising opportunity for students with diverse backgrounds to obtain more engaging and personalized training experiences [19]. They also help educators design a more progressive and adaptive learning strategy [10]. As shown in Fig. 2, we review a few empirical studies to elaborate on their effectiveness in cybersecurity education from three aspects, including student engagement and motivation, learning outcomes and skill acquisition, and comparison with traditional teaching methods.

2.1 Student Engagement and Motivation

A lot of research has studied different pedagogy in students' engagement and motivation, including the impact of instruction modes [20], assessment and feedbacks [26], technology integration [1], and so on [17,22,24], while LLMs can provide adaptive, interactive, and supportive learning to enhance their engagement and motivation. Among the four instruction modes, students prefer interactive learning followed by the constructive learning mode [20]. Learning engagement can also be prompted with collaborative and project-based assessments with timely and informative feedback [26], as well as the use of digital tools and multimedia content [1]. To meet the above preference in cybersecurity education,

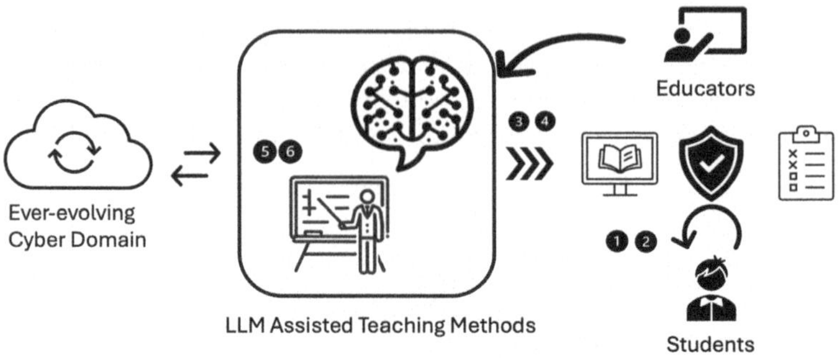

Fig. 2. Effectiveness of LLMs in Cybersecurity Education from the student engagement and motivation ①②, learning outcomes and skill acquisition ③④, and comparison with traditional teaching methods ⑤⑥. LLMs enable students to have adaptive and interactive learning ① with timely and informative feedback ②. Meanwhile, educators can utilize LLMs' advantages to create comprehension and personalized learning ③ for students at different levels and design teaching materials, cybersecurity scenarios, and assessments to ensure students' critical thinking and problem-solving skills④. Compared to traditional methods, LLMs increase the adaptability to the rapid evolution⑤ in cybersecurity education with rich teaching materials ⑥.

LLMs can not only simulate interactive dialogues with instant feedback but also provide comprehensive and diverse perspectives on various topics. We further explore the effectiveness of LLMs in designing adaptive and interactive learning and timely and informative feedback.

Adaptive and Interactive Learning: LLMs are effective in helping educators design adaptive and interactive instruction methods for cybersecurity education. Agrawal et al. [3] utilized LLMs with knowledge graphs to automatically generate a question-answer database for interactive learning in educating students and beginners about cybersecurity concepts and tools. This generator prevents educators from such labor-intensive material preparation, considering the rapid evolution of cybersecurity technology.

LLM-based interactive learning can also increase the student's engagement and motivation in cybersecurity awareness training. For instance, awareness training is essential in critical infrastructures due to their large impact on the surrounding population. Lee et al. [18] highlight the effectiveness of LLM-based interactive learning in ensuring the workers' engagement and motivation by prompting their cybersecurity thinking within the workplace.

Apart from the cybersecurity concepts and tools and cybersecurity awareness training, various cybersecurity exercise scenarios attract students' attention to enhance their engagement, clarify their motivations, and guide them in discussing the relationship between theoretical learning and real-world implementations. Yamin et al. [42] transformed the 'hallucination' potential inherent in LLMs into a potential advantage and leveraged the Retrieval-Augmented Generation

(RAG) approach to enrich the complexity and relevance of the scenarios. Further, they designed a pipeline with multiple LLMs for interactive communications, which provides a dynamic environment for cyber-attack scenarios. In this setup, educators can generate advanced and adaptive exercises by providing up-to-date and relevant information and querying the contextually rich cybersecurity scenarios.

Adaptive and interactive learning with LLMs significantly impacts another challenged education breach in cybersecurity—the Red and Blue teams' attack and defense training. LLMs proved their success in interaction and mutual feedback during the Red team and Blue team competition, which is the key factor in their training education [28].

Timely and Informative Feedback: Another positive implication factor of using LLMs in the Red team and Blue team competition training is their real-time and personalized feedback [28]. LLMs can provide explanations and indications adapted to specific threats and dangerous situations, which can increase students' engagement and clarify their motivations.

In a similar education scenario, the Catch-The-Flag (CTF) competition training can benefit from LLMs' timely and informative feedback. As indicated by Shao et al. [31], while the CTF game provides an interactive mode to display their security skill, convolutional LLMs like GPT-4 and PaLM-2 can help generate better outputs based on the previous inputs and outputs as feedback to learners. They can even further adjust feedback timely based on the learner's response [31], which increases learners' engagement as well. In other research, ChatGPT was chosen over other LLMs because its training with Reinforcement Learning from Human Feedback (RLHF) aligns its responses more closely with human expectations and minimizes hallucinations compared to open-source alternatives [3].

Yet, not all adaptive and interactive learning and timely and informative feedback from LLM can bring positive implementation in students' engagement and training. Some students misuse those LLM-generated answers to pass the quizzes and exams, which hinders authentic assessments and largely decreases students' engagement and mislead motivation. Therefore, recent research investigates some robust assessments against the misuse of AI-generated answers. We will explore those LLM-assisted assessment design methods in the next subsection while analyzing the LLMs' effectiveness in cybersecurity learning outcomes and skill acquisition.

2.2 Learning Outcomes and Skill Acquisition

With increased engagement and motivation, students' performance including their learning outcomes and skill acquisition could be significantly improved in cybersecurity education [40]. Apart from that, LLMs can benefit cybersecurity education from other perspectives to improve the student's learning outcomes and skill acquisition.

Cybersecurity is fundamentally an ongoing competition between attackers and defenders, where each side continuously evolves its strategies and techniques, especially with the fast development of AI tools. In this dynamic field, complex concepts emerged and techniques evolved rapidly, requiring students to comprehend timely and adapt to these changes in diverse workplace environments [15]. Meanwhile, critical thinking and problem-solving skills are essential, as students especially professional learners must navigate complex scenarios that demand thoughtful analysis and innovative solutions [27,42]. We further explore the effectiveness of LLMs in those two aspects.

Comprehension and Personalized Learning: Comprehension about swiftly understanding complex and rapidly evolving techniques can enhance the quality of students' learning and skill acquisition. LLMs allow students to interact with a conversational AI agent to break down some complex concepts into simpler ones, thereby, improving their understanding of cybersecurity and its implications. Lee et al. [18] have proven this statement especially involving XR in education. Similar to the CTF field, it is essential to understand cybersecurity concepts effectively to solve CTF challenges and win the competition. Shao et al. [31] show six different LLMs' proficiency in solving 26 diverse CTF problems and prove their effectiveness with both quantitative and qualitative results.

As for LLMs' effectiveness in personalized learning, Kallonas et al. [15] highlighted the significant potential of generative AI tools especially ChatGPT in various cybersecurity domains, and utilized their advent to tailor their learning content to specific jobs. Such a personalized learning strategy is based on industry needs, their long-term career goals, and their methodological progress with their feedback. It makes learning more efficient and acquiring skills more adaptable to their cybersecurity workforce.

In the tertiary education field, LLM-assist authentic assessments are designed to adapt to different year levels of students at university [27]. With different learning outcomes required by different year levels, answers for the same cybersecurity topic for undergraduate and postgraduate students should be different. It helps educators correctly evaluate whether students' answers align with their year level's learning outcome requirements. As for the Red team and Blue team competition training, LLMs can help involve human factors and perceptions in cybersecurity education especially in cyber kill chain understanding [28]. Herein, LLMs significantly enhance customized learning by offering detailed insights, thorough explanations, and tailored suggestions based on attackers' and defenders' behaviors and the way they interpret and comprehend their surroundings. Therefore, the learning outcomes and skill acquisition could be more accurate.

As for the security awareness training, Hafzullah [13] showed its effectiveness, especially in phishing defense. LLMs were used to not only adjust the training content based on participants' weaknesses but also tailor and dynamically update phishing simulations as training materials. This personalized interactive training method was proved to have significant improvement in their learning outcomes—the Phish Prone Percentage was dropped by more than 12%.

Critical Thinking and Problem-Solving: The ability to anticipate potential threats and devise effective countermeasures is crucial for maintaining robust security in an ever-changing digital landscape. While in cybersecurity training, corresponding exercise scenarios are essential to educate students on this ability.

Yamin et al. [42] proposed an exercise design framework with LLMs to generate dynamic and complex adaptable cybersecurity exercise scenarios for awareness training. In this framework, LLMs act in different roles including Chief Information Security Officer (CISO) and cybersecurity experts for generations, while the CISO LLM defines the training and learning objectives to ensure the training is aligned with the desired skills and knowledge outcomes. Meanwhile, six key criteria are used to evaluate the scenario quality including Details, Technical Soundness, Realism, Creativity, Usability in Exercises, and Expandability with Human Inputs [42]. These factors guarantee that exercises are realistic, engaging, and technically precise, mirroring the complexity of real-world cyber threats. They enhance creativity, prompting learners to think critically and innovatively. Usability ensures that exercises are accessible and beneficial to a diverse range of users, while expandability through human inputs allows for continuous updates and maintains relevance in the rapidly evolving cybersecurity landscape. One of the best-generated scenarios is assessed with above 4 out of 5 scores for each of the six criteria [42]. This method demonstrates that LLMs can enhance cybersecurity education by generating high-quality exercise scenarios that address a wide range of cyber threats. These scenarios simulate real-world challenges and cover both well-known and emerging cybersecurity issues.

In the tertiary education field, LLM-assist authentic assessment generator considers the type of questions and fact-checking strategy to prevent high scores awarded to AI-generated answers [27]. As mentioned above, misusing LLMs leads to low-quality skill acquisition while losing their critical thinking skill. Plapp et al. [27] proves that feeding LLMs generators with up-to-date information for long-form essay-type assessments, designing personalized prompts that go beyond surface-level knowledge, and training markers with fact-checking are the success keys to generating a robust assessment. Specifically, integrating fact-checking into assessments can ensure students' engagement with materials are meaningful with critical thinking instead of merely submitting AI-generated answers.

2.3 Comparison with Traditional Teaching Methods

Both LLM-assisted and traditional teaching methods follow a structured curriculum, utilize diverse educational resources, emphasize interactive learning, and value high-quality assessments and feedback [9]. Herein, LLMs are integrated into traditional teaching methods as supplementary tools to enhance the learning experience and deepen students' understanding of cybersecurity knowledge [23]. Apart from the effectiveness of LLM-assisted teaching methods mentioned above, we highlight two other advantages compared to traditional methods including their adaptability to the rapid evolution in cybersecurity education and their efficiency in preparing rich teaching materials from a vast array of digital resources [34].

Traditional education methods in cybersecurity education often struggle to keep pace with the rapidly evolving landscape of cybersecurity, making it challenging to maintain student engagement and motivation. In the ever-evolving cyber domain, educational materials for both cybersecurity awareness training and education towards various levels (K-12 to professional) require continuous updates [3,27,42]. Not even say those materials typically consist of unstructured text from diverse sources such as lecture notes, books, websites, and videos [27]. For instance, creating a comprehensive question-answer database from these varied sources is labor-intensive and time-consuming, as each source differs in writing style and content structure. This process not only demands significant effort but also incurs high costs.

In contrast, LLMs demonstrate remarkable adaptability to the fast-paced changes in cybersecurity technology. They can efficiently synthesize information from a wide array of digital sources, streamlining the creation of educational materials and enhancing the overall learning experience in cybersecurity education. Agrawal et al. [3] generated a cybersecurity-related question-answer dataset consisting of 4k questions with high readability and trained an open-ended question-answering model for K-12, undergraduate, graduate, and professional students. Similar to the paper [27], they utilized LLMs to generate high-quality assessments of different types including short answer questions, single answer multiple choice questions, and essay-type assessments. As for other educational materials like cybersecurity exercise simulation, Yamin et al. [42] showed the creativity of LLMs in the scenario design process and utilized LLMs to enrich the scenario variety.

3 Current Applications of LLMs in Cybersecurity Education

Below, we examine the current uses of LLMs in cybersecurity education, categorized into cybersecurity awareness education, security professional training, and tertiary cybersecurity education (Table 1).

3.1 Cybersecurity Awareness Education

Cybersecurity awareness education equips individuals with the knowledge and skills to protect themselves and their organizations from cyber threats [35]. It covers topics like recognizing phishing emails, creating strong passwords, practicing safe browsing habits, understanding social engineering tactics, and protecting sensitive data [45]. The goal is to foster a security-conscious culture where individuals can identify and mitigate risks, ultimately reducing the likelihood and impact of cyberattacks.

Table 1. Overview of Security Education Reviewed Papers

Paper	Category	Security Tasks	Target Audience	LLM-related Techniques	Other Techniques	Data Sources
[12]	Awareness Education	Phishing	Employees in organizations, security practitioners	LLM-generated training content	Simulated phishing campaigns, personality profiling	Psychological and demographic data
[30]		Policy formulation	Corporate employees	Prompt engineering, policy generation	Training program, feedback mechanisms	Internal network logs
[18]		Cyberattack simulation	Infrastructure operators	Attack description conversion	Extended reality, digital twins	Network scanning data
[13]		Phishing	Various backgrounds	LLM-generated content	Caniphish and KnowBe4 tools	Phishing simulation data
[42]	Security Professional Training	Cyber threats simulation	Security professionals	LLM hallucination utilization	Storytelling exercises	Cyber threats data
[29]		Cyber-physical defense	Industry 4.0 workforce	Learn-Apply-Create model	Hands-on experiments	Sensor data, MATLAB code
[31]		CTF tasks	CTF participants	Automated LLM workflow	Human-in-the-loop	Various CTF data
[28]		Red/Blue Team	SOC professionals	Real-time LLM feedback	MITRE ATT&CK framework	Incident reports
[37]	Tertiary Cybersecurity Education	Multiple-choice questions	Cybersecurity students	RAG, LLM validation	Benchmark datasets	Question banks
[27]		Assessment design	University students	Answer fact checking	Statistical analysis	35 security questions
[3]		Q&A generation	K-12 students	Prompt-chaining	Knowledge graph	Security knowledge base

Phishing: Papers [12,13] explore the use of LLMs to enhance the effectiveness of phishing education programs. Traditional Phishing Education, Training, and Awareness (PETA) programs often struggle with high costs and a lack of personalization, which limits their impact. To address this, the authors in paper [12] propose using LLMs to automatically generate customized training content that is tailored to individual employees based on their psychological traits, roles, and technical expertise. In paper [13], LLMs were used to create personalized and interactive training content, adjusting based on participants' weaknesses. LLMs can create realistic phishing emails for simulated campaigns that mimic real-world attacks, employing social engineering tactics such as persuasion principles (e.g., authority, scarcity) and emotional triggers (e.g., fear, curiosity). After employees fall for a simulated phishing attack, LLMs can generate personalized debriefing content that explains the phishing cues they missed and offers tailored tips on how to avoid future attacks. This approach allows for dynamic, engaging, and relevant training that is more closely aligned with each employee's specific vulnerabilities.

In addition to creating phishing emails, LLMs can also improve anti-phishing warnings by generating dynamic, context-aware messages that adapt to the employee's knowledge level and the specific threat they are facing. This helps prevent habituation, where users begin to ignore repetitive warnings. The paper [12] also highlights the efficiency of using LLMs to reduce the manual effort and high costs traditionally associated with creating PETA programs. By automating the customization of training materials and phishing simulations, LLMs make it possible to deliver personalized education at scale, ensuring that training is both effective and affordable for organizations. This automated, tailored approach promises to improve the overall cybersecurity awareness of employees, turning them into stronger defenders against phishing attacks.

Paper [30] demonstrates how LLMs, like Google Gemini, are used to enhance security awareness education against spear phishing. By analyzing past phishing incidents and internal data, the LLM helps generate adaptive training content tailored to the organization's specific threats. Through prompt engineering, the model creates dynamic and evolving educational materials that teach senior managers and employees to recognize and respond to phishing attacks. The LLM continuously refines the training based on real-time feedback, ensuring the education stays relevant and effective in tackling emerging cyber threats.

Cyberattack Simulation and Defense: Cyberattack simulation and defense in cybersecurity awareness education provides hands-on experience in a safe environment. Learners face simulated attacks, allowing them to practice identifying threats and implementing appropriate defensive measures. This interactive approach reinforces theoretical knowledge and builds practical skills to effectively respond to real-world cyberattacks. It helps bridge the gap between understanding concepts and applying them in practice, fostering a stronger security posture.

Lee et al., in the paper [18], use LLMs to enhance cybersecurity education by translating complex attack scenarios into understandable actions for non-experts. Specifically targeting water treatment plant employees, the system allows users to engage with a Conversational AI, which explains cybersecurity concepts and simulates attacks. LLMs interpret user inputs or preset scenarios—such as a slow network being a potential Denial of Service (DoS) attack—and generate appropriate commands that are executed in a simulated environment. This immersive approach helps users understand real-world cybersecurity threats and defense strategies through interactive, AI-driven learning.

3.2 Cybersecurity Professional Training

Security professionals typically undergo training through a combination of certification programs, hands-on labs, tabletop exercises, and continuous learning modules to stay current with emerging threats and defense strategies. These traditional training methods, while valuable, often struggle to provide real-time, personalized experiences that reflect the rapidly evolving nature of cybersecurity challenges [4]. The following section examines representative examples demonstrating how LLMs have been adapted and deployed to meet the unique requirements of security professional development.

Simulation of Cyber Attacks and Defense: The paper [42] utilizes LLMs to enhance the training of cybersecurity professionals by creating dynamic and realistic exercise scenarios. By using the generative capabilities of LLMs, the authors simulate a broad range of cyber threats, allowing security professionals to engage in more challenging and complex exercises. Rather than viewing the "hallucination" tendency of LLMs—where they might produce information that isn't factually accurate—as a limitation, the authors turn this into a creative tool for generating unpredictable and diverse cyber threat scenarios. This makes the training more engaging and less reliant on predictable, predefined threats.

In addition, in their approach, the authors use multiple LLMs in different roles [42]. For example, one LLM assumes the role of a CISO with organizational knowledge, while another acts as a Cybersecurity Expert familiar with current threats. These models interact through carefully crafted prompts, generating discussions that simulate real-world cyber attacks. This interaction, enhanced by a RAG technique, allows for the creation of rich, context-aware scenarios that challenge participants to think critically and respond to unexpected situations. The generated scenarios follow a structured design process, with elements like stories, scripts, events, and injects that simulate the progression of a cyber attack. This structure ensures that the scenarios are not only realistic but also varied, preventing predictability and keeping participants engaged. By using LLMs in this way, the paper proposes a significant advancement in cybersecurity training, making it more adaptable to evolving threats and better suited to preparing professionals for real-world incidents.

Red and Blue Team Simulations: The paper [28] uses Red and Blue Team simulations as a core method to enhance cybersecurity education, where the Red Team simulates attacks and the Blue Team defends against them. LLMs are integrated into these simulations to provide real-time, AI-driven support. For the Red Team, LLMs suggest attack strategies, identify vulnerabilities, and recommend effective delivery techniques during the Cyber Kill Chain (CKC) phases, such as reconnaissance and weaponization. On the Blue Team side, LLMs assist in detecting threats, analyzing suspicious activities, and guiding responses to attacks, particularly in later CKC phases like exploitation and command-and-control. The paper also highlights the use of the MITRE ATT&CK framework, a comprehensive knowledge base of real-world adversary tactics and techniques. LLMs help both teams navigate and apply this framework by offering tailored suggestions based on the specific tactics being employed. This AI-driven approach not only strengthens technical skills but also raises awareness of human factors, such as risk perception and decision-making under pressure. By combining AI, the CKC, and the MITRE ATT&CK framework, the paper proposes a holistic and more effective approach to cybersecurity education.

Cybersecurity Competition: A Capture The Flag (CTF) challenge is a type of cybersecurity competition where participants solve security-related puzzles to find hidden pieces of data, known as "flags". These flags are typically short strings of text that prove the participant has successfully completed a challenge. CTF challenges simulate real-world security scenarios and cover a wide range of topics, such as cryptography, reverse engineering, web exploitation, forensics, and more. Paper [31] examines how LLMs can be used to improve cybersecurity education through CTF challenges. Two approaches are explored: a human-in-the-loop (HITL) workflow, where participants guide the LLM by providing feedback on its responses, and a fully automated workflow, where the LLM solves challenges independently. In both cases, LLMs help solve security tasks like reverse engineering and web exploitation by generating code or instructions. This not only assists in solving challenges but also teaches participants how to interact with AI systems effectively, enhancing security awareness and problem-solving skills.

Understand Cyber-Physical Systems and Cyber Attacks: Paper [29] uses ChatGPT to enhance cybersecurity education through a three-phase "learn-apply-create" framework. In the learning phase, students interact with ChatGPT to understand cyber-physical systems (CPS) and cyberattacks, with a focus on real-world scenarios like attacks on industrial sensors. ChatGPT explains key concepts like false data injection and replay attacks, making it easier for students to grasp foundational knowledge. In the applying phase, ChatGPT helps students model these cyberattacks, simulating them on sensor data from a J-type thermocouple using MATLAB. This allows students to apply their knowledge to a practical, real-time scenario where they inject false data into the sensor readings to mimic a cyberattack. Finally, in the creating phase, ChatGPT assists in developing detection algorithms, guiding students in generating and refining code to identify and mitigate these attacks on the thermocouple sensor. This approach makes learning more interactive and accessible, helping students quickly acquire the skills needed for cybersecurity tasks in real industrial systems.

3.3 Tertiary Education in Cybersecurity

In tertiary cybersecurity education, traditional curricula combine theoretical concepts with practical labs, aligned with industry standards [38]. LLMs enhance this framework by helping educators develop dynamic course content that incorporates emerging trends and by generating diverse, scenario-based assessments. These models assist in creating customized learning pathways and provide automated feedback, enabling more effective evaluation of students' theoretical knowledge and practical skills.

CyberMetric: Tihanyi et al. [37] created the CyberMetric datasets, containing thousands of MCQs generated using LLMs, specifically GPT-3.5 with RAG. This method allows the model to generate questions based on a large corpus of cybersecurity documents, such as NIST standards, research papers, and books. The questions span nine critical domains of cybersecurity, including cryptography, network security, and penetration testing. By using LLMs to generate these questions, the authors aim to provide a consistent, large-scale method for evaluating both human and machine knowledge in cybersecurity.

To ensure the quality of the questions, the dataset underwent several rounds of validation, including checks by other LLMs like Falcon-180B and manual reviews by cybersecurity experts. The final dataset serves as a tool for both evaluating LLMs and educating humans in cybersecurity. The authors compared the performance of 25 state-of-the-art LLMs, such as GPT-4-turbo and Mixtral-8 × 7B-Instruct, against human experts who answered a smaller subset of the questions in a closed-book scenario. Interestingly, top-performing LLMs outscored many human participants, demonstrating the potential of LLMs not only to assist in cybersecurity education but also to serve as benchmarks for human expertise. In this way, LLMs help raise cybersecurity education by generating diverse, high-quality questions that can test a wide range of knowledge. This method provides an efficient and scalable tool for both teaching and evaluating cybersecurity expertise.

Cybersecurity Assessments Design: The paper [27] examines how LLMs, such as ChatGPT, affect the design of assessments in tertiary cybersecurity education. LLMs allow students to generate answers easily, which poses challenges for educators in ensuring assessments reflect genuine learning. The authors tested ChatGPT with various cybersecurity questions across undergraduate to PhD levels and analyzed the quality of its responses. To design more robust assessments, the paper suggests techniques like using long-form essay questions and requiring fact-checking. These methods make it harder for students to rely solely on AI-generated content, as LLMs struggle with in-depth analysis and providing accurate, up-to-date references. These techniques ensure students engage critically with the material, fostering better learning outcomes in cybersecurity education.

CyberQ: Paper [3] introduces CyberQ, a system that uses ChatGPT to generate question-answer pairs for cybersecurity education. To improve the factual accuracy of the generated content, the authors augment the LLM with a cybersecurity knowledge graph (AISecKG). This integration ensures that the LLM produces more reliable and contextually relevant questions. The system employs prompt-chaining techniques through three steps: Zero-Shot (ZS), Few-Shot (FS), and Ontology-Driven (OD) prompting. In Zero-Shot, the LLM generates questions based on a basic description of a concept. Few-Shot provides example questions to guide the model, while Ontology-Driven prompting uses specific knowledge graph relationships to create more complex and detailed questions.

This method reduces hallucinations and enhances the model's accuracy in generating cybersecurity-related questions. The resulting dataset, CyberQ, is validated by experts and used to train an interactive question-answering system for cybersecurity education.

The integration of KGs into the LLM's prompting process helps ground the model's responses in factual cybersecurity knowledge, thereby reducing errors commonly seen in LLM-generated content. The resulting question-answer pairs are then compiled into a dataset called CyberQ, which is evaluated by domain experts for accuracy and consistency. This dataset is not only used to train a generative model for question-answering tasks but also serves as a resource for building interactive educational tools in cybersecurity. The authors argue that this method enhances student engagement and provides a scalable solution for creating domain-specific educational content in cybersecurity.

4 Conclusions and Future Directions

This survey has provided a comprehensive examination of LLM's role in transforming cybersecurity education. Our analysis demonstrates that LLMs offer promising solutions to longstanding challenges in cybersecurity training, particularly in delivering personalized, engaging, and up-to-date educational experiences. The ability of LLMs to generate dynamic content, provide immediate feedback, and simulate real-world scenarios represents a significant advancement over traditional teaching methods.

From our analysis of empirical studies and existing applications, we have highlighted several significant benefits of integrating LLMs into cybersecurity education. Additionally, we conducted a systematic review of current methods for utilizing LLMs in areas such as cybersecurity awareness education, training for security professionals, and tertiary-level cybersecurity instruction. Looking forward, we identify several exciting avenues for future research and development.

4.1 Addressing Hallucinations and Ensuring Trustworthiness

Key challenges for LLMs in education include hallucinations and inadequate evaluation metrics. Research priorities should focus on developing methods to detect and mitigate hallucinations through fact-checking, knowledge base integration, and reinforcement learning. Integrating knowledge graphs can improve response accuracy for technical queries [25]. Creating specialized metrics to evaluate LLM-generated content for cybersecurity training is crucial, emphasizing accuracy, relevance, and security implications [44].

4.2 Leveraging Few-Shot and In-Context Learning

Recent advances in LLMs have brought attention to zero-shot learning capabilities, where models perform tasks without task-specific training examples.

Research by Wang et al. [39] indicates that zero-shot approaches constrain a model's learning potential. Their findings suggest that the absence of exemplars in zero-shot prompting can lead to suboptimal performance across various tasks. The limitations of zero-shot learning have motivated researchers to consider alternative approaches, such as few-shot and in-context learning paradigms, where models are provided with a small set of annotated examples. These methods show promise in enhancing model performance by providing context-specific guidance, potentially leading to more refined and task-appropriate responses.

4.3 Domian-Specific Performance

LLMs need to be trained to understand the specific terminology, jargon, and concepts used in cybersecurity [7]. This includes recognizing different types of attacks (e.g., phishing, malware, denial-of-service), security protocols (e.g., TLS, SSH), and network architectures [5]. Research should focus on developing specialized training datasets and fine-tuning techniques that enhance LLMs' comprehension of security-related language [41]. For example, training LLMs on vulnerability databases, security blogs, and incident reports can improve their ability to understand and respond to security-specific queries.

4.4 Ethical Considerations and Equity

The use of LLMs in educational contexts could raise ethical concerns, particularly around bias, fairness, and inclusivity [43]. To ensure the ethical and effective use of LLMs in cybersecurity education, bias mitigation is essential to guarantee fair representation of diverse populations, instructional styles, and teaching practices [8]. Transparency and explainability are crucial in helping students critically engage with LLM-generated feedback, allowing them to understand the reasoning behind suggestions and fostering deeper learning [6]. Hybrid models that incorporate both automated feedback and human oversight are necessary to maintain the quality of education, ensuring that students receive tailored and customized guidance. Finally, ethical safeguards must be implemented to prevent misuse of LLMs in this sensitive domain, ensuring that AI tools promote responsible behavior and adhere to ethical standards in cybersecurity training.

4.5 Combining LLMs with Other Technologies

Integrating LLMs with technologies like Virtual Reality (VR), Augmented Reality (AR), and the Metaverse enhances cybersecurity education [33]. VR enables immersive attack simulations, while AR provides real-time guidance during hands-on exercises like firewall configuration. The Metaverse fosters collaborative learning environments for tackling complex cybersecurity challenges, guided by LLMs. Additionally, combining LLMs with virtual labs and gamification platforms creates engaging, realistic, and interactive experiences, motivating students through challenges and rewards [16]. These integrations create immersive tools to better prepare students for cybersecurity complexities.

References

1. Abla, C., Fraumeni, B.R.: Student engagement: Evidence-based strategies to boost academic and social-emotional results. McREL International (2019)
2. Abrahams, T.O., Farayola, O.A., Kaggwa, S., Uwaoma, P.U., Hassan, A.O., Dawodu, S.O.: Cybersecurity awareness and education programs: a review of employee engagement and accountability. Comput. Sci. IT Res. J. **5**(1), 100–119 (2024)
3. Agrawal, G., Pal, K., Deng, Y., Liu, H., Chen, Y.C.: CyberQ: generating questions and answers for cybersecurity education using knowledge graph-augmented LLMs. In: Proceedings of the AAAI Conference on Artificial Intelligence. vol. 38, pp. 23164–23172 (2024)
4. AlDaajeh, S., Saleous, H., Alrabaee, S., Barka, E., Breitinger, F., Choo, K.: The role of national cybersecurity strategies on the improvement of cybersecurity education. Comput. Secur. **119**, 102754 (2022)
5. Aslan, Ö., Aktuğ, S.S., Ozkan-Okay, M., Yilmaz, A.A., Akin, E.: A comprehensive review of cyber security vulnerabilities, threats, attacks, and solutions. Electronics **12**(6), 1333 (2023)
6. Barman, K.G., Wood, N., Pawlowski, P.: Beyond transparency and explainability: on the need for adequate and contextualized user guidelines for LLM use. Ethics Inf. Technol. **26**(3), 47 (2024)
7. Butler, J.J., et al.: From jargon to clarity: improving the readability of foot and ankle radiology reports with an artificial intelligence large language model. Foot Ankle Surg. **30**(4), 331–337 (2024)
8. Chinta, S.V., et al.: FairAIED: Navigating fairness, bias, and ethics in educational AI applications. arXiv preprint arXiv:2407.18745 (2024)
9. Chowdhury, G., Chowdhury, S.: AI-and LLM-driven search tools: a paradigm shift in information access for education and research. J. Inf. Sci. 01655515241284046 (2024)
10. Dolmans, D., Schmidt, H.: The problem-based learning process. Lessons from problem-based learning, pp. 13–20 (2010)
11. Ferrag, M.A., Alwahedi, F., Battah, A., Cherif, B., Mechri, A., Tihanyi, N.: Generative AI and large language models for cyber security: All insights you need. arXiv preprint arXiv:2405.12750 (2024)
12. Greco, F., Desolda, G., Viganò, L.: Supporting the design of phishing education, training and awareness interventions: an LLM-based approach (2024)
13. Hafzullah, İ.: LLM-driven sat impact on phishing defense: a cross-sectional analysis. In: 2024 12th International Symposium on Digital Forensics and Security (ISDFS), pp. 1–5. IEEE (2024)
14. Jukes, I., McCain, T., Crockett, L.: Understanding the digital generation: Teaching and learning in the new digital landscape. Corwin Press (2010)
15. Kallonas, C., Piki, A., Stavrou, E.: Empowering professionals: a generative AI approach to personalized cybersecurity learning. In: 2024 IEEE Global Engineering Education Conference (EDUCON), pp. 1–10. IEEE (2024)
16. Lan, Y.J., Chen, N.S.: Teachers' agency in the era of LLM and generative AI. Educ. Technol. Soc. **27**(1), I–XVIII (2024)
17. Latif, E., Fang, L., Ma, P., Zhai, X.: Knowledge distillation of LLM for education. arXiv preprint arXiv:2312.15842 (2023)
18. Lee, A., King, K., Gračanin, D., Azab, M.: Experiential learning through immersive XR: cybersecurity education for critical infrastructures. In: International Conference on Human-Computer Interaction, pp. 56–69. Springer (2024)

19. Li, Q., et al.: Adapting large language models for education: Foundational capabilities, potentials, and challenges. arXiv preprint arXiv:2401.08664 (2023)
20. Menekse, M., Stump, G.S., Krause, S., Chi, M.T.: Differentiated overt learning activities for effective instruction in engineering classrooms. J. Eng. Educ. **102**(3), 346–374 (2013)
21. Mukherjee, M., Le, N.T., Chow, Y.W., Susilo, W.: Strategic approaches to cybersecurity learning: a study of educational models and outcomes. Information **15**(2), 117 (2024)
22. Neumann, A.T., Yin, Y., Sowe, S., Decker, S., Jarke, M.: An LLM-driven chatbot in higher education for databases and information systems. IEEE Transactions on Education (2024)
23. Ohm, M., Bungartz, C., Boes, F., Meier, M.: Assessing the impact of large language models on cybersecurity education: a study of ChatGPT's influence on student performance. In: Proceedings of the 19th International Conference on Availability, Reliability and Security, pp. 1–7 (2024)
24. Orenstrakh, M.S., Karnalim, O., Suarez, C.A., Liut, M.: Detecting LLM-generated text in computing education: Comparative study for ChatGPT cases. In: 2024 IEEE 48th Annual Computers, Software, and Applications Conference (COMPSAC), pp. 121–126. IEEE (2024)
25. Pan, S., Luo, L., Wang, Y., Chen, C., Wang, J., Wu, X.: Unifying large language models and knowledge graphs: a roadmap. IEEE Transactions on Knowledge and Data Engineering (2024)
26. Pan, X., Shao, H.: Teacher online feedback and learning motivation: Learning engagement as a mediator. Soc. Behav. Personal. Int. J. **48**(6), 1–10 (2020)
27. Plapp, A., Wu, J., Pan, L., Chen, C., Chua, C., Zhang, J.: Strengthening cyber security education: Designing robust assessments for ChatGPT-generated answers. In: International Conference on Machine Learning for Cyber Security, pp. 17–27. Springer (2023)
28. Santa Barletta, V., Caivano, D., Calvano, M., Curci, A., Piccinno, A.: CRASTE: Human factors and perception in cybersecurity education (2024)
29. Santhi, T.M., Srinivasan, K.: Chat-GPT based learning platform for creation of different attack model signatures and development of defense algorithm for cyberattack detection. IEEE Transactions on Learning Technologies (2024)
30. Shafee, S., Bessani, A., Ferreira, P.M.: Evaluation of LLM chatbots for OSINT-based cyber threat awareness. arXiv preprint arXiv:2401.15127 (2024)
31. Shao, M., et al.: An empirical evaluation of LLMs for solving offensive security challenges. arXiv preprint arXiv:2402.11814 (2024)
32. Shillair, R., Esteve-González, P., Dutton, W.H., Creese, S., Nagyfejeo, E., von Solms, B.: Cybersecurity education, awareness raising, and training initiatives: national level evidence-based results, challenges, and promise. Comput. Secur. **119**, 102756 (2022)
33. Soliman, M.M., Ahmed, E., Darwish, A., Hassanien, A.E.: Artificial intelligence powered metaverse: analysis, challenges and future perspectives. Artif. Intell. Rev. **57**(2), 36 (2024)
34. Stamper, J., Xiao, R., Hou, X.: Enhancing LLM-based feedback: insights from intelligent tutoring systems and the learning sciences. In: International Conference on Artificial Intelligence in Education, pp. 32–43. Springer (2024)
35. Sun, N., Zhang, J., Gao, S., Zhang, L.Y., Camtepe, S., Xiang, Y.: Data analytics of crowdsourced resources for cybersecurity intelligence. In: Network and System Security: 14th International Conference, NSS 2020, Melbourne, VIC, Australia, November 25–27, 2020, Proceedings 14, pp. 3–21. Springer (2020)

36. Sun, N., Zhang, J., Rimba, P., Gao, S., Zhang, L.Y., Xiang, Y.: Data-driven cybersecurity incident prediction: a survey. IEEE Commun. Surv. Tutorials **21**(2), 1744–1772 (2018)
37. Tihanyi, N., Ferrag, M.A., Jain, R., Bisztray, T., Debbah, M.: CyberMetric: a benchmark dataset based on retrieval-augmented generation for evaluating LLMs in cybersecurity knowledge. In: 2024 IEEE International Conference on Cyber Security and Resilience (CSR), pp. 296–302. IEEE (2024)
38. Towhidi, G., Pridmore, J.: Aligning cybersecurity in higher education with industry needs. J. Inf. Syst. Educ. **34**(1), 70–83 (2023)
39. Wang, R.E., Demszky, D.: Is ChatGPT a good teacher coach? Measuring zero-shot performance for scoring and providing actionable insights on classroom instruction. arXiv preprint arXiv:2306.03090 (2023)
40. Wen, Q., et al.: AI for education (AI4EDU): advancing personalized education with LLM and adaptive learning. In: Proceedings of the 30th ACM SIGKDD Conference on Knowledge Discovery and Data Mining, pp. 6743–6744 (2024)
41. Xu, H., et al.: Large language models for cyber security: A systematic literature review. arXiv preprint arXiv:2405.04760 (2024)
42. Yamin, M.M., Hashmi, E., Ullah, M., Katt, B.: Applications of LLMs for generating cyber security exercise scenarios (2024)
43. Yan, L., et al.: Practical and ethical challenges of large language models in education: a systematic scoping review. Br. J. Edu. Technol. **55**(1), 90–112 (2024)
44. Zhang, J., Bu, H., Wen, H., Chen, Y., Li, L., Zhu, H.: When LLMs meet cybersecurity: A systematic literature review. arXiv preprint arXiv:2405.03644 (2024)
45. Zhang-Kennedy, L., Chiasson, S.: A systematic review of multimedia tools for cybersecurity awareness and education. ACM Comput. Surv. (CSUR) **54**(1), 1–39 (2021)

A Comprehensive Survey on Deep Learning Solutions for 3D Flood Mapping

Wenfeng Jia[1], Bin Liang[2], Yuxi Lu[2], Muhammad Arif Khan[1], and Lihong Zheng[1(✉)]

[1] Charles Sturt University, New South Wales, Australia
lzheng@csu.edu.au
[2] University of Technology Sydney, New South Wales, Australia

Abstract. Flooding remains a major global challenge, worsened by climate change and urbanization, demanding advanced solutions for effective disaster management. While traditional 2D flood mapping techniques provide limited insights, 3D flood mapping, powered by deep learning (DL), offers enhanced capabilities by integrating flood extent and depth. This paper presents a comprehensive survey of deep learning-based 3D flood mapping, emphasizing its advancements over 2D maps by integrating flood extent and depth for effective disaster management and urban planning. The survey categorizes deep learning techniques into task decomposition and end-to-end approaches, applicable to both static and dynamic flood features. We compare key DL architectures, highlighting their respective roles in enhancing prediction accuracy and computational efficiency. Additionally, this work explores diverse data sources such as digital elevation models, satellite imagery, rainfall, and simulated data, outlining their roles in 3D flood mapping. The applications reviewed range from real-time flood prediction to long-term urban planning and risk assessment. However, significant challenges persist, including data scarcity, model interpretability, and integration with traditional hydrodynamic models. This survey concludes by suggesting future directions to address these limitations, focusing on enhanced datasets, improved models, and policy implications for flood management. This survey aims to guide researchers and practitioners in leveraging DL techniques for more robust and reliable 3D flood mapping, fostering improved flood management strategies.

Keywords: Deep Learning · Flood Mapping · 3D · Multi-source data

1 Introduction

Flooding has been a persistent threat to human society since its inception, and its impact has not diminished in modern times. Research indicates that flood frequency has shown an upward trend over the past 30 years [29]. Meanwhile, global urbanization has increased the concentration of populations and assets in flood-prone areas, leading to more severe damage when floods occur. According

X. Wu et al. (Eds.): PAKDD 2025, LNAI 15875, pp. 21–38, 2025.
https://doi.org/10.1007/978-981-96-8295-9_2

to the global hazards database, between 2018 and 2022, flood disasters caused more than 20,000 deaths or missing people, with direct economic losses exceeding $189.8 billion USD [30]. Currently, 1.81 billion people (approximately 23% of the global population) are directly exposed to flood risks [37].

Given the significant damage caused by floods, reliable flood mapping techniques are critical for disaster management and risk assessment. They offer the potential to significantly reduce the threat that floods pose to life and property. Traditional 2D flood mapping, which focuses solely on flood extent, is inadequate for capturing the full scope of flood risks. The inclusion of flood depth through 3D mapping provides critical insights into flood severity, enabling accurate risk assessments and targeted disaster management. Before deep learning (DL), physical models were the main tools for flood mapping, offering detailed simulations but requiring complex setups, high-quality data, and substantial computational resources [33]. With rapid advancements in computer technology, machine learning has greatly improved. To overcome the low computational efficiency of traditional physical models, some researchers have turned to physics-guided machine learning for flood mapping [25]. In recent years, DL, with its ability to automatically extract features and handle complex datasets [6,19,27], has emerged as a faster and more efficient alternative for 3D flood mapping [38].

Given the current lack of review articles on 3D flood mapping using DL techniques, this survey presents a comprehensive analysis of recent advancements in the field. By categorizing DL techniques into **task decomposition** and **end-to-end** methods, this paper explores their applicability to **static** and **dynamic** flood mapping. The review also explores the challenges of data scarcity, model generalizability, and practical implementation, offering insights into future directions for this rapidly evolving field. To identify high-quality literature and gain a comprehensive understanding of the research methods and current status of 3D flood mapping, this survey conducted a systematic search using WebofScience and Google Scholar. The search criteria were as follows:

- WebofScience: (TS=(flood mapping) AND TS=(deep learning) AND (TS=(3D) OR TS=(flood depth))) AND PY=(2020–2025)
- Google Scholar: “flood mapping” “deep learning” (OR 3D OR “flood depth”) -review -survey. (Since 2020)

The initial search retrieved 55 articles from WebofScience and 58 from Google Scholar. After removing duplicates and weakly relevant papers, 39 articles were ultimately selected for analysis in this survey.

Based on this curated collection, we outline the structure of the survey as follows: Sect. 2 covers the foundations of 3D flood mapping, including problem statements, method comparisons, and an introduction to data sources. Section 3 examines DL models categorized by their methodologies and the flood features they address. Section 4 highlights application directions, while Sect. 5 reviews model metrics. Section 6 discusses challenges and outlines potential future research directions. Finally, Sect. 7 concludes the survey.

2 The Foundations of DL-Based 3D Flood Mapping

This section provides a foundational understanding of 3D flood mapping, covering the problem statement, a comparison of existing methods, and the primary data sources used in DL-based 3D flood mapping.

2.1 Problem Statement

Flood mapping is essential for disaster management, aiming to predict flood extent and depth accurately to produce a 3D flood map. As shown in Fig. 1, two DL-based methodologies, task decomposition methods, and end-to-end methods offer distinct approaches to solving this problem. This section formalizes these methodologies with consistent notation and generalized loss functions.

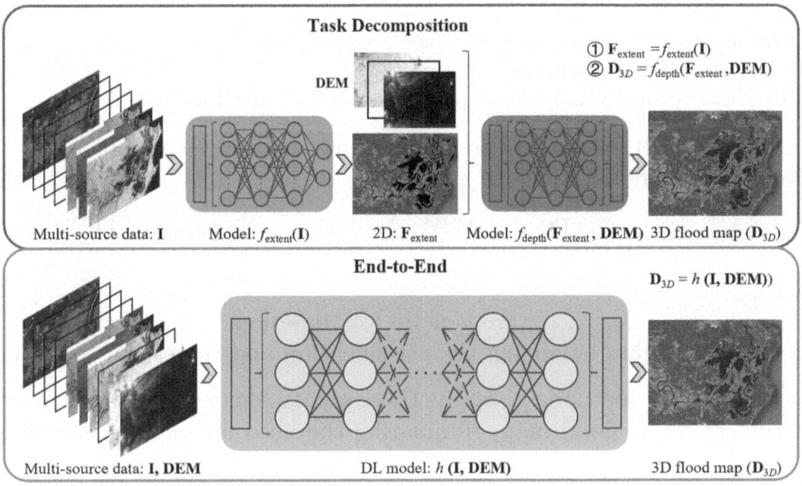

Fig. 1. Task decomposition and end-to-end methods for 3D flood mapping.

In task decomposition methods, the flood mapping process is divided into two steps. First, the flood extent detection step predicts the 2D flooded area. Second, the flood depth estimation step calculates the depth of flooding in the detected area and produces the final 3D flood map:

$$\mathbf{D}_{3D} = f_{\text{depth}}(\mathbf{F}_{\text{extent}}, \mathbf{DEM}) \tag{1}$$

where:

- f_{depth} is a flood depth estimation model predicting depth and generating 3D flood maps using $\mathbf{F}_{\text{extent}}$ and digital elevation model ($\mathbf{DEM}$).
- $\mathbf{F}_{\text{extent}}$ is a flood extent detection model predicting the 2D maps. It is defined as $\mathbf{F}_{\text{extent}} = f_{\text{extent}}(\mathbf{I})$, and $\mathbf{I}$ represents the input data, such as satellite, unmanned aerial vehicle (UAV), and rainfall data.

For the two separate steps, flood extent detection and flood depth estimation, each uses its independent loss function for the corresponding model training:

$$\begin{aligned}\mathcal{L}_{\text{extent}} &= \mathcal{L}_{\text{classification}}\left(\mathbf{F}_{\text{extent}}, \mathbf{F}_{\text{extent}}^{\text{true}}\right)\\ \mathcal{L}_{\text{depth}} &= \mathcal{L}_{\text{regression}}\left(\mathbf{D}_{\text{depth}}, \mathbf{D}_{\text{depth}}^{\text{true}}\right)\end{aligned} \tag{2}$$

As shown in the Eq. 2, $\mathcal{L}_{\text{extent}}$ is a general classification loss function, which measures how well the model predicts the flooded area. Common loss functions include Binary Cross-Entropy, Dice Loss, or Focal Loss. $\mathcal{L}_{\text{depth}}$ is a general regression loss function, which evaluates the accuracy of the predicted depth. Common options include Mean Squared Error and Mean Absolute Error.

In the end-to-end methods, a single unified model predicts the full 3D map:

$$\mathbf{D}_{3D} = h(\mathbf{I}, \mathbf{DEM}) \tag{3}$$

where:

- $\mathbf{I}$ is the input data (e.g., satellite, UAV or rainfall data).
- $\mathbf{DEM}$ is the digital elevation model.
- h is the end-to-end model that predicts the full 3D flood map.

The loss function for the end-to-end methods combines components for flood extent and depth prediction:

$$\mathcal{L}_{\text{e2e}} = \lambda\mathcal{L}_{\text{classification}}\left(\mathbf{D}_{\text{extent}}, \mathbf{F}_{\text{extent}}^{\text{true}}\right) + \gamma\mathcal{L}_{\text{regression}}\left(\mathbf{D}_{\text{depth}}, \mathbf{D}_{\text{depth}}^{\text{true}}\right) \tag{4}$$

where:

- $\mathcal{L}_{\text{classification}}$ and $\mathcal{L}_{\text{regression}}$ are general loss functions.
- λ and γ are weights balancing the classification and regression components.

The end-to-end methods simplify the process by using a single model to predict the entire 3D flood map, eliminating the need for intermediate steps.

2.2 Comparison of DL-Based 3D Flood Mapping Methods

Building on the problem statement, this subsection compares existing methodologies to highlight their strengths and limitations. Task decomposition methods break 3D flood mapping into two separate subtasks. In contrast, end-to-end methods use one model to learn flood extent and depth directly from multi-source data. Table 1 provides a comparison of the two methods.

As shown in Table 1, task decomposition methods allow for flexible optimization of each subtask and make it easy to integrate with current 2D flood mapping research. However, since these subtasks are performed in sequence, error propagation between them can reduce the accuracy of the final results. Additionally, the sequential nature increases computation time. In contrast, the end-to-end methods use a single model to directly learn patterns of flood extent and depth from multi-source data. This approach eliminates intermediate steps, avoiding human-induced errors and error propagation while improving computational efficiency. However, the lack of intermediate outputs reduces the model's flexibility, and re-training may be required for scenario changes.

Table 1. Comparison of task decomposition and end-to-end methods.

Aspect	Task Decomposition Methods	End-to-End Methods
Workflow	Two stages: $\mathbf{F}_{\text{extent}}$ and $\mathbf{D}_{\text{depth}}$	Single unified model $h(\mathbf{I}, \mathbf{DEM})$
Loss Functions	Separate: $\mathcal{L}_{\text{classification}}$ and $\mathcal{L}_{\text{regression}}$	Combined: $\mathcal{L}_{\text{e2e}}$
Interpretability	Intermediate outputs available	Direct output, less interpretable
Flexibility	Modular and adaptable	Requires re-training for changes
Performance	Errors can spread between subtasks	Joint optimization often improves accuracy
Efficiency	Sequential steps slow down inference	One model speeds up inference

2.3 Data Sources for 3D Flood Mapping

This subsection reviews the data sources that support these methods and presents the resolution, coverage, and information provided by different data types.

Common 3D data sources include satellites, UAVs, Light Detection and Ranging (LiDAR), DEM, rainfall, and water gauges. These data can be processed through DL and GIS technologies to create accurate flood models. Some typical data sources are shown in Table 2.

Table 2. Summary of key data sources

Sources	Spatial Res	Temporal Res	Coverage	Notes
Satellites	5–100 m	5–16 days	Global/Regional	Multispectral/Hyperspectral.
UAVs	<0.1 m	On-demand	Local	RGB, Multispectral.
DEM	1–30 m	>1 year	Global/Regional	Terrain data.
LiDAR	<1 m	On-demand	Local	Terrain data (higher cost).
Rainfall	1k–10km	Minutes/Hours	Global/Regional	Rain gauges or simulation.
Gauge	Point scale	–	Point level	Monitoring of water level

Certain data sources, such as satellite and UAV data, provide 2D information, while others, like DEMs, are crucial for estimating flood depth. Rainfall records contribute temporal features, and water gauge measurements can offer ground truth for flooding in specific regions.

However, these data sources differ not only in format but also in spatial coverage, resolution, and other characteristics. While the complementarity of different data sources offers multifaceted support for 3D flood mapping, it also introduces challenges in multi-data fusion.

3 DL Models for 3D Flood Mapping

This section focuses on reviewing DL models for 3D flood mapping and classifies these approaches into two types: task decomposition and end-to-end methods, both applicable to static and dynamic scenarios.

Static flood mapping focuses on predicting floods at a specific time, emphasizing spatial characteristics. In contrast, dynamic flood mapping captures the temporal features of floods, integrating both spatial and temporal dimensions. The proportion of two methods in DL-based 3D flood mapping is shown in Fig. 2.

Task Decomposition (36%) $\mathbf{D}_{3D} = f_{\text{depth}}(\mathbf{F}_{\text{extent}}, \mathbf{DEM})$		End-to-End (64%) $\mathbf{D}_{3D} = h\,(\mathbf{I}, \mathbf{DEM})$	
Static (31%)	5%	Static: 13%	Dynamic (51%)

Fig. 2. The proportion of the two methods in DL-based 3D flood mapping.

As shown in Fig. 2, among the classifications, studies based on end-to-end methods for dynamic scenarios are the most common, accounting for 51%, while dynamic mapping based on task decomposition is the least common, representing only 5%.

3.1 Task Decomposition Methods

According to Fig. 2, task decomposition can be categorized into Static Flood Mapping and Dynamic Flood Mapping based on the type of output mapping.

Static Flood Mapping. In static flood mapping, various models can handle the $\mathbf{F}_{\text{extent}}$ subtask. Convolutional Neural Networks (CNNs) are popular because they automatically extract spatial features from large flood datasets. For example, Gebrehiwot et al. [15] utilized a Fully Convolutional Neural Network (FCN-8 s) to extract $\mathbf{F}_{\text{extent}}$ and extended this to $\mathbf{D}_{\text{3D}}$ by incorporating DEM. They compared various methods, including Structure from Motion [14], and the Geomorphic Flood Index [16], concluding that integrating DL models with DEM and UAV imagery significantly enhances 3D flood mapping accuracy.

Building on this method, Tan et al. [41] employed the DeepLabV3 model with transfer learning for $\mathbf{F}_{\text{extent}}$ and used ResNet-34 to upscale low-resolution DEM data. This approach addressed the challenge of limited high-resolution data, enabling the generation of high-resolution $\mathbf{D}_{\text{3D}}$.

DL models can be used not only for detecting $\mathbf{F}_{\text{extent}}$ and enhancing flood data quality, but also for estimating $\mathbf{D}_{\text{depth}}$. Tokoya et al. [44] proposed a framework combining Attention U-Net and LinkNet to estimate flood depth and terrain deformation. Because of the powerful pattern learning capabilities of DL models, this approach can even compensate for omission or misclassification parts in remote sensing methods.

Additionally, DL models can be used for both $\mathbf{F}_{\text{extent}}$ detection and $\mathbf{D}_{\text{depth}}$ estimation. Generative Adversarial Networks (GANs) provide structural flexibility for complex scenarios in $\mathbf{D}_{\text{3D}}$. Based on this characteristic, do Lago et al. [28] proposed a dual-generator GAN model adhering to flood volume conservation principles. One generator identified $\mathbf{F}_{\text{extent}}$, while the other estimated $\mathbf{D}_{\text{depth}}$. This approach achieved simulation speeds 250 times faster than the hydrodynamic model HEC-RAS [28].

In addition to the above, DL models can also have other applications in task decomposition methods. For example, they can assist in data preprocessing to support hydrodynamic models. Fang et al. [12] developed a workflow using the RandLA-Net to classify UAV data, automatically assigning runoff parameters and grid resolutions to different areas. These classifications were subsequently utilized by the hydrodynamic model MIKE 21 to simulate floods. This framework reduces manual configuration efforts and improves simulation accuracy.

Although various DL models can be used, static mapping typically relies on spatial data alone, without temporal information. In contrast, the dynamic features of floods are often closely tied to temporal factors like rainfall [43], which can provide a better understanding of flood events.

Dynamic Flood Mapping. Recurrent Neural Networks (RNNs) are DL models designed for sequential data. However, traditional RNNs suffer from vanishing and exploding gradient problems. To address these challenges, the Long Short-Term Memory (LSTM) was proposed, which can learn long-term dependencies [40]. Zhou Y. et al. [49] introduced a Spatial Reduction and Reconstruction (SRR) framework combining LSTM, greatly enhancing computational efficiency by selecting and reconstructing key feature points. The SRR-DL framework achieved speeds 21 times faster than the hydrodynamic model (TUFLOW).

Though traditionally spatial, CNNs can incorporate temporal information by processing serialized temporal rainfall data. Zhou Y. et al. [48] enhanced their SRR framework [49] by using U-Net, achieving higher accuracy and computational efficiency (98 times faster than TUFLOW).

3.2 End-to-End Methods

While task decomposition methods could integrate with current 2D flood mapping research, the lack of ground truth for flood depth presents significant challenges in DL training and 3D evaluation. Therefore, many researchers prefer to generate large amounts of flood simulation data using hydrodynamic models, which enable the use of end-to-end approaches for DL-based 3D flood mapping, as described in Eq. 3.

Static Flood Mapping. CNNs could be used in end-to-end methods for static flood mapping. Hosseiny et al. [21] modified U-Net to predict $\mathbf{D}_{\text{3D}}$ flood discharge data, using hydraulic model (iRIC) simulations as ground truth. This approach reduced manual intervention, and achied a 29% improvement in flood depth prediction accuracy compared to task decomposition and MLP-based methods [22].

Guo et al. [17] treated flood depth prediction as an image transformation task, integrating spatial and temporal inputs to generate maximum flood depth predictions. This approach reduced computational time to 0.5% of that required by the cellular-automata flood model (CADDIES). The authors [18] further explored how catchment areas affect CNN-based flood mapping, finding that larger receptive fields improve accuracy.

Dynamic Flood Mapping. Building on LSTM models, researchers have demonstrated that DL methods are significantly more computationally efficient than hydraulic models in dynamic flood mapping. Zhou, Q et al. [47] combined rainfall time series data, DEM, LSTM, and Bayesian optimization to develop a framework for rapid $\mathbf{D}_{3D}$. This framework operates 19,585 times faster than the hydraulic model (MIKE 21) with a relative error of 9.5%.

Many other studies have shown that DL models offer high computational efficiency than hydraulic models. For instance, Kabir, S. et al. [23] utilized simulation data from a hydraulic model (LISFLOOD-FP) to estimate $\mathbf{D}_{3D}$ using a basic 1D CNN model. This model effectively simulates floods while significantly reducing computational costs. Several researchers have conducted similar studies [11,17,32], highlighting the efficiency of DL models.

To further enhance the extraction of dynamic flood features, some researchers have integrated attention mechanisms into CNN models. Shao Y et al. [39] introduced CRU-Net, a U-Net that integrates Convolutional Block Attention Module and Residual Block to predict $\mathbf{D}_{3D}$ in urban areas. Similarly, Chaudhery et al. [5] combined U-Net with temporal attention mechanisms, showcasing both the accuracy and computational efficiency of DL models.

In 3D flood mapping, the importance of DEM cannot be overstated due to its provision of crucial elevation information. To investigate the impact of DEM on $\mathbf{D}_{3D}$, Fereshtehpour et al. [13] investigated the influence of DEM types and resolutions using a 1D-CNN model. Their findings revealed that digital terrain models outperform digital surface models in accuracy. Additionally, while low-resolution DEMs reduce the precision of DL models, they remain highly effective for applications requiring rapid 3D flood mapping.

Graph Neural Networks (GNNs) represent data as graphs, where nodes represent entities like geographical locations and edges show their relationships. By propagating information across nodes, GNNs update node features and learn the overall graph structure. Bentivoglio et al. [1] used this approach to develop the Shallow Water Equation-Graph Neural Network (SWE-GNN) model for predicting water depth and discharges. Its propagation rule, based on shallow water equations, enhances the framework's interpretability.

3.3 Discussions

DL-based 3D flood mapping offers advantages over traditional methods, including better computational efficiency and scalability. Task decomposition methods are often used for static mapping and typically rely on measured data. These methods are modular, making it easier to integrate with other technologies. In contrast, end-to-end approaches are suited for dynamic mapping and primarily use simulated data. They simplify the process by employing a single unified model to generate 3D flood maps. More researchers are adopting end-to-end methods and simulated data for 3D flood mapping studies. However, their reliance on simulated data can limit their application in the real world.

Meanwhile, DL models demonstrate versatility across various stages of the 3D flood mapping workflow. CNNs are effective for capturing spatial and temporal features, while LSTMs are well-suited for modeling temporal dynamics. Advanced architectures, such as GANs and GNNs, further enhance the ability to handle complex scenarios. Moreover, DL models are not only used for flood extent and depth estimation but also play a critical role in data augmentation and preprocessing, highlighting their adaptability to diverse tasks and datasets.

4 Applications of 3D Flood Mapping with DL

This section highlights the applications of DL-based 3D flood mapping. The 3D flood maps provide not only flood extent but also key details on depth and hazard levels, enabling comprehensive risk assessments. These capabilities support near real-time disaster response, long-term flood prediction, and urban planning. Near real-time applications address flood events as they occur or shortly after, while long-term applications cover scenarios over years or decades. Figure 3 summarizes the applications of DL-based 3D flood mapping.

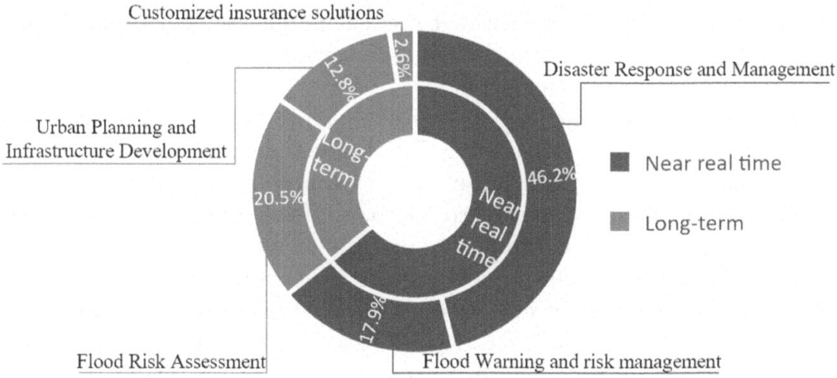

Fig. 3. The proportion of different types of applications in 3D flood mapping.

As shown in Fig. 3, most studies focus on near real-time applications, especially disaster response and flood management, which make up over 46% of the cases. This underscores the need for rapid 3D flood mapping, a key strength of DL models. In contrast, customized insurance solutions account for less than 3% of the studies. Although this approach is a passive response to flooding and is underexplored, it has potential to provide fair protection for residents and assets.

4.1 Near Real-Time Flood Prediction

The computational efficiency of DL models enables the generation of high-resolution 3D flood maps shortly after flood events [10,24]. However, due to the delays from data platforms, such mapping is typically classified as near real-time. To overcome this limitation, some studies have explored rainfall-driven real-time flood prediction as a solution [3,36].

By combining DL models with rainfall prediction, it is possible to perform 3D flood maps in advance for flood warnings [3,4,35,46]. For instance, Chaudhary et al. [5], Bentivoglio et al. [1], and El Baida et al. [11] can generate high-precision 3D flood maps 2 to several tens of hours in advance based on rainfall data. Combining predicted 3D flood maps for different regions with risk data, such as population density, decision-makers can make decisions more accurately.

Detailed 3D flood information is essential for disaster response, enabling targeted actions during flood disasters [20]. Tan et al. [41] demonstrated in Calgary, Canada, that DL models can generate high-resolution 3D flood maps even with low-resolution data. Furthermore, Li et al. [31] utilized GIS methods to develop an online portal for the automated processing and visualization of global 3D flood maps. This allows decision-makers to prioritize severely impacted regions and facilitate efficient rescue operations for people and property in danger zones.

4.2 Long Term Flood Prediction

Additionally, this survey explores the use of 3D flood mapping in strategic planning and long-term flood risk assessment. DL's ability to process large datasets makes it well-suited for 3D flood susceptibility maps based on historical flood data. These maps enable authorities to assess flood risk, customize insurance policies, and proactively plan urban flood control infrastructure, minimizing future economic losses and casualties.

Long-term historical data such as hydrology and flooding records can be used for flood risk assessment. Based on the 3D risk assessment, measures can be taken to reduce flood hazards [9,45]. Bui et al. [2] proposed a DL approach for flash flood risk assessment in Northwestern Vietnam. Their model provided valuable tools for land-use management, offering insights to reduce flood impacts.

Since 3D flood susceptibility maps provide richer information than 2D maps, customized insurance solutions for severely affected flood areas become possible. Chen et al. [7] studied rainfall, DEM, and historical flood events to develop 3D flood risk maps. Their study identified urban coastal areas as high-risk zones and emphasized the importance of adaptive measures, such as raising levee heights and improving insurance systems, to cope with increasing flood threats.

With the help of 3D flood susceptibility maps, targeted urban planning and infrastructure development for floods have seen increasing applications [42]. Panahi et al. [34] created flood probability maps for Golestan province in Iran, revealing that nearly 40% of the area is flood-prone and guiding planning efforts. Similarly, Khosravi et al. [26] mapped flood susceptibility for Iran, showing that 12% of the country is at risk. These maps provide essential data for flood control planning, helping high-risk regions mitigate future impacts.

5 Evaluation Metrics and Model Performance

This section details the criteria used to assess DL-based 3D mapping methods and provides a comparative analysis of evaluation metrics from the literature.

5.1 Evaluation Metrics

Accurately evaluating model performance is crucial, regardless of the type of research methods used. Compared to 2D flood maps, 3D maps include depth information, which requires additional factors to accurately assess both accuracy and precision. Fortunately, the geometric extent and elevation information of floods in 3D models can be decoupled and evaluated separately.

The evaluation of the 2D extent of floods can be treated as a typical binary classification task. There are various evaluation metrics, such as the Recall [23], and the Intersection over Union (IoU). Since the classification of flooded regions often demands high spatial accuracy, the IoU metric is commonly used to evaluate the spatial accuracy of flood predictions [41,44]. In 3D flood mapping, the model's prediction of flood depth is considered a regression task. The overall assessment of flood elevation results can be conducted using regression error metrics such as Root Mean Square Error (RMSE) [11,13,47] and Mean Absolute Error (MAE) [1,5].

In addition to examining the performance of 3D flood mapping models, it is essential to consider their computational efficiency, as this is vital for large-scale and near-real-time flood prediction. Since many researchers use hydrodynamic models to provide ground truth or as a benchmark for comparison, DL methods are often evaluated against these models in terms of simulation time to assess their computational efficiency quantitatively [28,48].

5.2 Model Performance of DL Based 3D Flood Mapping

Table 3 summarizes the evaluation metrics for DL-based 3D flood mapping and includes details on the methods, data, models, and more. Due to space constraints, only a subset of the literature is presented.

As shown in Table 3, 3D flood mapping relies heavily on simulated data, compared to 2D mapping, which frequently uses measured data. This reliance is due to the difficulty of obtaining ground truth for flood depth, resulting in a scarcity of publicly available 3D flood datasets based on measured data. Moreover, the lack of standardized datasets poses significant challenges for establishing baselines and comparing the performance of different methods and models. Despite limitations in datasets, DL models have a significant advantage over traditional models in computational efficiency. They can deliver predictions up to 100 times faster than hydraulic models. This enables real-time 3D flood mapping, which was unachievable due to the time-consuming nature of hydraulic models.

Table 3 also indicates that within task decomposition methods, studies focus more on static mapping. The measured data used can have centimeter-level resolutions, and the average RMSE for flood depth prediction is around 20 cm.

Table 3. Summary of the evaluation metrics for DL based 3D flood mapping

Method	Feature	Paper	Data: Sources	Data: Res[1] (m)	Models	Metrics: Extent(%) Recall	Metrics: Extent(%) IoU	Metrics: Depth (cm) MAE	Metrics: Depth (cm) RMSE	Time Efficiency[2]
Task Decomposition	Static	[16] 2021, M[3]	DEM, LiDAR, UAV, Gauge	0.026	FCN-8s (CNN)	-	-	-	26	-
		[12] 2024, M	UAV, MIKE 21	<0.1	RandLA-NET (CNN)	-	70.8	-	-	-
		[41] 2024, M	LR/HR DEM, HR UAV	0.1 – 18	ResNet, DeepLabV3 (CNN)	76.9	50.3	-	-	-
		[28] 2023, M	DEM, Rainfall HEC-RAS	3	cGAN-Flood (GAN)	-	82.5	-	10	50–250
		[22] 2020, S[4]	DEM iRIC	5	RF, ANN (MLP)	-	-	<5	<22	60
		[44] *[5] 2022, M	DEM, Satellite, UAV Numerical-model	5	Attention U-Net (CNN)	-	44.5	-	38.5	-
	Dynamic	[48] * 2022, M	DEM TUFLOW	10	1D-CNN U-Net (CNN)	97.6	-	-	3.4	98
		[49] * 2021, M	DEM TUFLOW	20	SRR-DL (LSTM)	99	-	-	80	21–86
End-to-End	Dynamic	[11] 2024, S	Rainfall Hydro-model	2	CNN	-	-	0.9	3.4	120
		[5] * 2024, S	DEM, Rainfall CADDIES	2	U-TAE (CNN)	-	-	0.75	-	-
		[39] 2024, M	DEM, Rainfall LISFLOOF-FP	3	CRU-Net (CNN)	94.2	87.1	1.1	5.4	-
		[23] * 2020, M	DEM (LiDAR) LISFLOOD-FP	5	CNN, SVR	93	-	-	18	38
		[32] 2021, S	DEM, Rainfall MIKE 21	5	U-FLOOD (CNN)	-	58.3	-	8	-
		[13] 2024, S	LiDAR, Rainfall LISFLOOD-FP	15 – 30	1D CNN	99.5	-	-	65	-
		[8] 2020, S	DEM TUFLOW	20	GRNN	-	-	-	51	100
		[1] * 2023, S	DEM Hydro-model	100	SWE-GNN (GNN)	-	80+	<5.8	<15.4	100+
		[47] 2023, S	DEM, Rainfall MIKE 21	-	LSTM	-	-	-	<5	19585
	Static	[17] * 2021, S	DEM, CADDIES	1	CNN	-	-	<10	-	200
		[38] * 2023, S	DEM TELEMAC	1	Unet (CNN)	-	60+	-	<10	-
		[18] * 2022, S	DEM CADDIES	2	Unet (CNN)	-	-	-	<40	6000+

Res: Resolution of the input data.
Time: How many times faster than hydrodynamic models.
M: Measured data.
S: Simulated data.
*: Open access datasets.

In contrast, end-to-end methods emphasize dynamic mapping. However, these methods use simulation data with a maximum resolution of 1 m and yield more accurate flood depth predictions, with most studies reporting RMSE below 20 cm. In both methods, CNNs are the most widely used DL models.

Figure 4 visualizes the depth performance from the literature (input data with resolutions below 20 m per pixel), showing the relationship between RMSE and resolution. The x-axis represents spatial resolution.

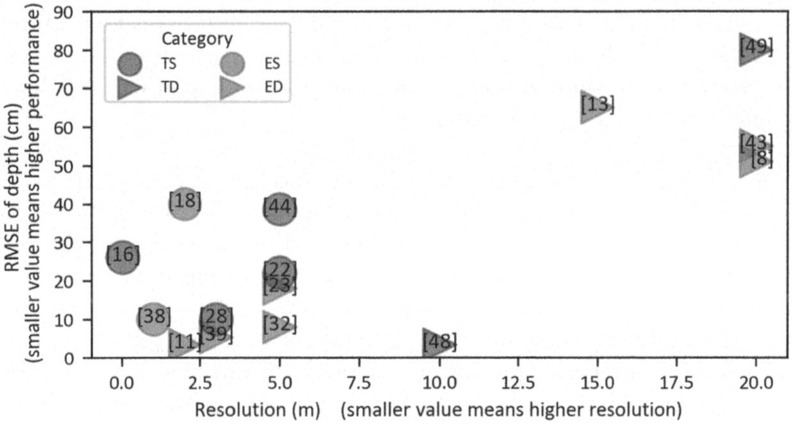

Fig. 4. Resolution vs Depth performance. (TS: Task decomposition Static; TD: Task decomposition Dynamic; ES: End-to-end Static; ED: End-to-end Dynamic)

As shown in Fig. 4, there is a clear correlation between RMSE and spatial resolution. Higher resolution generally yields a better RMSE, regardless of whether task decomposition or end-to-end methods are used. Moreover, the choice of data features also affects RMSE. At the same resolution, models using dynamic mapping (triangle markers) tend to perform better than those using static mapping (circle markers). This improved performance is likely due to the incorporation of temporal data in dynamic mapping, which offers a more accurate representation of flood characteristics.

6 Challenges and Future Directions

This section identifies the major obstacles in advancing DL-based 3D flood mapping and proposes potential solutions.

6.1 Challenges

Despite the advancements in 3D flood mapping using DL models, several challenges persist, particularly in data, models, and practical applications.

Challenges in Data. The scarcity of 3D datasets is a major obstacle, resulting in a lack of standard baselines for comparing methods. Researchers often have to create their own datasets, which raises the barrier to entry. Some use simulated data from hydraulic models, but this risks training on inaccurate flood characteristics. Moreover, ground truth for flood depth is limited—sparse water gauge records fail to capture comprehensive flood profiles, making model training, validation, and assessment difficult. In addition, satellite imagery often lacks the resolution required for detailed urban mapping, while UAV data, though high-resolution, is constrained by high costs, limited coverage, and logistical issues.

Integrating diverse data sources—such as DEM, multispectral, and hydrological datasets—is challenging due to variations in resolution, coverage, and accuracy. Reliance on simple interpolation methods can lead to data distortion, making effective fusion a persistent issue.

Challenges in DL Models. In terms of DL models, while advanced models like GANs and GNNs have emerged, traditional DL models, especially CNNs, still face limitations in robustness, generalization, and interpretability. Hybrid models that combine DL with hydrodynamic approaches show promise by merging computational efficiency with physical accuracy. However, poorly designed integrations may increase complexity without performance gains and can compromise stability. Furthermore, the "black-box" nature of DL models makes it difficult to discern what patterns are being learned. Techniques such as Grad CAM have improved explainability, yet they do not fully alleviate user concerns about model transparency, ultimately affecting trust in these systems.

Challenges in Applications. Integrating 3D flood mapping models into practical systems remains underexplored. Current research prioritizes mapping accuracy and speed over real-world deployment. Ethical issues add another challenge because high-resolution spatial data may reveal sensitive information. This raises privacy and consent concerns. Unrepresentative training data can introduce algorithmic bias and lead to flawed flood risk assessments. Consequently, public policies based on these maps must be transparent and accountable, and stakeholders must understand the models' limitations and assumptions.

6.2 Future Directions

Despite the challenges, advances in DL and related technologies promise more accurate and efficient models. These improvements are expected to generate higher-resolution maps and support a broader range of applications, from early warning systems to urban and land use planning.

Data scarcity remains a significant challenge. Future research should focus on developing concrete solutions to fill this gap. For instance, advanced data augmentation techniques can be developed to synthetically enrich benchmarked datasets. Moreover, establishing collaborative partnerships with governmental agencies and private organizations could facilitate access to more comprehensive, high-resolution data, thereby improving the robustness of flood mapping models.

Future research should explore advances in multi-data fusion that enable the integration of complementary information from diverse sources such as satellites, DEMs, and weather data. This approach enhances both spatial and temporal resolution while reducing uncertainties in mapping. In particular, fusing multiple data sources improves the accuracy of depth estimation, which is critical for boosting overall performance in 3D flood mapping.

Further work is needed to boost model performance. One promising approach is to merge various DL architectures—such as incorporating Vision Foundation Models like the Segment Anything Model (SAM)—to enhance mapping accuracy and usability. Beyond performance, improving DL model interpretability is crucial. Future research should explore advanced techniques that reveal the internal workings of these models, such as sensitivity analysis or integrating physical constraints into DL frameworks. These strategies can enhance both the interpretability and robustness of the results, foster greater user trust, and improve the practical applicability of 3D flood mapping.

Advances in 3D flood mapping, with its high-resolution and interpretable results, can significantly influence flood management strategies. These developments support not only early warning systems and targeted rescue but also urban planning, land use planning, and government flood policies. For instance, they can inform tailored flood insurance policies based on varying risk levels, offering stronger protection for vulnerable regions. While formulating such policies, it is essential to implement robust data governance frameworks, bias mitigation strategies, and transparent stakeholder engagement to address ethical concerns.

7 Conclusion

This survey emphasizes the potential of DL in 3D flood mapping, offering significant improvements in computational efficiency and scalability over traditional methods. Task decomposition approaches leverage modular workflows to build on existing 2D research, while end-to-end methods streamline the prediction process. Advanced models, including CNNs, LSTMs, GANs, and GNNs, have demonstrated their ability to address complex flood scenarios, though challenges in datasets, interpretability, model integration, and practical application remain.

Looking ahead, addressing data scarcity is paramount. Innovative solutions like few-shot synthesis and partnering with the government or industry can offer more publicly accessible benchmarked datasets. Researchers should also investigate hybrid models that merge DL with hydrodynamic methods or emerging architectures like SAM. Furthermore, enhancing model transparency through sensitivity analysis and incorporating physical constraints is equally important.

Finally, practical applications of 3D flood mapping must account for ethical and governance challenges. Policy makers should ensure that flood models support transparent and accountable decision-making. Robust data governance and bias mitigation are essential. By bridging current gaps, 3D flood mapping can better support disaster management, urban planning, and policy-making, ultimately reducing flood risks in vulnerable regions.

References

1. Bentivoglio, R., Isufi, E., Jonkman, S.N., Taormina, R.: Rapid spatio-temporal flood modelling via hydraulics-based graph neural networks. Hydrol. Earth Syst. Sci. **27**(23), 4227–4246 (2023)
2. Bui, D.T., et al.: A novel deep learning neural network approach for predicting flash flood susceptibility: a case study at a high frequency tropical storm area. Sci. Total Environ. **701**, 134413 (2020)
3. Chang, D.L., Yang, S.H., Hsieh, S.L., Wang, H.J., Yeh, K.C.: Artificial intelligence methodologies applied to prompt pluvial flood estimation and prediction. Water **12**(12), 3552 (2020)
4. Chaudhary, P., et al.: Flood uncertainty estimation using deep ensembles. Water **14**(19) (2022)
5. Chaudhary, P., Leitão, J.P., Schindler, K., Wegner, J.D.: Flood water depth prediction with convolutional temporal attention networks. Water **16**(9), 1286 (2024)
6. Chen, H., Shi, Z.: A spatial-temporal attention-based method and a new dataset for remote sensing image change detection. Remote Sens. **12**(10) (2020)
7. Chen, J., Huang, G., Chen, W.: Towards better flood risk management: assessing flood risk and investigating the potential mechanism based on machine learning models. J. Environ. Manage. **293**, 112810 (2021)
8. Chu, H., Wu, W., Wang, Q.J., Nathan, R., Wei, J.: An ANN-based emulation modelling framework for flood inundation modelling: application, challenges and future directions. Environ. Model. Softw. **124** (2020)
9. Costache, R., Ngo, P., Bui, D.T.: Novel ensembles of deep learning neural network and statistical learning for flash-flood susceptibility mapping. Water **12**(6), 1549 (2020)
10. Dong, S., Yu, T., Farahmand, H., Mostafavi, A.: A hybrid deep learning model for predictive flood warning and situation awareness using channel network sensors data. Comput. Aided Civil Infrastruct. Eng. **36**(4), 402–420 (2021)
11. El Baida, M., Boushaba, F., Chourak, M., Hosni, M.: Real-time urban flood depth mapping: convolutional neural networks for pluvial and fluvial flood emulation. Water Resour. Manage **38**(12), 4763–4782 (2024)
12. Fang, X., et al.: A rapid assessment method for flood risk mapping integrating aerial point clouds and deep learning. Water Resour. Manage **38**(5), 1753–1772 (2024)
13. Fereshtehpour, M., Esmaeilzadeh, M., Alipour, R.S., Burian, S.J.: Impacts of dem type and resolution on deep learning-based flood inundation mapping. Earth Sci. Inf. **17**(2), 1125–1145 (2024)
14. Gebrehiwot, A., Hashemi-Beni, L.: A method to generate flood maps in 3D using dem and deep learning. Int. Arch. Photogrammetry, Remote Sens. Spatial Inf. Sci. **XLIV-M-2-2020**, 25–28 (2020)
15. Gebrehiwot, A.A., Hashemi-Beni, L.: Three-dimensional inundation mapping using UAV image segmentation and digital surface model. ISPRS Int. J. Geo-Inf. **10**(3) (2021)
16. Gebrehiwot, A., Hashemi-Beni, L.: 3D inundation mapping: a comparison between deep learning image classification and geomorphic flood index approaches. Front. Remote Sens. **3** (2022)
17. Guo, Z., Leitão, J.P., Simões, N.E., Moosavi, V.: Data-driven flood emulation: speeding up urban flood predictions by deep convolutional neural networks. J. Flood Risk Manag. **14**(1) (2021)

18. Guo, Z., Moosavi, V., Leitão, J.P.: Data-driven rapid flood prediction mapping with catchment generalizability. J. Hydrol. **609** (2022)
19. Han, W., et al.: A survey of machine learning and deep learning in remote sensing of geological environment: challenges, advances, and opportunities. ISPRS J. Photogramm. Remote. Sens. **202**, 87–113 (2023)
20. He, J., Zhang, L., Xiao, T., Wang, H., Luo, H.: Deep learning enables super-resolution hydrodynamic flooding process modeling under spatiotemporally varying rainstorms. Water Res. **239**, 120057 (2023)
21. Hosseiny, H.: A deep learning model for predicting river flood depth and extent. Environ. Model. Softw. **145** (2021)
22. Hosseiny, H., Nazari, F., Smith, V., Nataraj, C.: A framework for modeling flood depth using a hybrid of hydraulics and machine learning. Sci. Rep. **10**(1) (2020)
23. Kabir, S., Patidar, S., Xia, X., Liang, Q., Neal, J., Pender, G.: A deep convolutional neural network model for rapid prediction of fluvial flood inundation. J. Hydrol. **590** (2020)
24. Kao, I.F., Liou, J.Y., Lee, M.H., Chang, F.J.: Fusing stacked autoencoder and long short-term memory for regional multistep-ahead flood inundation forecasts. J. Hydrol. **598**, 126371 (2021)
25. Karim, F., Armin, M.A., Ahmedt-Aristizabal, D., Tychsen-Smith, L., Petersson, L.: A review of hydrodynamic and machine learning approaches for flood inundation modeling. Water **15**(3) (2023)
26. Khosravi, K., et al.: Convolutional neural network approach for spatial prediction of flood hazard at national scale of Iran. J. Hydrol. **591**, 125552 (2020)
27. Kussul, N., Lavreniuk, M., Skakun, S., Shelestov, A.: Deep learning classification of land cover and crop types using remote sensing data. IEEE Geosci. Remote Sens. Lett. **14**(5), 778–782 (2017)
28. do Lago, C.A.F., Giacomoni, M.H., Bentivoglio, R., Taormina, R., Gomes, M.N., Mendiondo, E.M.: Generalizing rapid flood predictions to unseen urban catchments with conditional generative adversarial networks. J. Hydrol. **618** (2023)
29. Li, C., et al.: Review on urban flood risk assessment. Sustainability **15**(1), 765 (2022)
30. Li, H., Zhu, M., Li, F., Skitmore, M.: Solving flood problems with deep learning technology: Research status, strategies, and future directions. Sustainable Development (2024)
31. Li, S., et al.: High resolution 3D mapping of hurricane flooding from moderate-resolution operational satellites. Remote Sens. **14**(21), 5445 (2022)
32. Löwe, R., Böhm, J., Jensen, D.G., Leandro, J., Rasmussen, S.H.: U-flood – topographic deep learning for predicting urban pluvial flood water depth. J. Hydrol. **603**, 126898 (2021)
33. Mudashiru, R.B., Sabtu, N., Abustan, I., Balogun, W.: Flood hazard mapping methods: a review. J. Hydrol. **603** (2021)
34. Panahi, M., et al.: Deep learning neural networks for spatially explicit prediction of flash flood probability. Geosci. Front. **12**(3), 101076 (2021)
35. Pianforini, M., Dazzi, S., Pilzer, A., Vacondio, R.: Real-time flood maps forecasting for dam-break scenarios with a transformer-based deep learning model. J. Hydrol. **635** (2024)
36. Ravuri, S., et al.: Skilful precipitation nowcasting using deep generative models of radar. Nature **597**(7878), 672–677 (2021)
37. Rentschler, J., Salhab, M., Jafino, B.A.: Flood exposure and poverty in 188 countries. Nat. Commun. **13**(1) (2022)

38. Seleem, O., Ayzel, G., Bronstert, A., Heistermann, M.: Transferability of data-driven models to predict urban pluvial flood water depth in Berlin, Germany. Nat. Hazard. **23**(2), 809–822 (2023)
39. Shao, Y., Chen, J., Zhang, T., Yu, T., Chu, S.: Advancing rapid urban flood prediction: a spatiotemporal deep learning approach with uneven rainfall and attention mechanism. J. Hydroinf. **26**(6), 1409–1424 (2024)
40. Sherstinsky, A.: Fundamentals of recurrent neural network (RNN) and long short-term memory (LSTM) network. Physica D **404**, 132306 (2020)
41. Tan, W., Qin, N., Zhang, Y., McGrath, H., Fortin, M., Li, J.: A rapid high-resolution multi-sensory urban flood mapping framework via dem upscaling. Remote Sens. Environ. **301**, 113956 (2024)
42. Wu, Z., Zhou, Y., Wang, H., Jiang, Z.: Depth prediction of urban flood under different rainfall return periods based on deep learning and data warehouse. Sci. Total Environ. **716**, 137077 (2020)
43. Xie, S., Wu, W., Mooser, S., Wang, Q.J., Nathan, R., Huang, Y.: Artificial neural network based hybrid modeling approach for flood inundation modeling. J. Hydrol. **592** (2021)
44. Yokoya, N., et al.: Breaking limits of remote sensing by deep learning from simulated data for flood and debris-flow mapping. IEEE Trans. Geosci. Remote Sens. **60**, 1–15 (2022)
45. Yu, L., Qin, H., Huang, S., Wei, W., Jiang, H., Mu, L.: Quantitative study of storm surge risk assessment in an undeveloped coastal area of China based on deep learning and geographic information system techniques: a case study of double moon bay. Nat. Hazard. **24**(6), 2003–2024 (2024)
46. Zhang, W., Liu, Y., Tang, W., Chen, S., Xie, W.: Rapid spatio-temporal prediction of coastal urban floods based on deep learning approaches. Urban Climate **52** (2023)
47. Zhou, Q., et al.: A deep-learning-technique-based data-driven model for accurate and rapid flood predictions in temporal and spatial dimensions. Hydrol. Earth Syst. Sci. **27**(9), 1791–1808 (2023)
48. Zhou, Y., Wu, W., Nathan, R., Wang, Q.J.: Deep learning-based rapid flood inundation modeling for flat floodplains with complex flow paths. Water Resour. Res. **58**(12) (2022)
49. Zhou, Y., Wu, W., Nathan, R., Wang, Q.J.: A rapid flood inundation modelling framework using deep learning with spatial reduction and reconstruction. Environ. Model. Softw. **143** (2021)

A Survey of Foundation Models for Environmental Science

Runlong Yu[1], Shengyu Chen[1], Yiqun Xie[2], and Xiaowei Jia[1(✉)]

[1] University of Pittsburgh, Pittsburgh, USA
{ruy59,shc160,xiaowei}@pitt.edu
[2] University of Maryland, College Park, USA
xie@umd.edu

Abstract. Modeling environmental ecosystems is essential for effective resource management, sustainable development, and understanding complex ecological processes. However, traditional methods frequently struggle with the inherent complexity, interconnectedness, and limited data of such systems. Foundation models, with their large-scale pre-training and universal representations, offer transformative opportunities by integrating diverse data sources, capturing spatiotemporal dependencies, and adapting to a broad range of tasks. This survey presents a comprehensive overview of foundation model applications in environmental science, highlighting advancements in forward prediction, data generation, data assimilation, downscaling, model ensembling, and decision-making across domains. We also detail the development process of these models, covering data collection, architecture design, training, tuning, and evaluation. By showcasing these emerging methods, we aim to foster interdisciplinary collaboration and advance the integration of cutting-edge machine learning for sustainable solutions in environmental science.

Keywords: Foundation models · Knowledge-guided machine learning · Environmental informatics · Sustainable AI solutions

1 Introduction

Healthy environmental ecosystems are fundamental to human survival and well-being, providing essential resources such as clean air, water, food, and energy, which sustain life and fuel economic growth. Modeling these systems is essential for understanding underlying processes and generating reliable predictions to guide resource management. However, this is challenging because environmental systems are inherently complex, involving numerous interacting processes, and are often poorly observed due to the high costs of data collection. Traditionally, process-based physical models have been used to represent ecosystems across various environmental domains, including climate science, hydrology, agriculture, forestry, and geology [20,43]. These methods rely on mathematical or

R. Yu and S. Chen—contributed equally.

X. Wu et al. (Eds.): PAKDD 2025, LNAI 15875, pp. 39–57, 2025.
https://doi.org/10.1007/978-981-96-8295-9_3

physical equations to describe relevant processes. Given the complexity and limited knowledge in modeling certain processes, they often require approximations or parameterizations. Additionally, both calibration and inference for these models can be time-consuming and often demand significant domain expertise.

Advancements in data collection and processing in environmental science have sparked a growing interest in leveraging artificial intelligence (AI) and machine learning (ML) to better model environmental ecosystems [4,30]. These data-driven methods are particularly promising in scenarios where certain processes are not fully understood or require substantial computational resources. However, traditional ML models are typically designed for specific tasks, limiting their ability to capture the interconnectedness of various environmental processes. Integrating different ML models is challenging due to inconsistencies in model architectures, data inputs, and the scales of spatiotemporal processes involved. For instance, predicting water quality variables (e.g., water temperature and nutrient concentration) and water quantity variables (e.g., streamflow) is often handled by separate models, even though these variables are influenced by common processes. This siloed approach hinders understanding the relationships between tasks and limits effective information sharing across models. It may also introduce model bias due to the ignorance of related processes influencing the target variable. Additionally, clients (e.g., stakeholders and resource managers), who may lack expertise in AI and ML, often struggle to interpret and integrate results from multiple models to make informed decisions.

Recent ML studies have extensively explored different transfer learning and meta-learning strategies [23,73]. These works have demonstrated the potential of many ML models to be transferred from data-sufficient tasks to target data-sparse tasks. Building on the concept of transfer learning, there is a rising trend toward developing foundation models [9]. These models are pre-trained on a diverse set of tasks using either supervised or unsupervised methods to learn universal feature representations, enabling them to be fine-tuned for new tasks. Such power offers new opportunities for building data-driven models to address a broad range of environmental problems. In particular, many foundation models are able to harness large data from different sources and extract complex data patterns. They also offer flexibility in configuring input and output structures. For example, many large language models (LLMs) can accept user-specified inputs and generate different variables through proper prompt engineering. This is particularly useful in modeling environmental systems where we need to perform multiple related modeling tasks, but only a subset of them is observed in each data sample [40]. Additionally, foundation models can better adapt to new environments because they have been pre-trained on massive data [68].

Recognizing the transformative potential of foundation models, this survey comprehensively reviews their applications in environmental science. To ensure thorough coverage, we draw on research from major academic databases, leading journals (e.g., *Nature*, *Water Resources Research*, *Remote Sensing*, *TGRS*, *TPAMI*, *TKDE*), top conferences (e.g., *AAAI*, *IJCAI*, *NeurIPS*, *ICML*, *ICCV*, *CVPR*, *ACL*, *KDD*, *ICDM*, *SDM*, *WSDM*, *PAKDD*), specialized workshops (e.g., *KGML*), and preprints. We also review government reports from *NASA*, *USGS*, and *NOAA* to capture the latest advances in large-scale environmental AI.

The aim of this survey is to bring attention to the exciting advancements in the application of foundation models within environmental science, highlighting the opportunities for further research and development in this promising field. We hope that this survey will be valuable to both the machine learning community, by showcasing how foundation models are being developed to tackle complex environmental challenges, and to environmental scientists seeking to explore these cutting-edge models in their own work.

We organize the paper as follows: Sect. 2 provides an overview of the evolution from process-based models to foundation models, highlighting the paradigm shift in environmental modeling. Section 3 reviews the key objectives guiding the application of foundation models in environmental science. Section 4 explores innovative methods and architectures for building these models. Section 5 discusses future research opportunities, and the paper concludes with insights into the transformative potential of foundation models in environmental science.

2 Historical and Conceptual Overview

2.1 Environmental Modeling

The exponential growth of diverse environmental data has created unprecedented opportunities to advance our understanding of ecosystems. Traditional approaches to environmental computing have evolved through several stages [69], each reflecting a distinct paradigm for addressing environmental challenges:

- **Process-based models (1.0):** These models are grounded in fundamental scientific principles, including physics, chemistry, and biology, to analyze the underlying mechanisms of environmental phenomena. Often referred to as process-based or theory-driven models, they rely on mathematical formulations, such as differential equations, to simulate key processes. Although these models are interpretable, they often struggle with complex systems characterized by high variability and limited observational data.
- **Data-driven models (2.0):** With the rise of big data and AI, data-driven approaches emerged as a powerful alternative to traditional process-based models. These models prioritize pattern recognition, system characterization, and outcome prediction using large datasets. Aligning with the "Fourth Paradigm" of science, which emphasizes data-intensive methodologies, they excel at handling high-dimensional, complex problems. However, their "black-box" nature often limits interpretability and also requires large training data.
- **Hybrid physics-ML models (3.0):** To overcome the limitations of aforementioned single-paradigm approaches, process-guided or knowledge-guided ML integrates mechanistic insights into data-driven models [29,64]. This hybrid paradigm embeds physical laws and domain knowledge into machine learning workflows to improve accuracy, generalization, and consistency with fundamental principles such as conservation laws. For instance, in lake modeling, process-based components have been combined with recurrent neural

networks, yielding better performance for long-term trend predictions by constraining outputs with ecological principles [27,67].

Building on these computational paradigms, **foundation models (4.0)** represent the next leap forward. These models, which have already revolutionized domains such as natural language processing (NLP), are set to transform environmental science by enabling holistic, scalable, and integrated modeling approaches. Foundation models excel in their ability to assimilate data from multiple sources and domains, capturing intricate patterns across diverse systems. Their success is also attributed to the pre-training on extensive and diverse datasets, which enables scalability and adaptability from well-studied systems to data-scarce or unseen environments through effective knowledge transfer [9].

2.2 Emergence of Foundation Models

Foundation models are large, pre-trained machine learning systems capable of processing diverse modalities. These models have transformed the AI landscape by influencing the physical world, enabling complex reasoning, and facilitating meaningful human interaction. Their development builds on decades of advancements in ML, particularly the evolution of deep learning in the 2010s. Deep learning, driven by the concept of representation learning, allows systems to automatically extract meaningful features directly from raw data, significantly improving performance on complex tasks [2]. Despite their success, early deep learning models were still task-specific, which limited their broader applicability across multiple domains. The true breakthrough came with the advent of transfer learning, which allowed models trained on one task to be adapted to another. Transfer learning enabled the development of general-purpose models that could leverage pre-trained knowledge. Building on this, foundation models such as BERT and GPT expanded transfer learning by training on massive datasets, allowing them to generalize across tasks with minimal tuning [28].

A key driver of this transformation has been the rise of self-supervised learning, which enables models to learn representative patterns from vast amounts of unlabeled data with pre-defined pre-training tasks [70]. This is particularly valuable in environmental science, where labeled data is often scarce or incomplete [14,21,49]. Another critical factor in the success of foundation models is their scalability. Models such as GPT-3, with its 175 billion parameters, demonstrate the impact of scaling both data and model size. This scalability enables emergent capabilities like in-context learning, where the model adapts to new tasks simply by receiving natural language prompts [24,71]. In environmental science, this adaptability offers the potential to develop a unified model capable of handling multiple related tasks simultaneously, such as different water quality and quantity variables [34,40]. Such models could serve as powerful tools, offering a comprehensive perspective on diverse processes in dynamic environmental applications. Lastly, architectural innovations, particularly the Transformer architecture introduced by Vaswani et al. [60], have been instrumental in learning complex data patterns. Transformers are adept at capturing long-range dependencies in data, which is crucial for modeling spatial-temporal processes. Their ability to handle large datasets and integrate multi-modal data (e.g., text, images, sensors) makes them invaluable for unified ecosystem modeling [17,66].

3 Application-Centric Objectives

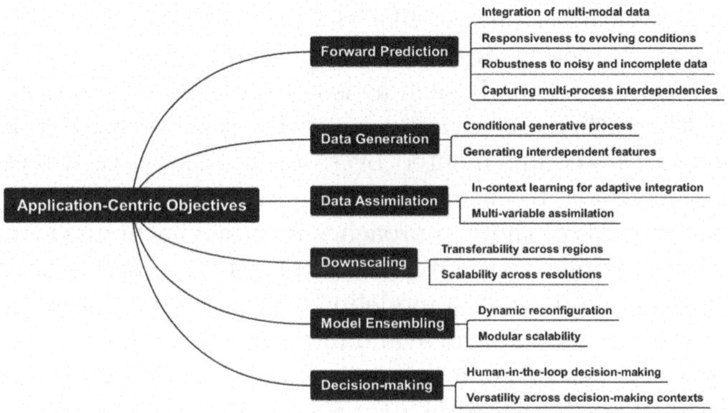

Fig. 1. Application-centric objectives and advancements enabled by foundation models.

This section examines the application of foundation models in environmental science, focusing on key objectives such as forward prediction, data generation, data assimilation, downscaling, model ensembling, and decision-making. Figure 1 highlights their advancements compared to traditional methods.

3.1 Forward Prediction

Forward prediction is a fundamental task in environmental science, focusing on modeling the relationships between input factors, such as climatic drivers or anthropogenic influences, and system characteristics to predict key environmental variables. This approach supports simulations of dynamic environmental processes over time, aiding in tasks like weather forecasting, greenhouse gas emissions modeling, water quality prediction, and tracking the spread of pests.

A specialized subset of forward prediction is forecasting, which emphasizes using historical and current data to predict future states of environmental variables. Forecasting often involves projecting temporal trends, such as long-term changes in climate patterns, air and water quality, or ecosystem health. Anomaly detection is another key aspect of forward prediction, focusing on identifying data points or patterns that significantly deviate from expected norms. In environmental science, it plays a crucial role in detecting early signs of critical or unusual changes, enabling timely interventions to prevent or mitigate potential negative impacts on ecosystems and human well-being.

Traditionally, process-based models based on physical principles have been key to environmental applications like climate, weather, and hydrology modeling. Parameterization is often used to handle complex or incompletely understood processes by approximating missing physics [3]. Parameter calibration approaches and reduced-order models have been developed to enhance model efficiency and reliability while reducing computational costs [51]. Data-driven models have emerged as effective alternatives where process-based models are computationally expensive or biased. Deep learning methods excel in extracting complex patterns, with techniques like variational autoencoders (VAEs) used for dimensionality reduction and feature extraction from large datasets [65]. For anomaly detection, methods such as isolation forests and clustering models effectively identify outliers in environmental datasets, such as pollutant levels. These methods can be complemented by traditional statistical techniques, like threshold-based methods, to ensure physical plausibility [7].

However, the success of these approaches is often shown on clean benchmark datasets for isolated applications. Without explicitly leveraging data from multiple sources, they remain limited in modeling real-world ecosystems as they often involve multiple processes that evolve and interact at different scales. In contrast, foundation models present significant opportunities to address these challenges: **(1) Integration of multi-modal data:** Foundation models excel in synthesizing diverse inputs, such as satellite imagery, sensor data, and historical records. For example, ClimaX by Microsoft [47] integrates visual, numerical, and textual data to predict climate variables across scales. Huawei's Pangu-Weather [6] provides accurate weather forecasts by processing global datasets, demonstrating the capability to model atmospheric dynamics holistically. **(2) Responsiveness to evolving conditions:** Foundation models are highly adaptable to changes in spatial and temporal conditions [68]. IBM's AI climate tools [44] use fine-tuning on limited regional datasets to model urban heat islands and shifts in precipitation patterns, showcasing rapid adaptability without extensive retraining. **(3) Robustness to noisy and incomplete data:** Environmental datasets often suffer from noise and missing values. Foundation models handle such issues using advanced imputation techniques and architectures capable of learning from incomplete data. For instance, the large sensor model employs self-supervised learning to effectively impute missing values in extensive sensor datasets [45]. **(4) Capturing multi-process interdependencies:** Environmental systems involve complex interdependencies among variables. Models like Prithvi [26] capture interactions between vegetation, soil, and climate. The physics-guided foundation model [68] predicts water temperature and dissolved oxygen in lakes, illustrating how temperature shifts affect oxygen solubility.

3.2 Data Generation

Significant data gaps often hinder environmental science applications. For example, process-based models depend on remote sensing data, which can be obscured by noise (e.g., clouds in satellite imagery). Similarly, field-based data, such as soil properties or land use, may be unavailable for certain regions due to the high labor and resource costs required for collection.

ML-based generative methods aim to address these gaps by generating realistic synthetic data that mimics real-world conditions. Generative models, such as generative adversarial networks (GANs), VAEs, and diffusion models can generate data simulations that closely resemble true observations for many environmental applications, such as forest and crop monitoring [42,48].

Foundation models, trained on vast and diverse datasets, present significant opportunities to enhance data generation capabilities in environmental science. These models excel in two key areas: **(1) Conditional generative process:** Foundation models generate synthetic data tailored to specific conditions by leveraging their ability to encode relationships across diverse features. This is particularly valuable for scenarios with limited observational data or where simulating data under targeted environmental policies is critical. For instance, Deng et al. [12] employed LLaMA to generate domain-specific datasets for data-scarce regions, enabling downstream applications in geoscience. By modifying prompts, LLMs can explore various environmental scenarios, facilitating the simulation of data under different management strategies. **(2) Generating interdependent features:** By incorporating multiple data sources, foundation models can effectively capture relationships among diverse variables and processes in the generative process. For example, Khanna et al. [31] proposed DiffusionSat, which captures internal relationships across multiple data sources, such as high-resolution remote sensing data and text-based inputs, improving performance in generative tasks like temporal generation and inpainting. Similarly, Li et al. [37] adapted the Segment Anything model for field mapping in agriculture, demonstrating its ability to integrate diverse environmental data. However, foundation models are prone to "hallucination", where plausible but incorrect data points are generated, especially when they encounter novel data combinations not seen during training. This can lead to misleading conclusions about environmental conditions.

3.3 Data Assimilation

Data assimilation integrates new observational data into physics-based models to improve the accuracy of simulations and predictions. Traditional methods like the Kalman filter have been fundamental for data assimilation in environmental science. For instance, the ensemble Kalman filter (EnKF), widely used for non-linear systems, refines initial conditions by incorporating diverse sources like satellite imagery and ground-based observations, leading to more accurate forecasts in water property prediction [50] and crop yield forecasting [46].

Foundation models have the potential to revolutionize data assimilation in environmental science by integrating diverse observational datasets, including satellite imagery, sensor data, physical simulations, and ground-based measurements. Their versatility and advanced architectures enable improved performance in various applications, such as weather forecasting [44,47] and streamflow prediction [34]. Key contributions of foundation models to data assimilation include: **(1) In-context learning for adaptive integration:** Foundation models can dynamically adapt to new observations through in-context learning, incorporating recent examples without retraining. This capability allows them to effectively respond to evolving environmental conditions and varying data availability, enhancing accuracy in dynamic scenarios such as extreme weather events or seasonal changes. **(2) Multi-variable assimilation:** Unlike traditional methods that typically focus on a single variable, foundation models excel at assimilating diverse environmental variables. For instance, in aquatic systems, they can integrate temperature observations at one time and dissolved oxygen measurements at another [68], or both field measurements and remote sensing estimations at different frequencies, by leveraging their multi-modal capabilities to synthesize and refine predictions. Models like ClimaX [47] illustrate this strength by combining satellite temperature data, weather station observations, and other inputs to produce robust and comprehensive climate predictions.

3.4 Downscaling

Downscaling refines large-scale environmental predictions to provide detailed local forecasts by transforming coarse-resolution data from global or regional models into high-resolution outputs. This approach is widely used to bridge the gap between broad-scale models and the specific needs of local-scale environmental management, such as the finer-scale prediction of local temperature and precipitation patterns. Traditional methods, including statistical techniques and dynamic modeling, have been extensively used. ML approaches like super-resolution further enhance this process by leveraging large datasets to capture complex patterns often missed by traditional methods [61]. However, ML-based methods often struggle to generalize across different spatial regions and scales.

Foundation models enhance downscaling by utilizing fine-tuning and prompt-based adjustments, allowing outputs to be tailored for specific regions without extensive retraining. The opportunities offered by foundation models include: **(1) Transferability across regions:** These models can seamlessly adapt knowledge from one region to another, allowing their application across diverse geographic and climatic contexts with minimal adjustments. For example, Dong et al. [13] proposed SMLFR, which incorporates sparse modeling and low-frequency information with learned general patterns to enable satellite image generation across regions. **(2) Scalability across resolutions:** These models effectively handle data across various spatial and temporal scales, providing both fine-grained local insights and broader regional analyses, demonstrating their versatility for diverse downscaling tasks. For example, Li et al. [35] explored the potential of GPT-4V in geo-localization, land cover classification, visual question answering, and basic image understanding by leveraging multi-modal and multi-scale data. These strengths of foundation models have also been demonstrated in other applications, such as agriculture [58] and remote sensing [10,22,57].

3.5 Model Ensembling

Model ensembling in environmental science integrates outputs from multiple predictive models to improve the accuracy and robustness of predictions [55,74]. This technique leverages the strengths of diverse models through approaches such as model blending, which combines various predictions, and post-processing, which refines outputs using observational data. Ensembling is widely applied in environmental disciplines, including weather forecasting, where it aggregates predictions from multiple numerical weather models to enhance forecast precision [15,19]. In hydrology, it synthesizes streamflow predictions from different hydrological models, ensuring more reliable and precise forecasts [25,59]. Similarly, in climate science, post-processing aligns model outputs with observed climate data to mitigate biases and improve long-term projections [1,32].

Foundation models capture complex interrelationships among diverse data types and provide greater flexibility in model integration. These models adeptly synthesize datasets such as atmospheric conditions, oceanographic measurements, and terrestrial observations, leveraging advanced algorithms to uncover and utilize relationships across these sources [39,54]. Foundation models excel in two key areas: **(1) Dynamic reconfiguration:** These models enable rapid adjustments to ensemble compositions by incorporating new data or refining underperforming components without extensive retraining. Li et al. [37] demonstrate the improvement of SAM's predictive performance in multiple agriculture and climate tasks by incorporating additional natural data. This adaptability ensures ensembles remain responsive to changing conditions and specific predictive challenges. **(2) Modular scalability:** The modular design of foundation models facilitates seamless integration with other predictive models and allows for easy adjustments. This scalability supports diverse applications, ranging from regional hydrological forecasts to global climate modeling, enhancing their effectiveness across various environmental domains. For instance, Kochkov et al. [32] proposed NeuralGCM, which integrates a differentiable solver for atmospheric dynamics with machine-learning components, demonstrating competitive performance in weather and climate prediction tasks.

3.6 Decision-Making

Decision-making in environmental science applies model-driven insights to develop strategies for natural resource management and environmental conservation, focusing on sustainability goals. This process often involves scenario analysis, where models simulate various conditions to evaluate the potential impacts of management strategies, aiding in risk and benefit assessment.

Foundation models streamline decision-making by integrating complex environmental data into actionable strategies. Unlike traditional ML models, which often require intricate reward designs, foundation models leverage extensive pretraining across diverse datasets, enabling them to adapt effectively to varied scenarios without relying on specific reward structures. Key advantages of foundation models include: **(1) Human-in-the-loop decision-making:** Foundation models enable meaningful interaction with stakeholders by producing human-interpretable outputs, such as visualizations that clarify predictions and recommendations. This capability facilitates transparent and collaborative decision-making processes, making the insights accessible to policymakers and community stakeholders [58]. **(2) Versatility across decision-making contexts:** Foundation models exhibit strong adaptability, handling diverse tasks such as linking global climate data to localized weather predictions or combining new environmental monitoring data with established models. For example, models like ClimaX [47] support a wide range of decision-making applications, from biodiversity conservation to water resource planning in dynamic conditions.

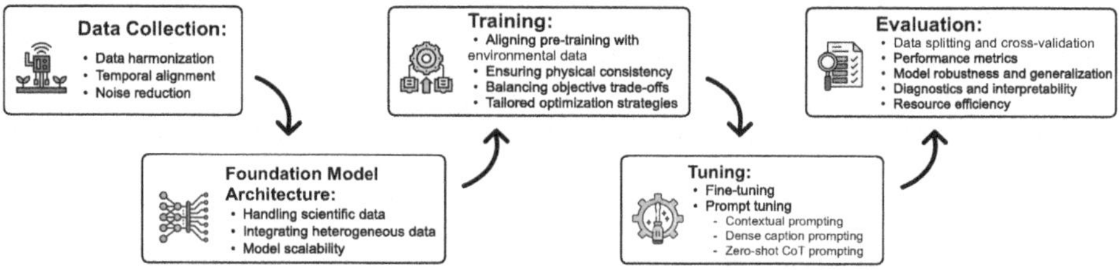

Fig. 2. Model design workflow for foundation models in environmental science.

4 Method Design

This section outlines the methodologies for designing foundation models in environmental science, as summarized in Fig. 2. It systematically examines key stages of the development process, including data collection, architecture design, training, tuning, and evaluation.

4.1 Data Collection

Building robust foundation models for environmental science requires addressing challenges in data collection and processing to ensure diverse, large-scale, and high-quality datasets. Capturing a broad range of environmental conditions supports generalization across varied contexts, necessitating datasets from ecosystems and climates such as tropical rainforests, arid deserts, and polar regions. Simulated data can effectively fill gaps where observational data is sparse, enabling models to incorporate rare or underrepresented phenomena.

Integrating heterogeneous datasets involves resolving variations in format, resolution, and temporal frequency. For example, satellite imagery provides high-resolution spatial data but often lacks temporal continuity, whereas ground sensors offer detailed time-series data with limited spatial coverage. Missing or noisy data, such as cloud-obscured satellite images or sensor malfunctions, further complicate integration [13, 47]. Key preprocessing steps include: ***Data harmonization:*** Standardizing units and aligning spatial resolutions to ensure consistency across datasets. ***Temporal alignment:*** Synchronizing datasets with different sampling frequencies to maintain temporal consistency. ***Noise reduction:*** Techniques such as wavelet transforms and median filtering remove distortions introduced during data collection.

Identifying relevant variables ensures an accurate representation of environmental processes. Variables such as temperature, precipitation, vegetation indices, and pollutant concentrations are commonly prioritized. The data collected are often unbalanced in diverse regions. To address spatial biases and overrepresented regions, techniques like spatial weighting are often adopted to ensure balanced and representative model performance.

Access to high-quality data sources enhances building foundation models:

Meteorological Data: The National Oceanic and Atmospheric Administration (NOAA) and the European Centre for Medium-Range Weather Forecasts (ECMWF) provide comprehensive weather and climate datasets for short-term forecasting and long-term climate modeling. **Geological and hydrological data:** The United States Geological Survey (USGS) offers datasets on geological conditions, natural hazards, and water quality. **Remote sensing data:** Resources like the National Land Cover Database (NLCD) and HydroLAKES provide high-resolution spatial data on land cover and water body distributions[1].

4.2 Foundation Model Architecture

Designing foundation model architectures for environmental science involves addressing challenges related to scientific data complexity, data heterogeneity, and scalability across regions and timescales.
Handling Scientific Data: Environmental data can be highly complex and domain-specific, requiring specialized model architectures. Pre-trained models, such as encoder-decoder transformers, are well-suited to text-based datasets; fine-tuning them on environmental data further enhances their ability to analyze reports, extract insights, and guide decision-making [12]. For example, T5 models fine-tuned on climate data can generate hypotheses or support adaptation strategies [33, 40]. When dealing with time-series data, temporal transformers excel at capturing temporal patterns and incorporating contextual factors, making them particularly effective for forecasting seasonal trends and weather patterns [47]. Additionally, physics-guided foundation models integrate physical principles, such as energy or water cycle laws, to ensure scientifically consistent predictions, especially in scenarios with limited data [29, 68].
Integrating Heterogeneous Data: Environmental datasets often differ in format, resolution, and frequency, requiring models that integrate diverse data

[1] URLs for the mentioned resources: NOAA (https://www.noaa.gov), ECMWF (https://www.ecmwf.int), USGS (https://www.usgs.gov), NLCD (https://www.mrlc.gov), HydroLAKES (https://www.hydrosheds.org/pages/hydrolakes).

sources. Multi-modal transformers combine inputs such as satellite imagery, sensor readings, and numerical data into unified representations for analysis. For example, these models can integrate satellite and ground-based data to assess land cover or air quality changes [37,39]. Spatial-temporal graph neural networks connect spatial and temporal data, such as linking weather station time-series data with geographic information to model regional dynamics [29,40]. Adaptive transformers streamline preprocessing by aligning datasets with different resolutions and frequencies, simplifying data integration [63].
Model Scalability: Scalability is essential for addressing multi-region and multi-scale challenges in environmental science. Fourier neural operators (FNO) [38] efficiently model cross-scale interactions, making them ideal for tasks like predicting regional rainfall while considering global climate patterns. For high-resolution data, vision Transformers and hierarchical models like Swin Transformers analyze satellite imagery to monitor deforestation or urbanization [58,63]. Transfer learning supports scalability by adapting models trained on global datasets to specific regions or contexts. For example, a climate model trained globally can be fine-tuned for regional biodiversity monitoring or flood prediction [58,62].

4.3 Training

Training or pre-training foundation models for environmental science requires strategies to capture the complexities of environmental systems and facilitate generalization across diverse tasks and datasets.
Aligning Pre-training with Environmental Data: Pre-training tasks should reflect the unique characteristics of environmental science. Instead of relying on generic objectives, tasks can focus on reconstructing physical variables (e.g., temperature, precipitation) or modeling multi-scale environmental interactions [68]. Self-supervised learning methods, such as masked autoencoders [18], allow models to leverage unlabeled datasets by predicting missing components. Contrastive learning, which distinguishes between similar and dissimilar contexts, enhances generalization to downstream tasks. These methods ensure that models learn representations aligned with environmental dynamics, including climate variability and ecosystem processes [35,68,72].
Ensuring Physical Consistency: Incorporating physical laws during training improves the reliability of predictions. Physical loss functions, such as enforcing energy or mass conservation, help maintain adherence to scientific principles [29]. For instance, in hydrology, these loss functions can ensure that predictions respect water balance equations [68]. Hybrid models, which combine machine learning with physics-based simulations, leverage the strengths of both approaches for enhanced consistency. Domain adaptation techniques transfer knowledge from data-rich to data-scarce regions [29,64].
Balancing Objective Trade-Offs: Training should balance predictive accuracy with adherence to physical principles. Multi-objective optimization techniques, such as Pareto optimization, optimize multiple criteria simultaneously,

ensuring a balance between error minimization and physical consistency. Regularization methods, like physics-informed priors, help maintain this balance without overfitting. Advanced optimizers, such as AdamW, stabilize training by effectively handling large parameter spaces and preventing weight decay issues. ***Tailored Optimization Strategies:*** Training strategies tailored to environmental data improve performance: **Curriculum learning:** Start with simpler tasks, such as single-variable predictions, and progressively incorporate complex tasks, such as multi-variable interactions. This helps models build foundational knowledge before addressing advanced scenarios. **Multi-scale training:** Environmental processes operate at varying spatial and temporal scales. Training on datasets spanning local, regional, and global scales ensures generalization across different resolutions [38, 47]. **Gradient clipping:** Stabilize training and prevent exploding gradients in large-scale models by limiting the magnitude of updates.

4.4 Tuning

Tuning adapts pre-trained foundation models to specific tasks and datasets in environmental science, aligning them with domain-specific challenges for practical applications. Two main approaches are fine-tuning and prompt tuning, each targeting distinct aspects of model customization.
Fine-Tuning: Fine-tuning involves retraining pre-trained models on task-specific datasets to improve their performance for specialized applications, such as climate modeling, air quality analysis, or biodiversity studies. A common strategy is layer-wise adaptation, where earlier layers—responsible for general features—are minimally updated, while later task-specific layers are fine-tuned more aggressively. Freezing base layers during the initial training phases reduces computational costs and prevents overfitting, especially when working with smaller task-specific datasets. Fine-tuning also requires careful hyperparameter optimization to balance generalization and accuracy [35, 72]. Beyond standard adaptations, domain knowledge can be integrated to ensure physically valid predictions [68]. Lastly, models are often fine-tuned on localized datasets to enhance predictions for specific regions. For instance, a model pre-trained on global climate data can be fine-tuned with regional meteorological datasets to optimize predictions for localized phenomena, such as drought forecasting or flood risk assessment [25].
Prompt Tuning: Prompt tuning adapts pre-trained models by structuring or modifying input prompts to guide their outputs without altering model parameters. This approach is especially valuable for domain-specific challenges, such as land use classification, climate forecasting, or species identification, where minimal retraining is preferred. **Contextual Prompting:** Contextual prompting adds task-specific context to the input to guide the model's understanding. For example, a prompt for species classification might include, "Each species has a unique scientific name composed of genus and species," enhancing the model's interpretive accuracy while preserving its original structure. **Dense Caption Prompting:** This technique generates detailed descriptions of input features,

such as satellite imagery, and incorporates them into the task prompt. For example, in land cover analysis, the model first describes features like vegetation and water bodies before addressing questions on environmental changes. This approach enriches data understanding for complex tasks. **Zero-Shot Chain of thought (CoT) Prompting:** CoT prompting encourages step-by-step reasoning, which is critical for multi-factor analyses. For example, in evaluating deforestation impacts, a prompt might guide the model to consider sequential effects on carbon storage, local temperatures, and biodiversity.

4.5 Evaluation

Effective evaluation of foundation models in environmental science focuses on their ability to address unique challenges, such as multi-modal data integration, scalability, and generalization across diverse environmental tasks and datasets.
Data Splitting and Cross-Validation: Time-based and region-based splits ensure models account for temporal dynamics and spatial heterogeneity without introducing data leakage. K-fold cross-validation assesses transferability across varied datasets, while domain-specific methods, such as leave-one-region-out validation, evaluate performance on unobserved geographic areas [29].
Performance Metrics: Beyond conventional regression metrics like root mean square error, multi-modal alignment scores, such as contrastive loss in joint embeddings, assess the model's ability to integrate text, imagery, and numerical data. For downstream environmental applications, extrinsic evaluation focuses on metrics such as F1 scores for species classification or intersection-over-union (IoU) for land-cover segmentation, ensuring task-specific adaptation [35, 72].
Model Robustness and Generalization: Foundation models should demonstrate robustness across diverse environmental conditions and unseen data domains. Transferability tests evaluate performance when the model is fine-tuned or adapted to new regions or phenomena, such as applying a global weather model to predict localized flood patterns [25]. Out-of-distribution testing assesses generalization to novel modalities or sparsely observed datasets, a critical capability for applications like rare species identification or extreme event prediction [72].
Diagnostics and Interpretability: Evaluating foundation models involves diagnosing biases and interpreting decision-making processes. Feature importance tools help interpret the contribution of variables across multi-modal inputs [41, 68]. They can reveal whether vegetation indices or precipitation data dominate predictions in drought forecasting tasks. Residual analysis uncovers systematic biases, such as overgeneralizing patterns across diverse geographic areas [5].
Resource Efficiency: Foundation models are expected to efficiently process large-scale environmental datasets, such as high-resolution satellite imagery or climate simulations. Evaluation should include scalability metrics, such as runtime per dataset size or memory usage for multi-modal tasks. Efficient adaptation techniques, such as parameter-efficient tuning (e.g., LoRA or prompt tuning),

are evaluated based on their ability to minimize computational overhead while maintaining performance on environmental applications [9,35].

Table 1. Application-centric objectives and methods for foundation models.

3 Objectives	4.1 Data Collection	4.2 Foundation Model Architecture	4.3 Training	4.4 Tuning	4.5 Evaluation
3.1 Forward Prediction	[5,8,13,16,32,37, 47,53,56]	[5,8,10,13,16,22,33,35, 39,47,53,54,56,68]	[5,12,13,16,32,33, 35,47,53,56,68]	[5,13,16,32,33,35, 40,47,53,56,57,68]	[5,32,33,47,56,68]
3.2 Data Generation	[5,13,16,32,36,37, 47,56]	[12,13,16,31,32,36,56]	[11,13,31,32,52,56]	[13,56]	
3.3 Data Assimilation	[5,37,47]	[44,47,68]	[47,68]	[34,68]	[68]
3.4 Downscaling	[5,13,32,37,47]	[13,16,22,31,32,47,58]	[11,13,31,32,52,57]	[13,31,32]	
3.5 Model Ensembling		[32,39,54]		[37]	
3.6 Decision-making		[47,58]			

5 Discussions

Table 1 summarizes the applications of foundation models in environmental science. Areas with extensive research activity indicate current hotspots, while those with fewer contributions highlight challenges or opportunities for further exploration. This section reflects on the current state of foundation models in this field, their limitations, and potential directions for future research.

Foundation models offer immense potential for environmental science but remain in the early stages of their application. By integrating diverse data sources such as satellite imagery, sensor measurements, and historical records, these models provide a holistic understanding of environmental systems. Pre-training on large-scale datasets enables these models to quickly adapt to new tasks. Their adaptability, driven by techniques like prompt engineering and tuning, enables meaningful insights even from incomplete or variable data. Emerging technologies like retrieval-augmented generation (RAG) and in-context learning also show promise for incorporating external knowledge into real-time decision-making.

However, foundation models face notable challenges that need to be addressed to fully unlock their potential. One key obstacle in scientific discovery is the lack of trust in these models due to their "black-box" nature. Addressing this challenge requires advancements in the explainability of their predictions. Uncertainty quantification for these models is also underdeveloped, yet it is vital for high-stakes applications such as disaster preparedness, climate mitigation, and resource management. Modeling rare extreme events, such as floods and heatwaves, is particularly difficult due to sparse data and the complexity of integrating physical constraints. Computational efficiency poses a significant barrier, especially for high-resolution simulations over large regions and long periods.

Looking forward, several opportunities exist to advance the development and application of foundation models in environmental science. Embedding domain knowledge can improve physical consistency and enhance model interpretability, addressing concerns around explainability and trust. Open-access datasets and pre-trained models can democratize research and foster interdisciplinary collaboration, broadening participation and accelerating innovation. Active learning offers a practical way to enhance model performance by prioritizing data collection in regions with high uncertainty or limited representation. Integrating decision-making frameworks into foundation models has the potential to transform them from predictive tools into actionable systems, supporting adaptive management strategies for water resources, disaster mitigation, and more. Addressing long-term error accumulation through recalibration, ensemble modeling, and hybrid approaches can further improve the reliability of extended forecasts. Finally, foundation models hold significant potential for scientific discovery by uncovering novel patterns, generating hypotheses, and synthesizing interdisciplinary knowledge, advancing our understanding of complex environmental systems.

6 Conclusion

In this survey, we explored the transformative potential of foundation models in environmental science, reviewing their applications and detailing development methodologies. These models, with their ability to integrate diverse data, capture complex spatiotemporal dynamics, and adapt flexibly, represent a significant leap forward in addressing environmental challenges. They also promote scientific discovery and support sustainable decision-making. We hope this work lays the foundation for further research toward a more resilient global community.

Acknowledgments.. This work was supported by NSF grant No. 2239175 and NASA award No. 80NSSC24K1061.

References

1. Ahmed, K., et al.: Multi-model ensemble predictions of precipitation and temperature using machine learning algorithms. Atmospheric Research (2020)
2. Bengio, Y., et al.: Representation learning: a review and new perspectives. IEEE Transactions on Pattern Analysis and Machine Intelligence (2013)
3. Bennett, A., et al.: Deep learned process parameterizations provide better representations of turbulent heat fluxes in hydrologic models. WRR (2021)
4. Bergen, K.J., et al.: Machine learning for data-driven discovery in solid earth geoscience. Science (2019)
5. Bi, K., et al.: Pangu-weather: A 3D high-resolution model for fast and accurate global weather forecast. arXiv (2022)
6. Bi, K., et al.: Accurate medium-range global weather forecasting with 3D neural networks. Nature (2023)
7. Blázquez-García, A., et al.: A review on outlier/anomaly detection in time series data. ACM Comput. Surveys (CSUR) (2021)

8. Bodnar, C., et al.: Aurora: A foundation model of the atmosphere. arXiv (2024)
9. Bommasani, R., et al.: On the opportunities and risks of foundation models. arXiv preprint arXiv:2108.07258 (2021)
10. Chen, K., et al.: RSMamba: remote sensing image classification with state space model. IEEE Geoscience and Remote Sensing Letters (2024)
11. Cong, Y., et al.: SatMAE: pre-training transformers for temporal and multi-spectral satellite imagery. NeurIPS (2022)
12. Deng, C., et al.: K2: a foundation language model for geoscience knowledge understanding and utilization. In: WSDM (2024)
13. Dong, Z., et al.: Generative convnet foundation model with sparse modeling and low-frequency reconstruction for remote sensing image interpretation. IEEE Transactions on Geoscience and Remote Sensing (2024)
14. Ghosh, R., et al.: Robust inverse framework using knowledge-guided self-supervised learning: an application to hydrology. In: KDD (2022)
15. Grönquist, P., et al.: Deep learning for post-processing ensemble weather forecasts. Philosophical Transactions of the Royal Society A (2021)
16. Guo, X., et al.: SkySense: a multi-modal remote sensing foundation model towards universal interpretation for earth observation imagery. In: CVPR (2024)
17. Han, X., et al.: A survey of transformer-based multimodal pre-trained modals. Neurocomputing (2023)
18. He, K., et al.: Masked autoencoders are scalable vision learners. In: CVPR (2022)
19. Hewage, P., et al.: Deep learning-based effective fine-grained weather forecasting model. Pattern Analysis and Applications (2021)
20. Hipsey, M.R., et al.: A general lake model (GLM 3.0) for linking with high-frequency sensor data from the global lake ecological observatory network (GLEON). Geoscientific Model Development (2019)
21. Hoffmann, S., Lessig, C.: AtmoDist: self-supervised representation learning for atmospheric dynamics. Environmental Data Science (2023)
22. Hong, D., et al.: SpectralGPT: spectral remote sensing foundation model. IEEE Transactions on Pattern Analysis and Machine Intelligence (2024)
23. Hospedales, T., et al.: Meta-learning in neural networks: a survey. IEEE Transactions on Pattern Analysis and Machine Intelligence (2021)
24. Huang, Y., et al.: EnviroExam: Benchmarking environmental science knowledge of large language models. arXiv preprint arXiv:2405.11265 (2024)
25. Islam, A.R.M.T., et al.: Flood susceptibility modelling using advanced ensemble machine learning models. Geoscience Frontiers (2021)
26. Jakubik, J., et al.: Foundation Models for Generalist Geospatial Artificial Intelligence. Preprint Available on arxiv:2310.18660 (2023)
27. Jia, X., et al.: Physics guided RNNs for modeling dynamical systems: a case study in simulating lake temperature profiles. In: SDM (2019)
28. Kalyan, K.S.: A survey of GPT-3 family large language models including ChatGPT and GPT-4. Natural Language Processing Journal (2023)
29. Karpatne, A., Jia, X., Kumar, V.: Knowledge-guided machine learning: Current trends and future prospects. arXiv preprint arXiv:2403.15989 (2024)
30. Karpatne, A., et al.: Machine learning for the geosciences: Challenges and opportunities. IEEE Transactions on Knowledge and Data Engineering (2018)
31. Khanna, S., et al.: DiffusionSat: A generative foundation model for satellite imagery. arXiv preprint arXiv:2312.03606 (2023)
32. Kochkov, D., et al.: Neural general circulation models for weather and climate. Nature (2024)

33. Lacoste, A., et al.: GEO-bench: Toward foundation models for earth monitoring. NeurIPS (2024)
34. Li, H., et al.: LITE: Modeling environmental ecosystems with multimodal large language models. arXiv preprint arXiv:2404.01165 (2024)
35. Li, J., et al.: Foundation models in smart agriculture: Basics, opportunities, and challenges. Computers and Electronics in Agriculture (2024)
36. Li, N., et al.: Using LLMs to build a database of climate extreme impacts. In: ACL Workshop ClimateNLP (2024)
37. Li, W., et al.: Segment anything model can not segment anything: assessing AI foundation model's generalizability in permafrost mapping. Remote Sensing (2024)
38. Li, Z., et al.: Fourier neural operator for parametric partial differential equations. arXiv (2020)
39. Lin, Z., et al.: GeoGalactica: A scientific large language model in geoscience. arXiv preprint arXiv:2401.00434 (2023)
40. Luo, S., Ni, J., Chen, S., et al.: FREE: The foundational semantic recognition for modeling environmental ecosystems. arXiv preprint arXiv:2311.10255 (2023)
41. Mangalathu, S., et al.: Failure mode and effects analysis of RC members based on machine-learning-based shapley additive explanations (SHAP) approach. Engineering Structures (2020)
42. Martinez, J., et al.: A comparison of cloud removal methods for deforestation monitoring in amazon rainforest. ISPRS (2022)
43. Moorcroft, P.R., et al.: A method for scaling vegetation dynamics: the ecosystem demography model (ed). Ecological monographs (2001)
44. Mukkavilli, S.K., et al.: AI foundation models for weather and climate: Applications, design, and implementation. arXiv preprint arXiv:2309.10808 (2023)
45. Narayanswamy, G., et al.: Scaling wearable foundation models. arXiv preprint arXiv:2410.13638 (2024)
46. Nguyen, L.H., et al.: Spatial-temporal multi-task learning for within-field cotton yield prediction. In: PAKDD. Springer (2019)
47. Nguyen, T., et al.: ClimaX: A foundation model for weather and climate. ICML (2025)
48. Nie, J., et al.: Prediction of liquid magnetization series data in agriculture based on enhanced CGAN. Frontiers in plant science (2022)
49. Pantazis, O., et al.: Focus on the positives: Self-supervised learning for biodiversity monitoring. In: ICCV (2021)
50. Park, S., et al.: Variable update strategy to improve water quality forecast accuracy in multivariate data assimilation using the ensemble kalman filter. Water Res. **176**, 115711 (2020)
51. Quarteroni, A., Rozza, G. (eds.): Reduced Order Methods for Modeling and Computational Reduction. MMSA, vol. 9. Springer, Cham (2014). https://doi.org/10.1007/978-3-319-02090-7
52. Reed, C.J., et al.: Scale-MAE: a scale-aware masked autoencoder for multiscale geospatial representation learning. In: CVPR (2023)
53. Ren, Y., et al.: WaterGPT: Training a large language model to become a hydrology expert. Water (2024)
54. Rezayi, S., et al.: AgriBERT: knowledge-infused agricultural language models for matching food and nutrition. In: IJCAI (2022)
55. Shahhosseini, M., Hu, G., Archontoulis, S.V.: Forecasting corn yield with machine learning ensembles. Front. Plant Sci. **11**, 1120 (2020)
56. Stevens, S., et al.: BioCLIP: a vision foundation model for the tree of life. In: CVPR, pp. 19412–19424 (2024)

57. Sun, X., et al.: RingMo: A Remote Sensing Foundation Model with Masked Image Modeling. IEEE Trans, Geosci Remote (2022)
58. Tan, C., et al.: On the promises and challenges of multimodal foundation models for geographical, environmental, agricultural, and urban planning applications. arXiv preprint arXiv:2312.17016 (2023)
59. Troin, M., et al.: Generating ensemble streamflow forecasts: A review of methods and approaches over the past 40 years (2021)
60. Vaswani, A., et al.: Attention is all you need. NeurIPS **30** (2017)
61. Wang, F., et al.: Deep learning for daily precipitation and temperature downscaling. Water Resources Research (2021)
62. Wang, S., et al.: GPT, large language models and generative artificial intelligence models in geospatial science: a systematic review. Digital Earth (2024)
63. Wang, Z., et al.: SiamHRnet-OCR: a novel deforestation detection model with high-resolution imagery and deep learning. Remote Sensing (2023)
64. Willard, J., et al.: Integrating scientific knowledge with machine learning for engineering and environmental systems. ACM Computing Surveys (2022)
65. Wu, Q., et al.: A novel deep learning framework with variational auto-encoder for indoor air quality prediction. Frontiers of Environmental Sci & Eng (2024)
66. Xu, P., et al.: Multimodal learning with transformers: a survey. IEEE Trans. Pattern Anal. Mach. Intell. **45**(10), 12113–12132 (2023)
67. Yu, R., et al.: Adaptive process-guided learning: an application in predicting lake do concentrations. In: ICDM (2024)
68. Yu, R., et al.: Physics-guided foundation model for scientific discovery: An application to aquatic science. arXiv preprint arXiv:2502.06084 (2025)
69. Zeyang, W., et al.: Environmental computing: concept, evolution, and challenges. J. Tsinghua Univ. (Science and Technology) **62**(12), 1839–1850 (2022)
70. Zhang, M., et al.: Foundation model for generalist remote sensing intelligence: potentials and prospects. Science Bulletin (2024)
71. Zhu, J.J., et al.: ChatGPT and environmental research. Environmental Science & Technology (2023)
72. Zhu, X.X., et al.: On the foundations of earth and climate foundation models. arXiv preprint arXiv:2405.04285 (2024)
73. Zhuang, F., et al.: A comprehensive survey on transfer learning. Proceedings of the IEEE (2020)
74. Zounemat-Kermani, M., et al.: Ensemble machine learning paradigms in hydrology: a review. Journal of Hydrology (2021)

A Survey on Efficient Graph Reachability Queries

Huangleshuai He[1], Zhengyi Yang[1(✉)], Dong Wen[1], Wenqian Zhang[1], Michael Yu[2], Wenke Yang[1], and Wenjie Zhang[1]

[1] The University of New South Wales, Sydney, NSW, Australia
{huangleshuai.he,zhengyi.yang,dong.wen, wenqian.zhang,wenke.yang,wenjie.zhang}@unsw.edu.a

[2] The Chinese University of Hong Kong, Sha Tin, Hong Kong SAR
mryu@cse.cuhk.edu.hk

Abstract. Graph reachability queries, which determine whether a path exists between two vertices in a graph, are a foundational problem in graph analytics. This survey provides a comprehensive review of techniques for efficient graph reachability querying, including static and dynamic approaches, indexed and online methods, and recent advancements. Detailed discussions explore various design choices for efficiently handling graph reachability queries, emphasising their applicability, limitations, and performance trade-offs. Additionally, we outline key open challenges and potential future directions to advance this field. This survey aims to guide researchers in navigating and advancing the state-of-the-art in graph reachability queries.

Keywords: Graph Reachability · Indexing · Query Optimisation

1 Introduction

Graphs provide a versatile framework for modeling real-world data with vertices as entities and edges as relationships, making them essential in fields like biology, finance, social networks, and knowledge graphs. Among graph analytics tasks, reachability queries are fundamental, determining if a directed path exists between two vertices to capture network connectivity.

Applications. Reachability queries have a wide range of applications in multiple domains, covering everything from financial security to data analytics, including but not limited to the following key scenarios. In financial systems, it is used for fraud detection, where potential frauds are detected by analysing the reachability in the transaction network [28]; In social networks, reachability queries can be used for access control to determine whether a user has access to a certain resource or community [13]; In the field of program analysis, it is used in data flow and control flow analysis to help the compilers optimise code or perform security checks [29]; In user input dependency analysis, reachability queries

X. Wu et al. (Eds.): PAKDD 2025, LNAI 15875, pp. 58–77, 2025.
https://doi.org/10.1007/978-981-96-8295-9_4

can determine how input data is propagated through a program to assess security risks [32]; And in XML query processing, reachability queries can be used to optimise query execution and improve retrieval efficiency [42]. These applications highlight that reachability queries are not just a fundamental graph algorithm problem but also a crucial tool for solving complex real-world challenges.

Objectives of the Survey. Graph reachability has been previously surveyed in the literature [49,51], but existing reviews are either outdated or incomplete. Specifically, [49] discusses algorithms developed only up to 2010, while a more recent tutorial [51] focuses mainly on reachability indexes, overlooking the broader design of reachability queries and online methods that do not rely on indexes. This survey aims to fill these gaps by providing a comprehensive review of graph reachability query, summarising existing methods, offering an in-depth analysis, and discussing challenges and future research directions. Note that while reachability queries have led to extended problems, such as k-reach queries [11,45] and label-constrained reachability [9,23,44], this paper focuses on the original reachability query problem in directed graphs, which serves as the foundation for all these variants. Meanwhile, this paper focuses exclusively on traditional algorithms, deliberately excluding learning-based methods. While learning-based approaches have been explored in reachability analysis [48], they inherently provide only approximate results, limiting their applicability to reachability queries.

Contributions. Our main contributions are summarised as follows.

- *Comprehensive Review*: We present a thorough and up-to-date survey of graph reachability query methods, incorporating both indexed and online approaches, and highlighting advancements that were not covered in earlier reviews.
- *In-Depth Analysis*: We provide an extensive evaluation of existing methods, examining their strengths, limitations, and practical applications to offer a nuanced understanding of current techniques.
- *Future Directions*: We outline key challenges in the field and propose promising directions for future research to further improve the efficiency and scalability of graph reachability queries.

2 Background

This paper focuses on simple directed graphs $G = (V, E)$, where V is the vertex set and $E \subseteq V \times V$ is the edge set. We define $n = |V|$ as the number of vertices and $m = |E|$ as the number of edges. A vertex without any incoming edges is referred to as a *source*, while a vertex with no outgoing edges is referred to as a *sink*. The reverse graph G^R is constructed by retaining all vertices from the original graph G and reversing the direction of each edge in E. We use $\deg_{\text{in}}$ and $\deg_{\text{out}}$ to represent the in-degree and out-degree of vertices.

Definition 1 (Reachability). *A sequence of vertices* $s = v_0 \rightarrow \cdots \rightarrow v_k = t$*, where* $k \geq 0$*, is called an* $s - t$ *path if for every consecutive pair of vertices* $v_i \rightarrow v_{i+1}$*, there exists an edge* $(v_i, v_{i+1}) \in E$*. We denote* $s \rightsquigarrow t$ *to mean that* s *can reach* t *via at least one path.*

Definition 2 (Reachability Query). *Given two vertices* u *and* v *in a graph, a reachability query determines whether* $u \rightsquigarrow v$*. If* u *can reach* v*,* $Query(u, v)$ *is* `true`*; otherwise, it is* `false`*.*

We denote a reachability query that returns `true` as a *positive* query and one that returns `false` as a *negative* query.

Example 1. In the graph shown in Fig. 1, the following reachability queries hold $Query(a, b) =$ `true`, $Query(b, a) =$ `false`, and $Query(a, c) =$ `true`.

Definition 3 (Strongly Connected Components). *A strongly connected component (SCC) of graph* G *is a maximal set of vertices* $S \subseteq V$ *such that for every pair of vertices* $u, v \in S$*, both* $u \rightsquigarrow v$ *and* $v \rightsquigarrow u$ *hold.*

Definition 4 (Directed Acyclic Graph). *A Directed Acyclic Graph (DAG) is a directed graph that contains no cycles.*

This means that no pathd starts and ends at the same vertex while following the direction of the edges. In a DAG, all edges point in a single direction, making it impossible to traverse the graph in a way that loops back to the starting vertex.

Definition 5 (Spanning Tree). *A spanning tree of an undirected graph* $G = (V, E)$ *is a subgraph* $T = (V, E_T)$ *such that: 1)* T *is a tree (i.e., it is connected and contains no cycles), 2)* T *spans* G*, meaning it includes all the vertices of* $G(V_T = V)$*, and 3)* $E_T \subseteq E$*, meaning the edges of* T *are a subset of the edges of* G*. 4) For a graph with* n *vertices, a spanning tree contains exactly* $n - 1$ *edges.*

Definition 6 (Tree Edge and Non-Tree Edge). *Given a directed acyclic graph* $G = (V, E)$ *and its spanning tree* $T = (V, E_T)$*, an edge* $(u, v) \in E_T$ *is classified as a Tree Edge, while any edge* $(u, v) \in E \setminus E_T$ *is classified as a Non-Tree Edge.*

Example 2. A possible spanning tree of the graph shown in Fig. 1 is illustrated in Fig. 2. The edge (a, b) is a tree edge, while the edge (d, c) is a non-tree edge.

Definition 7 (Topological Sorting [20]). *Topological Sorting is a technique for arranging the vertices of a Directed Acyclic Graph (DAG) in a linear order such that for every directed edge* (u, v)*, vertex* u *appears before vertex* v*.*

A topological sort may not be unique as there can be multiple valid orderings.

Example 3. In the graph shown in Fig. 1, $[a, d, b, f, g, c, e]$ and $[a, b, e, d, f, g, c]$ are the two possible topological order of the vertices.

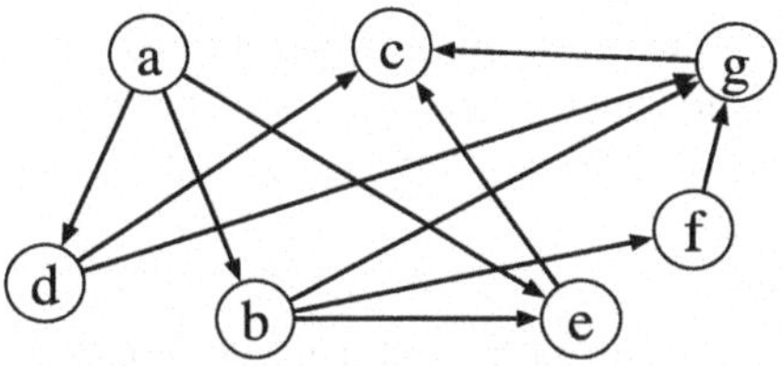

Fig. 1. An Example DAG

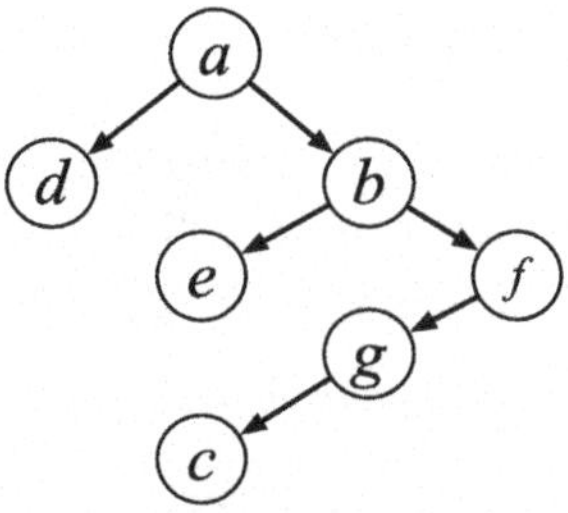

Fig. 2. Spanning Tree

Table 1. Features of Different Approaches

Method	Year	Framework	Index Type	DAG Prep.	Dyna.	Supp. Index	Query Prop.
BFS/DFS	–	Other	No index	✗	✓	✗	Pos.
FullMatrix	–	Transitive Closure	Label-Only	✗	✗	✗	–
TCBase [30,36]	1988	Transitive Closure	Label-Only	✓	✗	✗	–
TreeCoverBase [2]	1989	Tree Cover	Label-Only	✓	✗	✗	–
2-HopBase [12]	2003	2-Hop	Label-Only	✗	✗	✗	–
Tree + SSPI [8]	2005	Tree Cover	Label+Graph	✓	✗	✗	Neg.
HOPI [31]	2005	2-Hop	Label-Only	✗	✓	✗	–
Dual-Labelling [42]	2006	Tree Cover	Label-Only	✓	✗	✓	–
GRIPP [40]	2007	Tree Cover	Label+Graph	✗	✗	✗	–
PathTree [19]	2008	Tree Cover	Label-Only	✓	✗	✗	–
3-Hop [18]	2009	2-Hop	Label-Only	✓	✗	✗	–
U2-Hop [4]	2010	2-Hop	Label-Only	✓	✓	✗	–
PathHop [6]	2010	2-Hop	Label-Only	✓	✗	✗	–
GRAIL [50]	2012	Tree Cover	Label+Graph	✓	✗	✗	Neg.
TF [10]	2013	2-Hop	Label-Only	✓	✗	✗	–
DL [17]	2013	2-Hop	Label-Only	✗	✗	✗	–
HL [17]	2013	Other	Label-Only	✓	✗	✗	–
DAGGER [47]	2013	Tree Cover	Label+Graph	✓	✓	✗	Neg.
PPL [46]	2013	2-Hop	Label-Only	✓	✗	✗	–
Ferrari [34]	2013	Tree Cover	Label+Graph	✓	✗	✗	Neg.
Feline [41]	2014	Other	Label+Graph	✓	✗	✗	Neg.
PReaCH [26]	2014	Other	Label+Graph	✓	✗	✓	–
TOL [56]	2014	2-Hop	Label-Only	✗	✓	✗	–
BFL [37]	2017	Transitive Closure	Label+Graph	✓	✗	✗	Neg.
IP [43]	2018	Transitive Closure	Label+Graph	✓	✓	✗	Neg.
ARROW [33]	2019	Other	No index	✗	✓	✗	Pos.
DBL [25]	2021	2-Hop	Label+Graph	✗	✓	✓	–
O'Reach [15]	2022	2-Hop	Label+Graph	✓	✗	✓	Neg.
IFCA [27]	2023	Other	No index	✗	✓	✗	Pos.

3 Overview

In this section, we categorise reachability algorithms from various perspectives, identify their commonalities and differences, and discuss the complexity of them.

3.1 Categorisation

The details of graph reachability algorithms are summarised in Table 1. We categorise existing algorithms based on the following criteria: framework, index type, whether they require a preprocessing to a DAG (DAG Prep.), whether they support dynamic graphs, whether they construct supportive indexes (Supp. Index), and their query propensity (Query Prop.).

Framework. Framework provides a valuable perspective for understanding and comparing different reachability methods. Algorithms that share the same framework often exhibit similar characteristics, making it easier to analyse their design and performance. As shown in Table 1, frameworks can be categorised into four major types: *Tree Cover-based*, *2-Hop based*, *Transitive Closure-based*, and *others*. The details of each framework will be described in Sect. 4.

Index Type. To efficiently answer graph reachability queries, most existing methods rely on constructing index structures to accelerate the querying process. We categorise the index into two types: *Label-Only* and *Label+Graph*. In a reachability index, *labels* encode local information (e.g., vertices that are reachable or can be reached) assigned to each vertex in the graph. By comparing these labels, one can efficiently determine whether two vertices are reachable or unreachable. Each reachability algorithm employs a unique label design, as will be discussed in detail in Sect. 4.

The *Label-Only* approach excels in query performance by encoding the structural information of the graph into labels. This method constructs an index as a complete search space using these labels, enabling it to answer reachability queries without needing to access the original data graph. By leveraging the precomputed labels, the *Label-Only* approach facilitates rapid determination of reachability between vertices, eliminating the need for full graph traversal. Although the index construction and preprocessing is typically time-consuming, it provides significant advantages during the querying phase.

In contrast, the *Label+Graph* approach combines a label-based index with the original data graph. Specifically, when a query can be answered using the index, the method directly utilises the constructed index to provide a result, ensuring efficient query processing. However, if the index alone cannot resolve the query, the method performs graph traversal, typically using a depth-first search, to determine a path in the graph in an online manner. The key advantage of the *Label+Graph* approach is its ability to reduce index construction time and index size, as it avoids building extensive index structures as in the *Label-Only* approach. However, it may result in longer query times, especially in the case where graph traversal becomes necessary.

In addition to these methods, there are also purely online methods. One such example is IFCA [27], a representative of recent online methods. IFCAis designed to optimise query speed without imposing storage space requirements or requiring preprocessing time. This makes it particularly well-suited for dynamic graphs, where frequent structural updates make maintaining an index challenging. The adaptability of online methods like IFCAprovides an efficient alternative for cases where graph updates are frequent, and preprocessing is impractical.

DAG Preprocessing. Some methods require converting the original graph into a Directed Acyclic Graph (DAG) during preprocessing. For static graphs, this transformation involves identifying strongly connected components (SCCs) and merging all vertices within each SCC into a single new vertex, effectively preserving reachability while simplifying the graph. Commonly used algorithms for SCC detection include Tarjan's [39] and Kosaraju's [35] algorithms, both of which operate in linear time. Moreover, [52] proposed an I/O-efficient method for finding SCCs in massive graphs that cannot be held in memory, while [53,55] introduced DAG reduction and later improved it with RCN reduction [54]. DAG preprocessing serves multiple purposes, with the most notable being information compression. Additionally, it is often necessary because certain algorithms cannot handle SCC structures efficiently or require excessive computational time to do so. However, challenges arise when extending this approach to dynamic graphs. For dynamic graphs, the inability to maintain SCCs efficiently becomes a significant bottleneck. Currently, there is no robust solution for the dynamic maintenance of SCCs. The overhead of maintaining SCCs can sometimes render these methods less effective than simpler approaches (e.g. BFS/DFS). This underscores the limitation for algorithms that rely on DAG preprocessing in dynamic graphs.

Dynamic. Dynamic graphs, where vertices and edges are updated over time through insertion and deletion operations, are increasingly being used to model real-world applications such as financial systems [21], social networks [1], and epidemic network [16,22]. However, most existing approaches to reachability queries are designed for static graphs. While state-of-the-art indexing techniques can efficiently construct indexes for large graphs, these methods often struggle to handle the challenges posed by dynamic graphs. For example, while DBL [25] supports insertion operations, it cannot handle deletions. Both IP [43] and TOL [56] support insertions and deletions but rely on a DAGGER-based [47] approach (discussed in more detail later). However, these methods require the graph to be preprocessed into a DAG, as discussed previously, which can lead to performance degradation, sometimes failing to surpass the efficiency of a straightforward DFS/BFS approach. The limited number of solutions for dynamic graph reachability queries and the inherent limitations of existing approaches underscore the need for efficient dynamic reachability indexing techniques. These methods must effectively handle reachability queries while simultaneously supporting graph updates, including both insertions and deletions.

Table 2. Complexity of Different Approaches

Method	Index Size	Init Time	Query Time
BFS/DFS	0	0	$O(n+m)$
FullMatrix	$n^2/8$ (Byte)	$O(n(n+m))$	$O(1)$
TCBase	$O(n^2)$	$O(nm)$	$O(1)$
TreeCoverBase	$O(n^2)$	$O(nm)$	$O(\log n)$
2-HopBase	$O(nm^{1/2})$	$O(n^3\|TC\|)$	$O(m^{1/2})$
Tree + SSPI	$O(n+m)$	$O(n+m)$	$O(m-n)$
HOPI	$O(nm^{1/2})$	$O(n^3\|TC\|)$	$O(m^{1/2})$
Dual-Labelling	$O(n+t^2)$	$O(n+m+t^3)$	$O(1)$
GRIPP	$O(n+m)$	$O(n+m)$	$O(m-n)$
PathTree	$O(nk_{pt}))$	$O(mk_{pt}) or O(nm)$	$O(\log^2 k_{pt})$
3-Hop	$O(nk_{3-hop})$	$O((k_{3-hop}n^2) \times \|Con(G)\|)$	$O(\log n + k_{3-hop})$
U2-Hop	$O(n^2)$	$O((n+m)n\log n)$	$O(\|L(s)\|+\|L(t)\|)$
PathHop	$O(n^2)$	$O(k_{ph}n^2)$	$O(k_{ph}h_{ph})$
GRAIL	$O(k_g n)$	$O(k_g(n+m))$	$O(k_g)$ or $O(m+n)$
TF	$O(n \log L_{tf})$	$O\left(\sum_{v\in V_{G_i^*}} (\deg_{\text{in}}(v,G_i^*)\cdot \deg_{\text{out}}(v,G_i^*))\right)$	$O(\|L(s)\|+\|L(t)\|)$
DL	$O(n^2)$	$O(n(n+m)L_{dl})$	$O(\|L(s)\|+\|L(t)\|)$
HL	$O(n^2)$	$O(n+m+ML_{hl})$	$O(\|L(s)\|+\|L(t)\|)$
DAGGER	$O(nk_{dagger})$	$O(n+m)$	$O(k_{dagger})$ or $O(bk_{dagger})$
PPL	$O(n\log n)$	$O(n\log n+m)$	$O(\log n)$
Ferrari	$O((k_f+s_f)n)$	$O(k_f^2 m+S_f)$	$O(k_f)$ or $O(m+n)$
Feline	$O(n)$	$O(n\log n+m)$	$O(1)$ or $O(m+n)$
PReaCH	$56n$ (Byte)	$O(n\log n+m)$	$O(1)$ or $O(n+m)$
TOL	$O(nm^{1/2})$	$O(n^3\|TC\|)$	$O(\|L(s)\|+\|L(t)\|)$
BFL	$(2\lceil s_{bfl}/8\rceil n)$ (Byte)	$O(s_{bfl}(n+m))$	$O(s_{bfl})$ or $O(s_{bfl}n+m)$
IP	$O((k_{ip}+h_{ip})n)$	$O((k_{ip}+h_{ip})(n+m))$	$O(k)$ or $O(kn\rho^2)$
ARROW	0	0	$O\left(\sqrt[3]{n^2 \ln n} \times \text{diam}\right)$
DBL	$O((k_d+k_b)n)$	$O((k_d+k_b)(m+n))$	$O(k_d+k_b)$ or $O((k_d+k_b)(m+n))$
O'Reach	$(12+12d_o+2\lceil k_o/8\rceil)n$ (Byte)	$O((d_o+k_o p_o)(n+m))$	$O(k_o+d_o+1)$ or $O(n+m)$
IFCA	0	0	$SubLinear(n+m)$

Supportive Index. Some methods construct *supportive indexes* in answering reachability queries. A supportive index selectively indexes a subset of vertices rather than the entire graph, reflecting a shift in focus from indexing all vertices to prioritising only the important ones. This approach addresses inefficiencies that arise when labels are constructed uniformly for all vertices, which can lead to redundant or oversized labels for certain vertices. Supportive indexing maintains pruning efficiency while reducing index size and can be tailored to optimise either positive or negative queries, depending on the specific requirements. Recent research indicates the growing utility of supportive indexes. For instance, both static methods like O'Reach [15] and dynamic methods like DBL [25] have adopted supportive indexing, showcasing its potential. By indexing only a subset of vertices, supportive indexes offer better control over index size and provide opportunities to integrate different methodological frameworks to offset their respective weaknesses. This makes supportive indexing a promising direction for further exploration and development.

Query Propensity. Algorithms can also be classified based on their query propensity, as some are designed to favor either positive or negative queries. Positive query propensity indicates that an algorithm is faster at resolving queries that return positive results, while negative query propensity means the algorithm performs better with negative results. *Label+Graph* algorithms often exhibit negative query propensity because their indexes are typically designed to efficiently handle negative queries. When the index cannot resolve a query, these algorithms rely on DFS/BFS-like approaches, which may require traversing the entire graph for positive queries. This design demonstrates a good trade-off between index size and performance, particularly in real-world scenarios where graphs are sparse. The sparsity of real-world graphs increases the likelihood of reachability queries returning negative results, aligning well with the algorithms optimised for negative queries. On the other hand, online methods generally show positive query propensity. For positive queries, they can terminate as soon as the target vertex is found, ensuring faster resolution. However, for negative queries, they must traverse the entire graph to confirm the absence of a path, resulting in longer processing times. This inherent difference leads to contrasting query propensities between *Label+Graph* and online methods, highlighting the potential to combine them.

3.2 Complexity

Typically, the performance of a reachability algorithm is evaluated based on three key metrics: index size, initialisation time, and query time, as summarised in Table 2.

Symbols. In addition to the number of vertices (n) and edges (m), the complexities of reachability algorithms depend on various parameters specific to each algorithm. $|TC|$ denotes the total number of reachable pairs in the graph, and t is the number of non-tree edges. k_{pt} represents the number of paths into which a DAG is decomposed [19], while k_{3-hop}, k_{ph}, and k_g represent the number of chains in the chain decomposition [6,18,50]. $|CON(G)|$ indicates the number of pseudo-diagonal cells in the transitive closure profile [18], reflecting the index compression size. $|L(s)|$ and $|L(t)|$ are the number of elements in the labels of vertices s and t, respectively. h_{ph} denotes the Path-Hop Count [6]. L_{tf} is the number of graph levels [10], and G_i^* is the intermediate graph after the Topological Folding operation [10]. L_{dl} and L_{hl} are the maximum label sizes, and M is the maximal Backbone Vertex Set size [17]. k_{dagger} and b denote the number of SCCs and their average size, respectively. k_f is the number of intervals per vertex [34]. s_f is the number of seed vertices, and S_f is the complexity of finding the top-s largest degree vertices. s_{bfl} represents the size of the Bloom filter (in bits). k_{ip} the number of permutations, and h_{ip} the number of vertices with precomputed positives. ρ denotes the graph's reachability rate. $diam$ means its diameter. k_d and k_b is the average number of landmark and leaf vertices per vertex [25]. d_o is the number of topological orderings, k_o is the number of supportive vertices, and p_o is the number of candidates per supportive vertex.

Comparative Analysis. These metrics often require trade-offs during algorithm design to achieve a balance among them. The traditional BFS/DFS methods offer linear query time with almost no initialisation time or additional index space, making them suitable for small graphs or infrequent queries. In contrast, the full-matrix approach provides extremely fast query times but incurs significant costs in initialisation time and index size, limiting its applicability to large-scale graph data. Further, index-based methods such as GRAIL, Ferrari, and IPaccelerate queries by constructing indexes during the initialisation phase, trading off initialisation time and space complexity for improved query efficiency. Notably, methods like Ferrari and IP offer flexibility, allowing parameters to be adjusted to balance query performance and the cost of index construction. In addition, methods like Dual-Labellingand PReaCHachieve a refined balance between query time and index size. Notably, PReaCHdemonstrates strong potential for large-scale graph data by ensuring low query time while significantly reducing index size. Among all methods, O′Reach [15] stands out as the most flexible and adaptable due to its innovative supportive index design. This design enables it to integrate seamlessly with existing methods, effectively balancing index size, initialisation time, and query time.

Selecting the appropriate reachability query method depends on the specific application, data size, and query frequency. Achieving efficient query times often requires higher initialisation time and space costs, and for dynamic graphs, the cost of index updates must also be considered. It is not entirely fair to assess methods based solely on complexity, as this represents the worst-case scenario. In practice, factors such as graph structure, density, and other characteristics significantly influence real-world performance.

4 Framework

In this section, we introduce the three main types of frameworks: *Tree Cover*, *2-Hop*, and *Transitive Closure*. Other methods will be introduced in later sections.

4.1 Tree Cover

Tree cover-based methods leverage the concept of a *spanning tree*, which is a subgraph that connects all the vertices in a graph without forming cycles, and *intervals*, which are assigned to vertices to compactly represent reachability information. The main idea is to use the spanning tree as the backbone of the graph structure and encode reachability relationships using intervals associated with each vertex. This approach ensures compressed storage while supporting efficient query answering. During the index construction process, a spanning tree is first constructed for the original graph. The vertices in the spanning tree are then post-ordered based on the reverse topological order [20], and intervals such as $[l, r]$ are assigned to them according to this ordering, where l encodes the smallest ordering number of its children and r encodes its own ordering number.

To handle edges such as (u, v) that are not part of the spanning tree (*non-tree edges*), intervals of v will be added directly to intervals of u and checked with the original intervals to remove redundancy. To answer a reachability query $Query(s,t)$, the method checks whether the subtree in the spanning tree rooted at s contains t. The intervals assigned to the vertices encode all necessary information to perform this check efficiently. This framework is particularly effective for graphs such as trees, directed acyclic graphs (DAGs), and sparse real-world graphs, though challenges arise when the graph has a high density of non-tree edges. An example of a tree cover for Fig. 1 is illustrated in Fig. 3, which involves three main steps: 1) identify a spanning tree of the DAG to serve as the tree cover, 2) define reachability intervals for the vertices based on the spanning tree, 3) the non-tree edges are incorporated and utilized to propagate the reachability information.

Optimisations. Two primary designs are used to reduce the number of intervals: *accurate recording of k intervals* (as in GRAIL [50]) and *recording up to k intervals* (as in Ferrari [34]), where k is an input parameter. In GRAIL, k intervals are generated by constructing k random spanning trees, ensuring accurate interval representation. Conversely, in Ferrari, intervals are merged even if they are not adjacent, limiting the number of intervals per vertex to k, which provides a trade-off between accuracy and index size. Both methods do not calculate complete indexes, and the results of a query using index lookups may contain false positives (but no false negatives). Therefore, these partial indexes can be used to guide online traversal to compute the correct query results. Although partial indexes require additional graph traversal to process the query, their index construction time and index size are linearly related to the size of the input graph, making them eligible for large graphs.

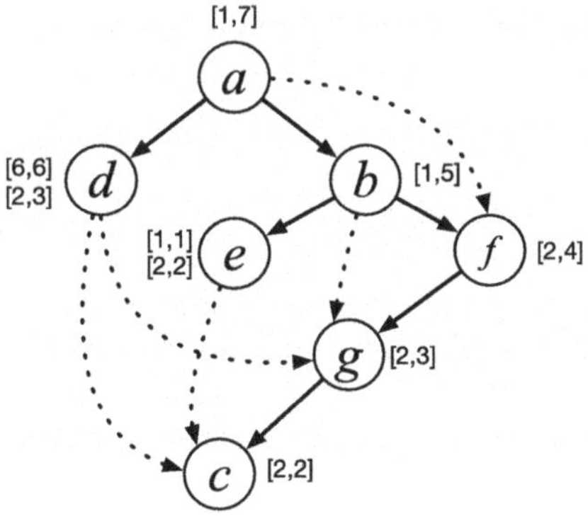

Fig. 3. An Example of Tree Cover

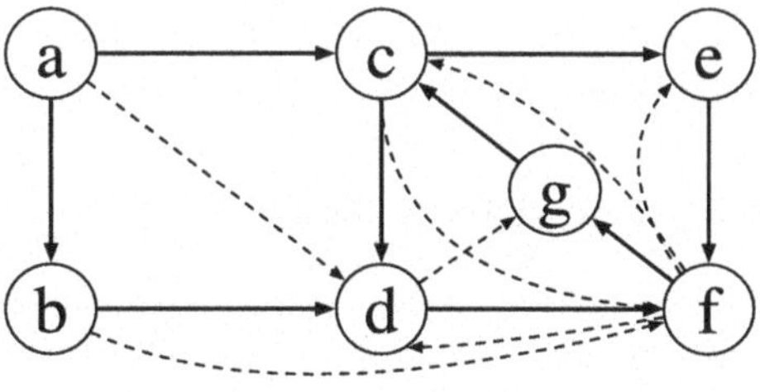

u	*a*	*b*	*c*	*d*	*e*	*f*	*g*
Lin(v)			*f*	*f,c*	*c,f*	*d,c*	*d,f*
Lout(v)	*b,c,d*	*f,d*	*d,f*	*g,f*	*f*	*g,c,e*	*c*

Fig. 4. An Example for 2-hop Labelling with Table of Vertex Sets L_{in} and L_{out}

4.2 2-Hop Labelling

2-hop labelling encodes the graph's connectivity using two sets of labels for each vertex: an *in-label*, denoted as $L_{\text{in}}(v)$, and an *out-label*, denoted as $L_{\text{out}}(v)$. The in-label of a vertex v represents the set of intermediate vertices from which v can be reached, while the out-label represents the set of intermediate vertices that can be reached from v. These labels capture the graph's structure, allowing reachability between any two vertices to be determined in at most two hops. During the label construction process, intermediate vertices are selected such that for any pair of reachable vertices u and v, there exists at least one common vertex in $L_{\text{out}}(u) \cap L_{\text{in}}(v)$. To determine if vertex u can reach vertex v, a query checks whether $L_{\text{out}}(u) \cap L_{\text{in}}(v) \neq \emptyset$. If the labels have a common vertex, u can reach v; otherwise, it cannot. 2-hop labelling provides fast query performance due to the simplicity of set intersection operations and has a relatively small index size, particularly for sparse graphs. However, the label sizes are not uniform, as some vertices may have large labels, which can impact query efficiency. Additionally, constructing the labels can be computationally expensive for large or dense graphs. Figure 4 shows an example for 2-hop reachability labelling.

Optimisations. There are several main approaches to improving the 2-hop technique, two of which are highlighted here: *compressing the index size by leveraging special graph structures* (as in PPL [46]), and *optimising the construction order to achieve smaller index sizes* (as in TOL [56]). PPLenhances 2-hop performance by using a path-based structure. The graph is split into non-overlapping paths, transforming the original approach of vertices reaching vertices through other vertices into vertices reaching vertices through paths. This significantly reduces the number of indexes, thereby improving query time. TOLintroduces a novel reachability index called the "Butterfly" to improve upon 2-hop. By refining the vertex sorting process, TOL reduces construction time, index size, and query time, achieving notable performance gains. Additionally, TOL supports insertion and deletion operations, making it suitable for dynamic graph scenarios.

4.3 Transitive Closure

The transitive closure compresses the Boolean matrix by means of special structures in the graph such as chains. Boolean matrix stores the answers to all possible reachability queries in a graph, as in FullMatrix. The transitive closure provides excellent query time performance. However, its index construction time and index size are typically large.

Optimisations. To reduce storage space, an *approximate transitive closure* method can be used with an approximation function $AP()$. This approach creates an approximate set $AP(Out(v))$ with significantly fewer elements than the original set $Out(v)$, reducing index size while preserving query efficiency. The core idea is to use set inclusion to determine reachability. If vertex t is unreachable from s, then $Out(t)$ will not be a subset of $Out(s)$. Similarly, if $AP(Out(t))$ is not a subset of $AP(Out(s))$, t is not reachable from s. This method is effective for verifying negative reachability, while actual reachability can be confirmed using traditional methods like DFS or BFS. There are two main approaches to implementing approximate transitive closures: BFL [37] and IP [43]. The BFL approach uses Bloom filters to create the approximation function $AP()$, while the IP approach employs k-min-wise independent permutations for this purpose. Both methods effectively reduce index size and construction time while maintaining query accuracy. However, they may produce false positives (incorrectly identifying unreachable vertices as reachable) which can be verified and corrected through further graph traversal. In summary, by using approximate transitive closures and the $AP()$ function, it is possible to reduce index construction time and size while maintaining query efficiency. This makes these approaches particularly suitable for reachability queries on large-scale graph data.

5 Static Methods

In this section, we introduce each of the static methods.

Tree+SSPI. Tree + SSPI [8] constructs indexes for reachability queries by creating a spanning tree for the DAG and assigning interval encodings to vertices, enabling quick checks of ancestor-descendant relationships within the tree. Additionally, it designs the Surrogate and Surplus Predecessor Index (SSPI) to store information about edges outside the tree cover. By combining the spanning tree and SSPI, the algorithm can fully derive reachability relations in the DAG. Furthermore, the paper employs a bit-vector encoding prefiltering technique to select vertices that satisfy structural constraints and uses stack operations along with partial solution pools to efficiently process path and twig pattern queries.

Dual-Labelling. Dual-Labelling [42] decomposes a graph into a spanning tree and non-tree edges, leveraging interval labels to record reachability within the spanning tree and a link table for non-tree edge reachability information. Interval labels enable fast queries between vertices in the spanning tree, while the link table precomputes its transitive closure for efficient multi-hop path queries.

GRIPP. GRIPP [40] proposes graph indexing based on preorder and postorder numbering to construct an index for reachability queries. By performing a depth-first traversal, each vertex in the graph is assigned preorder and postorder numbers. For graphs with cycles and non-tree edges, multiple instances are created for target vertices to accommodate additional numbering. During queries, reachability is assessed by checking number ranges. If the initial range is insufficient, a hop technique recursively expands the search to ensure all paths are covered.

PathTree. PathTree [19] proposes a new graph structure called Path-Tree Cover for efficiently answering reachability queries in large directed graphs. Path-Tree Cover is an index construction method for covering directed acyclic graphs

(DAGs) by decomposing the original graph into tree-like paths. Specifically, Path-Tree Cover forms a tree cover subgraph structure by dividing the graph into a number of paths and generating a tree based on these paths. Under its tree structure, each vertex is assigned a ternary label. In addition, it also reduces storage space and computation overhead by compressing the passing closure.

3-hop and PathHop. Both 3-Hop [18] and PathHop [6] look for new graph structures based on 2-hop to further compress the information. In the 3-hop method, a chain structure and a multiple-hop strategy are used to describe the reachability between the source and target vertices in a directed graph. Combines the use of entry and exit points to compress the delivery closure and avoids storing complete reachability information. In contrast, in the Path-hop method, the reachable paths in the graph are first covered by generating an optimal spanning tree, and then the remaining transitive closure information is further compressed using a path-hopping strategy. In this method, the paths in the tree are used as jump points for indexing on the basis of the spanning tree.

GRAIL. GRAIL [50] constructs the index for reachability queries using randomised interval labelling. Specifically, the graph is traversed multiple times using randomised DFS, generating multiple postorder interval labels for each vertex to form a multi-dimensional interval set. It uses interval containment to quickly rule out non-reachability (if the interval set of the target vertex is not contained in that of the source vertex, they are deemed non-reachable). For cases that cannot be resolved directly, a DFS with pruning verifies reachability. This method achieves linear time and space complexity for index construction.

TF. TF(Topological Folding) [10] builds a reachability index by assigning topological levels to vertices and dividing the graph into hierarchical layers. Vertices from odd levels are recursively removed, and direct edges are created between their in-neighbors and out-neighbors to preserve reachability. This process halves the levels iteratively, producing a reduced DAG. TF-label maintains two sets of labels for each vertex: one for reachable vertices and one for vertices that can reach it, generated by merging neighboring labels. To handle high-degree vertices, only directly related vertices are kept in labels. Reachability queries are answered by checking label intersections.

HL and DL. HL(Hierarchical Labelling) [17] constructs a reachability index by hierarchically decomposing directed acyclic graphs (DAGs). Each layer extracts a backbone graph that retains essential reachability information. Vertices are assigned to different levels based on their importance, with higher-level vertices being more critical. Labels are then generated for each vertex, including directly related vertices in the current layer and propagated labels from higher levels. DL(Distribution Labelling) [17] creates a reachability index through vertex traversal, sorting vertices to prioritise those covering more reachability pairs. For each vertex, reverse and forward searches are conducted to add it to the $Lout$ and Lin sets of reachable vertices, using pruning to avoid redundancy. when querying, both HLand DLwill check their labels for intersections.

PPL. PPL [46] introduces a 2-hop indexing method for reachability queries on directed graphs by precomputing labels. Unlike traditional 2-hop approaches, it uses paths instead of individual vertices, storing interval information to compress label sizes. Pruning techniques are applied during label construction to stop redundant searches, prioritising critical vertices and paths to enhance efficiency.

Ferrari. Ferrari [34] utilises postorder numbering based on tree cover to generate initial tree interval representations. By combining adaptive compression strategies, the transitive closure of the graph is compactly represented as a mix of exact and approximate intervals. During index construction, vertices are processed in reverse topological order, merging interval sets from child vertices and optimising interval allocation under a given size constraint to balance storage overhead and query performance.

Feline. Feline [41] maps each vertex in the graph to a unique coordinate pair (x, y) on a two-dimensional plane, utilising the geometric "weak dominance relation" to quickly determine reachability. Specifically, if the coordinates of a vertex u are less than or equal to those of another vertex v, there may be a path from u to v; otherwise, non-reachability can be ruled out in constant time. The index is constructed by generating x coordinates through topological sorting and optimising y coordinates using a heuristic algorithm to reduce false positives.

PReaCH. PReaCH [26] achieves reachability queries by combining Contraction Hierarchies (CH) and pruning techniques. The index is constructed by iteratively contracting vertices with zero in-degree or out-degree based on their total degree, marking the remaining edges with directional labels to support bidirectional search. Additionally, topological levels and DFS-based vertex numbering are utilised to derive *full ranges* and *empty ranges* for pruning, significantly reducing the search space during queries. The entire process ensures linear space complexity and near-linear preprocessing time.

BFL. BFL [37] uses Bloom filters [3] to build reachability indexes by labelling each vertex's *out-set* and *in-set* as Bloom filter labels. The probabilistic Bloom filter efficiently checks set membership, providing efficient pruning to determine reachability without complex traversals. During initialisation, BFL constructs a Bloom filter label, for each vertex. In the query phase, Bloom filter labels are compared to quickly determine reachability; if uncertain, a DFS is performed.

O'Reach. O'Reach [15] preprocesses graphs using supportive vertices and extended topological orderings to answer most reachability queries in constant time. When preprocessing is insufficient, it falls back on methods like bidirectional BFS. With a linear-sized index and combined hop labelling and pruning, O'Reach effectively reduces query times. It can be used standalone or integrated with other algorithms to balance query time, initialisation, and index size.

6 Dynamic Methods

In this section, we introduce each of the static methods. Online methods are also included, as they can be directly used to answer queries on dynamic graphs.

HOPI and U2-Hop. HOPI [31] and U2-Hop [4] both use 2-hop cover indexing for reachability queries. HOPI constructs a compact graph representation for XML documents, enabling efficient querying and incremental updates. U2-Hop proposes an incremental update algorithm for large graphs, using two-hop labelling to maintain efficient connectivity during data changes. Both methods face limitations in maintenance efficiency, which are not discussed further here.

DAGGER. DAGGER [47] proposes a lightweight indexing method for reachability in large dynamic graphs. It uses interval labelling for vertices and edges, enabling insertions and deletions. When adding new elements, DAGGERupdates SCCs and their DAG, adjusting interval labels as needed. For deletions, it splits SCCs and reallocates interval labels to maintain consistency. This dynamic maintenance of SCCs and labels ensures efficient real-time query handling with minimal overhead. Although DAGGER can independently perform SCC maintenance and supports both insertions and deletions, its maintenance cost and query time typically fail to meet real-time demands on large dynamic graphs.

TOL. TOL [56] proposes a framework base on 2-hop for constructing reachability indices that are applicable to large dynamic graphs. The TOLframework constructs indices by assigning a total order to the vertices in the graph. This ordering enables efficient support for reachability queries. Insertions and deletions are handled without altering the order of other vertices, thus maintaining the efficiency of the index. Specifically, when a new vertex is inserted, the algorithm selects the optimal position to minimise the index size and updates the affected label sets. When deleting a vertex, the algorithm adjusts the corresponding label sets to ensure the index remains correct while minimising redundancy.

IP. IP [43] introduces an efficient method for reachability indexing using random permutation-based labels for vertices. The minimum k-value set describes vertex reachability, and queries are resolved by comparing labels, with DFS as a fallback when necessary. In dynamic, it enables insertion and deletion.

Remark. Both IP [43] and TOL [56] are unable to process SCC maintenance. In practice, they need to be completed with the help of DAGGER [47]. This can also cause them to not work on large actual graphs.

ARROW. ARROW [33] uses random walks answering reachability queries. It conducts multiple random walks from both source and target vertices, building two sets: one reachable from the source and the other that can reach the target. If these sets intersect, a path exists. This method efficiently estimates reachability in large, dynamic graphs without needing precomputed indexes.

DBL. DBL [25] constructs two complementary indexes: Dynamic Landmark (DL) labels and Bidirectional Leaf (BL) labels for efficient reachability queries. DL labels use high-centrality landmarks to quickly determine reachability, while BL labels index leaf vertices to prune unreachable pairs. During edge insertion, pruned BFS updates affected DL and BL labels, propagating labels as needed. DBLavoids high maintenance costs related to SCC changes and, though it doesn't support deletions, remains practical for dynamic graph reachability.

IFCA. IFCA [27] introduces an index-free, community-aware framework for reachability in large dynamic graphs. It identifies community structures using an approximation algorithm inspired by Personalised PageRank for efficient within-community searches. For vertices in different communities, IFCA uses community contraction to merge communities into super-vertices, reducing graph complexity. It also employs a cost-based mechanism to choose between guided search and bidirectional BFS. IFCA's lack of constraints with SCC and index maintenance makes it well-suited for large, dynamic graphs compared to traditional methods.

7 Analysis and Discussion

7.1 Recommendations for Existing Methods

Static Graphs. For static graphs, O'Reach [15] demonstrates strong performance, especially for real-world sparse graphs. It can be combined with other methods to achieve efficient query times while balancing index size, initialisation, and query time, making it a practical choice for various applications.

Dynamic Graphs. For dynamic graphs, algorithm selection varies. When the SCC structure changes and both insertions and deletions are required, DAGGER [47] is recommended, though it has limitations for large graphs. For cases with a stable SCC structure, IP [43] and TOL [56] are strong candidates, but SCC maintenance challenges limit their scalability. DBL [25] supports insertions but not deletions. The online method IFCA [27] offers flexibility without preprocessing, but its query performance is weaker than preprocessed approaches. In summary, while several promising methods exist for dynamic graphs, each has limitations, and further research and technological breakthroughs are needed to enhance their applicability to large-scale dynamic scenarios.

7.2 Open Challenges

Dynamic Graph Update and Maintenance. Dynamic graphs pose significant challenges, as frequent updates to vertices and edges require index-based methods to synchronise with graph changes. High maintenance costs in dynamic environments can negate the efficiency gains of indexing, making practical implementation difficult.

Complexity of Heterogeneous Networks. Heterogeneous information networks [7,14,38], with multiple types of vertices and edges, increase query complexity by requiring the consideration of inter-vertex relationships and path pattern matching, which traditional methods struggle to address.

Reachability with Label Constraints. Applications requiring label sequence constraints on paths demand methods that account for edge label orders, a challenge traditional approaches are not equipped to handle efficiently.

7.3 Future Directions

Preprocessing and Indexing. Efficient indexing structures are key to rapid reachability queries. Future work aims to improve preprocessing time, index size, and query performance for various graph types.

Dynamic Graphs. Dynamic updates, such as vertex and edge insertions or deletions, require adaptable indexing methods. Advances in maintaining SCCs could eliminate DAG constraints, enabling existing algorithms to handle large dynamic graphs more effectively.

Extended Queries and Attribute Graphs. Extended query types, such as k-reach [11], can be adapted to better meet practical needs. Incorporating attributes into reachability queries, as seen in temporal [44] or labelled [9] graphs, is also a growing research area.

Distributed and Parallel Processing. For large-scale graphs, distributed systems and parallel computing offer solutions to scale efficiently, with a focus on data partitioning and communication optimisation.

Learning-based Approaches. Graph embedding techniques [5,24] for predicting reachability provide fast query responses, making them suitable for real-time applications, though challenges in accuracy remain.

8 Conclusion

This paper offers a comprehensive review on understanding and addressing the problem of directed graph reachability queries. It details various existing techniques and methods, compares their advantages and limitations, and explores future directions in the field.

References

1. Aggarwal, C.C., Subbian, K.: Evolutionary network analysis: a survey. ACM Comput. Surv. (CSUR) **47**(1), 10 (2014)
2. Agrawal, R., Borgida, A., Jagadish, H.V.: Efficient management of transitive relationships in large data and knowledge bases. In: Proceedings of ACM SIGMOD International Conference on Management Data, pp. 253–262 (1989)
3. Bloom, B.H.: Space/time trade-offs in hash coding with allowable errors. Commun. ACM **13**(7), 422–426 (1970)
4. Bramandia, R., Choi, B., Ng, W.K.: Incremental maintenance of 2-hop labeling of large graphs. IEEE Trans. Knowl. Data Eng. **22**(5), 682–698 (2010)
5. Cai, H., Zheng, V.W., Chang, K.: A comprehensive survey of graph embedding: problems, techniques, and applications. IEEE Trans. Knowl. Data Eng. **30**(9), 1616–1637 (2018)
6. Cai, J., Poon, C.K.: Path-hop: Efficiently indexing large graphs for reachability queries. In: Proceedings of the 19th ACM International Conference on Information and Knowledge Management, pp. 119–128 (2010)

7. Carranza, A.G., Rossi, R.A., Rao, A., Koh, E.: Higher-order clustering in complex heterogeneous networks. In: Proceedings of the 26th ACM SIGKDD International Conference on Knowledge Discovery & Data Mining, pp. 25–35. KDD '20, Association for Computing Machinery, New York, NY, USA (2020)
8. Chen, L., Gupta, A., Kurul, M.E.: Stack-based algorithms for pattern matching on DAGs. In: Proceedings of the 31st International Conference on Very Large Data Bases (VLDB 2005) (2005)
9. Chen, X., Peng, Y., Wang, S., Yu, J.X.: DLCR: efficient indexing for label-constrained reachability queries on large dynamic graphs. Proc. VLDB Endowment **15**(8), 1645–1657 (2022)
10. Cheng, J., Huang, S., Wu, H., Fu, A.W.C.: TF-label: a topological-folding labeling scheme for reachability querying in a large graph. In: Proceedings of the ACM SIGMOD International Conference on Management of Data. Kenneth A. Ross, Divesh Srivastava, and Dimitris Papadias (Eds.), ACM, pp. 193–204 (2013)
11. Cheng, J., Shang, Z., Cheng, H., Wang, H., Yu, J.X.: K-reach: who is in your small world. Proc. VLDB Endow. **5**(11), 1292–1303 (2012)
12. Cohen, E., Halperin, E., Kaplan, H., Zwick, U.: Reachability and distance queries via 2-hop labels. SIAM J. Comput. **32**(5), 1338–1355 (2003)
13. Dhia, I.B.: Access control in social networks: a reachability-based approach. In: Proceedings of the 2012 Joint EDBT/ICDT Workshops, pp. 227–232 (2012)
14. Fu, C., Zheng, G., Huang, C., Yu, Y., Dong, J.: Multiplex heterogeneous graph neural network with behavior pattern modeling. In: Proceedings of the 29th ACM SIGKDD Conference on Knowledge Discovery and Data Mining, pp. 482–494. KDD '23 (2023)
15. Hanauer, K., Schulz, C., Trummer, J.: O'Reach: even faster reachability in large graphs. J. Exp. Algorithmics **27**(4), 4.2 (2022)
16. Holme, P., Saramäki, J.: Temporal networks. Physics Reports (2012)
17. Jin, R., Wang, G.: Simple, fast, and scalable reachability oracle. Proc. VLDB Endowment **6**(14), 1978–1989 (2013)
18. Jin, R., Xiang, Y., Ruan, N., Fuhry, D.: 3-hop: a high-compression indexing scheme for reachability query. In: Proceedings of the 2009 ACM SIGMOD International Conference on Management of Data, pp. 813–826. SIGMOD '09 (2009)
19. Jin, R., Xiang, Y., Ruan, N., Wang, H.: efficiently answering reachability queries on very large directed graphs. In: Proceedings of ACM SIGMOD International Conference on Management Data, pp. 595–608 (2008)
20. Kahn, A.B.: Topological sorting of large networks. Commun. ACM **5**(11), 558–562 (1962)
21. Kaminski, M., Mahoney, M.W.: Dynamic financial networks and applications. Annu. Rev. Financ. Econ. **9**, 259–299 (2017)
22. Kiss, I.Z., Miller, J.C., Simon, P.L.: Mathematics of Epidemics on Networks: From Exact to Approximate Models. Springer (2017)
23. Li, X., et al.: Hop-constrained s-t simple path enumeration in billion-scale labelled graphs. In: Web Information Systems Engineering – WISE 2022: 23rd International Conference, pp. 49–64. Springer-Verlag, Berlin, Heidelberg (2022)
24. Luo, L., Fang, Y., Cao, X., Zhang, X., Zhang, W.: Detecting communities from heterogeneous graphs: a context path-based graph neural network model. In: CIKM '21: The 30th ACM International Conference on Information and Knowledge Management, Virtual Event, Queensland, Australia, November 1 - 5, 2021, pp. 1170–1180. ACM (2021)
25. Lyu, Q., Li, Y., He, B., Gong, B.: DBL: efficient reachability queries on dynamic graphs, pp. 761–777 (2021)

26. Merz, F., Sanders, P.: PReaCH: a fast lightweight reachability index using pruning and contraction hierarchies. In: Proceedings of the European Symposium on Algorithms, pp. 701–712 (2014)
27. Pang, Y., Zou, L., Liu, Y.: IFCA: index-free community-aware reachability processing over large dynamic graphs. In: 2023 IEEE 39th International Conference on Data Engineering (ICDE) (2023)
28. Qiu, X., et al.: Real-time constrained cycle detection in large dynamic graphs. Proc. VLDB Endowment **11**(12), 1876–1888 (2018)
29. Reps, T.: Program analysis via graph reachability. Inf. Softw. Technol. **40**(11–12), 701–726 (1998)
30. van Schaik, S.J., de Moor, O.: A memory efficient reachability data structure through bit vector compression. In: Proceedings of ACM SIGMOD International Conference on Management Data, pp. 913–924 (2011)
31. Schenkel, R., Theobald, A., Weikum, G.: Efficient creation and incremental maintenance of the HOPI index for complex XML document collections. In: Proceedings of the 21st International Conference on Data Engineering, pp. 360–371 (2005)
32. Scholz, B., Zhang, C., Cifuentes, C.: User-input dependence analysis via graph reachability. In: Proceedings of the 2008 8th IEEE International Working Conference on Source Code Analysis and Manipulation, pp. 25–34 (2008)
33. Sengupta, N., Bagchi, A., Ramanath, M., Bedathur, S.: ARROW: approximating reachability using random walks over web-scale graphs. In: 35th IEEE International Conference on Data Engineering, ICDE 2019, pp. 470–481. IEEE (2019)
34. Seufert, S., Anand, A., Bedathur, S.J., Weikum, G.: FERRARI: flexible and efficient reachability range assignment for graph indexing. Proc. IEEE Int. Conf. Data Eng. **2013**, 1009–1020 (2013)
35. Sharir, M.: A strong-connectivity algorithm and its applications in data flow analysis. Comput. Math. Appl. **7**(1), 67–72 (1981)
36. Simon, K.: An improved algorithm for transitive closure on acyclic digraphs. Theoretical Comput. Sci. **58**(1–3), 325–346 (1988)
37. Su, J., Zhu, Q., Wei, H., Yu, J.X.: Reachability querying: can it be even faster? IEEE Trans. Knowl. Data Eng. **29** (2017)
38. Sun, Y., Han, J., Yan, X., Yu, P.S., Wu, T.: PathSim: meta path-based top-k similarity search in heterogeneous information networks. Proc. VLDB Endowment **4**(11), 992–1003 (2011)
39. Tarjan, R.: Depth-first search and linear graph algorithms. SIAM J. Comput. **1**(2), 146–160 (1972)
40. Trißl, S., Leser, U.: Fast and practical indexing and querying of very large graphs. In: Proceedings of the 2007 ACM SIGMOD International Conference on Management of Data (SIGMOD 2007). ACM (2007)
41. Veloso, R.R., Cerf, L., Jr., W.M., Zaki, M.J.: Reachability queries in very large graphs: a fast refined online search approach. In: Proceedings of 17th International Conference on Extending Database Technology, pp. 511–522 (2014)
42. Wang, H., He, H., Yang, J., Yu, P.S., Yu, J.X.: Dual labeling: answering graph reachability queries in constant time. In: Proceedings of IEEE 29th International Conference on Data Engineering (2006)
43. Wei, H., Yu, J.X., Lu, C., Jin, R.: Reachability querying: an independent permutation labeling approach. VLDB J. **27** 1–26 (2018)
44. Wen, D., Yang, B., Zhang, Y., Qin, L., Cheng, D., Zhang, W.: Span-reachability querying in large temporal graphs. VLDB J. 1–19 (2021). https://doi.org/10.1007/s00778-021-00715-z

45. Xu, Z., Yang, Z., Chen, X., He, H., Cao, X.: Efficient answering of k-reachability on temporal bipartite graphs. In: Databases Theory and Applications, pp. 224–238. Springer Nature Singapore, Singapore (2025)
46. Yano, Y., Akiba, T., Iwata, Y., Yoshida, Y.: Fast and scalable reachability queries on graphs by pruned labeling with landmarks and paths. 22nd CIKM 2013: San Francisco, CA, USA (2013)
47. Yildirim, H., Chaoji, V., Zaki, M.J.: DAGGER: A scalable index for reachability queries in large dynamic graphs (2013)
48. Yu, D., Ma, H., Li, S., Chen, J.: Reachability constrained reinforcement learning. In: International Conference on Machine Learning, 17-23 July 2022, Baltimore, Maryland, USA. Proceedings of Machine Learning Research
49. Yu, J.X., Cheng, J.: Graph reachability queries: a survey. In: Managing and Mining Graph Data, Advances in Database Systems, vol. 40, pp. 181–215. Springer (2010)
50. Yıldırım, H., Chaoji, V., Zaki, M.J.: GRAIL: a scalable index for reachability queries in very large graphs. VLDB J. **21**(4), 509–534 (2012)
51. Zhang, C., Bonifati, A., Özsu, M.T.: An overview of reachability indexes on graphs. In: Companion of the 2023 International Conference on Management of Data, SIGMOD/PODS 2023, pp. 61–68. ACM (2023)
52. Zhang, Z., Yu, J.X., Qin, L., Chang, L., Lin, X.: I/O efficient: computing SCCs in massive graphs. In: SIGMOD Conference, pp. 181–192 (2013)
53. Zhou, J., Yu, J.X., Li, N., Wei, H., Chen, Z., Tang, X.: Accelerating reachability query processing based on DAG reduction. VLDB J. **27**(2), 271–296 (2018)
54. Zhou, J., Yu, J.X., Qiu, Y., Tang, X., Chen, Z., Du, M.: Fast reachability queries answering based on RCN reduction. IEEE Trans. Knowl, Data Eng (2023)
55. Zhou, J., Zhou, S., Yu, J.X., Wei, H., Chen, Z., Tang, X.: Dag reduction: fast answering reachability queries. In: SIGMOD Conference, pp. 375–390 (2017)
56. Zhu, A.D., Lin, W., Wang, S., Xiao, X.: Reachability queries on large dynamic graphs: A total order approach, pp. 1323–1334 (2014)

Machine Learning

Disentangled Representation Learning for Geospatial-Temporal Data Modeling

Siyuan Ren[1], Guannan Chang[1], Luqi Jing[1], Zhenghong Wu[1], Shuailin Chen[1], and Qianru Wang[2(✉)]

[1] Xi'an Modern Control Technology Research Institute, Xi'an, China
[2] Xidian University, Xi'an, China
wangqianru@xidian.edu.cn

Abstract. While the air quality prediction has been extensively studied, existing methods often struggle with stability due to complex data and modeling challenges. Data heterogeneity from various city areas can lead to significant information loss during representation generation, which involves the *data complexity*. Additionally, capturing geospatial-temporal dependencies amidst multiple uncertain factors adds to the *model complexity*. To tackle these dual challenges, effectively disentangling mixed impact factors from heterogeneous geospatial-temporal data is crucial for generating robust representations. In this paper, we propose a novel geospatial-temporal disentangled representation learning (GT-DRL) method for accurate air quality prediction. Our deep predictive model not only disentangles mixed data dependencies into independent geospatial-temporal dimensions for uniform reconstruction, but also models complex coupling relationships from multi-dimensional uncertain factors. Evaluations on two real-world datasets demonstrate GT-DRL's superiority over baseline methods, highlighting its robustness and efficiency in handling data and modeling complexities.

Keywords: Geospatial-temporal data · disentangled representation learning · complex coupling relationship · air quality prediction

1 Introduction

Accurate air quality prediction is essential to improve public health and urban living. With the advent of advanced deep neural networks, various data-driven research in urban computing has emerged, aimed at enhancing the construction of smart cities. Recurrent neural networks [16,25], attention mechanisms [10,18], temporal convolution networks [22,28], spatial-temporal graph convolution networks [3,9], transformer [27,31,32], and their variants, are powerful in capturing non-linear relationships and have achieved great success in general spatial-temporal correlations learning.

This work is supported by the National Natural Science Foundation of China (No. 62302367).

X. Wu et al. (Eds.): PAKDD 2025, LNAI 15875, pp. 81–93, 2025.
https://doi.org/10.1007/978-981-96-8295-9_5

Although existing approaches have demonstrated considerable prowess in terms of their results and predictive capabilities, directly applying deep models to infer inter-dependencies from raw data inadequately captures the complex nature of geospatial-temporal data and their complex coupling relationships [5,6,19], which contain three aspects: 1) the temporal data comprises sequential information alongside timestamps for each step, is often processed using sequence modeling techniques that weaken the significance of timestamps and grapple with time alignment issues in cases of irregular sampling; 2) the authentic relative geographical positions of monitoring sites harbor invaluable information, such as relative distances and directions, while existing methods rarely utilize this information and simply obtain topological spatial relationships rather than the true essence of geospatial relationships; 3) the geospatial-temporal mutual dependencies are jointly caused by multiple impact factors, each of which is unfixed and sparse after being disentangled, making it difficult to model such uncertain factors adaptively and concurrently.

Driven by these challenges and identified gaps, the aim of this paper is to elevate both the stability and performance of air quality modeling by tackling the inherent complexities of the data and the modeling process. In contrast to prevailing methods, we propose the following strategies to address the data characteristics and modeling challenges: 1) explicitly decompose raw temporal data into components with clear, inherent meanings, and generate multi-dimensional temporal position representations for each timestamp to resolve the data heterogeneity issue, i.e. time alignment, 2) project geospatial data onto a rectangular coordinate system and quantify mutual dependencies based on relative geographical locations, thereby disentangling geospatial couplings into two independent dimensions, and 3) model geospatial-temporal couplings using a stacked encoder-decoder architecture to effectively manage the joint impact of multiple uncertain factors. Accordingly, we propose a novel **geospatial-temporal disentangled representation learning (GT-DRL)** method for estimating air quality changes. Our key contributions can be summarized as follows:

- We incorporate a novel geospatial-temporal representation method into the deep predictive model, effectively addressing the challenge of data complexity. To the best of our knowledge, our approach marks the pioneering application of such a method to geospatial-temporal data modeling, particularly in the realm of air quality prediction.
- We devise an efficient stacked encoder-decoder framework that adeptly assembles disentangled factors, significantly bolstering our ability to handle complex geospatial-temporal coupling relationships.
- Comprehensive experiments conducted on real-world air quality datasets from two representative cities demonstrate that our model surpasses existing state-of-the-art methods. Additionally, we assess the stability of GT-DRL under conditions of irregular sampling, further validating its robustness and reliability.

2 Related Work

2.1 Deep Learning for Air Quality Prediction

Efficient and accurate air quality prediction is of great importance for urban regulatory applications and public health protection. Deep neural networks have become a powerful tool for air quality prediction tasks [11,23]. These methods typically rely on linear or nonlinear transformations for automatic extraction of high-dimensional features. Further, Transformer-based approaches, including informer [31], autoformer [27], FEDformer [32], learn complex inter-dependencies between variables by reconstructing and stacking multiple layers of multi-head attentions. However, these methods generally capture temporal dependencies based on sequence learning, rather than time series modeling, which lacks specific design for temporal information, such as timestamps. Moreover, they rely on predefined graph structures and omit geographical location information, making them unsuitable for air quality prediction tasks. Despite substantial research in spatial-temporal prediction, existing methods still have gaps in learning geographical-related spatial coupling relationships. Our work is the first to apply disentangled representation learning to geospatial-temporal modeling, emphasizing the importance of representation to improve the efficiency of air quality prediction.

2.2 Disentangled Representation Learning

Disentangled representation learning has become an increasingly popular approach for modeling complex data in a more interpretable and effective manner. The core idea is to separate the underlying factors of variation in the data into multiple independent components that can be explicitly controlled and manipulated. Many prior works have demonstrated the effectiveness of disentangled representation learning on image data through Variational Autoencoders (VAE) [14] based methods, such as β-VAE [12], Factor-VAE [13], and β-TCVAE [7]. In the field of sequential data analysis, S3VAE [33] and SV2P [2] are two approaches that have been proposed to separate static and dynamic factors of sequential data, allowing for better modeling of time-varying dynamics. However, such methods have primarily been applied to purely sequential data and lack attention to the modeling of geospatial-temporal data. The problem of generating disentangled representation from multi-sourced heterogeneous data, while simultaneously modeling several latent geospatial-temporal couplings relationships, remains largely unexplored. Therefore, there is a pressing demand for innovative methodologies capable of effectively modeling complex, multivariate geospatial-temporal data and deriving efficient representations from it. This paper aims to address these challenges by incorporating a disentangled representation learning method into a deep predictive model, to effectively learn and decouple the complex geospatial-temporal interactions inherent in air quality forecasting.

3 Methodology

To efficiently model complex coupling relations among multi-sourced data, we propose a novel approach to deeply explore GT-data by first introducing a geospatial-temporal disentangled representation learning method. This method decomposes mixed data couplings into multi-dimensional independent representations, which are then assembled to model complex relationships via a stacked encoder-decoder architecture (Fig. 1). Detailed module descriptions follow.

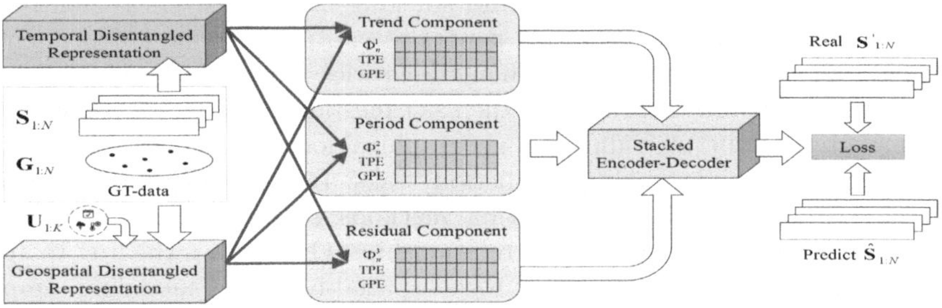

Fig. 1. The architecture of geospatial-temporal disentangled representation learning.

3.1 Geospatial-Temporal Disentangled Representation

Geospatial Disentangling. Since air quality in each area depends on both historical and neighboring observations [21], it is critical to obtain site geo-locations and learning their geospatial relationships, including relative distances and relative directions. However, existing methods neglect directional information and focusing solely on distances. Our approach models more complex geospatial relationships by considering time-varying pollutant diffusion directions, site locations, and observation data changes.

To simplify the modeling difficulty, the site geo-locations are projected into a coordinate system, which disentangling their relative distances and directions into two independent dimensions. Specifically, confusion matrices X, Y calculate relative locations between sites from 2 perpendicular directions (i.e. horizontal coordinates and vertical coordinates), which are represented as:

$$\mathrm{X}=\begin{bmatrix} 0 & x_{\Delta 12} & \cdots & x_{\Delta 1N} \\ x_{\Delta 21} & 0 & \cdots & x_{\Delta 2N} \\ \vdots & \vdots & \ddots & \vdots \\ x_{\Delta N1} & x_{\Delta N2} & \cdots & 0 \end{bmatrix} \quad \mathrm{Y}=\begin{bmatrix} 0 & y_{\Delta 12} & \cdots & y_{\Delta 1N} \\ y_{\Delta 21} & 0 & \cdots & y_{\Delta 2N} \\ \vdots & \vdots & \ddots & \vdots \\ y_{\Delta N1} & y_{\Delta N2} & \cdots & 0 \end{bmatrix} \tag{1}$$

where $x_{\Delta mn} \in \mathrm{X}$ and $y_{\Delta mn} \in \mathrm{Y}$ denote the relative location from source site m to the target site n ($m, n \in \{1, 2, \cdots, N\}$ and $m \neq n$). And their values can be obtained by:

$$\begin{cases} x_{\Delta mn} = g_n^x - g_m^x \\ y_{\Delta mn} = g_n^y - g_m^y \end{cases} \tag{2}$$

Specifically, $|x_{\Delta mn}|$ denotes the horizontal relative distance from m to n, and $|y_{\Delta mn}|$ is the vertical relative distance. And the relative direction is record by its positive/negative value, e.g. $x_{\Delta mn} > 0$ indicates that m is west of n, and $y_{\Delta mn} < 0$ means that m is north of n. When modeling relationships between one target monitoring site with others, and $\mathbf{x}_n, \mathbf{y}_n$ are two vectors that record all pair-wise relative distances for target site n in two perpendicular dimensions, i.e. $\mathbf{x}_n = [x_{\Delta 1n}, x_{\Delta 2n}, \ldots, x_{\Delta Nn}]^\top$ and $\mathbf{y}_n = [y_{\Delta 1n}, y_{\Delta 2n}, \ldots, y_{\Delta Nn}]^\top$, where $x_{\Delta nn} = 0$, $y_{\Delta nn} = 0$.

Considering that the pollutants' diffusion is also related to time-varying meteorological conditions (i.e. wind, temperature, humidity, etc.) and dynamically changes over time [4]. We use deep neural networks with residual blocks to learn non-linear dependencies and generate timely parameters from exogenous features $\mathbf{u}_{1:K}^t$:

$$\varepsilon_x, \varepsilon_y = ResBlock(\mathbf{u}_{1:K}^t) \tag{3}$$

where the implementation of **ResBlock** is formalized as:

$$\mathbf{Out} = LayerNorm(\mathbf{In} + FeedForward(\mathbf{In})) \tag{4}$$

The parameters $\varepsilon_x, \varepsilon_y$ from **ResBlock** reflect pollutant diffusion and shared across sites, where their values and directions are related to atmospheric stability and weather. Besed on this, we designed a geospatial position encoding (GPE) method to generate geospatial representations for each disentangled dimension based on relative locations:

$$GPE(1:N)(n) = \begin{cases} \exp(-\varepsilon_x \cdot \mathbf{x}_n) \otimes \sigma(\varepsilon_x \cdot \mathbf{x}_n) \\ \exp(-\varepsilon_y \cdot \mathbf{y}_n) \otimes \sigma(\varepsilon_y \cdot \mathbf{y}_n) \end{cases} \tag{5}$$

$GPE(1:N)(n)$ contains geo-information for all sites referred to the target site n from both horizontal and vertical directions. For example, the operation $\exp(-\varepsilon_x \cdot \mathbf{x}_n)$ quantifies the attenuation of effect along one specific direction, where their effects decrease as relative distance increased. Besides, with considering the effect of direction, we also employ a gating $\sigma(\bullet)$ to filter out invalid connections. For instance, some sites far away from the target site is unnecessary to maintain.

Temporal Disentangling. Despite the existing deep learning methods, such as RNN [24], TCN [30], TPA-LSTM [20], and FEDformer [32] have considered the seasonal changes, but they purely consider the seasonal relations from a single scale. Since we predict air quality based on hourly (24 h), daily (7 days) and weekly (52 weeks) historical data, we designed a multi-scale temporal position embedding approach for timestamps. Motivated by the idea of Sinusoidal position embedding, the timestamps can be expressed by the pair-wise sin/cos transformation with the following equations:

$$TPE(t) = \begin{cases} \exp\{-ReLU(W_*(h-t))\} \\ \left[\sin\left(\frac{2\pi}{24}t\right);\ \cos\left(\frac{2\pi}{24}t\right)\right] \\ \left[\sin\left[\frac{2\pi}{7}(t-1)\right];\ \cos\left[\frac{2\pi}{7}(t-1)\right]\right] \\ \left[\sin\left(\frac{2\pi}{52}t\right);\ \cos\left(\frac{2\pi}{52}t\right)\right] \end{cases} \tag{6}$$

For each position t of time series, the temporal position embedding $TPE(t)$ consists of two parts: **relative position embedding** (first row) is used to represent the temporal decay effects because the influence will decay over time. And **absolute position embedding** (last three rows) is designed to uniquely identify a specific timestamp. We use multi-scale: hour, day and week to embed the position of one timestamp, respectively. In this way, any position in the sequential data can be embedded by the multi-scale seasonal changes.

Besides, to extract latent impact factors hidden in temporal data, we propose to decomposing raw sequences into discriminative components across multiple scales. Specifically, the decomposed latent impact factors requires to make different components discriminative in multiple temporal scales and easy to calculate. The generic temporal decomposer $Decomposer(\bullet)$ is defined. For input temporal data $\mathbf{s}_n$ of length h at the nth site, the detailed decomposition is formalized as:

$$\mathbf{\Phi}_n = [\Phi_n^1, \Phi_n^2, \cdots, \Phi_n^D]^\top = Decomposer(\mathbf{S}_n) \tag{7}$$

where $d \in \{1, 2, \ldots, D\}$ denotes the number latent impact factors extracted by the decomposer. $\Phi_n^d = [\varphi_n^{d,1}, \varphi_n^{d,2}, \cdots, \varphi_n^{d,t}, \cdots, \varphi_n^{d,h}] \in \mathbb{R}^h$ is a 1D tensor with length h, it represents a component of factors extracted by the decomposer, and $\varphi_n^{d,t}$ represents the value of the dth component of the nth site's observation at time t. Thus $\mathbf{\Phi}_n \in \mathbb{R}^{D\times h}$ is a 2D tensor that contains all extracted factors for site n, similar express also applied on other monitoring sites temporal data.

To inductively disentangle impact factors, we employed the seasonal and trend decomposition using Loess (STL) [26] to decompose temporal data into periodic, trend, and residual. Although residuals are typically white noise, they may contain abrupt data indicating uncertain changes, which we preserve. Therefore, in Eq. 7, the number of latent impact factors (D) is set to 3 when the STL adopted.

Geospatial-Temporal Representation Generation. After obtaining disentangled multi-dimensional geospatial-temporal information, representations are generated by fusing the position embedding vectors with each decomposed temporal impact factors. For example, the impact factor from source m to target n of the dth component at time t can be expressed as:

$$r_{m\rightarrow n}^{d,t} = Concat[\varphi_m^{d,t}; TPE(t); GPE_m(n)]^\top \tag{8}$$

By meaning of concatenating operation $Concat[\bullet]$, the geospatial position $GPE_m(n)$ and temporal position $TPE(t)$ uniquely marked the representation

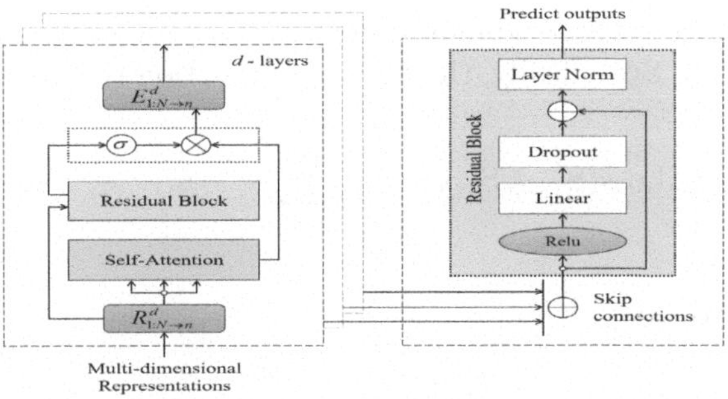

Fig. 2. The framework of stacked encoder-decoder and residual block.

of current dimension. And then, the entire representation corresponding to target n of dth component is merged by:

$$R^d_{1:N\to n} = \sum_{m=1}^{N}\sum_{t=1}^{h} r^{d,t}_{m\to n}, d \in \{1, 2, \ldots, D\} \tag{9}$$

where $R^d_{1:N\to n} \in \mathbb{R}^{3\times Nh}$ is a 2D tensor that contains all generated representation elements for site n, and similar operation also applied on other sites to generate their representations.

3.2 Stacked Encoder-Decoder Framework Implement

Our stacked encoder-decoder framework leverages multi-dimensional representations for geospatial-temporal coupling modeling (Fig. 2). The GT encoder adapts to these couplings, with each encoder targeting a dimension, including self-attention block for automatically weighting units and a gated residual block for controlling uncertainty:

$$E^d_{1:N\to n} = \mathrm{Self-attn}\left(R^d_{1:N\to n}\right) \otimes \sigma(\mathrm{ResBlock}(R^d_{1:N\to n})) \tag{10}$$

where $E^d_{1:N\to n}$ represents the output of the encoder corresponding to d^th dimension representation. Several layers of the GT encoders are stacked paralleled, followed by a predictive decoder to estimate the future changes of air qualities. The decoder first merges the encoder output of all layers, and then employs a residual block to adaptively select the most relevant data for prediction:

$$\hat{S}_n = \mathrm{ResBlock}\left(\sum_{d=1}^{D} E^d_{1:N\to n}\right) \tag{11}$$

where $\hat{S}_n$ denotes the predicted result of nth site, and the $\hat{S}_{1:N}$ can be obtained by repeat this framework in N times with different representations $R^d_{m\to n}$.

In model training process, the ground truth is given by: $\mathbf{S}'_{1:N} = [\mathbf{s}'_1, \mathbf{s}'_2, \ldots, \mathbf{s}'_N]^\top$, where $\mathbf{s}'_n = \left[\tau^{t+1}_n, \tau^{t+2}_n, \ldots, \tau^{t+l}_n\right]$, and $\hat{\mathbf{S}}_{1:N}$ denotes the predicted result. Since the model is smooth and differentiable, it can be trained using standard back-propagation. We adopt average Mean Square Error (MSE) as the loss function. Specifically, the MSELoss function can be calculated by:

$$\text{MSELoss}(\mathbf{S}'_{1:N}, \hat{\mathbf{S}}_{1:N}) = \frac{1}{N}\sum_{i=1}^{N}(\mathbf{s}'_i - \hat{\mathbf{s}}_i)^\top(\mathbf{s}'_i - \hat{\mathbf{s}}_i) \quad (12)$$

4 Evaluation

4.1 Datasets

The proposed model was evaluated using real-world datasets from two cities: Beijing and Xi'an, noted for frequent pollution and high concentrations, enhancing result representativeness. Both cities' datasets include sub-datasets on air quality and meteorology, as depicted in Table 1.

Table 1. Details of the datasets.

Datasets	Time Spans	Interval	# Variables	# Instance	Indicators	# E-features
Beijing	05/12/2013 - 18/01/2021	1 h	17	12,517	AQI, PM2.5	6
Xi'an	13/05/2014 - 20/08/2022	1 h	12	15,881	AQI, PM2.5	4

Air quality dataset sourced from multiple urban monitoring sites, includes site coordinates and historical observations. Specifically, we have selected two measurements: AQI and PM2.5 as representatives, and their hourly observations are formed as multivariate time series. Preprocessing involves removing anomalies and imputing missing values using neighboring averages.

Meteorological dataset serving as exogenous features (E-features), encompasses several types of air quality-related meteorological features: air temperature, dew point, wind direction (angular degrees), wind speed rate, sea level pressure, and sky conditions (0–19, higher values=more clouds). Beijing's dataset includes all, while Xi'an's uses the first four. These time-varying features are synchronized with air quality data for training.

4.2 Experimental Setting

We implement GT-DRL in Pytorch, and trained by Adam optimizer ($1e^{-2}$ initial learning rate, batch size 256). The dataset is split into non-overlapped training, validation and test sets by a ratio of 7:2:1. All experiments are conducted on an RTX-3080Ti GPU and i7-11800H CPU. Two criteria are employed to evaluate our proposed model, including Root Mean Squared Error (RMSE) and Mean Absolute Error (MAE), both of which are commonly used in regression tasks, with lower values indicating better performance. To be fair, all deep learning models are trained five times and their average performance is compared.

Specifically, given $\hat{\mathbf{S}}$ (predicted values) and $\mathbf{S}'$ (ground truth), averaged RMSE and MAE are calculated in Eq. 13, based on output tensor dimension Nl.

$$\mathrm{RMSE}(\mathbf{S}', \hat{\mathbf{S}}) = \sqrt{\frac{1}{Nl}\sum_{i=1}^{N} ||\mathbf{s}_i - \hat{\mathbf{s}}_i||_2^2}; \qquad \mathrm{MAE}(\mathbf{S}', \hat{\mathbf{S}}) = \frac{1}{Nl}\sum_{i=1}^{N} ||\mathbf{s}_i - \hat{\mathbf{s}}_i||_1 \quad (13)$$

4.3 Prediction Performance

We compare the proposed method with several baselines, including classical methods: Auto-regressive Integrated Moving Average (ARIMA) [1] and Vector Auto-Regressive (VAR) [8]. The deep learning-based methods include LSTNet [15], GeoMAN [17], STGCN [29], MTGNN [28] and a state-of-art long sequence time-series forecasting method FEDformer [32]. Table 2 compares GT-DRL to baseline methods for 12-hour prediction on two datasets, with GT-DRL consistently achieving competitive results.

Table 2. Performance comparison of GT-DRL and the baselines.

Models	Beijing dataset				Xi'an dataset			
	AQI		PM2.5		AQI		PM2.5	
	MAE	RMSE	MAE	RMSE	MAE	RMSE	MAE	RMSE
ARIMA	30.75	46.17	29.79	44.35	30.05	46.88	28.95	40.36
VAR	28.51	44.44	27.89	43.52	29.77	43.39	26.81	40.54
LSTNET	26.57	47.53	24.46	42.09	29.05	43.05	26.25	38.76
GeoMAN	25.72	44.84	22.70	39.1	27.75	40.47	22.15	36.55
DCRNN	23.24	38.04	20.84	37.34	25.53	39.29	21.90	33.96
GMAN	22.09	37.31	19.15	38.08	25.25	38.92	21.50	33.92
STGCN	24.22	40.91	20.15	39.06	25.98	38.73	22.46	32.85
MTGNN	20.76	36.18	19.61	35.45	22.62	37.98	19.82	31.47
FEDformer	21.63	37.87	20.46	35.23	23.36	38.52	19.76	30.97
GT-DRL	**19.94**	**32.77**	**18.12**	**34.05**	**21.54**	**37.68**	**17.93**	**30.49**

GT-DRL exhibits a larger margin over traditional methods, as ARIMA solely models univariate temporal dependencies, and VAR has limited capability in modeling complex inter-variable dependencies, resulting in their poorer performance.

GT-DRL outperforms deep models relying on single-view data, including LSTNet, GeoMAN, and GMAN. LSTNet combines graph convolution networks with recurrent neural networks in a relatively simple structure to extract long- and short-term temporal features. GeoMAN and GMAN, compared to LSTNet, can adaptively extract features from different sequences using multi-level attention mechanisms. However, these methods solely consider the feature information from the temporal view, falls short in considering spatial relationships within the data.

GT-DRL surpasses graph-learning-based deep models such as DCRNN, STGCN, and MTGNN. STGCN's fixed graph structure hinders its adaptability to dynamic spatial relationships, while DCRNN and MTGNN construct directed graph features using bidirectional random walks and automatic relationship learning, respectively. Although they capture feature information from different spatial-temporal views, they neglect the importance of timestamp information in time series.

GT-DRL also demonstrates superior performance compared to the state-of-the-art FEDformer model. FEDformer employs a self-attention mechanism to extract features from any variable at any time in data and adaptively allocates feature weights. Additionally, its design for seasonal trend decomposition is similar to the decoupling approach in our model, enabling effective extraction of features from potential influencing factors. However, this method ignores relative distance and direction information in the spatial view during feature extraction.

4.4 Ablation Study

We conduct the ablation study by removing the key components manually and comparing it with the original model. We name GT-DRL without different components as follows:

- GT-DRL w/o GD : We simply replace the geospatial disentangling component with geographical coordinates.
- GT-DRL w/o TD: We simply replace the temporal disentangling component with a Sinusoidal position embedding.
- GT-DRL w/o GTD : We remove both the geospatial and temporal disentangling representations.

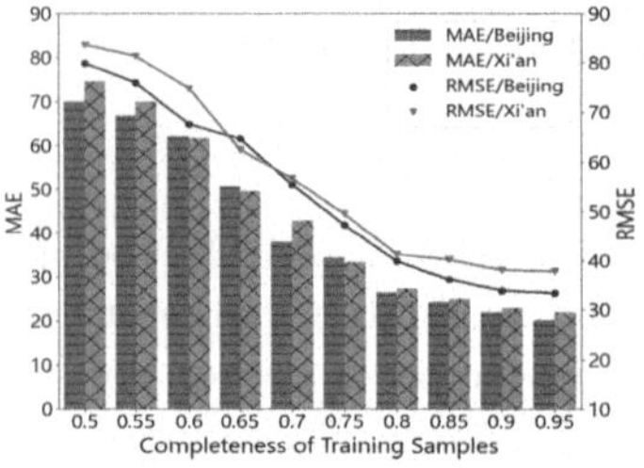

Fig. 3. The effect of GT-DRL with different missing data ratios.

Table 3. Experimental results on ablation study.

Dataset	Methods	AQI		PM2.5	
		MAE	RMSE	MAE	RMSE
Beijing	GT-DRL	19.94	32.77	18.12	34.05
	w/o GD	22.43	35.79	20.13	35.42
	w/o TD	20.76	33.31	19.46	34.61
	w/o GTD	23.52	39.28	21.17	35.94
Xi'an	GT-DRL	21.54	37.68	17.93	30.49
	w/o GD	22.81	38.75	20.26	31.84
	w/o TD	20.25	38.62	19.56	33.68
	w/o GTD	24.79	39.27	22.19	32.98

The experimental results in Table 3 underscores the necessity of each component in GT-DRL. Removing any component leads to a notable decline in performance, emphasizing the importance of disentangled representation learning for both geospatial and temporal aspects. Specifically, GT-DRL w/o GTD yields the worst results due to neglecting vital geospatial-temporal couplings, resulting in loss of crucial information during feature extraction. Similarly, GT-DRL w/o GD exhibits compromised performance by omitting significant geospatial information, which undermines the capture of directional pollutant diffusion relationships. While GT-DRL w/o TD fares better than these variations, it still lags behind the original GT-DRL, suggesting that the design of disentangling multiple latent impact factors and the inclusion of timestamps further enhance model performance. In conclusion, the results of the ablation study clearly demonstrate that all components are indispensable for our proposed model.

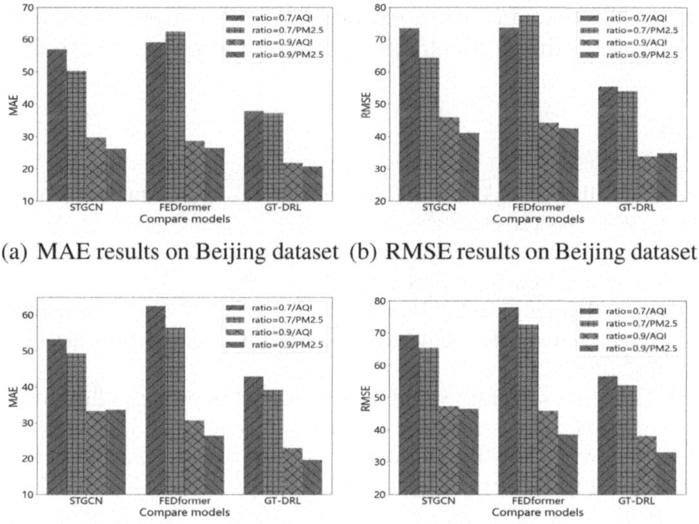

(a) MAE results on Beijing dataset (b) RMSE results on Beijing dataset

(c) MAE results on Xi'an dataset (d) RMSE results on Xi'an dataset

Fig. 4. Performance comparison on irregularly sampling.

4.5 Effect of the Irregular Sampling

To showcase GT-DRL's effectiveness with irregularly sampled datasets, we conducted experiments with constant model parameters and varying completeness of training samples (from 0.5 to 0.95), simulating irregular sampling by randomly marking values as missing. From the results in Fig. 3, we observe that the model's performance steadily improves as the training sample size increases. This underscores the model's ability to effectively learn the intricate coupling relationships among multi-sourced geospatial-temporal data from larger datasets.

Additionally, we compared GT-DRL with STGCN and FEDformer at training sample completeness of 0.7 and 0.9 (Fig. 4). Notably, even with <0.7 completeness, GT-DRL performs competitively in MAE and RMSE, highlighting its applicability and strong predictive performance on irregular datasets due to timestamp utilization.

5 Conclusion

This paper proposes a data-driven approach, GT-DRL, incorporating two innovative concepts into geospatial-temporal data modeling: 1) A novel disentangled representation is designed to preserve critical information from raw data and address data complexity issues. 2) A stacked encoder-decoder is employed to learn complex coupling relationships adaptively. Experiments conducted on real-world air quality datasets from two representative cities demonstrate the effectiveness of GT-DRL. In future research, we will extend GT-DRL accessible for general use and evaluate its efficiency across various applications.

References

1. Asteriou, D., Hall, S.G.: Arima models and the box-jenkins methodology. Appl. Econo. **2**(2), 265–286 (2011)
2. Babaeizadeh, M., Finn, C., Erhan, D., Campbell, R.H., Levine, S.: Stochastic variational video prediction. In: ICLR (2017)
3. Bai, L., Yao, L., Kanhere, S.S., Wang, X., Sheng, Q.Z.: Stg2seq: spatial-temporal graph to sequence model for multi-step passenger demand forecasting. In: IJCAI, pp. 1981–1987 (2019)
4. Battista, G., de Lieto Vollaro, R.: Correlation between air pollution and weather data in urban areas: Assessment of the city of Rome (Italy) as spatially and temporally independent regarding pollutants. Atmos. Environ. **165**, 240–247 (2017)
5. Cao, L.: Coupling learning of complex interactions. Inform. Process. Manag. **51**(2), 167–186 (2015)
6. Cao, L.: Beyond iid: non-iid thinking, informatics, and learning. IEEE Intell. Syst. **37**(4), 5–17 (2022)
7. Chen, R.T., Li, X., Grosse, R.B., Duvenaud, D.K.: Isolating sources of disentanglement in variational autoencoders. NeurIPS **31** (2018)
8. Dowell, J., Pinson, P.: Very-short-term probabilistic wind power forecasts by sparse vector autoregression. IEEE Trans. Smart Grid **7**(2), 763–770 (2015)
9. Geng, X., et al.: Spatiotemporal multi-graph convolution network for ride-hailing demand forecasting. In: AAAI, vol. 33, pp. 3656–3663 (2019)
10. Guo, S., Lin, Y., Feng, N., Song, C., Wan, H.: Attention based spatial-temporal graph convolutional networks for traffic flow forecasting. In: AAAI, vol. 33, pp. 922–929 (2019)
11. Han, J., Liu, H., Zhu, H., Xiong, H., Dou, D.: Joint air quality and weather prediction based on multi-adversarial spatiotemporal networks. In: AAAI, vol. 35, pp. 4081–4089 (2021)
12. Higgins, I., et al.: beta-vae: learning basic visual concepts with a constrained variational framework. In: ICLR (2017)

13. Kim, H., Mnih, A.: Disentangling by factorising. In: ICML, pp. 2649–2658. PMLR (2018)
14. Kingma, D.P., Welling, M.: Auto-encoding variational bayes. arXiv preprint arXiv:1312.6114 (2013)
15. Lai, G., Chang, W.C., Yang, Y., Liu, H.: Modeling long-and short-term temporal patterns with deep neural networks. In: ACM SIGIR, pp. 95–104 (2018)
16. Li, Y., Zhu, Z., Kong, D., Han, H., Zhao, Y.: Ea-lstm: evolutionary attention-based lstm for time series prediction. Knowl.-Based Syst. **181**, 104785 (2019)
17. Liang, Y., Ke, S., Zhang, J., Yi, X., Zheng, Y.: Geoman: multi-level attention networks for geo-sensory time series prediction. In: IJCAI, pp. 3428–3434 (2018)
18. Lin, H., Bai, R., Jia, W., Yang, X., You, Y.: Preserving dynamic attention for long-term spatial-temporal prediction. In: ACM SIGKDD, pp. 36–46 (2020)
19. Ren, S., Guo, B., Li, K., Wang, Q., Yu, Z., Cao, L.: Coupledmuts: coupled multivariate utility time series representation and prediction. IEEE IoT-J (2022)
20. Shih, S.Y., Sun, F.K., Lee, H.y.: Temporal pattern attention for multivariate time series forecasting. Mach. Learn. **108**(8), 1421–1441 (2019)
21. Tobler, W.R.: A computer movie simulating urban growth in the detroit region. Econ. Geogr. **46**(sup1), 234–240 (1970)
22. Wan, R., Mei, S., Wang, J., Liu, M., Yang, F.: Multivariate temporal convolutional network: a deep neural networks approach for multivariate time series forecasting. Electronics **8**(8), 876 (2019)
23. Wang, C., Zhu, Y., Zang, T., Liu, H., Yu, J.: Modeling inter-station relationships with attentive temporal graph convolutional network for air quality prediction. In: Proceedings of the 14th ACM International Conference on Web Search and Data Mining, pp. 616–634 (2021)
24. Wang, J., Sun, T., Liu, B., Cao, Y., Zhu, H.: Clvsa: a convolutional lstm based variational sequence-to-sequence model with attention for predicting trends of financial markets. In: IJCAI, pp. 3705–3711 (2019)
25. Wang, J., Sun, T., Liu, B., Cao, Y., Zhu, H.: Clvsa: a convolutional lstm based variational sequence-to-sequence model with attention for predicting trends of financial markets. arXiv preprint arXiv:2104.04041 (2021)
26. Wen, Q., Gao, J., Song, X., Sun, L., Xu, H., Zhu, S.: Robuststl: a robust seasonal-trend decomposition algorithm for long time series. In: AAAI, vol. 33, pp. 5409–5416 (2019)
27. Wu, H., Xu, J., Wang, J., Long, M.: Autoformer: decomposition transformers with auto-correlation for long-term series forecasting. NeurIPS **34**, 22419–22430 (2021)
28. Wu, Z., Pan, S., Long, G., Jiang, J., Chang, X., Zhang, C.: Connecting the dots: Multivariate time series forecasting with graph neural networks. In: ACM SIGKDD, pp. 753–763 (2020)
29. Yu, B., Yin, H., Zhu, Z.: Spatio-temporal graph convolutional networks: a deep learning framework for traffic forecasting. In: IJCAI, pp. 3634–3640 (2018)
30. Yu, F., Koltun, V.: Multi-scale context aggregation by dilated convolutions. In: ICLR (2016)
31. Zhou, H., et al.: Informer: beyond efficient transformer for long sequence time-series forecasting. In: AAAI, vol. 35, pp. 11106–11115 (2021)
32. Zhou, T., Ma, Z., Wen, Q., Wang, X., Sun, L., Jin, R.: Fedformer: frequency enhanced decomposed transformer for long-term series forecasting. In: ICML, pp. 27268–27286 (2022)
33. Zhu, Y., Min, M.R., Kadav, A., Graf, H.P.: S3vae: self-supervised sequential vae for representation disentanglement and data generation. In: CVPR, pp. 6538–6547 (2020)

Treatment Effect Estimation for Graph-Structured Targets

Shonosuke Harada(✉), Ryosuke Yoneda, and Hisashi Kashima

Kyoto University, Kyoto, Japan
harada.shonosuke.r66@kyoto-u.jp

Abstract. Treatment effect estimation, which helps understand the causality between treatment and outcome variable, is a central task across various domains. While most studies focus on treatment effect estimation on individual targets, in specific contexts, there is a necessity to comprehend the treatment effect on a group of targets, especially those that have relationships represented as a graph structure between them. In such cases, the focus of treatment assignment is prone to depend on a particular node of the graph, such as the one with the highest degree, thus resulting in an observational bias from a small part of the entire graph. Whereas a bias tends to be caused by the small part, straightforward extensions of previous studies can not provide efficient bias mitigation owing to the use of the entire graph information. In this study, we propose *Graph-target Treatment Effect Estimation (GraphTEE)*, a framework designed to estimate treatment effects specifically on graph-structured targets. GraphTEE aims to mitigate observational bias by focusing on confounding variable sets and consider a new regularization.

Keywords: Treatment effect estimation · Causal inference · Graph

1 Introduction

The task of treatment effect estimation aims to understand the causal effect of a treatment or intervention on the outcome of a target object. It plays an crucial role in various fields such as advertisement, education, economic policy and healthcare [12,22,27]. Most studies focus on treatment effect on a individual target; however, in some scenarios, treatment effect on graph-structured including set-structured targets need to be considered. For instance, an influencer, who has a large number of followers on a social network or social media because of their social position, could be involved in a form of marketing called influencer marketing [5]. Figure 1 describes an example in a social network. In influencer marketing, advertisement companies leverage influencers' brands to promote their products, and this enables businesses to reach targeted individuals. Another practical application is group recommendation [3], which aims to recommend items for a group, rather than individuals. These groups of targets can be represented as graphs because they can capture complex relationships between entities and it becomes crucial to comprehend the treatment effect on

X. Wu et al. (Eds.): PAKDD 2025, LNAI 15875, pp. 94–109, 2025.
https://doi.org/10.1007/978-981-96-8295-9_6

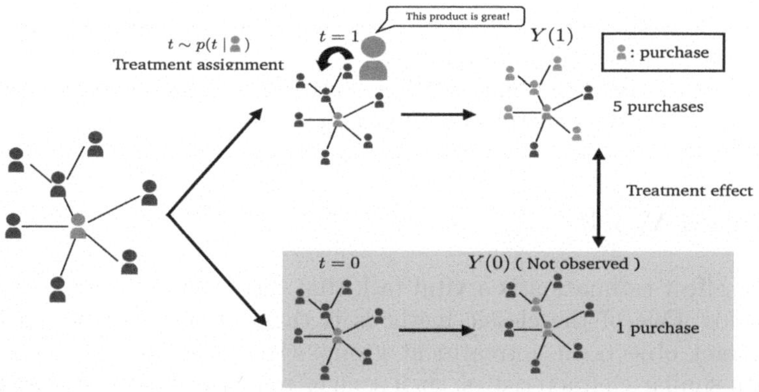

Fig. 1. An illustrative example of treatment effect estimation on graph-structured targets like a social network. The shaded outcome is a counterfactual outcome and never observed. For example, if a central user promotes a product, the entire graph may have positive impression on the product, i.g., the majority of users (5 users) may buy it; however, if no-advertisement is given, only 1 user may have interests in the product and the others may not buy it. In this example, the treatment effect on this network is 4. Our interest is to understand how interventions affect on entire graphs.

graph-structured targets, rather than focusing solely on individual targets. In these applications, understanding the treatment effect on graph-structured individuals becomes vital, not on individual. There are mainly two fundamental challenges in treatment effect estimation, namely (i) observational bias and (ii) the counterfactual nature of observational data. Observational bias is a form of bias that occurs in the process of treatment assignment on target objects based on their covariates. For example, the elderly are more likely to receive medical treatment than younger ones. Regarding the counterfactual nature of observational data, treatment effect is often defined as the difference between outcomes of a graph with and without treatment; however, we can only observe one of the possible outcomes. Hence, studies have been focusing on understanding counterfactual outcomes. Moreover, in treatment effect estimation for graph-structured targets, observational bias becomes a more severe technical difficulty to be addressed because a policymaker does not necessarily pay attention to all the nodes in a graph but, rather, to some specific nodes, such as a central node, often referred to as a hub node, that has a large number of connections compared to other nodes. In social networks, for example, an influencer who has a large number of followers is prone to attract a significant amount of attention for promoting products in influencer marketing. While the sizes of input graphs are large, the space for nodes which affect both treatment assignments and outcomes is relatively small. From the perspective of invariant learning on graph data, these features are well-known as invariant features that are essential for prediction [14]. Therefore, efficient mitigation of biases in graphs is a major technical challenge. The contributions of this study are summarized as:

- We propose *Graph-target Treatment Effect Estimation (GraphTEE)*, a novel algorithm that efficiently mitigates bias based on theoretical analysis.

- We provide a theoretical analysis that shows how existing approaches can suffer from inefficient bias mitigation.
- Through extensive experiments using synthetic and semi-synthetic datasets, we demonstrate the effectiveness of GraphTEE.

2 Related Work

Treatment effect estimation is a vital task that has been extensively studied in various fields. One of the classic methods is the matching [19] which involves pairing target objects in a treatment group with target objects in a control group with similar characteristics, such as age, gender or health status. Propensity score matching [18] is a class of the matching that uses a predictive model to estimate the probability of targets receiving treatments based on their covariates. The estimated probability, called propensity score, is used for matching targets. Deep learning-based methods have also been successfully applied to the treatment effect estimation problem [9,20]. One of the promising methods is a CounterFactual Regression (CFR) [20] that is characterized by a regularizer designed to minimize the distribution discrepancy between treatment and control groups. CFR without any regularization is called Treatment-Agnostic Representation Network (TARNet), which simply has representation sharing and outcome prediction layers for each group. The key idea behind representation balancing is profoundly linked with domain adaptation analysis [2,15]. In recent times, the problems of treatment effect estimation using graph-structured treatments like molecules have also been introduced [8,11]; however their goal is to estimate the effect of graph-structured treatments on individuals, rather than the treatment effect on graph-structured targets, and completely different from our focus. Several studies have also considered treatment effect estimation on graph-structured data [6,28]; however, all these studies have considered treatment effect estimation on node level targets, rather than at the level of entire graphs.

3 Problem Statement

Suppose we have a dataset $\mathcal{D} = \{G_i, t_i, Y_i^{t_i}\}_{i=1}^N \in \mathcal{G} \times \mathcal{T} \times \mathcal{Y}$, where $G_i = \{V_i, A_i\}_{i=1}^N$ is a graph and, $V_i \in \mathcal{V}$, A_i represents a set of nodes and an adjacency matrix, respectively. Set V_i consists of $|V_i|$ nodes as $V_i = \{v_{i,0}, ..., v_{i,|V_i|}\}$ and each node $v_{i,j} \in R^d$ has d dimensional covariates. We assume the treatment space $\mathcal{T}$ is binary; i.e., $\mathcal{T} = \{0, 1\}$ and the outcome space is one-dimensional and continuous-valued; i.e., $\mathcal{Y} = R$. Thus, our goal is to learn a predictive function $g : \mathcal{G} \times \mathcal{T} \to \mathcal{Y}$. We make standard assumptions, i.e., (i) Positivity, i.e., any graph-structured target has positive probability of receiving all treatments as $1 > p(t_i \mid G_i) > 0$; (ii) Stable unit treatment value assumption, i.e., the potential outcomes of each target is independent of the treatment assignment of other targets; (iii) Unconfoundedness, i.e., the treatment assignment to an instance is independent of the outcome given covariates.

Definition 1. *We call a node a confounding node when it has an effect on both treatment assignment and an outcome. We define $V_i(c)$ a set of confounding nodes and $V_i(y)$ a set of non-confounding nodes in G_i.*

Assumption 1. *We assume the decomposition function that decomposes graphs, $V_i = V_i(c) \cup V_i(y)$, is the same function across different graphs and $\mid V_i(c) \mid \ll \mid V_i \mid$, unless $\mid V_i \mid = 1$, i.e., a small part of a graph has an effect on treatment assignment.*

This assumption enables us to learn such a function from observational data to decompose V_i into $V_i(c)$ and $V_i(y)$. We wish to emphasize that this assumption is reasonable, as particular nodes or sub-substructures play a crucial role in the context of invariant learning [14,23]. We sometimes express graphs without i for convenience.

4 Methodology

The proposed framework consists of two steps. In the first step, our aim is to learn the decomposition function ϕ and find nodes or a sub-graph that have an effect on treatment assignment and an outcome. We achieve this task using GNNs and self-attention graph (SAG) pooling [13]. In the second step, using ϕ learned in the first step, we decompose the input graph into confounding node set $V(c)$ and non-confounding node set $V(y)$. This decomposition facilitates the model in mitigating bias with greater efficiency, as detailed in the theoretical analysis.

4.1 Confounding Nodes Selecting Step

In the first step, we perform a propensity score learning using GNNs. The goal of this step is to build a function that takes a graph as an input and returns confounding nodes and not-confounding nodes. To select confounding nodes in a graph, we apply SAG-pooling technique [13]. SAG-pooling aims to learn graph representation while giving larger weights on more important nodes that seem to have larger effects on a target value. In this context, we need to specify a set of nodes that have effects on treatment assignment process. We aim to find these nodes by selecting nodes based on their weights through applying SAG-pool in the final step of GNNs in ϕ. The loss function L_t for selecting confounding nodes are given as:

$$\mathcal{L}_t = \sum_{i=1}^{N} -t_i \log \sigma(g_t(G_i)) - (1 - t_i) \log(1 - \sigma(g_t(G_i))), \tag{1}$$

where $g_t(G_i) = h_t\big(\sum_{i \in V_i}^{|V_i|} \mathrm{SAGP}(\Phi_t(G_i)) \odot (\Phi_t(G_i))\big)$, and SAGP denotes the SAG-pooling and returns weights of each node given node representations and adjacency matric, $\odot$ represents the Hadamard product, h_t is standard feed-forward neural networks, and σ denotes the sigmoid function, respectively.

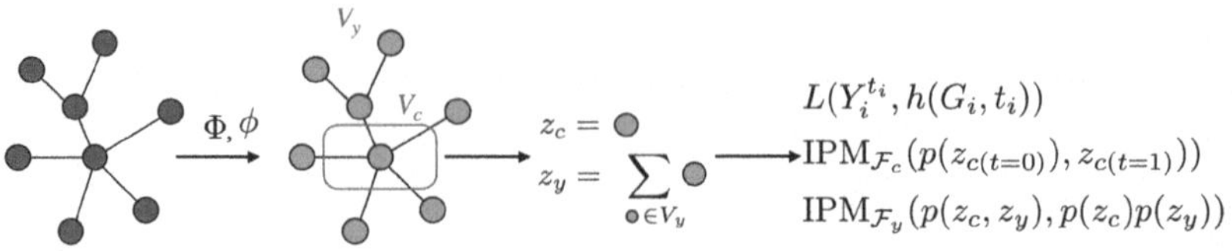

Fig. 2. Overview of the proposed algorithm. An input graph is first mapped into representations using GNNs Φ and then decomposed into two types of nodes using ϕ: (i) confounding and (ii) other nodes. Outcome prediction is conducted using the entire nodes' representations with TARNet-based feedforward neural networks as Eq. (2). Representations of nodes of each type are used for bias mitigation as Eq. (4).

We use SAGP to select the confounding nodes as $\phi : \mathrm{top}(\mathrm{SAGP}(\Phi_t(G_i)), k) = \{V_i(c), V_i(y)\}$ where top is a function that return the top k% nodes in terms of their weights in V_i as $V_i(c)$ and the others as $V_i(y)$.

4.2 Outcome Prediction Step

In the second step of the proposed method, we implement a methodology for treatment effect estimation. Let $\Phi : \mathcal{G} \to \mathcal{Z}$ where $z \in \mathcal{Z}$ is a space of sets of representations, be a mapping function and $h : \mathcal{Z} \times \mathcal{T} \to \mathcal{R}$ be a hypothesis. We employ GNNs as the mapping function Φ for an input graph. As discussed in the following section, several conditions have to be satisfied to fully use some theoretical benefits. For instance, it is imperative for the GNNs to have a one-to-one mapping between input graphs and their representations, although neural networks that do not necessarily satisfy one-to-one mapping functions empirically can show satisfactory performances. Recent theoretical analysis provided remarkable insights into the theoretical properties of GNNs [24,26]. Particularly, we utilize a GIN-based GNN [24] to map graph-structured targets into representations $z \in \mathcal{Z}$. After performing the mapping, we obtain representations of each node for a input graph. As a predictive model, we employ a TARNet-based model [20], which takes the graph representation and treatment indicator as inputs. The supervised loss $\mathcal{L}_{\mathrm{supervised}}$ is given as:

$$\mathcal{L}_{\mathrm{supervised}} = \sum_{i=0}^{N} L(Y_i^{t_i}, g(G_i, t_i)), \tag{2}$$

where $g(G_i, t_i) := h(\Phi(G_i), t_i)$. Due to the observational problem, the supervised loss itself does not give desirable results for the test distribution, necessitating the implementation of a bias mitigation technique. We introduce a popular metric between probability distributions, the *Integral Probability Metric (IPM)* for bias mitigation. The technique of IPM has been extensively utilized to measure the dissimilarity between probability distributions [17,21].

Definition 2. *The IPM for two probability distributions* p, q *over* $\mathcal{X}$ *and for a function family* $\mathcal{F}$ *is defined as:*

$$IPM_{\mathcal{F}}(p, q) = \sup_{f \in \mathcal{F}} \left| \int_{\mathcal{X}} f(x)(p(x) - q(x))dx \right|. \tag{3}$$

Depending on the choice of function family $\mathcal{F}$, the IPM can be reduced to a widely recognized and utilized distance between two distributions. For example, the IPM yields the Wasserstein distance by choosing $\mathcal{F}$ as a 1-Lipschitz function. However, as we will discuss in the theoretical analysis, using entire node representations in a graph makes empirical IPM computation inefficient. In order to avoid such a situation, we apply the confounding nodes select function $\phi : \mathcal{G} \rightarrow \mathcal{V}, \mathcal{V}$, which is built in the first step, to decompose the sets of nodes into necessary and not necessary nodes for bias mitigation. We compute the two IPMs, i.e., $\text{IPM}_{\mathcal{F}}(p(z_{c(t=0)}), p(z_{c(t=1)}))$ and $\text{IPM}(p(z_c, z_y), p(z_c)p(z_c))$, where $z_c := \{z \in z_j \mid j \in V(c)\} \in \mathcal{Z}_c$, $z_y := \{z \in z_j \mid j \in V(y)\} \in \mathcal{Z}_y$, and $\mathcal{Z} = \mathcal{Z}_c \times \mathcal{Z}_y$. The first term aims to mitigate selection bias and the second one tries to decrease the dependency between z_c and z_y. Using these terms, the bias mitigation regularizer loss $\mathcal{L}_{\text{reg}}$ is given as:

$$\mathcal{L}_{\text{reg}} = \text{IPM}_{\mathcal{F}_c}(p(z_{c(t=0)}), p(z_{c(t=1)})) + \text{IPM}_{\mathcal{F}_y}(p(z_c, z_y), p(z_c)p(z_c)). \tag{4}$$

We use the sum function as the aggregation function of nodes. Note that the sum function ensures that $z_c + z_y = z = \Phi(G)$ and the mapping function has one-to-one mapping property [24,26]. Finally, the loss function to be minimized in the second step is given as:

$$\mathcal{L} = \mathcal{L}_{\text{supervised}} + \lambda \cdot \mathcal{L}_{\text{reg}}, \tag{5}$$

where λ is a hyper-parameter that controls the strength of regularization. We describe the whole training process in Fig. 2.

5 Theoretical Analysis

In this section, we provide a theoretical analysis with regard to our approach. We define the expected the risk ϵ_{train} and ϵ_{test} using data from the train data distribution $p(G, t)$ and the test data distribution $p(G)p(t)$ as follows:

Definition 3. *The expected risks on the train data distribution and the test data distribution are defined as:*

$$\epsilon_{\text{train}}(h, \Phi) = \mathbb{E}_{(G,t)\sim p(G,t)}[L(h(\Phi(G), t), Y^t)], \tag{6}$$

$$\epsilon_{\text{test}}(h, \Phi) = \mathbb{E}_{(G,t)\sim p(G)p(t)}[L(h(\Phi(G), t), Y^t)], \tag{7}$$

where L is a loss function, such as squared loss.

Note that we assume the test data distribution we aim to evaluate our method on is $p(G, t) :- p(G)p(t)$, i.e., a treatment assignment is randomized. Empirically, we approximate the IPM using finite samples. For example, within deep learning-based framework, IPM is typically computed in a mini-batch manner. We refer to $\text{IPM}_{\mathcal{F}}(\hat{p}, \hat{q})$ as such an empirical estimate of the IPM. Sriperumbudur et al. [21] analyzed the following empirical risk of IPMs.

Proposition 1. *Given m, n samples from p, q respectively and for any $\mathcal{F}$ such that $\nu := \sup_{x \in \mathcal{X}} f(x) < \infty$, we have with a probability at least $1 - \delta$,*

$$\left| IPM_{\mathcal{F}}(p, q) - IPM_{\mathcal{F}}(\hat{p}, \hat{q}) \right| \leq 2R_m(\mathcal{F}) + 2R_n(\mathcal{F}) + \sqrt{18\nu^2 \log \frac{4}{\delta}} (\frac{1}{\sqrt{m}} + \frac{1}{\sqrt{n}}), \tag{8}$$

where $R_m(\mathcal{F}) := \mathbb{E} \sup_{f \in \mathcal{F}} \left| \frac{1}{m} \Sigma_{i=1}^m \rho_i f(x_i) \right|$ and $\{\rho_i\}_{i=1}^m$ is the Rademacher complexity and independent Rademacher random variables that take 1 or -1 at the probability 0.5, respectively.

The IPM has been successfully applied in recent treatment effect estimation methods that were based on representation learning to mitigate the selection bias [9,20]. Combining Proposition 1 with the expected risks, we can detive the following bound in a similar manner as the previous study [10].

Proposition 2. *Let $\Phi : \mathcal{G} \to \mathcal{Z}$ be a one-to-one mapping function. Let $h: \mathcal{Z} \times \{0, 1\} \to \mathcal{Y}$ be a hypothesis. Let $\mathcal{F}$ be a family of functions and $f \in \mathcal{F}: \mathcal{Z} \to \mathcal{Y}$. Assume there exists a constant B_Φ such that $\frac{1}{B_\Phi} \int_{\mathcal{Y}} L(Y^t, h(\Phi(G), t)) p(Y^t \mid G) p(G, t) dY^t \in \mathcal{F}$. Supposing $f = h(z) \in \mathcal{F}$, we have with a probability at least $1 - \delta$,*

$$\begin{aligned} \epsilon_{\text{test}}(h, \Phi) &\leq \epsilon_{\text{train}}(h, \Phi) + B_\Phi \cdot IPM_{\mathcal{F}}(\hat{p}(z_{(t=0)}), \hat{p}((z_{(t=1)}))) \\ &+ 2R_m(\mathcal{F}) + 2R_n(\mathcal{F}) + \sqrt{18\nu^2 \log \frac{4}{\delta}} (\frac{1}{\sqrt{m}} + \frac{1}{\sqrt{n}}). \end{aligned} \tag{9}$$

Proof. See the original papers [10,20]. This regularization still provides a theoretically correct upper bound for counterfactual outcomes; however, if we naively apply this approach to a graph-structured targets setting, this regularization computes the IPM using all the node representations in the input sets and cannot avoid an inefficient computation. Based on the assumption that a small part of an input graph has an effect on the treatment assignment, rather than on the entire graph; i.e., confounding nodes account for small part of large graphs, we aim to establish a more efficient way to mitigate bias. When each variable and node is independent of each other, we only need to use a confounding node for bias mitigation. However, the technical difficulty when we focus on graph-structured targets is confounding nodes and other structures are basically relevant and dependent on each other. Therefore, we try to achieve efficient bias mitigation by focusing only on the confounding nodes. If there are no relationship between z_c and z_y, i.e., $p(z) = p(z_c)p(z_y)$, only minimizing $\text{IPM}_{\mathcal{F}}(p(z_{c(t=0)}), p(z_{c(t=1)}))$ is sufficient to mitigate bias; however, due to the dependency between z_c and z_y, i.e., $p(z_c, z_y) = p(z_c \mid z_y)p(z_y)$, just minimizing the IPM with regard to the confounding node representation is insufficient and we need to remove the dependency by minimizing $\text{IPM}_{\mathcal{F}_c}(p(z_c, z_y), p(z_c)p(z_y))$. We give a theoretical analysis on the proposed method between this approach and the original regularization $\text{IPM}(p(z_{(t=0)}), p(z_{(t=1)}))$.

Table 1. Performance comparisons on the synthetic and Reddit datasets. We conduct experiments 20 times for different seeds, and show the average results and standard errors. We set the number of node as 100 for the synthetic dataset. The bold results indicate the best results in terms of average, and † and ‡ indicate the proposed method perform e statistically significantly better than the baseline methods by the paired t-test with $p < 0.05$ and $p < 0.01$. Lower values are better.

Method	Synthetic		Reddit	
	$\sqrt{\epsilon_{\text{PEHE}}}$	ϵ_{ATE}	$\sqrt{\epsilon_{\text{PEHE}}}$	ϵ_{ATE}
Mean	$^\ddagger 65.477_{\pm 7.447}$	$^\ddagger 29.918_{\pm 4.834}$	$^\ddagger 64.458_{\pm 5.347}$	$^\ddagger 12.786_{\pm 2.425}$
DeepSets	$^\ddagger 43.372_{\pm 1.292}$	$^\ddagger 17.415_{\pm 0.938}$	$^\ddagger 34.586_{\pm 1.316}$	$^\ddagger 7.239_{\pm 0.405}$
DeepSets+CFR	$^\ddagger 43.905_{\pm 1.174}$	$^\ddagger 18.917_{\pm 0.893}$	$^\ddagger 34.692_{\pm 1.610}$	$^\ddagger 6.672_{\pm 1.187}$
GNN	$^\ddagger 40.481_{\pm 1.706}$	$^\ddagger 12.248_{\pm 1.605}$	$^\ddagger 28.105_{\pm 4.738}$	$^\dagger 4.419_{\pm 0.848}$
GNN+CFR	$^\ddagger 16.703_{\pm 0.637}$	$4.556_{\pm 0.541}$	$^\dagger 16.212_{\pm 0.628}$	$2.523_{\pm 0.321}$
GraphTEE	$\mathbf{15.644}_{\pm 0.644}$	$\mathbf{4.137}_{\pm 0.519}$	$\mathbf{15.317}_{\pm 0.633}$	$\mathbf{2.331}_{\pm 0.352}$

Theorem 1. *Suppose we have an error as* $IPM(p(t)p(z_c \mid z_y)p(z_y), p(t)p(z_c) p(z_y)) = \Delta$. *We have with probability at least* $1 - \delta$,

$$\epsilon_{\text{test}}(h, \Phi) \leq \epsilon_{\text{train}}(h, \Phi) + B_\Phi \cdot (IPM_{\mathcal{F}_c}(\hat{p}(z_{c(t=0)}), \hat{p}(z_{c(t=1)})) + IPM_{\mathcal{F}_y}(\hat{p}(z_c, z_y), \hat{p}(z_c)p(z_y)))$$
$$+ 2R_m(\mathcal{F}_c) + 2R_n(\mathcal{F}_c) + \sqrt{18C_\nu^2 \log \frac{8}{\delta}}(\frac{1}{\sqrt{m}} + \frac{1}{\sqrt{n}}) + 4R_{m+n}(\mathcal{F}_y) + 2\sqrt{\frac{18\nu^2}{m+n} \log \frac{8}{\delta}} + 2\Delta. \quad (10)$$

Proof. Please refer to Appendix.

Remark 1. The main benefit of Theorem 1 comes from the analysis that the empirical error for the IPM terms can become smaller than the existing approach (Proposition 2) mainly because the constant C_c is smaller than 1 and using $m+n$ samples when we can minimize Δ enough. Hence, our regularization enables us to compute the regularization more efficiently than the existing one, particularly when $\mathcal{Z}_c$ is small compared to $\mathcal{Z}$. Note that we cannot necessarily claim the upper bound of Theorem 1 is always tighter than the existing one (Proposition 2); however, we expect that space of set of confounding nodes representation $\mathcal{Z}_c$ is sufficiently smaller than the entire representation space to give computationally efficient bias mitigation, and moreover, by using both of them, with probability at least $1-2\delta$, we can provide a tighter bound. We demonstrate the effectiveness of this regularization empirically.

6 Experiments

Synthetic Dataset: We generated a synthetic graph dataset using NetworkX [7][1] following the Barabási–Albert algorithm [1]. Each node has d-dimensional

[1] https://networkx.org/.

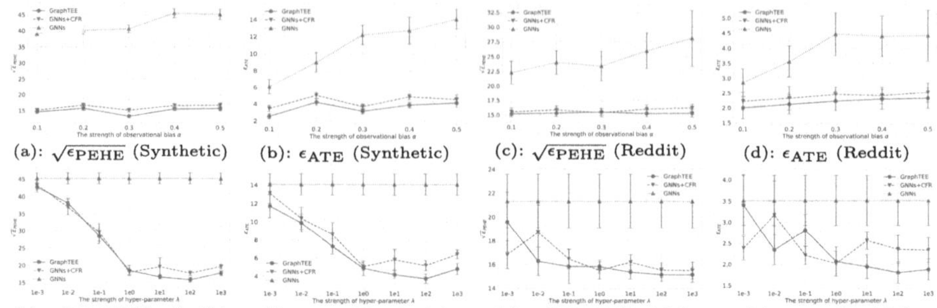

(a): $\sqrt{\epsilon_{\text{PEHE}}}$ (Synthetic) (b): ϵ_{ATE} (Synthetic) (c): $\sqrt{\epsilon_{\text{PEHE}}}$ (Reddit) (d): ϵ_{ATE} (Reddit)

(e): $\sqrt{\epsilon_{\text{PEHE}}}$ (Synthetic) (f): ϵ_{ATE} (Synthetic) (g): $\sqrt{\epsilon_{\text{PEHE}}}$ (Reddit) (h): ϵ_{ATE} (Reddit)

Fig. 3. Sensitivity of the results to the strength of observational bias α (upper) and sensitivity of the results to the strength of hyper-parameter λ (lower). Lower values are better.

covariates $v_{ij} \in R^d \sim \mathcal{N}(0, \text{Degree(i, j)})$, where Degree(i, j) is a degree of node j in G_i. We assume treatment is assigned by a node who has the highest degree in a graph as $t_j \sim \text{Bern}(\sigma(\alpha \sum_{d=1}^{d} \text{HighestDegree}(G_i)))$, where Bern represents the Bernoulli distribution, $\text{HighestDegree}(G_i) : \mathcal{G} \rightarrow R^d$ is a function that returns covariates of node who has the highest degree, and α is a constant value that controls the strength of observational bias. If there are more than one node which has the highest degree, HighestDegree returns the covariates of node whose sum of the covariates are the largest. We use widely employed GNNs in generating outcomes as GNNs_0 and GNNs_1 [16] which are defined as $\{z_{i0}, (1) \dots z_{i,|V_i|}(1)\} = \text{gnn}_1^w(V_i, A_i)$, where $z_{ij}(1) = W_1^w(1)v_{ij} + W_2^w(1)\sum_{k \in \mathcal{N}_{A_i}(j)} v_{ik}$, $z_{ij}(2) = \text{gnn}_2^w(\{z_{i0}, (1) \dots z_{i,|V_i|}(1)\}, A_i, j) = W_1^w(2)z_{ij}(1) + W_2^w(2)\sum_{k \in \mathcal{N}_{A_i}(j)} z_{ik}(1)$, $\text{GNNs}_w(G_i = (V_i, A_i), j) := \text{gnn}_2^{\text{w}}(\text{gnn}_1^{\text{w}}(V_i, A_i)), j)$, where $\mathcal{N}_{A_i}(j)$ represents a set of neighbors of j in A_i. Outcomes are generated as $Y_i^0 = \frac{1}{d}\frac{1}{|V_i|}C_s \sum_j^{|V_i|} z_{ij}^0 + \tau, Y_i^1 = \frac{1}{d}\frac{1}{|V_i|}C_s \sum_j^{|V_i|} z_{ij}^1 + \tau$ where $z_{ij}^0 = \text{GNNs}_0(G_i, j)$, $z_{ij}^1 = \text{GNNs}_1(G_i)$, and C_s is a scaling parameter, and we set C_s as 1 for the synthetic dataset. τ is a noise sampled from $\mathcal{N}(0, 0.01)$. We set parameters of GNNs_0; i.e., $W_1^0(1), \dots W^2(2)$, using random variables sampled from $\mathcal{U}(-3, 3)$ and similarly GNNs_1 from $\mathcal{U}(0, 3)$ so that graphs whose central nodes are larger values tend to have larger treatment effect and vice versa. $\mathcal{U}$ denotes a uniform distribution. Note that when we set α as 0, the treatment assignment reduces to random assignment and unless otherwise stated, we report experimental results when we set α as 0.5. This setting reflects a situation in which treatment is assigned only by a small part of an entire graph on which people are prone to focus. We set each node as 20-dimensional covariates and 2,000 graphs are used in one experiment.

Reddit Dataset [25]**:** We use Reddit binary dataset [25] for a semi-synthetic dataset. In Reddit dataset, each graph represents an online discussion thread. Each node represents a user and edges are given between two nodes if at least one of them correspond to the other comments. In this paper, we make use of graphs included in the dataset as a real world data and generate a tuple of covariates, treatments and outcomes in the same manner as the synthetic data. We set C_s as 0.5 for the synthetic dataset. We also set each node as 20-dimensional covariates. We remove graphs which have more than 500 nodes for computational convenience and have 1,617 graphs.

6.1 Experimental Settings

Note that due to the nature of input data, popular methods such as the TARNet or CFR [20] cannot be applied directly to this experiment in the original form and hence, none of previous studies have handled treatment effect estimation on graph-structured targets. Therefore we extend these methods [20,26] to this setting by replacing the neural networks with DeepSets or GNNs because they are unable to handle graph-structured data without modifications to the model architectures. (i) *Mean* predicts the outcome using the mean of train dataset. (ii) *DeepSets* [26] predicts the outcome by combining DeepSets and TARNet. (iii) *DeepSets-CFR* predicts the outcome by the combination of DeepSets and TARNet with the Wasserstein regularization as the IPM for bias mitigation. (iv) *GNNs* predicts the outcomes using GNNs without any regularization. (v) *GNNs-CFR* predicts the outcomes using GNNs with the Wasserstein regularization as the IPM for bias mitigation. We split the dataset into 20%, 20%, and 60% for train, validation, and test datasets, respectively. Note that we particularly focus on a relatively small number of data because the error of bias mitigation becomes severer. As a evaluation metric, we employ $\sqrt{\epsilon_{\text{PEHE}}}$ that measures regression error in terms of CATE and is defined as $\sqrt{\epsilon_{\text{PEHE}}} = \frac{1}{N_{\text{test}}} \sum_i^N ((Y_i^1 - Y_i^0) - (\hat{Y}_i^1 - \hat{Y}_i^0))^2$, where N_{test} represents the number of samples in test data and $\hat{Y}_i^t$ is a predicted outcome given as $\hat{Y}^t = g(G_i, t)$. We also employ ϵ_{ATE} that evaluates performances in terms of ATE defined as $\epsilon_{\text{ATE}} = \left| \frac{1}{N_{\text{test}}} \sum_{i=1}^N (Y_i^1 - Y_i^0) - \frac{1}{N_{\text{test}}} \sum_{i=1}^N (\hat{Y}_i^1 - \hat{Y}_i^0) \right|$. The regularization hyper-parameter λ is selected from $\{10^{-3},\ 10^{-2}, \ldots 10^2,\ 10^3\}$ based on the loss for the validation dataset. We set the k in selecting the confounding node in the SAG-pooling as 10; i.e., 10% nodes in each graph are treated as the confounding nodes. For the computation of Wasserstein distance as the IPM regularization, we employ Sinkhorn's algorithm [4]. We use the *elu* function as an activation function in all the models. We set the number of layers to 3 and the dimension of representations as 32 through the experiments.

6.2 Results

Table 1 summarizes the entire results in comparison with several baseline methods. First, whereas mean or the DeepSets-based methods do not work well for the both datasets, methods making use of GNNs outperform these baseline methods significantly. Second, the bias mitigation techniques remarkably improve the predictive performances for the both datasets compared to GNNs without any such technique. The naive methods without any bias mitigation technique suffer severe observational bias and fail to predict treatment effect. Conversely, GraphTEE significantly outperforms all the baseline methods in both the datasets. This result shows the effectiveness of efficient bias regularization and GraphTEE could mitigate bias. Figure 3 shows the performance comparisons when we change the strength of observational bias (upper figures) and the strength of regularization λ (lower figures). GraphTEE consistently outperforms the baseline methods even when strong bias is given. Although we need to somewhat exercise discretion in selecting λ to give the best performance, GraphTEE still performs better than the baseline methods.

7 Conclusion

In this study, we considered treatment effect estimation on graph-structured targets. We proposed GraphTEE, a novel two-step approach that aims to mitigate bias more efficiently. In the first step, we identify confounding nodes which have effects on treatment assignment. In the second step, we employ GNNs to predict outcomes while we perform bias mitigation by using confounding nodes. In this study, we mainly considered 1-Lipschitz function space. We cannot claim similar results as Theorem 1 currently for other function spaces and this might limit the range of application to some extent. Future work includes theoretical analyses towards that direction.

Acknowledgments. This work was supported by JST CREST Grant. Number JPMJCR21D1.

A Appendix

Lemma 1. *Let $\Phi : \mathcal{G} \rightarrow \mathcal{Z}$ be a one-to-one mapping function. Let $h: \mathcal{Z} \times \{0,1\} \rightarrow \mathcal{Y}$ be a hypothesis. Let $\mathcal{F}$ be a family of functions and $f \in \mathcal{F}: \mathcal{Z} \rightarrow \mathcal{Y}$. Assume there exists a constant B_Φ such that $\frac{1}{B_\Phi}\int_{\mathcal{Y}} L(Y^t, h(\Phi(G), t))p(Y^t \mid G)p(G,t)dY^t \in \mathcal{F}$. Then, we have*

$$\epsilon_{\text{test}}(h, \Phi) \leq \epsilon_{\text{train}}(h, \Phi) + B_\Phi \cdot IPM_{\mathcal{F}}(p(z_{(t=0)}), p((z_{(t=1)}))), \tag{11}$$

where $z_{(t=0)}$ and $z_{(t=1)}$ stand for representations of control and treatment group, respectively.

Proof. See the original papers [10,20]. We also used the following inequality.

$$\begin{aligned} &\mathrm{IPM}_{\mathcal{F}}(p(x,t), p(x)p(t)) \\ &= \sup_{f\in\mathcal{F}}\left|\int_{\mathcal{X}} f(x,t=0)p(t=0)(p(x \mid t=0) - p(x))dx\right. \\ &\left. + \int_{\mathcal{X}} f(x,t=1)p(t=1)(p(x \mid t=1) - p(x))dx\right| \end{aligned} \tag{12}$$

$$\leq \sup_{f \in \mathcal{F}} \left| \int_{\mathcal{X}} f(x, t=0) p(t=0) p(t=1) (p(x \mid t=0) - p(x \mid t=1)) dx \right|$$

$$+ \sup_{f \in \mathcal{F}} \left| \int_{\mathcal{X}} f(x, t=1) p(t=0) p(t=1) (p(x \mid t=1) - p(x \mid t=0)) dx \right| \tag{13}$$

$$= p(t=0)p(t=1) \mathrm{IPM}_{\mathcal{F}}(p(x \mid t=0), p(x \mid t=1))$$

$$+ p(t=0)p(t=1) \mathrm{IPM}_{\mathcal{F}}(p(x \mid t=0), p(x \mid t=1)) \tag{14}$$

$$\leq \frac{1}{2} \mathrm{IPM}_{\mathcal{F}}(p(x \mid t=0), p(x \mid t=1)) \quad (\because p(t=0)p(t=1) \leq \frac{1}{4}) \tag{15}$$

$$\leq \mathrm{IPM}_{\mathcal{F}}(p(x \mid t=0), p(x \mid t=1)). \tag{16}$$

Proposition 2. *Let $\Phi : \mathcal{G} \to \mathcal{Z}$ be a one-to-one mapping function. Let $h \colon \mathcal{Z} \times \{0,1\} \to \mathcal{Y}$ be a hypothesis. Let $\mathcal{F}$ be a family of functions and $f \in \mathcal{F} \colon \mathcal{Z} \to \mathcal{Y}$. Assume there exists a constant B_Φ such that $\frac{1}{B_\Phi} \int_{\mathcal{Y}} L(Y^t, h(\Phi(G), t)) p(Y^t \mid G) p(G, t) dY^t \in \mathcal{F}$. Supposing $f = h(z) \in \mathcal{F}$, we have with a probability at least $1 - \delta$,*

$$\epsilon_{\text{test}}(h, \Phi) \leq \epsilon_{\text{train}}(h, \Phi) + B_\Phi \cdot IPM_{\mathcal{F}}(\hat{p}(z_{(t=0)}), \hat{p}((z_{(t=1)})))$$

$$+ 2R_m(\mathcal{F}) + 2R_n(\mathcal{F}) + \sqrt{18\nu^2 \log \frac{4}{\delta}} (\frac{1}{\sqrt{m}} + \frac{1}{\sqrt{n}}). \tag{9}$$

Proof. Combining Proposition 1 and Lemma 1, we have the result.

Lemma 2. *Suppose we have the following error:*

$$IPM_{\mathcal{F}_{ty}}(p(t)p(z_c, z_y), p(t)p(z_c)p(z_y)) = IPM_{\mathcal{F}_y}(p(z_c, z_y), p(z_c)p(z_y)) = \Delta, \tag{17}$$

then the following inequality holds.

$$IPM_{\mathcal{F}_{ty}}(p(t \mid z_c)p(z_c, z_y), p(t)p(z_c)p(z_y)) \leq IPM_{\mathcal{F}_{ty}}(p(t \mid z_c)p(z_c)p(z_y), p(t)p(z_c)p(z_y)) + 2\Delta. \tag{18}$$

Proof. We can rewrite $p(z_c, z_y)$ as $p(z_c, z_y) = p(z_c)p(z_y) + \Delta_{z_c, z_y}$ using a difference Δ_{z_c, z_y} between $p(z_c, z_y)$ and $p(z_c)p(z_y)$ at z_c and z_y. Then, we can derive the following equation.

$$\mathrm{IPM}_{\mathcal{F}_y}(p(z_c, z_y), p(z_c)p(z_y))$$

$$= \sup_{f \in \mathcal{F}_y} \left| (\int_{\mathcal{Z}_c \times \mathcal{Z}_y} f(z_c, z_y) \Big(p(z_c, z_y) - p(z_c)p(z_y) \Big) dz_c dz_y \right| \tag{19}$$

$$= \sup_{f \in \mathcal{F}_y} \left| (\int_{\mathcal{Z}_c \times \mathcal{Z}_y} f(z_c, z_y) \Big((p(z_c)p(z_y) + \Delta_{z_c, z_y} - p(z_c)p(z_y) \Big) dz_c dz_y \right| \tag{20}$$

$$= \sup_{f \in \mathcal{F}_y} \left| (\int_{\mathcal{Z}_c \times \mathcal{Z}_y} f(z_c, z_y) \Delta_{z_c, z_y dz_c dz_y} \right| \tag{21}$$

$$= \Delta. \tag{22}$$

Then, the claim is proved as follow.

$$
\begin{aligned}
&\mathrm{IPM}_{\mathcal{F}_{ty}}(p(t \mid z_c)p(z_c, z_y), p(t)p(z_c)p(z_y))) &(23)\\
&= \sup_{f \in \mathcal{F}_{ty}} \left|\left(\int_{\mathcal{T}\times\mathcal{Z}_c\times\mathcal{Z}_y} f(t, z_c, z_y)\Big(p(t \mid z_c)p(z_c, z_y) - p(t)p(z_c)p(z_y)\Big)dtdz_cdz_y\right| &(24)\\
&= \sup_{f \in \mathcal{F}_{ty}} \left|\left(\int_{\mathcal{T}\times\mathcal{Z}_c\times\mathcal{Z}_y} f(t, z_c, z_y)\Big(p(t \mid z_c)(p(z_c)p(z_y) + \varDelta_{z_c,z_y})\right.\\
&\qquad \left. - p(t)p(z_c)p(z_y)\Big)dtdz_cdz_y\right| &(25)\\
&\leq \mathrm{IPM}_{\mathcal{F}_{ty}}(p(t \mid z_c)p(z_c)p(z_y), p(t)p(z_c)p(z_y))\\
&\qquad + \sup_{f \in \mathcal{F}_{ty}} \left|\int_{\mathcal{T}\times\mathcal{Z}_c\times\mathcal{Z}_y} f(t, z_c, z_y)p(t \mid z_c)\varDelta_{z_c,z_y}dtdz_cdz_y\right| &(26)\\
&\quad (\because \text{triangle inequality})\\
&\leq \mathrm{IPM}_{\mathcal{F}_{ty}}(p(t \mid z_c)p(z_c)p(z_y), p(t)p(z_c)p(z_y)) + 2\varDelta. &(27)\\
&\quad (\because \text{Eq. (22)},\ 0 < p(t) < 1,\ 0 < p(t \mid z_c) < 1).
\end{aligned}
$$

Theorem 1. *Suppose we have an error as* $IPM(p(t)p(z_c \mid z_y)p(z_y), p(t)p(z_c)p(z_y)) = \varDelta$. *We have with probability at least* $1 - \delta$,

$$
\begin{aligned}
&\epsilon_{\text{test}}(h, \varPhi) \leq \epsilon_{\text{train}}(h, \varPhi) + B_{\varPhi} \cdot \big(IPM_{\mathcal{F}_c}(\hat{p}(z_{c(t=0)}), \hat{p}(z_{c(t=1)}))\\
&+ IPM_{\mathcal{F}_y}(\hat{p}(z_c, z_y), \hat{p}(z_c)p(z_y))\big) + 2R_m(\mathcal{F}_c) + 2R_n(\mathcal{F}_c)\\
&+ \sqrt{18C_\nu^2 \log\frac{8}{\delta}}\left(\frac{1}{\sqrt{m}} + \frac{1}{\sqrt{n}}\right) + 4R_{m+n}(\mathcal{F}_y) + 2\sqrt{\frac{18\nu^2}{m+n}\log\frac{8}{\delta}} + 2\varDelta. \qquad (10)
\end{aligned}
$$

Proof. Using Lemma 2, we have

$$
\begin{aligned}
&\mathrm{IPM}_{\mathcal{F}_{ty}}(p(t \mid z_c)p(z_c, z_y), p(t)p(z_c)p(z_y))\\
&\quad \leq \mathrm{IPM}_{\mathcal{F}_{ty}}(p(t \mid z_c)p(z_c)p(z_y), p(t)p(z_c)p(z_y)) + 2\varDelta. \qquad (28)
\end{aligned}
$$

Therefore,

$$
\begin{aligned}
&\mathrm{IPM}_{\mathcal{F}_{ty}}(p(t \mid z_c)p(z_c, z_y), p(t)p(z_c, z_y))\\
&\leq \mathrm{IPM}_{\mathcal{F}_{ty}}(p(t \mid z_c)p(z_c, z_y), p(t)p(z_c)p(z_y))) + \mathrm{IPM}_{\mathcal{F}_{ty}}(p(t)p(z_c, z_y), p(t)p(z_c)p(z_y)) &(29)\\
&\leq \mathrm{IPM}_{\mathcal{F}_{ty}}(p(t \mid z_c)p(z_c)p(z_y), p(t)p(z_c)p(z_y)) + 2\varDelta\\
&+ \mathrm{IPM}_{\mathcal{F}_{ty}}(p(t)p(z_c, z_y), p(t)p(z_c)p(z_y)) &(30)\\
&= \mathrm{IPM}_{\mathcal{F}_c}(p(t \mid z_c)p(z_c), p(t)p(z_c)) + 2\varDelta + \underbrace{\mathrm{IPM}_{\mathcal{F}_y}(p(z_c, z_y), p(z_c)p(z_y))}_{=\varDelta}. &(31)
\end{aligned}
$$

Using the theoretical analysis in [21], at least probability 1-δ,

$$
\begin{aligned}
&\mathrm{IPM}_{\mathcal{F}}(p(t \mid z_c)p(z_c, z_y), p(t)p(z_c, z_y)) - \mathrm{IPM}_{\mathcal{F}}(\hat{p}(t \mid z_c)\hat{p}(z_c, z_y), \hat{p}(t)\hat{p}(z_c, z_y))\\
&\leq 2R_m(\mathcal{F}) + 2R_n(\mathcal{F}) + \sqrt{18\nu^2 \log\frac{4}{\delta}}\left(\frac{1}{\sqrt{m}} + \frac{1}{\sqrt{n}}\right) \qquad (32)
\end{aligned}
$$

Here, we claim there exists a constant C_c ($1 \geq C_c > 0$) such that

$$C_c \sup f(z_c, z_y,, t) = \sup g(z_c, t). \tag{33}$$

This is because

$$\begin{aligned} \forall f \in \mathcal{F}, \sup g(z_c, t) &= \sup \mathbb{E}_{z_y}[f(z_c, z_y,, t)] &(34)\\ &\leq \mathbb{E}_{z_y}[\sup f(z_c, z_y, t)] &(35)\\ &\leq \sup f(z_c, z_y, t). &(36) \end{aligned}$$

Therefore, we focus on an empirical error with regard to the following IPM.

$$\mathrm{IPM}_{\mathcal{F}}(p(z_{c(t=1)}), p(z_{c(t=0)})) - \mathrm{IPM}_{\mathcal{F}}(\hat{p}(z_{c(t=1)}), \hat{p}(z_{c(t=0)})). \tag{37}$$

By performing similar operation, we have with probability at least $1 - \frac{\delta}{8}$,

$$\begin{aligned} \sup_{f \in \mathcal{F}} \left| \mathbb{P}_m f - Pf \right| &\leq \mathbb{E} \sup_{f \in \mathcal{F}} \left| \mathbb{P}_m f - Pf \right| + \sqrt{\frac{2C_c^2 \nu^2}{m} \log \frac{8}{\delta}} &(38)\\ &\leq 2\mathbb{E} \sup_{f \in \mathcal{F}} \left| \frac{1}{m} \sum_{i=1}^{m} \sigma_i f(z_c, t) \right| + \sqrt{\frac{2C_c^2 \nu^2}{m} \log \frac{8}{\delta}}. &(39) \end{aligned}$$

we have with probability at least $1 - \frac{\delta}{8}$,

$$\mathbb{E} \sup_{f \in \mathcal{F}} \left| \frac{1}{m} \sum_{i=1}^{m} \sigma_i f(z_c, t) \right| \leq \mathbb{E}_\sigma \sup_{f \in \mathcal{F}} \left| \frac{1}{m} \sum_{i=1}^{m} \sigma_i f(z_c, t) \right| + \sqrt{\frac{2C_c^2 \nu^2}{m} \log \frac{8}{\delta}}. \tag{40}$$

we have that with probability at least $1 - \frac{\delta}{4}$

$$\sup_{f \in \mathcal{F}} \left| \mathbb{P}_m f - Pf \right| \leq 2R_m(\mathcal{F}) + \sqrt{\frac{18C_c^2 \nu^2}{m} \log \frac{8}{\delta}}. \tag{41}$$

Then we have that with probability at least $1 - \frac{\delta}{2}$,

$$\begin{aligned} &\left| \mathrm{IPM}_{\mathcal{F}_c}(p(z_{c(t=1)}), p(z_{c|t=0})) - \mathrm{IPM}_{\mathcal{F}_c}(\hat{p}(z_{c(t=1)}), \hat{p}(z_{c(t=0)})) \right| \\ &\leq 2R_m(\mathcal{F}_c) + 2R_n(\mathcal{F}_c) + \sqrt{\frac{18C_c^2 \nu^2}{m} \log \frac{8}{\delta}} + \sqrt{\frac{18C_c^2 \nu^2}{n} \log \frac{8}{\delta}}. &(42) \end{aligned}$$

Next we perform similar analysis on $\mathrm{IPM}_{\mathcal{F}}(p(t)p(z_c, z_y), p(t)p(z_c)p(z_y))$. Then we have that with probability at least $1 - \frac{\delta}{2}$,

$$\begin{aligned} &\left| \mathrm{IPM}_{\mathcal{F}_y}(p(z_c, z_y), p(z_c)p(z_y)) - \mathrm{IPM}_{\mathcal{F}_y}(\hat{p}(z_c, z_y), \hat{p}(z_c)p(z_y)) \right| \\ &\leq 2R_{m+n}(\mathcal{F}_y) + 2R_{m+n}(\mathcal{F}_y) + \sqrt{\frac{18\nu^2}{m+n} \log \frac{8}{\delta}} + \sqrt{\frac{18\nu^2}{m+n} \log \frac{8}{\delta}}. &(43) \end{aligned}$$

Finally, by combining these results, we have with probability at least $1-\delta$,

$$\begin{aligned}
&\left|\mathrm{IPM}_{\mathcal{F}_c}(p(z_{c(t=1)}), p(z_{c(t=0)})) - \mathrm{IPM}_{\mathcal{F}_c}(\hat{p}(z_{c(t=1)}), \hat{p}(z_{c(t=0)}))\right| \\
&+\left|\mathrm{IPM}_{\mathcal{F}_y}(p(z_c, z_y), p(z_c)p(z_y)) - \mathrm{IPM}_{\mathcal{F}_y}(\hat{p}(z_c, z_y), \hat{p}(z_c)\hat{p}(z_y))\right| \\
\leq& 2R_m(\mathcal{F}_c) + 2R_n(\mathcal{F}_c) + \sqrt{18C_c^2\nu^2 \log\frac{8}{\delta}}(\frac{1}{\sqrt{m}} + \frac{1}{\sqrt{n}}) \\
&+2R_{m+n}(\mathcal{F}_y) + 2R_{m+n}(\mathcal{F}_y) + 2\sqrt{\frac{18\nu^2}{m+n}\log\frac{8}{\delta}} \qquad (44) \\
\leq& 2R_m(\mathcal{F}_c) + 2R_n(\mathcal{F}_c) + \sqrt{18C_c^2\nu^2 \log\frac{8}{\delta}}(\frac{1}{\sqrt{m}} + \frac{1}{\sqrt{n}}) \\
&+4R_{m+n}(\mathcal{F}_y) + 2\sqrt{\frac{18\nu^2}{m+n}\log\frac{8}{\delta}}. \qquad (45)
\end{aligned}$$

References

1. Barabási, A.L., Albert, R.: Emergence of scaling in random networks. Science **286**(5439), 509–512 (1999)
2. Blitzer, J., Crammer, K., Kulesza, A., Pereira, F., Wortman, J.: Learning bounds for domain adaptation. In: Advances in Neural Information Processing Systems, vol. 20 (2007)
3. Cao, D., He, X., Miao, L., An, Y., Yang, C., Hong, R.: Attentive group recommendation. In: Proceedings of the 41st International ACM SIGIR Conference on Research & Development in Information Retrieval, pp. 645–654 (2018)
4. Cuturi, M.: Sinkhorn distances: lightspeed computation of optimal transport. In: Advances in Neural Information Processing Systems, vol. 26 (2013)
5. De Veirman, M., Hudders, L., Nelson, M.R.: What is influencer marketing and how does it target children? A review and direction for future research. Front. Psychol. **10**, 2685 (2019)
6. Guo, R., Li, J., Liu, H.: Learning individual causal effects from networked observational data. In: Proceedings of the 13th International Conference on Web Search and Data Mining (2020)
7. Hagberg, A., Swart, P., S Chult, D.: Exploring network structure, dynamics, and function using networkx. Technical report, Los Alamos National Lab. (LANL), Los Alamos, NM (United States) (2008)
8. Harada, S., Kashima, H.: Graphite: estimating individual effects of graph-structured treatments. In: Proceedings of the 30th ACM International Conference on Information & Knowledge Management, pp. 659–668 (2021)
9. Johansson, F., Shalit, U., Sontag, D.: Learning representations for counterfactual inference. In: Proceedings of the 33rd International Conference on Machine Learning (2016)
10. Johansson, F.D., Kallus, N., Shalit, U., Sontag, D.: Learning weighted representations for generalization across designs. arXiv preprint arXiv:1802.08598 (2018)
11. Kaddour, J., Zhu, Y., Liu, Q., Kusner, M.J., Silva, R.: Causal effect inference for structured treatments. In: Advances in Neural Information Processing Systems, vol. 34, pp. 24841–24854 (2021)

12. LaLonde, R.J.: Evaluating the econometric evaluations of training programs with experimental data. Am. Econ. Rev. (1986)
13. Lee, J., Lee, I., Kang, J.: Self-attention graph pooling. In: Proceedings of the 36th International Conference on Machine Learning, pp. 3734–3743. PMLR (2019)
14. Ma, Y., Wang, Y., Tresp, V.: Causal inference under networked interference. arXiv preprint arXiv:2002.08506 (2020)
15. Mansour, Y., Mohri, M., Rostamizadeh, A.: Domain adaptation: learning bounds and algorithms. arXiv preprint arXiv:0902.3430 (2009)
16. Morris, C., et al.: Weisfeiler and leman go neural: higher-order graph neural networks. In: Proceedings of the 33rd AAAI Conference on Artificial Intelligence, vol. 33, pp. 4602–4609 (2019)
17. Müller, A.: Integral probability metrics and their generating classes of functions. Adv. Appl. Probab. (1997)
18. Rosenbaum, P.R., Rubin, D.B.: The central role of the propensity score in observational studies for causal effects. Biometrika (1983)
19. Rubin, D.B.: Matching to remove bias in observational studies. Biometrics (1973)
20. Shalit, U., Johansson, F.D., Sontag, D.: Estimating individual treatment effect: generalization bounds and algorithms. In: Proceedings of the 34th International Conference on Machine Learning (2017)
21. Sriperumbudur, B.K., Fukumizu, K., Gretton, A., Schölkopf, B., Lanckriet, G.R.: On integral probability metrics,\phi-divergences and binary classification. arXiv preprint arXiv:0901.2698 (2009)
22. Wood, L., et al.: Empirical evidence of bias in treatment effect estimates in controlled trials with different interventions and outcomes: meta-epidemiological study. BMJ **336**(7644), 601–605 (2008)
23. Wu, Y.X., Wang, X., Zhang, A., He, X., Chua, T.S.: Discovering invariant rationales for graph neural networks. In: Proceedings of the 10th International Conference on Learning Representations (2022)
24. Xu, K., Hu, W., Leskovec, J., Jegelka, S.: How powerful are graph neural networks? In: Proceedings of the 6th International Conference on Learning Representations (2018)
25. Yanardag, P., Vishwanathan, S.: Deep graph kernels. In: Proceedings of the 21th ACM SIGKDD International Conference on Knowledge Discovery & Data Mining, pp. 1365–1374 (2015)
26. Zaheer, M., Kottur, S., Ravanbakhsh, S., Poczos, B., Salakhutdinov, R.R., Smola, A.J.: Deep sets. In: Advances in Neural Information Processing Systems, vol. 30 (2017)
27. Zhao, S., Heffernan, N.: Estimating individual treatment effect from educational studies with residual counterfactual networks. In: Proceedings of the 10th International Conference on Educational Data Mining (2017)
28. Zheleva, E., Arbour, D.: Causal inference from network data. In: Proceedings of the 27th ACM SIGKDD International Conference on Knowledge Discovery & Data Mining, pp. 4096–4097 (2021)

Dynamic DropConnect: Enhancing Neural Network Robustness Through Adaptive Edge Dropping Strategies

Yuan-Chih Yang and Hung-Hsuan Chen(✉)

National Central University, Taoyuan, Taiwan
hhchen1105@acm.org

Abstract. Dropout and DropConnect are well-known techniques that apply a consistent drop rate to randomly deactivate neurons or edges in a neural network layer during training. This paper introduces a novel methodology that assigns dynamic drop rates to each edge within a layer, uniquely tailoring the dropping process without incorporating additional learning parameters. We perform experiments on synthetic and openly available datasets to validate the effectiveness of our approach. The results demonstrate that our method outperforms Dropout, DropConnect, and Standout, a classic mechanism known for its adaptive dropout capabilities. Furthermore, our approach improves the robustness and generalization of neural network training without increasing computational complexity. The complete implementation of our methodology is publicly accessible for research and replication purposes at https://github.com/ericabd888/Adjusting-the-drop-probability-in-DropConnect-based-on-the-magnitude-of-the-gradient/

1 Introduction

Dropout [3,12] and DropConnect [13] are prominent techniques designed to mitigate overfitting in deep neural networks. Dropout functions by independently zeroing each neuron within a layer with a predetermined fixed probability p. In contrast, DropConnect takes a slightly different approach by randomly eliminating an edge within the network with a fixed probability p. This makes DropConnect a generalization of Dropout; specifically, removing a single neuron as performed in Dropout equates to eliminating all incoming and outgoing edges associated with that neuron, which DropConnect facilitates.

Both Dropout and DropConnect uniformly apply a fixed dropping rate across all neurons or edges within a layer. However, this universal dropping rate might not always represent the optimal strategy; ideally, a model should utilize available data-driven insights to tailor the dropping rate for each individual edge or neuron based on their specific characteristics.

To address this, we introduce a novel methodology termed DynamicDropConnect (DDC). This approach dynamically assigns a drop probability to each edge based on the magnitude of the gradient associated with that edge. The

X. Wu et al. (Eds.): PAKDD 2025, LNAI 15875, pp. 110–121, 2025.
https://doi.org/10.1007/978-981-96-8295-9_7

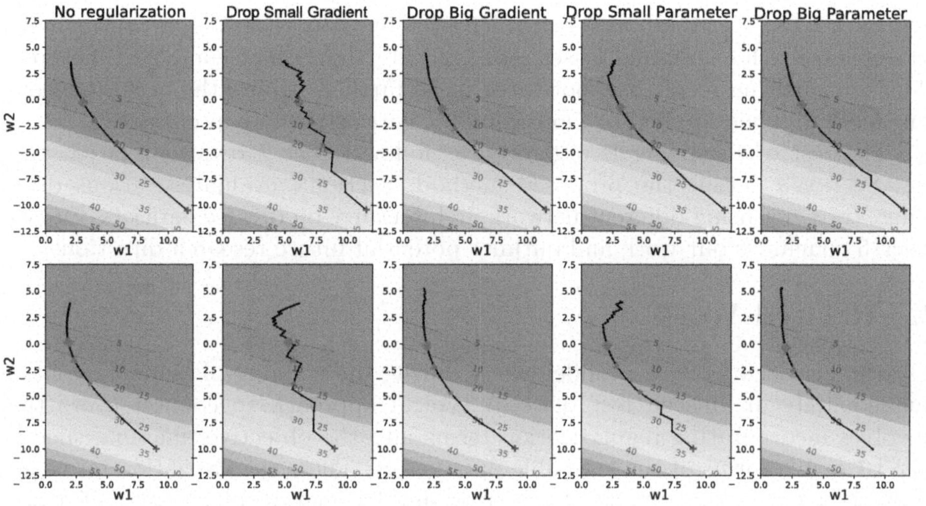

Fig. 1. The contour plot of the loss and the parameter update process. The two rows represent two sets of initial values and their updating process. The blue pluses denote the initial values. The red stars, triangles, and diamonds represent the values of w_1 and w_2 after training for 7, 13, and 19 epochs. The method that tends to drop the edges with small gradients (the second column) reaches a small error faster. (Color figure online)

underlying principle is that edges with larger gradients are crucial for learning and should be retained, whereas it might be acceptable to omit those with minimal impact on the model's output occasionally.

DDC offers several advantages over other methods such as Standout [1], which also employs a dynamic dropping rate. Firstly, DDC does not require any additional learning parameters, which simplifies the model architecture and reduces memory requirements during training. Secondly, DDC provides a more deterministic and transparent approach to deciding the dropping rate, as opposed to the potentially opaque and unpredictable learning processes used by methods like Standout. Moreover, empirical evidence from our experiments demonstrates that DDC achieves superior accuracy compared to Standout.

Our experimental analysis extends beyond theoretical discussions. We conduct rigorous experiments on both synthetic and real open datasets to validate the efficacy of our methodology. Figure 1 illustrates the trajectory of parameter values and their corresponding losses using a synthetic dataset. The first column in the figure depicts a model without any regularization (labelled as "No regularization" in Fig. 1, whereas the second column shows the impact of training the model by preferentially dropping edges associated with smaller gradients ("Drop Small Gradient"). For comparative purposes, we also trained the same model using different variants ("Drop Big Gradient", "Drop Small Parameter", and "Drop Big Parameter"). The plots clearly demonstrate that training a linear regressor using our methodology, which preferentially drops edges with smaller gradients, reaches minimal losses more quickly than other compared methods. This encouraging result has motivated us to further explore the performance of our method when applied to more complex networks on open datasets.

The rest of the paper is organized as follows: Sect. 2 reviews related studies. Section 3 details the proposed method. Section 4 evaluates various dropping strategies using different network architectures on diverse datasets. Finally, Sect. 5 concludes our work and outlines potential future research directions.

2 Related Work

Overfitting occurs when a model excels during the training phase but performs inadequately on unseen test data. A typical approach to counter overfitting involves incorporating regularization terms into the objective function, such as the L1 or L2 norms of the learnable parameters. These regularization terms are designed to encourage the development of simpler relationships between features and targets, thus mitigating the risk of overfitting. Early stopping is another frequently adopted strategy to prevent overfitting. This technique involves ceasing the parameter updates once the loss on the validation set begins to show signs of increase, thereby avoiding the overfitting of the training data. Additional well-established methods to combat overfitting include data augmentation [7], employing model ensembles [4], and modifying the architecture of neural networks by reducing their depth or width. These methods have been proven to be effective in enhancing the generalizability of models across unseen datasets.

Dropout is another widely recognized method specifically utilized in neural networks to curb overfitting [12]. During training, Dropout randomly deactivates a subset of neurons, ensuring that the neural network does not become overly dependent on any particular set of neurons. This technique effectively simulates the use of multiple neural network models and aggregates their predictions, which enhances the robustness of the model. Notably, the random neuron deactivation is only applied during the training phase and not during inference. To maintain consistent expected values between training and inference, Inverted Dropout is commonly utilized.

DropConnect extends the idea of Dropout by randomly omitting connections, or edges, between neurons during the training phase. This approach is similar to Dropout in that masking a neuron in Dropout is analogous to masking both the incoming and outgoing connections of that neuron in DropConnect. Empirically, both Dropout and DropConnect often result in models with comparable performance levels.

From a Bayesian perspective, Dropout can also be interpreted as performing an approximated Bayesian inference in deep Gaussian processes [2]. Notable techniques within this framework include Monte Carlo Dropout [2], fast dropout [14], and variational dropout [5], each offering unique approaches to implementing Dropout within a Bayesian inference context.

Some recent studies have explored the idea of varying the dropping rate throughout the training process. A notable example is Standout [1], which dynamically adjusts the dropping rate based on the values of the model parameters. Unlike other variations of Standout, which often necessitate additional learning parameters, our approach simplifies the process by avoiding extra learning mechanisms. This not only reduces training time and memory requirements but also provides a more transparent method for determining dynamic dropping rates. Furthermore, our research has explored the efficacy of utilizing gradient values rather than parameter values for adjusting drop rates, which has shown promising results in our experimental evaluations.

For a comprehensive review of Dropout and its various adaptations, please refer to the survey by Labach et al. [8].

3 Methodology

This section introduces DropConnect as the background knowledge, followed by our methodology to assign dynamic dropping rates.

3.1 Preliminary: DropConnect

Let $l \in \{1, \ldots, L\}$ be the L hidden layers in a neural network and $\boldsymbol{y}^{(l)}$ be the output of layer l (and therefore the input of layer $l+1$), each layer of a neural network transforms $\boldsymbol{y}^{(l-1)}$ to $\boldsymbol{y}^{(l)}$ by Eq. 1.

$$\boldsymbol{y}^{(l)} = f_l\left(\boldsymbol{z}^{(l)}\right) = f_l\left(\boldsymbol{W}^{(l)}\boldsymbol{y}^{(l-1)}\right), \tag{1}$$

where $\boldsymbol{W}^{(l)}$ is the set of parameters in layer l, and $f_l(\cdot)$ is the activation function of the same layer.

DropConnect assigns a universal dropping rate p such that each edge has a probability p of being turned off during training. Therefore, a neural network, with DropConnect, transforms $\boldsymbol{y}^{(l-1)}$ to $\boldsymbol{y}^{(l)}$ by the equation below during training.

$$\boldsymbol{y}^{(l)} = f_l\left(\boldsymbol{z}^{(l)}\right) = f_l\left(\left((1-\boldsymbol{M}^{(l)}) \odot \boldsymbol{W}^{(l)}\right)\boldsymbol{y}^{(l-1)}\right), \tag{2}$$

where $\boldsymbol{M}^{(l)} = [m_{i,j}]$ is a $(0,1)$-matrix whose shape is the same as $\boldsymbol{W}^{(l)}$; each entry $m_{i,j}$ is sampled from a Bernoulli distribution with a fixed hyper-parameter p, and $\odot$ performs the Hadamard product (i.e., element-wise product) of matrices. Thus, $\boldsymbol{M}^{(l)}$ is a mask matrix that decides which parameters in $\boldsymbol{W}^{(l)}$ (edges) to omit.

3.2 DDC - Mask Generation

The proposed DDC advances DropConnect by allowing variable dropping probabilities for distinct neural network edges.

Algorithm 1: Generating a mask for $\boldsymbol{W}^{(l)}$

Input : gradients $\boldsymbol{G}^{(l)} = \left[g_{i,j}^{(l)}\right] \in \mathcal{R}^{n_1 \times n_2}$,
hyperparameters p, p_g, and τ

Output: mask $\widetilde{\boldsymbol{M}}^{(l)} = \left[\widetilde{m}_{i,j}^{(l)}\right] \in \mathcal{R}^{n_1 \times n_2}$

1 **for** $i \leftarrow 1$ **to** n_1 **do**
2   **for** $j \leftarrow 1$ **to** n_2 **do**
3         $v_{i,j}^{(l)} \leftarrow \left|g_{i,j}^{(l)}\right|$;
4   **end**
5 **end**
6 $\mu^{(l)} \leftarrow \dfrac{\sum_{i=1}^{n_1} \sum_{j=1}^{n_2} v_{i,j}^{(l)}}{n_1 \times n_2}$;
7 $\sigma^{(l)} \leftarrow \dfrac{\sum_{i=1}^{n_1} \sum_{j=1}^{n_2} (v_{i,j}^{(l)} - \mu^{(l)})^2}{n_1 \times n_2}$;
8 **for** $i \leftarrow 1$ **to** n_1 **do**
9   **for** $j \leftarrow 1$ **to** n_2 **do**
10         $z_{i,j}^{(l)} \leftarrow \dfrac{v_{i,j}^{(l)} - \mu^{(l)}}{\sigma^{(l)}}$;
11   Update $q_{i,j}^{(l)}$ by Eq. 3;
12   Update $p_{i,j}^{(l)}$ by Eq. 4;
13         $\widetilde{m}_{i,j}^{(l)} = \text{Bernoulli}(p_{i,j}^{(l)})$;
14   **end**
15 **end**

We create a mask matrix, $\widetilde{\boldsymbol{M}}^{(l)}$ to drop edges. Each entry $\widetilde{m}_{i,j}^{(l)}$ in $\widetilde{\boldsymbol{M}}^{(l)}$ is sampled from a Bernoulli distribution with a unique parameter, $p_{i,j}^{(l)}$. An edge is omitted if $\widetilde{m}_{i,j}^{(l)} = 1$.

Algorithm 1 shows a pseudocode to generate $\widetilde{\boldsymbol{M}}^{(l)}$ during training. For every edge $e_{i,j}^{(l)}$ that bridges neuron i at layer $l-1$ and neuron j at layer l, the algorithm decides the value of the corresponding mask cell $\widetilde{m}_{i,j}^{(l)}$ based on its current gradient $g_{i,j}^{(l)}$ at every training iteration. The algorithm first computes the absolute gradient value of $g_{i,j}^{(l)}$, represented as $v_{i,j}^{(l)}$ (line 1 to line 5). DDC needs gradient normalization to harmonize the gradient magnitudes across distinct layers. Therefore, the mean and standard deviation of the set of $v_{i,j}^{(l)}$ values for layer l are calculated (line 6 and line 7). Subsequently, a z-score normalization is performed on $v_{i,j}^{(l)}$, yielding normalized gradient magnitudes $z_{i,j}^{(l)}$-s. DDC then calculates $q_{i,j}^{(l)}$ for each edge based on the normalized gradient magnitude $z_{i,j}^{(l)}$. This candidate dropping probability positively correlates with the edge's final dropping probability. Eventually, we define the candidate dropping probability based on Eq. 3, so the model tends to keep the edge with a larger gradient magnitude.

Algorithm 2: One-layer forward re-calibration during training

Input : weights $\boldsymbol{W}^{(l)} = \left[w_{i,j}^{(l)}\right] \in \mathcal{R}^{n_1 \times n_2}$,
gradients $\boldsymbol{G}^{(l)} = \left[g_{i,j}^{(l)}\right] \in \mathcal{R}^{n_1 \times n_2}$,
outputs from previous layer $\boldsymbol{y}^{(l-1)} \in \mathcal{R}^{n_2 \times 1}$
Output: $\boldsymbol{y}^{(l)} \in \mathcal{R}^{n_1 \times 1}$

Compute $\widetilde{\boldsymbol{M}}^{(l)}$ by Algorithm 1 using $\boldsymbol{W}^{(l)}$ and $\boldsymbol{G}^{(l)}$;
$\widetilde{\boldsymbol{W}}^{(l)} \leftarrow \left(1 - \widetilde{\boldsymbol{M}}^{(l)}\right) \odot \boldsymbol{W}^{(l)}$;
Compute $r^{(l)}$ by Equation 5;
Compute $\boldsymbol{y}^{(l)}$ by Equation 6;

$$q_{i,j}^{(l)} \leftarrow \begin{cases} 1 - \sigma\left(z_{i,j}^{(l)}\right) & \text{if } 1 - \sigma\left(z_{i,j}^{(l)}\right) \geq \tau \\ 0 & \text{otherwise,} \end{cases} \tag{3}$$

where $\sigma()$ is the sigmoid function. Since $\sigma()$ is monotonically increasing and the output is always between 0 and 1, a larger $z_{i,j}^{(l)}$ results in a smaller candidate dropping probability. We set hyperparameter τ to 0.5. So, half of the $q_{i,j}^{(l)}$-s are zero on average because the expected value of $\sigma\left(z_{i,j}^{(l)}\right)$ is 0.5.

The final dropping rate $p_{i,j}^{(l)}$ for each edge $e_{i,j}^{(l)}$ is a function of three variables: the candidate dropping probability $q_{i,j}^{(l)}$, a base dropping probability p (a hyper-parameter), and a gradient unit dropping rate p_g (another hyper-parameter). We limit the value of $p_{i,j}^{(l)}$ to be no greater than 1 to ensure $p_{i,j}^{(l)}$ as a proper probability value. The computation of $p_{i,j}^{(l)}$ is shown in Eq. 4.

$$p_{i,j}^{(l)} \leftarrow \begin{cases} p + p_g \times q_{i,j}^{(l)} & \text{if } p + p_g \times q_{i,j}^{(l)} \leq 1 \\ 1 & \text{otherwise.} \end{cases} \tag{4}$$

Finally, the DDC generates each $\widetilde{m}_{i,j}^{(l)}$ in $\widetilde{\boldsymbol{M}}^{(l)}$ by sampling from a Bernoulli distribution with the parameter $p_{i,j}^{(l)}$ (line 13).

3.3 DDC - Training Re-calibration and Testing

Since $\widetilde{\boldsymbol{M}}^{(l)}$ is the mask for the parameters in layer l, the effective parameters in this layer go from $\boldsymbol{W}^{(l)}$ to $(1 - \widetilde{\boldsymbol{M}}^{(l)}) \odot \boldsymbol{W}^{(l)}$. The dropping mechanism is only applied in training but not in prediction. However, if we set the value of $\boldsymbol{y}^{(l)}$ as $f_l\left((1 - \widetilde{\boldsymbol{M}}^{(l)}) \odot \boldsymbol{W}^{(l)}\boldsymbol{y}^{(l-1)}\right)$ in training but predict $\boldsymbol{y}^{(l)}$ as $f_l\left(\boldsymbol{W}^{(l)}\boldsymbol{y}^{(l-1)}\right)$ in inference, the learned $\boldsymbol{W}^{(l)}$ from training cannot be used directly in the tests. To fix the inconsistency, we need to recalibrate $(1 - \widetilde{\boldsymbol{M}}^{(l)}) \odot \boldsymbol{W}^{(l)}$ during training, so that the learned $\boldsymbol{W}^{(l)}$ in training can be used directly during inference by $f_l\left(\boldsymbol{W}^{(l)}\boldsymbol{y}^{(l-1)}\right)$.

Since $\widetilde{\boldsymbol{M}}^{(l)}$ is a $(0,1)$-matrix, $\sum_{i=1}^{n_1}\sum_{j=1}^{n_2}\widetilde{m}_{i,j}^{(l)}$ is the number of 1-s in $\widetilde{\boldsymbol{M}}^{(l)}$. So, we can compute the expected value of the dropping rate of layer l by Eq. 5.

$$r^{(l)} = \frac{\sum_{i=1}^{n_1}\sum_{j=1}^{n_2}\widetilde{m}_{i,j}^{(l)}}{n_1 \times n_2}, \tag{5}$$

where n_1 and n_2 are the number of rows and the number of columns of the masking matrix $\widetilde{\boldsymbol{M}}^{(l)}$.

We re-calibrate the output of layer l during training such that the learned weights $\boldsymbol{W}$ can be used during inference. The re-calibration is achieved by dividing the output by the keep rate:

$$\boldsymbol{y}^{(l)} \leftarrow f_l\left(\widetilde{\boldsymbol{W}}^{(l)}\boldsymbol{y}^{(l-1)} \times \frac{1}{1-r^{(l)}}\right). \tag{6}$$

The mask matrix $\widetilde{\boldsymbol{M}}^{(l)}$ and the masked parameter matrix $\widetilde{\boldsymbol{W}}^{(l)}$ are only used in training. At inference, we use only the unmasked parameter matrix $\boldsymbol{W}^{(l)}$ for forwarding: $\boldsymbol{y}^{(l)} \leftarrow f_l\left(\boldsymbol{W}^{(l)}\boldsymbol{y}^{(l-1)}\right)$.

Algorithm 2 shows one-layer forward re-calibration in training.

4 Experiments

This section introduces the experiment settings and results. All the models are implemented by PyTorch and trained on the NVIDIA GTX 3090. We conducted experiments based on one synthetic dataset and four open datasets: MNIST, CIFAR-10, CIFAR-100, and NORB. We split the labeled instances for each open dataset into the training set, the validation set, and the test set. Detailed settings, such as parameter initialization, learning rate, and batch size, are included in the experimental code.

4.1 Experiments on the Synthetic Dataset

We generate a synthetic dataset to analyze the learning process of various edge-dropping strategies. For each instance i, we sample two independent features $x_1^{(i)}$ and $x_2^{(i)}$, each $x_j^{(i)} \sim N(0,1)$. We fix the values of w_1 and w_2 and let the target $y^{(i)}$ be $w_1x_1^{(i)} + w_2x_2^{(i)} + \xi$, where $\xi \sim N(0,1)$. We only use two parameters, w_1 and w_2, because it is easier to visualize the parameter updating/learning process in a 2-dimensional contour plot.

Figure 1 shows the parameter learning process of various edge-dropping strategies: the first column (labeled "No regularization") represents the scenario

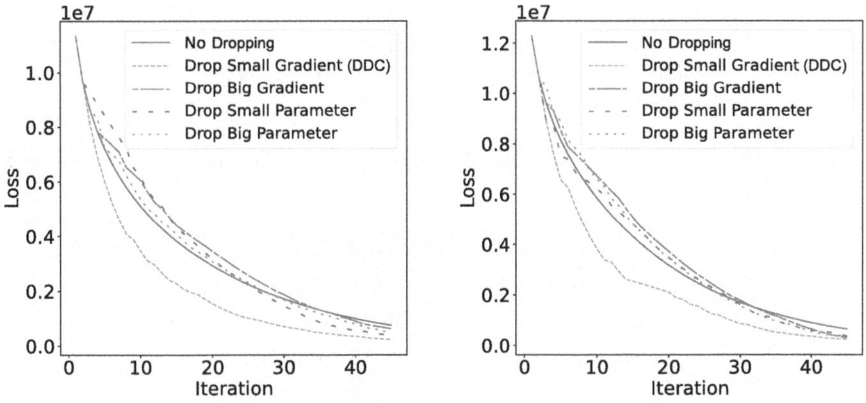

(a) $w_1 = 11.5, w_2 = -10$ (b) $w_1 = 9, w_2 = -10.5$

Fig. 2. Loss vs. epochs of different dropping strategies.

of the no-dropping mechanism. The second column (labeled "Drop Small Gradient") corresponds to the DDC method. As for the other methods, "Drop Big Gradient" tends to drop the edges with more significant gradients; "Drop Small Parameter" favors dropping edges with small weights; "Drop Big Parameter" tends to drop edges with large weights. The horizontal and vertical axes denote the values of the estimated w_1 and w_2 at different epochs, and the contour lines represent the loss. We highlight the initial values of w_1 and w_2 and their values at the 7th, 13th, and 19th epochs by blue pluses, red stars, red triangles, and red diamonds, respectively. As shown, updating the parameters with DDC reaches small losses faster than the compared baselines. We randomly initialize the values of w_1 and w_2 and repeat the process several times. Figure 1 shows two cases with different initializations.

We show the relationship between iteration and the losses of the compared methods in Fig. 2. The results show that DDC gets the lowest loss when giving a fixed iteration count. The methodologies "Drop Small Parameter", "Drop Big Parameter", and "No Dropping" are in the middle, and "Drop Big Gradient" yields the largest loss given a fixed iteration count. We demonstrate the results using two sets of initial values for w_1 and w_2. Other sets of initial values give similar patterns. It appears that the parameter values per se are unlikely to be influential factors in deciding the dropping rate; the gradients are more influential.

4.2 Experiments on Open Datasets

The previous section shows that DDC helps linear models learn faster. This section explores DDC's capacity for more complicated deep learning models using four open datasets: MNIST [9], CIFAR-10, CIFAR-100 [6], and NORB [10].

Table 1. The test accuracies (%) of applying SimpleCNN on MNIST. We repeat each experiment 5 times and report the mean $\pm$ standard deviation. Let the test accuracies of "No Dropping" and a dropping method m be $a \pm b$ and $c \pm d$, respectively, if $a+b < c-d$, we label the method m with symbol $\upuparrows$ (much better). If $a < c$ but $a+b \not< c-d$, we denote m with $\uparrow$ (better). If $a-b > c+d$, we denote $\downdownarrows$ (much worse). Finally, if $a > c$ but $a-b \not> c+d$, we label $\downarrow$ (worse). We also highlight the method with the largest average accuracy with a boldface.

Method	SimpleCNN
No Dropping	99.08 ± 0.04
DDC	$\mathbf{99.25 \pm 0.01}(\upuparrows)$
Dropout	$99.03 \pm 0.08(\downarrow)$
DropConnect	$99.22 \pm 0.03(\upuparrows)$
Standout	$99.01 \pm 0.02(\downdownarrows)$
Drop Small Parameter	$99.24 \pm 0.01(\upuparrows)$
Drop Big Parameter	$99.10 \pm 0.01(\uparrow)$
Drop Big Gradient	$99.12 \pm 0.01(\uparrow)$

In addition to the baselines introduced in Sect. 4.1, we include Dropout, DropConnect, and Standout for comparison. We adjust the dropping rate $r^{(l)}$ in Algorithm 2 to be close to the dropping probability of the compared baselines whenever possible.

We test the convolutional neural network used in the DropConnect paper [13] (called SimpleCNN below). In addition, we add two more complicated networks –AlexNet [7] and VGG [11] – for comparison.

MNIST contains grayscale images; the size of each image is 28×28. The dataset is simple, so we tested only different dropping strategies on SimpleCNN, which includes three convolutional layers and two fully connected layers. The detailed structure is given in [13]; the hyperparameters and detailed settings are included in the experimental code.

The results are presented in Table 1. After repeating each experiment five times, we report the average accuracy $\pm$ standard deviation. As demonstrated, most test accuracies improve when dropping strategies are applied, and the proposed DDC method achieves the best accuracy among all compared methods. Although all methods exhibit accuracies above 99%, DDC is significantly better because it consistently shows very small standard deviations across all repeated trials.

CIFAR-10 is a more challenging dataset: it contains color images; each image's size is 32×32. We use AlexNet [7] and VGG [11] network structures for the experiments because CIFAR-10 is more complex than MNIST.

Table 2 shows the results of applying AlexNet and VGG using the CIFAR-10 dataset. For both the AlexNet and VGG networks, the dropping methods mostly increase test accuracies, and our DDC performs the best among all methods in both networks.

Table 2. The test accuracies (%) of applying AlexNet and VGG on the CIFAR-10 dataset, the symbols (⇈), (↑), (↓), and (⇊) are the same as in Table 1.

Method	Network Structure	
	AlexNet	VGG
No Dropping	81.04 ± 0.04	90.46 ± 0.12
DDC	**84.25 ± 0.09**(⇈)	**90.94 ± 0.11**(⇈)
Dropout	83.86 ± 0.09(⇈)	90.66 ± 0.07(⇈)
DropConnect	83.52 ± 0.17(⇈)	90.68 ± 0.08(⇈)
Standout	83.75 ± 0.04(⇈)	90.64 ± 0.08(↑)
Drop Small Parameter	83.00 ± 0.10(⇈)	89.88 ± 0.01(⇊)
Drop Big Parameter	83.81 ± 0.06(↑)	90.45 ± 0.02(↓)
Drop Big Gradient	83.19 ± 0.16(↑)	90.88 ± 0.03(⇈)

We also test VGG and AlexNet on CIFAR-100 by changing the output softmax layer to 100 categories; all other settings are identical to the previous setup.

Table 3 shows the results of applying the AlexNet and VGG on the CIFAR-100 dataset. The results are similar to those reported earlier: methods with dropping rates usually perform better than without dropping. Our DDC performs best on the VGG network and second on AlexNet.

Finally, we apply the AlexNet and the VGG network to the NORB dataset. The outcomes of these experiments are displayed in Table 4. Our DDC method still consistently outperforms the other techniques on both datasets. It is noteworthy that, in some cases when using the AlexNet, methodologies that involve dropping rates occasionally yield lower performance compared to methods that do not employ any dropping strategy on the NORB dataset. However, the difference is not statistically significant. By examining the mean accuracies plus or minus one standard deviation, we observe that the performance of the poorer performing methods with dropping rates overlaps with the "No Dropping" method.

5 Conclusion

This paper presents a straightforward yet effective methodology for dynamically adjusting the dropping rate of the edges in a neural network. This approach avoids the complexity of introducing additional learning parameters and simplifies the implementation process. We demonstrate the parameter updating process through a series of experiments using a synthetic dataset and multiple open datasets and compare our proposed methodology against Dropout and its various adaptations. The results of these experiments indicate that our method surpasses traditional approaches in nearly all scenarios.

The efficacy of gradient magnitude as a reliable indicator for setting the dropping rate has been validated through our results, suggesting a promising direction for future enhancements. Building on this foundation, we propose to develop a

Table 3. The test accuracies (%) of applying AlexNet and VGG on the CIFAR-100 dataset, the symbols (↑↑), (↑), (↓), and (↓↓) are the same as in Table 1.

	Network Structure	
Method	AlexNet	VGG
No Dropping	62.53 ± 0.50	71.38 ± 0.03
DDC	64.06 ± 0.37(↑↑)	**72.09 ± 0.04**(↑↑)
Dropout	**66.53 ± 0.27**(↑↑)	71.69 ± 0.06(↑↑)
DropConnect	63.72 ± 0.11(↑↑)	71.60 ± 0.21(↑)
Standout	63.97 ± 0.38(↑↑)	71.53 ± 0.31(↑)
Drop Small Parameter	63.10 ± 0.13(↑)	71.71 ± 0.10(↑↑)
Drop Big Parameter	63.02 ± 0.47(↑)	71.82 ± 0.03(↑↑)
Drop Big Gradient	62.92 ± 0.22(↑)	71.13 ± 0.07(↓↓)

Table 4. The test accuracies (%) of applying AlexNet and VGG on the NORB dataset, the symbols (↑↑), (↑), (↓), and (↓↓) are the same as in Table 1.

	Network Structure	
Method	AlexNet	VGG
No Dropping	92.08 ± 0.11	93.36 ± 0.03
DDC	**92.93 ± 0.09**(↑↑)	**94.20 ± 0.16**(↑↑)
Dropout	92.04 ± 0.07(↓)	93.79 ± 0.25(↑↑)
DropConnect	92.09 ± 0.14(↑)	93.94 ± 0.19(↑↑)
Standout	92.08 ± 0.07	93.81 ± 0.08(↑↑)
Drop Small Parameter	92.42 ± 0.10(↑↑)	94.02 ± 0.05(↑↑)
Drop Big Parameter	92.02 ± 0.04(↓)	93.82 ± 0.15(↑↑)
Drop Big Gradient	91.85 ± 0.13(↓)	94.17 ± 0.07(↑↑)

model that algorithmically uses gradients as features to determine optimal dropping rates. We hypothesize that this advanced strategy could potentially elevate the model's predictive accuracy even further. However, it's important to note that this approach would introduce additional parameters that require learning, which could extend the training duration and increase computational demands.

Furthermore, we are interested in delving into a more rigorous theoretical analysis of our methodology. Previous studies, such as those referenced in [2], have explored the connection between Dropout techniques and Bayesian learning. Inspired by this, we are interested in investigating potential theoretical links between our dynamic edge dropping approach and Bayesian inference principles. This exploration could offer deeper insights into the probabilistic foundations of our method and possibly reveal new theoretical reasoning that could explain why our Dynamic DropConnect method enhances model robustness and generalizability more effectively than traditional Dropout techniques.

In summary, our current methodology provides a significant improvement over traditional techniques. Our future efforts will focus on refining this approach by incorporating more sophisticated learning strategies and gaining deeper theoretical insight. Our goal is to further boost the robustness and accuracy of neural network training.

Acknowledgement. We acknowledge partial support from the National Science and Technology Council of Taiwan under grant number 113-2221-E-008-100-MY3. We thank to National Center for High-performance Computing (NCHC) of National Applied Research Laboratories (NARLabs) in Taiwan for providing computational and storage resources.

References

1. Ba, J., Frey, B.: Adaptive dropout for training deep neural networks. Adv. Neural Inform. Process. Syst. **26** (2013)
2. Gal, Y., Ghahramani, Z.: Dropout as a bayesian approximation: representing model uncertainty in deep learning. In: International Conference on Machine Learning, pp. 1050–1059. PMLR (2016)
3. Hinton, G.E., Srivastava, N., Krizhevsky, A., Sutskever, I., Salakhutdinov, R.R.: Improving neural networks by preventing co-adaptation of feature detectors. arXiv preprint arXiv:1207.0580 (2012)
4. Ke, G., et al.: Lightgbm: a highly efficient gradient boosting decision tree. Adv. Neural Inform. Process. Syst. **30** (2017)
5. Kingma, D.P., Salimans, T., Welling, M.: Variational dropout and the local reparameterization trick. Adv. Neural Inform. Process. Syst. **28** (2015)
6. Krizhevsky, A., Hinton, G., et al.: Learning multiple layers of features from tiny images. University of Toronto, ON, Canada, Tech. rep. (2009)
7. Krizhevsky, A., Sutskever, I., Hinton, G.E.: Imagenet classification with deep convolutional neural networks. Commun. ACM **60**(6), 84–90 (2017)
8. Labach, A., Salehinejad, H., Valaee, S.: Survey of dropout methods for deep neural networks. arXiv preprint arXiv:1904.13310 (2019)
9. LeCun, Y., Cortes, C., Burges, C.: Mnist handwritten digit database. ATT Labs, vol. 2 (2010). http://yann.lecun.com/exdb/mnist
10. LeCun, Y., Huang, F.J., Bottou, L.: Learning methods for generic object recognition with invariance to pose and lighting. In: Proceedings of the 2004 IEEE Computer Society Conference on Computer Vision and Pattern Recognition, CVPR 2004, vol. 2, pp. II–104. IEEE (2004)
11. Simonyan, K., Zisserman, A.: Very deep convolutional networks for large-scale image recognition. arXiv preprint arXiv:1409.1556 (2014)
12. Srivastava, N., Hinton, G., Krizhevsky, A., Sutskever, I., Salakhutdinov, R.: Dropout: a simple way to prevent neural networks from overfitting. J. Mach. Learn. Res. **15**(1), 1929–1958 (2014)
13. Wan, L., Zeiler, M., Zhang, S., Le Cun, Y., Fergus, R.: Regularization of neural networks using dropconnect. In: International Conference on Machine Learning, pp. 1058–1066. PMLR (2013)
14. Wang, S., Manning, C.: Fast dropout training. In: international Conference on Machine Learning, pp. 118–126. PMLR (2013)

The Brownian Integral Kernel: A New Kernel for Modeling Integrated Brownian Motions

Béla H. Böhnke[1(✉)], Edouard Fouché[2], and Klemens Böhm[1]

[1] Karlsruhe Institute of Technology (KIT) IPD, Kaiserstraße 12, 76131 Baden-Würtemberg, Karlsruhe, Germany
bela.boehnke@kit.edu

[2] Siemens Deutschland – Work originated while at KIT, Werner-von-Siemens-Straße 1, München, 80333 Bayern, Germany

Abstract. Brownian motion is widely used to model random processes across various domains. However, many practical scenarios only provide aggregated data over time intervals, rather than direct measurements of the underlying process. This poses significant challenges for accurate modeling, as conventional Brownian kernels are not designed to account for the uncertainty introduced by these aggregates. We introduce the *Brownian integral kernel* (BIK), the first analytical kernel specifically developed to model aggregated data from Brownian motion. Through extensive experiments on synthetic and real-world datasets, we demonstrate the BIK's superiority in prediction accuracy, uncertainty estimation, and data synthesis compared to existing Kernels. To support adoption, we provide a Python implementation (git: https://github.com/bela127/Brownian-Integral-Kernel.) compatible with GPy, along with all code and data to reproduce our experiments.

Keywords: Integral Measurements · Learning from Aggregated Data · Integrated Brownian Motion · Gaussian Process Regression · Kernels

1 Introduction

Brownian motion (Fig. 1a) is central to modeling various physical and technological processes, such as: (1) The movement of particles in a fluid [7]. (2) The movement and loosening of machine elements due to vibration [13]. (3) The behavior of financial and other markets [17], population behavior and effects. (4) The load profile of electrical grids where producers and consumers with varying loads are plugged in or out of the grid at any time. These processes often involve stochastic uncertainties, which are effectively modeled using Gaussian processes and Brownian kernels [4], enabling synthesis of process data, regression of real-world data with associated uncertainty, and the combination of both, i.e., synthesis of data from partially conditioned models.

However, in many real-world scenarios, data collection pipelines contain "integrators" that implicitly or explicitly aggregate data over time intervals [19]

X. Wu et al. (Eds.): PAKDD 2025, LNAI 15875, pp. 122–134, 2025.
https://doi.org/10.1007/978-981-96-8295-9_8

(Fig. 1b). Direct observation of the quantity of interest is not possible in such scenarios. Examples include sensors with inherent integration properties (e.g., temperature sensors with heat capacity) or practical constraints (e.g., smart meters providing 15-minute aggregated load data for value privacy). This aggregation obscures the underlying behavior and increases uncertainty, which conventional Brownian kernels cannot model accurately (Fig. 1c). For instance, load forecasting requires precise variance estimation to manage short-term peaks in energy consumption [8,18,28]. Without this, providers risk grid instability or legal penalties due to insufficient capacity [3]. Similar problems arise in other domains like disease monitoring and simulation [34]. As emphasized by [5], the importance of integral kernels for modeling integrating processes is undisputed in many fields.

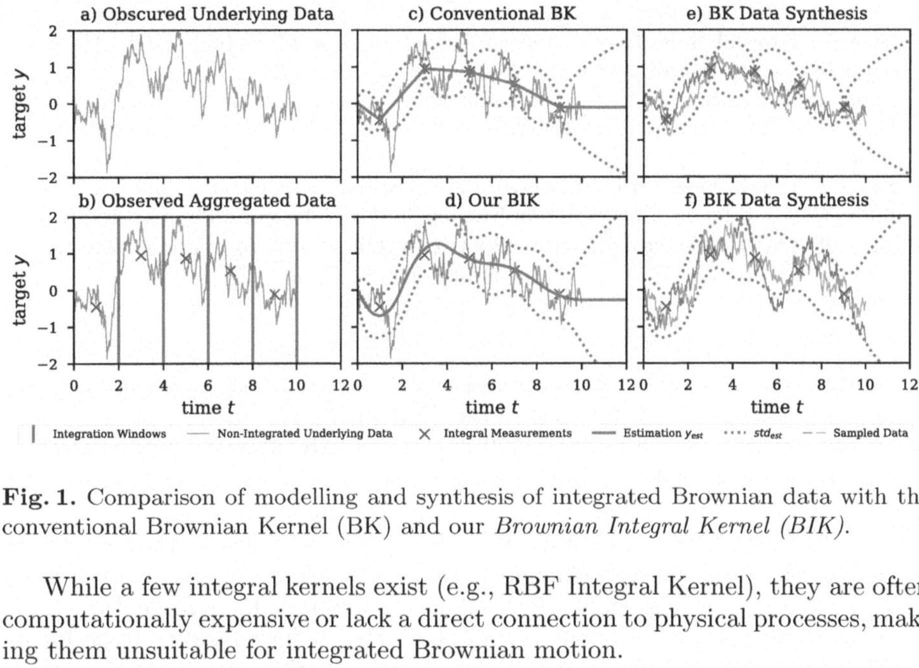

Fig. 1. Comparison of modelling and synthesis of integrated Brownian data with the conventional Brownian Kernel (BK) and our *Brownian Integral Kernel (BIK).*

While a few integral kernels exist (e.g., RBF Integral Kernel), they are often computationally expensive or lack a direct connection to physical processes, making them unsuitable for integrated Brownian motion.

To address these challenges, we derive the *Brownian Integral Kernel (BIK)*, a novel analytical solution for modeling aggregated data from Brownian motions. The BIK accurately estimates variance and supports Gaussian process regression, enabling better predictions and uncertainty quantification (Fig. 1d), as well as data synthesis (Fig. 1f). We validate its performance through extensive experiments on synthetic and real-world datasets. Further, to foster accessibility, we provide a Python implementation of the BIK compatible with the widely used [9] framework for Gaussian Process modeling. For brevity of this paper, we provide proofs, theoretical findings, and additional experiments in our extensive Supplementary material.

2 Related Work

Gaussian processes (GPs) are a standard tool for modeling continuous processes, offering flexible kernels that enable accurate regression, uncertainty quantification, and data generation [4]. Their robustness to overfitting and scalability make them well-suited for both small and large datasets [6]. A key strength of GPs is their use of kernel functions, which allow for modeling complex, non-linear relationships without explicit parametric assumptions. The general workflow for using GPs is outlined in Algorithm 1:

Algorithm 1. Generic Use-Case of a Gaussian Processes

1: **Input:** Training data $\mathbf{t}_{\text{train}}$, $\mathbf{y}_{\text{train}}$, Test data $\mathbf{t}_{\text{test}}$
2: **Output:** Predicted mean $\mu(\mathbf{t}_{\text{test}})$, uncertainty $\sigma(\mathbf{t}_{\text{test}})$, generated data $\mathbf{y}_{\text{gen}}$
3: **Step 1: Kernel Selection.** Choose a kernel function $k(\cdot,\cdot)$ (e.g., RBF, Matérn, our BIK, etc., or combination).
4: **Step 2: Training.** Train the Gaussian process model on $\mathbf{t}_{\text{train}}$, $\mathbf{y}_{\text{train}}$ using the kernel $k(\cdot,\cdot)$. Learn hyperparameters θ of the kernel by maximizing the log marginal likelihood.
5: **Step 3: Prediction and Uncertainty Estimation.** For test points $\mathbf{t}_{\text{test}}$, compute the posterior mean $\mu(\mathbf{t}_{\text{test}})$ and variance $\sigma^2(\mathbf{t}_{\text{test}})$ using the trained model.
6: **Step 4: Data Generation from Posterior Distribution.** Sample from the posterior $\mathcal{N}(\mu(\mathbf{t}_{\text{test}}), \sigma^2(\mathbf{t}_{\text{test}}))$ to generate new data $\mathbf{y}_{\text{gen}}$.

Despite their strengths, conventional kernels face limitations when modeling aggregated data. The RBF Integral Kernel [5,29] lacks physical grounding, while numeric integration-based methods [20,33] and MCMC approximations [34] are computationally expensive. Spectral approximations [10,30] succeed for smooth kernels but fail for discontinuous ones like the Brownian kernel [22].

The Brownian kernel, in contrast, is closely tied to physical processes such as load profiles and market behavior. However, it assumes direct observations, limiting its utility for aggregated data. Efforts to adapt Kalman filters [12,15,24, 25,31,32,35,36] and integrate numeric methods [16,20] have resulted in problem-specific or approximate solutions. For more details, see the extended related work in Supp.Mat. 2

Some approaches are out of scope for this work. Time-discrete methods, such as Kalman filters, lack the continuous modeling required for integrated data, while extensions for time-continuous modeling [15,32] require problem-specific transition functions. Methods like Support Vector Machines (SVMs) do not support uncertainty estimation, and neural networks lack efficient mechanisms for data generation and uncertainty quantification. These limitations make them unsuitable as baselines for this study.

This paper introduces the *Brownian Integral Kernel (BIK)*, an analytical solution for modeling aggregated data. The BIK directly relates to common physical processes and provides exact covariance estimation for integrated Brownian motion, enabling efficient computation within standard GP frameworks while maintaining a direct connection to physical processes.

3 Problem Statement

For clarity and completeness, the detailed notation and fundamental definitions are provided in Supp. Mat. 1. Although the notation is designed to be intuitive and should become clear from the context, readers are encouraged to refer to Supp. Mat. 1 as needed for additional details.

We assume a data-generating process $B(t)$ that behaves like a Brownian motion. Some real processes that behave like this do not allow observation of their actual value $b(t)$ (realization) at time t. One can only observe average or integral measurements of $B(t)$, while the true value $b(t)$ remains unknown.

Definition 1. *An integral measurement $\mathcal{B}(s,e)$ is the integration over time t of measurements from $b(t)$ from start time s to end time e:*

$$\mathcal{B}(s,e) = \int_s^e b(t)\delta t$$

Even if $b(t)$ follows a Brownian motion which is per definition erratic, integral measurements $\mathcal{B}(s,e)$ behave differently, this is because integration smoothes the values leading to a smoother function. In consequence the measured value $\mathcal{B}(s,e)$ is most certainly not the true value of the underlying process $b(t)$. However, a Brownian motion model, e.g., a Gaussian process with Brownian kernel $k_{ff'}(t,t') = v_b \cdot min(t,t')$ assumes that the observed data points are the true values. Fitting such a model on integral measurement $\mathcal{B}(s,e)$ gives the wrong predictive variance of zero at measurement locations. To obtain the correct covariance, i.e., to correctly model the additional variance due to integration, a new kernel is necessary.

4 The Brownian Integral Kernel

We propose to model the additional variance (described in the previous section) directly in a new kernel to solve the problem of mis-estimating the variance. Thus, this calls for a kernel that yields the covariance between two integrated time intervals of a Brownian motion. Following [5] we can derive this kernel by integrating the Brownian kernel:

Definition 2. *The Brownian integral kernel (BIK) is the two-time integration over time intervals (s,e) and (s',e') of the Brownian kernel:*

$$k_{\mathcal{F}\mathcal{F}'}((s,e),(s',e')) = v_b \cdot \int_{s'}^{e'} \int_s^e min(t,t')\delta t \delta t'.$$

This kernel generalizes the Brownian kernel to integral measurements by modeling the higher variance due to integration, which the Brownian kernel neglects. For intuition and proof of how integrating a kernel is equivalent to integrating the underlying process, please refer to Supp.Mat. 3. Further, a kernel needs to fulfill some properties, namely, it needs to be positive-semidefinite [26]. We provide proof for this property in Supp.Mat. 4.

Theorem 1. *The Brownian integral kernel admits the following solution:*

$$
k_{\mathcal{F}\mathcal{F}'}((s,e),(s',e')) = \\
v_b \cdot \begin{cases}
\textit{if } s \le s' < e' \le e: \\
\quad \mathcal{F}_{tt}((e',t),(s',e')) + \mathcal{F}_{tt'}(s',e') + \mathcal{F}_{t't'}((s,s'),(s',e')) \\
\textit{if } s' < e' \le s < e: \\
\quad \mathcal{F}_{tt}((s,e),(s',e')) \\
\textit{if } s' \le s < e' \le e: \\
\quad \mathcal{F}_{tt}((s,e),(s',s)) + \mathcal{F}_{tt}((e',e),(s,e')) + \mathcal{F}_{tt'}(s,t') \\
\textit{if } s \le s' < e \le e': \\
\quad \mathcal{F}_{t't'}((s,s'),(s',e')) + \mathcal{F}_{t't'}((s',e),(e,e')) + \mathcal{F}_{tt'}(s',e) \\
\textit{if } s < e \le s' < e': \\
\quad \mathcal{F}_{t't'}((s,e),(s',e')) \\
\textit{if } s' \le s < e \le e': \\
\quad \mathcal{F}_{tt}((s,e),(s',s)) + \mathcal{F}_{tt'}(s,e) + \mathcal{F}_{t't'}((s,e),(e,e'))
\end{cases},
$$

with $\mathcal{F}_{tt}, \mathcal{F}_{t't'}, \mathcal{F}_{tt'}$ being sub-parts of the primitive function:

$$\mathcal{F}_{tt}((l',u'),(l,u)) = \frac{1}{2}u^2 \cdot u' - \frac{1}{2}u^2 \cdot l' - \frac{1}{2}l^2 \cdot u' + \frac{1}{2}l^2 \cdot l', \tag{1}$$

$$\mathcal{F}_{t't'}((l',u'),(l,u)) = \frac{1}{2}u'^2 \cdot u - \frac{1}{2}l'^2 \cdot u - \frac{1}{2}u'^2 \cdot l + \frac{1}{2}l'^2 \cdot l, \textit{ and} \tag{2}$$

$$\mathcal{F}_{tt'}(l,u) = \frac{1}{3}(u-l)^3 + (u-l)^2 \cdot l. \tag{3}$$

We provide a detailed proof of this derivation in Supp.Mat. 8.

The resulting kernel $k_{\mathcal{F}\mathcal{F}'}$ calculates the covariance between training data intervals, i.e., between two integrated time intervals. However, during inference, most users are interested in the predictive variance $k_{ff'}^{post}$.

Definition 3. *The predictive variance $k_{ff'}^{post}$ of a Gaussian process is a variance within the original space of the non-integrated Brownian motion. For a given point t', $k_{ff'}^{post}$ quantifies how likely a prediction $y_{t'}$ is equal to the unobserved ground truth $\hat{y}_{t'}$.*

The predictive variance $k^{post}_{ff'}$ is often used to quantify the uncertainty associated with predictions $y_{t'}$ [27]. To calculate it, we need to calculate the *partly integrated* covariance $k_{\mathcal{F}f'}$.

Definition 4. *The* partly integrated *covariance $k_{\mathcal{F}f'}$ is the covariance between an inference point t' (point in the original space of non-integrated Brownian motion) and an integration time interval (s, e). It is calculated as a one-time integration of $k_{ff'}$:*

$$k_{\mathcal{F}f'}((s,e),t') = v_b \cdot \int_s^e min(t,t')\delta t.$$

Theorem 2. *The* partly integrated *kernel admits the following solution:*

$$k_{\mathcal{F}f'}((s,e),t') = v_b \cdot \begin{cases} if\, t' \le s < e: \\ \quad \mathcal{F}_t(t',s,e) \\ if\, s < t' < e: \\ \quad \mathcal{F}_{t'}(s,t') + \mathcal{F}_t(t',t',e) \\ if\, s < e \le t': \\ \quad \mathcal{F}_{t'}(s,e) \end{cases}$$

with $\mathcal{F}_t, \mathcal{F}_{t'}$ as follows:

$$\mathcal{F}_t(t,l',u') = t \cdot u' - t \cdot l', \tag{4}$$

$$\mathcal{F}_{t'}(l',u') = \frac{1}{2}u'^2 - \frac{1}{2}l'^2. \tag{5}$$

We obtain this partially integrated covariance by one-time integration of $k_{ff'}(t,t')$. The proof follows as a direct result from the proof of Theorem 1, see Supp.Mat. 8. Here, Eqs. 4 and 5 are an intermediate result (compare with Eqs. 1 and 2) from Supp.Mat. 8.

We can now calculate the predictive variance $k^{post}_{ff'}$ of the Gaussian process, according to [23] with the standard formula for Gaussian processes. Here we use matrix notation, where K_{ff}, K_{Ff}, K_{FF} is the matrix obtained by using the appropriate kernel $k_{ff'}(*,*), k_{\mathcal{F}f'}(*,*), k_{\mathcal{F}\mathcal{F}'}(*,*)$ and K^{post}_{ff} is the result matrix that corresponds to $k^{post}_{ff'}$:

Corollary 1.

$$K^{post}_{ff} = K_{ff} - K_{fF}K^{-1}_{FF}K_{Ff}.$$

5 Experimental Design

We derived the Brownian Integral Kernel (BIK) and highlighted its advantages over existing kernels.

To validate our findings, we compare the BIK in various scenarios against the Brownian Kernel (BK), the RBF Integral Kernel (RBFIK) from [5], and the Brownian Kernel with added white noise (BNK) [2,14]. The comparisons assess prediction quality, variance estimation, and the plausibility of data generated by a GP conditioned on integral measurements.

5.1 Used Data

For our evaluation, we use data from multiple sources: **Synthetic data:** *Synth* from a Gaussian process with a Brownian Kernel and subsequent numeric integration. **Simulated data:** *Load* generated with a realistic and widely used data generator [21]. **Real-world data:** *Real* provided by a private household, *HIPE* using the High-resolution Industrial Production Energy (HIPE) data set [1], and *Stock*, using daily stock market prices from [11]. More details on the used data can be found in our Supp. Mat. 5. We provide all used datasets which are otherwise not easily accessible, together with the experiment code within our GitHub repository to facilitate reproducibility.

It is important to note that, contrary to the intuition of the name *load profile*, load profiles can have both positive and negative values due to generating nodes such as photovoltaic systems. Especially for short-term modeling of such load profiles, they behave Brownean-like because of randomly moving cloud cover [37]. For long-term modeling, one almost always has additional periodicity (day and night). However, the periods are usually much longer than 15 min, which is why they are not relevant for modeling uncertainty due to integration. Our kernel can be used for long-term modeling with standard kernel composition by just adding a periodic kernel. Yet such an extension is not necessary for evaluating the short-term uncertainty effects we are targeting.

5.2 Metrics

We evaluate multiple application-relevant aspects of the kernels. One is often interested in the prediction error. However, in our scenario (where one can only observe integrals of the ground truth but not the ground truth itself) predictions will always be close to the mean within the observed integration interval. Therefore, standard metrics such as Mean Square Error (MSE) are not meaningful when comparing the kernels. Instead, this scenario requires an evaluation that combines prediction and prediction uncertainty. For this, we use the Weighted Mean Absolute Error (WMAE):

Definition 5. *The WMAE quantifies the estimation quality considering the estimated likelihood $l_{GP}(t')$ of an estimation $y_{t'}$ vs the ground truth $\hat{y}_{t'}$:*

$$WMAE = \frac{1}{|\mathbf{T}|} \sum_{t' \in \mathbf{T}} l_{GP}(t') \cdot |y_{t'} - \hat{y}_{t'}|$$

with $\mathbf{T}$ *being the set of test points.*

The WMAE is an intuitive measure: The estimated likelihood $l_{GP}(t')$ quantifies for a prediction $y_{t'}$ how likely this prediction is. If the model is confident ($l_{GP}(t') \approx 1$) and the prediction is accurate, the WMAE is low. WMAE penalizes prediction errors more when the model is (wrongly) confident, and less when it is (correctly) uncertain. Note that we are interested in the likelihood per data point, thus it is not required to normalize the likelihood across all data points.

While the WMAE is relevant for the prediction accuracy, we can also directly evaluate the estimated uncertainty. For this, we calculate the variance of the ground truth within an integration interval and compare it with the estimated variance, which should be similar. We use the Mean Square Error MSE_{var} between true and estimated variance. Please note, that this metric is only meaningful in combination with the prediction error.

Finally, we evaluate whether the generated data (which is partially conditioned on observed integral measurements) matches the process assumptions: We know that the integral of the ground truth results in the observed integral measurement. Therefore, the integral of generated data should also result in the observed integral. Here, we calculate the Mean Absolute Error MAE_{int} between the integral measurement and the integral value of generated data. We calculate the integral value of generated data by sampling data at the same points in time used to calculate the ground truth integrals and then apply the same windowing procedure, i.e., summing all the resultant values within the integration time interval.

5.3 Experiment Procedure and Model

We evaluate each kernel using the same Gaussian process model. As a model, we use the standard GPy implementation [9]. We also train all configurations in the same way, using the GPy gradient-based maximum likelihood optimizer in its standard configuration [9]. In this configuration, the optimizer performs 5 independent optimization runs, each time with random start kernel parameters. The model then uses the parameters from the best run.

For each dataset, we repeat this procedure 20 times, each time with a different ground truth time series. We then calculate the mean and the variance across the runs of the respective metrics.

6 Evaluation

For brevity, we focus here on the quantitative experiments. However, we feature additional qualitative kernel comparisons and visualizations in Supp.Mat. 6. These are especially useful for domain experts who like to attain an intuitive visual understanding of the superiority of the BIK against other kernels. Further, we provide an ablation study of process and kernel parameters in Supp.Mat. 7 , and a theoretical runtime analysis in Supp.Mat. 3.

A comparison of the different approaches, in Table 1, shows that BIK is superior to its competitors in every regard evaluated: The integral of generated data

Table 1. Evaluation metrics for different kernels and data sets. Bold entries mark the best result within one metric

		Metrics		
Data	Kernel	MSE_{var}	MAE_{int}	WMAE
Synth	BIK	**5.82 ± 12.**	**7.6 ± .45**	**3.79 ± 2.9**
w=25	RBFIK	24.9 ± 25.	—	44.2 ± 60.
	BK	15.6 ± 20.	72.8 ± 4.6	6.9 ± 133
	BNK	15.6 ± 20.	73.0 ± 6.0	6.9 ± 133
Synth	BIK	**5.87 ± 14.**	**3.67 ± .26**	**2.33 ± 2.0**
w=50	RBFIK	24.6 ± 27.	—	13.5 ± 19.
	BK	20.8 ± 25.	71.1 ± 5.8	3.38 ± 42.
	BNK	20.8 ± 25.	72.4 ± 4.3	3.38 ± 42.
Synth	BIK	**4.7 ± 13.**	**2.0 ± .14**	**4.35 ± 3.3**
w=100	RBFIK	20.6 ± 25.	—	44.8 ± 62.
	BK	17.8 ± 23.	68.5 ± 5.2	5.95 ± 89.
	BNK	17.8 ± 23.	70.0 ± 4.7	5.95 ± 89.
Synth	BIK	**5.41 ± 9.2**	**1.12 ± .09**	**4.27 ± 3.5**
w=200	RBFIK	22.4 ± 19.	—	50.9 ± 71.
	BK	20.6 ± 18.	63.2 ± 5.4	5.50 ± 51.
	BNK	20.6 ± 18.	62.9 ± 4.2	5.50 ± 51.
Load	BIK	**.50 ± 1.2**	**.39 ± .006**	**.35 ± .74**
	RBFIK	.64 ± 1.6	—	3.5 ± 7.4
	BK	.66 ± 1.6	3.3 ± .07	12. ± 26.
	BNK	.56 ± 1.4	3.7 ± .08	.80 ± 1.7
HIPE	BIK	**12.5 ± 26.**	**.69 ± .031**	**0.42 ± 0.3**
	RBFIK	43.1 ± 43.	—	8.02 ± 6.5
	BK	45.5 ± 44.	5.84 ± .21	518. ± 429
	BNK	45.5 ± 44.	5.69 ± .19	476. ± 395
Stock	BIK	**1.13 ± 2.9**	**14.8 ± 1.1**	**0.32 ± 0.3**
	RBFIK	3.65 ± 5.1	—	4.48 ± 3.9
	BK	3.98 ± 5.3	63.1 ± 3.2	46k ± 38k
	BNK	3.98 ± 5.3	60.9 ± 2.6	45k ± 38k

has at least 10 times lower MAE_{int} compared to all baselines. (The RBFIK from [5] cannot generate data.) For our BIK, we provide a more detailed analysis of MAE_{int} in Supp.Mat. 7.

Regarding the prediction with uncertainty, BIK has at least 2 times better WMAE on *Load* and *Synth* $w = 25$ than with all competitors. On *Synth* $w = 50, 100, 200$ the baselines catch up a bit, but BIK still beats them by 30% less error. Also, the WMAE variance suggests that the baselines are very unstable across data sets.

The MSE_{var} between predicted variance and actual variance during measurement on *Synth*, *HIPE*, *Stock* is 2.6 times better for BIK than any baseline. On *Load* BIK is still 10% better.

Note that the results of BK and BNK are similar on synthetic data. This similarity arises because BNK learns a noise of zero. Indeed, BNK only gives good uncertainty estimates when trained on several different observations of the same data point (or with a smaller time step). With integral measurements, such data cannot be observed, leading to BNK learning incorrect small variances.

One may also notice that the variance in predictive variance (measured with MSE_{var}) is high across the experimental runs for all kernels. This is to be expected, since by chance the ground truth is sometimes simply close to the average within the integration interval, leading to a high deviation. Even with this high deviation, MSE_{var} is still meaningful for comparing the quality of the uncertainty quantification as long as the prediction error is similar. Even though BNK estimates a similar uncertainty as BIK, the metrics MAE_{int} and WMAE show that the uncertainty estimation of BNK is useless because of its much higher prediction error.

7 Conclusions

Integrated Brownian motions are crucial in many physical processes and data collection techniques. The conventional Brownian kernel, while effective for modeling Brownian motion, falls short in capturing the uncertainties associated with integrated data.

This study has tackled modeling integrated Brownian motions precisely, by providing an analytical solution to the novel Brownian integral kernel (BIK). The BIK enables precise estimation of variance associated with the underlying quantity of interest. Further, the BIK is a valuable tool for tasks like regression with uncertainty estimation and for data synthesis partially conditioned on measurements, as shown in our experiments.

Our contributions bridge a significant gap in modeling integrated processes. Our Brownian integral kernel enhances the accuracy and reliability of Gaussian process modeling in such uncertain and dynamic environments by a factor of at least 2 on every dataset and against every baseline. Data synthesis with our integral kernel is better by a factor of at least 10 compared to all baselines.

While our Brownian integral kernel is a substantial advancement, there are avenues for further exploration: First, the challenge of concept drift, i.e., changes of behavior of the underlying ground truth stream, and how to handle it, is a relevant research direction. Additionally, investigating the properties of other integrated processes besides the Brownian integrated process could result in additional new kernels useful for modeling such processes. Finally, using our kernel in applied research could lead to advances in several directions, such as estimating privacy violations and disaggregating load data from smart meters.

Supplementary Material. We include detailed notation, additional experiments, and proofs in our Supp. Mat. published on Git Hub[1]

[1] git: https://github.com/bela127/Brownian-Integral-Kernel.

Acknowledgments. This work was supported by the German Research Foundation (DFG) as part of the Research Training Group GRK 2153: Energy Status Data - Informatics Methods for its Collection, Analysis, and Exploitation and by the Baden-Württemberg Foundation via the Elite Program for Postdoctoral Researchers.

Disclosure of Interests. The authors have no competing interests to declare that are relevant to the content of this article.

References

1. Bischof, S., Trittenbach, H., Vollmer, M., Werle, D., Blank, T., Böhm, K.: HIPE: an energy-status-data set from industrial production. In: e-Energy, pp. 599-603. ACM (2018)
2. Boyle, P.: Gaussian Processes for Regression and Optimisation. Thesis at WGTN (2007)
3. Cutsem, T., Vournas, C.: Voltage Stability of Electric Power Systems. Springer US (1998)
4. Eric, S., Maarten, S., Andreas, K.: A tutorial on gaussian process regression: modelling, exploring, and exploiting functions. J. Math., Psychol. (2018)
5. Yousefi, F., Smith, M.T., Álvarez, M.: Multi-task learning for aggregated data using gaussian processes. In: Advances in Neural Information Processing Systems. Curran Associates, Inc. (2019)
6. Ferris, B., Hähnel, D., Fox, D.: Gaussian processes for signal strength-based location estimation. In: Robotics: Science and Systems (2006)
7. Friz, P.K., Gassiat, P., Lyons, T.: Physical brownian motion in a magnetic field as a rough path. AMS **367**, 7939–7955 (2015)
8. Gilanifar, M., Wang, H., Sriram, L., Ozguven, E.E., Arghandeh, R.: Multitask Bayesian spatiotemporal gaussian processes for short-term load forecasting. IEEE Trans. Ind., Elect. (2020)
9. GPy: GPy: a gaussian process framework in Python. http://github.com/SheffieldML/GPy (since 2012)
10. Jidling, C., Wahlström, N., Wills, A., Schön, T.B.: Linearly constrained gaussian processes. In: NIPS, vol. 30. Curran Associates, Inc. (2017)
11. Kumar, S.: 2019-2024 us stock market data (2024). https://doi.org/10.34740/KAGGLE/DSV/7553516
12. Kurtz, V., Lin, H.: Kalman filtering with gaussian processes measurement noise (2019)
13. Lanoiselée, Y., Briand, G., Dauchot, O.: Statistical analysis of random trajectories of vibrated disks: towards a macroscopic realization of brownian motion. Phys. Rev. **98** (2018)
14. Li, Z., Guo, F., Chen, L., Hao, K., Huang, B.: Hybrid kernel approach to gaussian process modeling with colored noises. Comput. Chem. Eng. **143**, 107067 (2020)
15. Liu, F., Gao, Z., Yang, C., Ma, R.: Extended Kalman filters for continuous-time nonlinear fractional-order systems involving correlated and uncorrelated process and measurement noises. Int. J. Control Autom. Syst. **18**, 2229–2241 (2020)
16. Longi, K., et al.: Sensor placement for spatial gaussian processes with integral observations. In: UAI, vol. 124, pp. 1009–1018. PMLR (2020)
17. Meng, X., Zhang, J., Guo, H.: Quantum brownian motion model for the stock market. Phys. A Stat. Mech. Appl. **452**, 281–288 (2016)

18. Mori, H., Ohmi, M.: Probabilistic short-term load forecasting with gaussian processes. In: Proceedings of the 13th International Conference on, Intelligent Systems Application to Power Systems. IEEE (2005)
19. Musicant, D.R., Christensen, J.M., Olson, J.F.: Supervised learning by training on aggregate outputs. In: ICDM, pp. 252–261. IEEE (2007)
20. O'Callaghan, S., Ramos, F.: Continuous occupancy mapping with integral kernels. AAAI **25**, 1494–1500 (2011)
21. Pflugradt, N., Stenzel, P., Kotzur, L., Stolten, D.: LoadProfileGenerator: an agent-based behavior simulation for generating residential load profiles. J. Open Source Softw. **7**, 3574 (2022)
22. Purisha, Z., Jidling, C., Wahlström, N., Schön, T.B., Särkkä, S.: Probabilistic approach to limited-data computed tomography reconstruction. Inverse Prob. **35**, 105004 (2019)
23. Rasmussen, C.E., Williams, C.: Gaussian Processes for Machine Learning. The MIT Press, Adaptive computation and machine learning (2006)
24. Reece, S., Roberts, S.: An introduction to gaussian processes for the Kalman filter expert. In: 2010 13th International Conference on Information Fusion, pp. 1–9 (2010)
25. Sarkka, S., Hartikainen, J.: Infinite-dimensional Kalman filtering approach to spatio-temporal gaussian process regression. In: Lawrence, N.D., Girolami, M. (eds.) Proceedings of the Fifteenth International Conference on Artificial Intelligence and Statistics. Proceedings of Machine Learning Research, vol. 22, pp. 993–1001. PMLR (2012)
26. Schölkopf, B., Smola, A.J.: Learning with Kernels: support vector machines, optimization, and beyond. MIT, Regularization (2002)
27. Settles, B.: Active Learning. Springer Cham (2012)
28. Shepero, M., der Meer, D.V., Munkhammar, J., Widén, J.: Residential probabilistic load forecasting: a method using gaussian process designed for electric load data. Appl. Energy (2018)
29. Smith, M.T., Alvarez, M.A., Lawrence, N.D.: Gaussian process regression for binned data (2019)
30. Solin, A., Särkkä, S.: Hilbert space methods for reduced-rank gaussian process regression. Stat. Comput. **30**, 419–446 (2019)
31. Sun, L., Alkhatib, H., Kargoll, B., Kreinovich, V., Neumann, I.: Ellipsoidal and gaussian Kalman filter model for discrete-time nonlinear systems. Mathematics **7**(12) (2019)
32. Taghvaei, A., de Wiljes, J., Mehta, P.G., Reich, S.: Kalman filter and its modern extensions for the continuous-time nonlinear filtering problem. J. Dyn. Syst., Measur. Control **140**(3), 030904 (2017)
33. Tanaka, Y., et al.: Spatially aggregated gaussian processes with multivariate areal outputs. In: NEURIPS, vol. 32. Curran Associates, Inc. (2019)
34. Tanskanen, V., Longi, K., Klami, A.: Non-linearities in gaussian processes with integral observations. In: MLSP (2020)
35. Todescato, M., Carron, A., Carli, R., Pillonetto, G., Schenato, L.: Efficient spatio-temporal gaussian regression via Kalman filtering. Automatica **118**, 109032 (2020)

36. Xie, J., Dubljevic, S.: Discrete-time Kalman filter design for linear infinite-dimensional systems. Processes **7**(7) (2019)
37. Xu, J., Yang, C., Zhao, J., Wu, L.: Fast modeling of realistic clouds. In: CNMT (2009)

Fed-ARIMA-OPARBFN: An Ensemble Model for Cross-Domain Crop Yield Time Series Prediction Based on Federated Learning

Shiqi Xu, Shuqin Li(✉), and Tianyi Song

College of Information Engineering, Northwest A&F University, Shaanxi, China
{xusq,songtianyi}@nwafu.edu.cn, lsq_cie@nwsuaf.edu.cn

Abstract. Accurate crop yield prediction has become increasingly critical for global food security in the face of climate change and population growth. Traditional prediction methods are constrained by their dependence on costly equipment and multisource data collection, while data silos and the complexity of agricultural data characterized by long prediction cycles and extensive temporal spans present significant obstacles to effective forecasting. Our key innovations address these challenges through an integrated solution. First, we propose Fed-ARIMA-OPARBFN, a novel lightweight model that combines horizontal federated learning with historical yield data analysis, effectively breaking down data silos while reducing deployment costs. Most significantly, we introduce a hybrid time series modeling mechanism that integrates ARIMA analysis with an optimized OPARBFN neural network, enhancing the capture of complex temporal dependencies across extended prediction horizons. Validation using multiregional crop data from Nepal demonstrates that Fed-ARIMA-OPARBFN achieves a coefficient of determination (R2) 6% to 29% higher than centralized learning models while enabling cross-regional data collaboration. Compared with existing multisource approaches, our model attains comparable accuracy with enhanced generalization using only historical yield data, offering developing countries a cost-effective solution for reliable agricultural decision support.

Keywords: Federated Learning · Agriculture · Time Series Data · Data Silos · Yield Prediction

1 Introduction

Crop yield prediction plays a strategic role in agricultural production. Accurate predictions help farmers optimize planning and resource allocation while supporting agricultural enterprises' decision-making, ultimately contributing to national food security. Despite advances in big data and deep learning applications [1], challenges remain in achieving accurate predictions while addressing privacy concerns and resource constraints in cross-organizational settings.

Recent studies have demonstrated substantial advancements in crop yield estimation techniques, particularly in machine learning and deep learning. Prasad et al. utilized

X. Wu et al. (Eds.): PAKDD 2025, LNAI 15875, pp. 135–146, 2025.
https://doi.org/10.1007/978-981-96-8295-9_9

remote sensing data of land surface temperature and soil moisture [11], while Marshall et al. employed hyperspectral satellite data for global crop predictions [12]. Mateo-Sanchis et al. integrated optical and microwave sensing [13], Alibabaei et al. developed a climate-based model [2], and Sakamoto implemented random forest with multi-source data [3]. However, these methods are generally characterized by the following issues: (1) The reliance on expensive sensing equipment, remote sensing systems, and satellite technology presents a significant financial challenge for SMEs, particularly in terms of the economic and technical costs involved; (2) the limited availability of agricultural data, coupled with the difficulty in obtaining it, further exacerbates the challenges faced by these producers; (3) The spatiotemporal limitations of the data restrict the generalization ability of deep learning models and make it challenging to overcome the limitations of data silos, as this is the core intellectual property of the relevant entities [4].

Time series-based prediction techniques have gained prominence as they utilize historical yield data accumulated by farmers, addressing equipment and data acquisition challenges [7]. Several deep learning approaches have been developed: Bali et al. proposed an RNN-LSTM model for wheat yield prediction [26], Wang et al. designed a CNN-GRU structure for winter wheat evaluation [27], and Sun et al. developed a CNN-LSTM model for soybean yield prediction [28]. However, these deep learning methods still have the following limitations: (1) the model structure is complex and requires many training samples to obtain optimal results; (2) the lack of interpretability of the prediction results makes providing clear guidance for agricultural production decisions challenging; and (3) the training process depends on large-scale computing resources, which increases implementation costs.

To address these issues, the ARIMA-ANN hybrid model presents a novel solution for time series prediction. The model combines ARIMA for linear data characteristics and neural networks for nonlinear components, achieving an optimal balance between complexity and accuracy. Compared with pure deep learning methods, it offers three key advantages: (1) ARIMA preprocessing effectively alleviates neural network overfitting [14]; (2) it has a clear statistical theoretical basis, providing better interpretability [15]; and (3) the model structure is simple and intuitive, facilitating implementation. The model's effectiveness has been demonstrated across various domains, including wind speed prediction [16], solar energy forecasting [17], and handling seasonal and long-term trends [18].

Data privacy and cross-organizational collaboration challenges impede predictive model development. Federated learning, first proposed by McMahan et al. at Google Research [5], addresses this by transmitting only model parameters rather than raw data [6]. Among its three categories [9], horizontal federated learning is particularly suitable for agricultural production prediction, as it handles scenarios with overlapping data features but different sample IDs [8, 10].

Conventional ARIMA-ANN models face challenges with agricultural time series data characterized by long prediction cycles and complex external factors, including climate change and seasonal fluctuations. While RBFN offers advantages in resource-constrained scenarios through its nonlinear approximation ability and fast training, its standard Gaussian kernel function inadequately addresses the multidimensional characteristics of agricultural data.

Based on the above analysis, this paper proposes a Fed-ARIMA-OPARBFN model, integrating an enhanced ARIMA-RBFN with horizontal federated learning. The main innovations are: (1) An offset power adaptive RBFN (OPARBFN) that introduces learnable offset and adaptive power parameters, optimizing the traditional radial basis function for agricultural data characteristics. (2) To the best of our knowledge, a novel Fed-ARIMA-OPARBFN model, the first to combine federated learning with enhanced ARIMA-RBFN for agricultural time series prediction, enabling privacy-preserving cross-organizational collaboration while maintaining model interpretability and statistical foundations.

2 Method

This study proposes a hybrid prediction methodology founded upon a federated learning framework. Figure 1 illustrates the complete algorithmic process and model framework. First, an autoregressive integrated moving average (ARIMA) model is employed for local historical yield time series modeling. ARIMA captures time-series correlation through autoregressive (AR) terms, handles nonstationarity via differencing (I), and describes random fluctuations using moving average (MA) terms. Given ARIMA's linear limitations and the data's nonlinear characteristics, we developed paired sequences of ARIMA predictions and actual values to train an offset power adaptive radial basis function network (OPARBFN) for capturing nonlinear relationships.

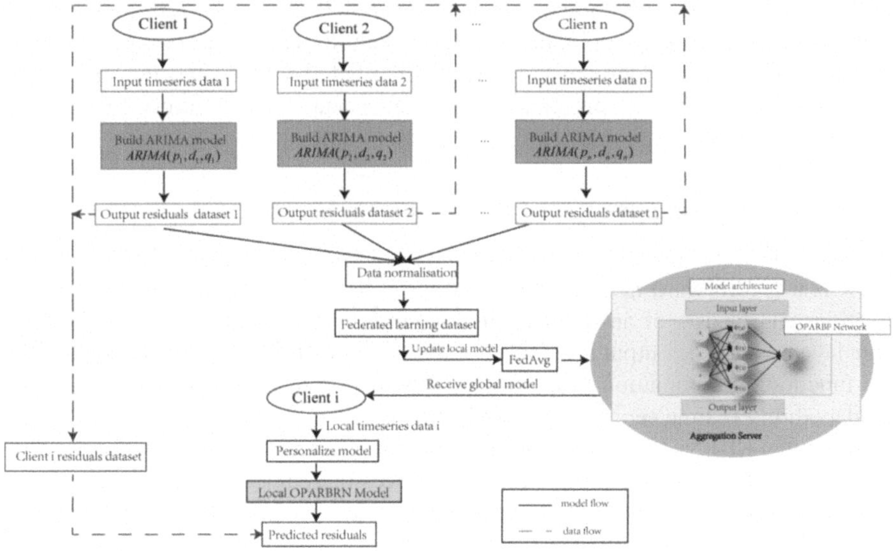

Fig. 1. The flowchart shows the proposed Fed-ARIMA-OPARBFN framework.

Through its unique network configuration, OPARBFN enables adaptive adjustments to the radial basis function. In the federated framework, clients train OPARBFN using local paired data of ARIMA predictions and actual values to learn nonlinear correction mappings. The trained parameters (weights, offsets, and power parameters) are uploaded to the server, which applies federated averaging (FedAvg) to generate global parameters for redistribution. This iterative process enhances model performance while maintaining data privacy through local storage.

In the prediction stage, ARIMA generates initial predictions, which the federally trained OPARBFN model then corrects. This hybrid approach enables OPARBFN to capture nonlinear patterns in residuals while maintaining prediction dimensionality and scale information. Experimental results confirm that this integrated linear-nonlinear method improves prediction accuracy while protecting data privacy through federated learning, making it suitable for privacy-sensitive distributed predictions.

2.1 Autoregressive Integrated Moving Average Model

The autoregressive integrated moving average (ARIMA) model, also known as the Box–Jenkins model [19], is widely used in time series forecasting. The nonseasonal ARIMA model is expressed as follows: where p, d, and q are nonnegative integers representing the autoregressive order, differencing number, and moving average order, respectively.

In the context of a time series $\{y_t\}$, where t represents an integer time point and where y_t is a real number, the model is defined as (1):

$$\phi(B)(1 - B)^d y_t = \theta(B)\varepsilon_t \tag{1}$$

In this model, the polynomial $\phi(B) = 1 - \phi_1 B - \phi_2 B^2 - ... - \phi_p B^p$ represents the autoregressive part, the polynomial $\theta(B) = 1 - \theta_1 B - \theta_2 B^2 - ... - \theta_p B^p$ represents the moving average part, and the random error B represents the error at the t time point, which is assumed to be independently and identically distributed with mean 0 and variance σ^2. The polynomial $(1 - B)^d y_t$ represents the time series data after d differencing steps. The model parameters ϕ_i and θ_j must be estimated by training data.

2.2 Offset Power Adaptive Radial Basis Function Network

Blumhead and Lore introduced the Radial Basis Function Network (RBFN) in 1988 [20]. It is a feedforward neural network using radial basis functions as hidden layer activations. It consists of an input layer, a nonlinear RBF hidden layer, and a linear output layer, where the output is a linear combination of RBF-transformed weights. Due to its function approximation properties, RBFN demonstrates superior learning speed and adaptability in time series prediction [29].

Traditional RBFNs face limitations in agricultural time series prediction due to long cycles and extensive periods. This paper introduces the Offset Power Adaptive Radial Basis Function (OPARBFN) to improve model adaptability and better capture nonlinear characteristics of ARIMA residuals. Integration with federated learning enhances the model's prediction capability for complex time series data.

This novel function (2) is derived through enhancements to the conventional radial basis function.

$$\phi(x) = \exp(-\gamma((\sum_{i=1}^{n}(x_i - c_i)^2 + b)^p)) \tag{2}$$

In addition to the scaling parameters typically included in radial basis functions, the OPARBFN incorporates additional trainable parameters, including the offset parameter b and the power parameter p. These enhancements significantly enhance the flexibility of the function. Incorporating the offset parameter b enables the baseline of the kernel function to be adaptively calibrated, thereby improving the model's resilience in addressing data shifts. The power parameter p nonlinearly transforms the input distance, enabling the model to more accurately control the shape of the kernel function and facilitating the fitting of complex patterns and relationships. Introducing these new parameters enhances the flexibility and expressiveness of the OPARBFN while preserving its rapid learning speed and robust adaptability. These enhancements impart greater expressiveness and predictive capability to OPARBFN in applications such as time series prediction while maintaining the efficiency and stability of model training.

In contrast, traditional radial basis functions typically assume the form of a Gaussian function, as illustrated in Eq. (3).

$$\phi(r) = e^{-\varepsilon r^2} \tag{3}$$

The squared Euclidean distance between input vector and center vectors is given by (4), where x_i and c_i are their i-th components, and ε controls the function width.

$$r^2 = \sum_{i=1}^{n} (x_i - c_i)^2 \tag{4}$$

3 Experiment and Results

This study presents Fed-ARIMA-OPARBFN (FAOAR), a federated learning framework for crop yield prediction that enables regional collaboration through model structure sharing while preserving data privacy. Experiments conducted on TensorFlow Federated with NVIDIA RTX 4060 GPU, Intel i9-13980HX CPU, and 16GB RAM compare FAOAR against independently and centrally trained models. Results demonstrate significant improvements in prediction accuracy.

3.1 Dataset Construction

Nepal's diverse terrain, ranging from southern plains to northern mountains, creates an ideal environment for agricultural production, making its agrarian output highly representative. The production of its principal crops - rice, wheat, and maize - affects national food security and reflects agricultural development patterns in similar regions. Thus, this study focuses on these three major crops in Nepal to provide insights for global agrarian forecasting.

The dataset utilized in this paper is derived from the International Centre for Integrated Mountain Development (ICIMOD). It encompasses annual rice, wheat, and maize production data for each district in Nepal, extending from 1979/80 to 2013/14 fiscal years [21]. In light of the absence of pertinent data, additional literature and data reports were consulted to supplement the research process. Through a comprehensive examination of these datasets, this article investigates the evolving trends in Nepal's agricultural production and endeavors to extend these insights and models to the global level, particularly concerning the agricultural development of analogous regions. This study provides theoretical support for global food security, agricultural forecasting, and other fields.

Table 1. Descriptive statistics of raw data

ID	Species	Count	Mean	Min	Max
1	Rice	3431	2455.9161	562	6600
2	Wheat	3525	1693.7543	315	5120
3	Maize	3478	1929.3485	103	6400

This study maintains Nepal's original 75-district administrative division (pre-2015) to ensure data consistency. Data anomalies were excluded for accuracy, including zero rice production in the Manang district and limited rice data (1979/80-1982/83) in the Mustang district. The dataset spans from 1975/76 to 2021/22 across all districts, providing a robust foundation for agricultural analysis, as validated by references [22–24]. Table 1 presents the statistical summary of annual production for Nepal's three main staple crops, illustrating the dataset's methodology and characteristics.

3.2 Evaluation Indicators

The selection of appropriate performance evaluation indicators is essential in agricultural production forecasting. This experiment employs five evaluation indicators, namely, the root mean square error (RMSE), coefficient of determination (R^2), symmetric mean absolute percentage error (SMAPE), median absolute error (MedAE), and root mean square logarithmic error (RMSLE), to comprehensively evaluate forecasting performance.

The RMSE is a commonly used statistical measure that quantifies the average difference between predicted and actual values. As such, it reflects the prediction accuracy of the model in question. The calculation formula for the RMSE is provided in Eq. (5).

$$RMSE = \sqrt{\sum_{i=1}^{n} \frac{(\hat{y}_i - y_i)^2}{n}} \tag{5}$$

Second, R^2 (6) is a pivotal metric for evaluating the efficacy of a specific model. It quantifies the model's capacity to elucidate the fundamental drivers of fluctuations in the dependent variable, reflecting its overall predictive accuracy.

$$R^2 = 1 - \frac{\sum_{i=1}^{n} (y_i - \overline{y})^2}{\sum_{i=1}^{n} (y_i - \hat{y}_i)^2} \tag{6}$$

Subsequently, the SMAPE (7) is employed to quantify the discrepancy between the predicted and actual values. The SMAPE is expressed as a percentage, facilitating comparisons across disparate magnitudes of data. Furthermore, by normalizing the predicted and actual values to the mean, the SMAPE enhances adaptability to data of varying magnitudes.

$$SMAPE = \frac{100\%}{n}\sum_{i=1}^{n}\frac{|\hat{y}_i - y_i|}{\left(|\widehat{y}_i| + |y_i|\right)/2} \tag{7}$$

The MedAE(8) is employed to quantify the median prediction error and assess the stability of the model predictions. In contrast to the root mean square error, the MedAE is not significantly affected by extreme values, thereby providing a more reliable assessment of the model's prediction error and reflecting its robustness in most cases.

$$MedAE = med(|y_1 - \hat{y}_1|, \ldots, |y_n - \hat{y}_n|) \tag{8}$$

The RMSLE (9) is employed to evaluate exponential growth data. The RMSLE enables the reduction of outliers' impact by applying a logarithmically transformed function to the predicted and actual values. Furthermore, it can effectively handle disparate orders of magnitude deviations, rendering it particularly well-suited to datasets exhibiting exponential growth characteristics.

$$RMSLE = \sqrt{\frac{1}{n}\sum_{i=1}^{n}(\log(\hat{y}_i + 1) - \log(y_i + 1))^2} \tag{9}$$

In this context, the following definitions are provided: $\hat{y}_i$ represents the predicted value, y_i represents the observed value, and $\overline{y}$ represents the average observed value.

These indicators enable a comprehensive evaluation of model accuracy, trend analysis, and adaptability. This assessment method objectively validates the FAOAR model's superior performance in agricultural production forecasting.

3.3 Comparison of the Results of the Training Set Fitting via Three Different Methods

This study evaluates model performance through three approaches: single data source, centralized processing, and our proposed federated learning model. While the single-source method uses local data and centralized processing to combine all regional data, our federated learning model enables efficient prediction with data privacy by training global models through regional client collaboration.

Table 2 demonstrates the three algorithms' fitting performance across the crop datasets. The federated learning model shows superior performance compared to single-source and centralized processing approaches. This analysis specifically evaluates the models' fitting capabilities on the training set, which is distinct from prediction accuracy assessment.

The FAOAR algorithm demonstrates notable advantages over traditional centralized and single-region models in predicting yields for rice, wheat, and corn. The algorithm

Table 2. Comparison of prediction results

	Rice			Wheat			Maize		
	Conventional		**FAOAR**	Conventional		**FAOAR**	Conventional		**FAOAR**
	individual	central		individual	central		individual	central	
RMSE	298.6117	300.9144	**288.3724**	308.3556	304.7213	**288.6745**	271.2933	275.0283	**255.3460**
R^2	0.6524	0.6534	**0.6827**	0.6405	0.6569	**0.6845**	0.6462	0.6458	**0.6802**
SMAPE	9.6824%	9.6632%	**9.1518%**	14.2100%	13.1858%	**11.9678%**	10.4763%	10.2470%	**9.5750%**
MedAE	148.1550	140.7416	**135.3020**	144.5466	123.6005	**126.4216**	127.5986	121.3366	**109.1176**
RMSLE	0.1388	0.1391	**0.1345**	0.1890	0.1840	**0.1790**	0.1512	0.1504	**0.1443**

significantly improves error metrics (RMSE, SMAPE, MedAE, RMSLE) and accuracy indicators (R^2), confirming closer alignment between predicted and observed values. This enhanced accuracy improves model reliability and practicality and strengthens its capability to identify data patterns, thereby leading to more accurate predictions.

Furthermore, the FAOAR algorithm effectively integrates multisource data while maintaining data privacy in distributed processing, enhancing model performance and accuracy. The algorithm achieves both privacy protection and efficient prediction through the collaborative processing of distributed data, demonstrating its practical viability for agricultural forecasting applications.

3.4 Cross-Validation Results for the Test Set Using Three Different Methods

As shown in Table 3, this study utilizes the final year's data as the test set to evaluate production forecasting capability. This testing phase is crucial for assessing model generalization to novel data, with test set results providing more practical insights into predictive accuracy than training set performance.

Table 3. Leave-one-year-out cross validation

	Rice			Wheat			Maize		
	Conventional		**FAOAR**	Conventional		**FAOAR**	Conventional		**FAOAR**
	individual	central		individual	central		individual	central	
RMSE	405.3435	330.1487	**314.4849**	439.2381	249.1063	**229.8575**	589.3497	366.3392	**313.8415**
R^2	0.4865	0.6048	**0.6421**	0.4155	0.7823	**0.8170**	0.2585	0.5006	**0.6461**
SMAPE	13.8682%	10.6052%	**10.2167%**	18.4709%	9.9176%	**9.7230%**	22.4959%	12.4619%	**11.3527%**
MedAE	247.8567	223.3682	**220.6824**	371.4817	148.9719	**138.9633**	495.8794	293.8873	**279.2689**
RMSLE	0.1399	0.1068	**0.1028**	0.1871	0.0997	**0.0980**	0.2283	0.1251	**0.1139**

Experimental results validate the FAOAR model's superior performance over traditional single-source and centralized models in predicting rice, wheat, and corn yields. While limited data and regional adaptability constrain single-source models, centralized models, despite improved flexibility, face potential data privacy risks.

The FAOAR model enhances prediction accuracy while preserving regional data privacy through federated learning integration. Specifically, it improves the R^2 values for rice, wheat, and maize by 32%, 97%, and 350% compared to single-source models and by 6%, 4%, and 29% versus centralized models. These improvements demonstrate federated learning's effectiveness in enhancing model generalization and privacy protection for agricultural decision support.

3.5 MEA Index Analysis Under Three Different Training Methods

Beyond quantitative analysis, Fig. 2 presents the time series mean absolute error (MAE) curves for regional predictions across the three ARIMA-OPAGRBFN models. These visualizations demonstrate prediction accuracy and temporal variations, providing additional insights into model performance.

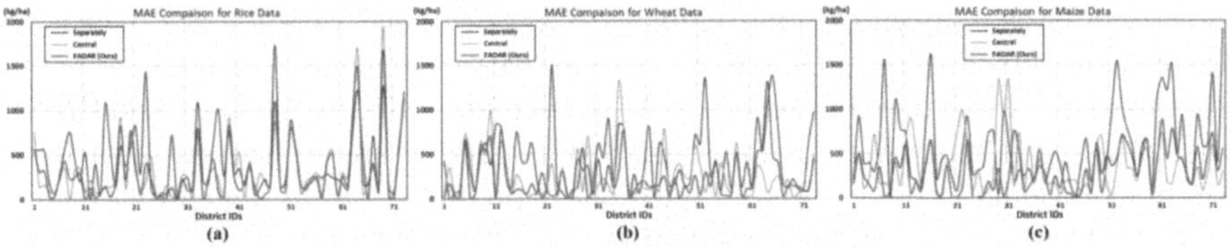

Fig. 2. The following figures present the MEA prediction indicators for each region under the three methods: (a) rice dataset, (b) wheat dataset, (c) maize dataset.

As illustrated in the graph, the error curve of the FAOAR model is predominantly situated at a lower level, indicating that it achieves more modest prediction errors at most points in time. Conversely, the error curve of the traditional single-source model is generally positioned at a higher level, reflecting its relatively substantial prediction errors. The performance of the centralized model is intermediate; however, the overall effect remains inferior to that of the FAOAR model.

Notably, the FAOAR model has apparent advantages in dealing with outliers and extreme cases: the error curves are smoother and more continuous, with fewer fluctuations, indicating better stability and reliability of the prediction capability. The error curves of the traditional models clearly peak at specific points in time, reflecting their limitations in predicting challenging scenarios. Overall, the FAOAR model achieves more accurate prediction results through the federated learning approach, fully demonstrating its advantages in dealing with complex challenges.

3.6 Results of Ablation Experiments

This section verifies the effect of improving the traditional RBFN kernel function in the FAOAR model by conducting comparative experiments on the Nepalese dataset. The experiments compare the performance of the conventional RBFN model with that of the FAOAR model under the same initial model parameters. Table 4 shows the experimental results of the FAOAR model with the conventional RBFN model and the FAOAR model on the Nepalese cereal yield dataset under the same conditions.

Table 4. Results of the ablation experiment

	Rice		Wheat		Maize	
	Conventional	**FAOAR**	Conventional	**FAOAR**	Conventional	**FAOAR**
RMSE	843.2251	**314.4849**	1011.0638	**229.8575**	1224.3221	**313.8415**
R^2	−0.5464	**0.6421**	−1.2225	**0.8170**	−2.8869	**0.6461**
SMAPE	29.5611%	**10.2167%**	35.6392%	**9.7230%**	47.9374%	**11.3527%**
MedAE	840.9473	**220.6824**	1011.0638	**138.9633**	1302.7018	**279.2689**
RMSLE	0.2992	**0.1028**	0.4592	**0.0980**	0.4913	**0.1139**

The experimental results revealed that the FAOAR model outperformed the traditional model in all the evaluation indicators. The improved model's root mean square error (RMSE) was reduced by approximately 62%, 77%, and 74% for rice, wheat, and maize, respectively. In addition, the R^2 value improved from negative to positive, significantly improving the prediction accuracy. Moreover, the SMAPE, MedAE, and RMSLE metrics also significantly improved. These results validate the effectiveness of RBFN kernel function enhancement, improve the model's predictive ability, and provide a more accurate tool for agricultural yield prediction.

4 Discuss

The Ljung–Box Q test [25], a standard time series autocorrelation test, was used to ensure each region's yield series had a particular autocorrelation structure. The test examines observation correlations, as autocorrelation can reduce forecasting accuracy and cause data interpretation errors. A p-value ≥ 0.05 suggests no significant autocorrelation under the null hypothesis. As shown in Fig. 3, over 98% of the time series data exhibit autocorrelation. Given the independent and homogeneous distribution of same-crop time series, this indicates inherent autocorrelation patterns. Since perfect time series are unrealistic, perturbation groups were introduced to simulate actual conditions and enhance system robustness. The strong autocorrelation validates using ARIMA to capture linear relationships in the time series data.

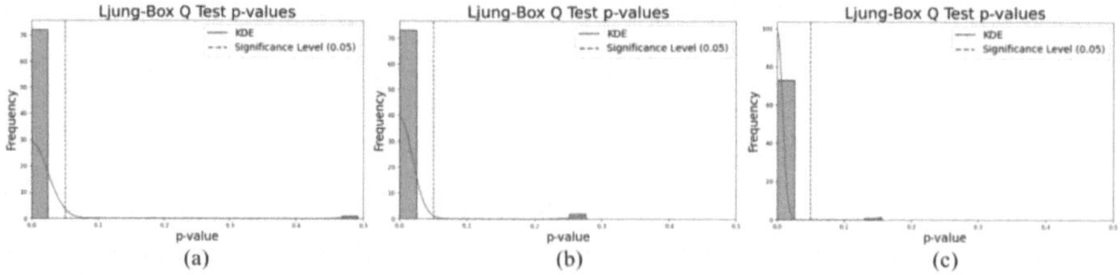

Fig. 3. Ljung-Box Q test results for rice (a), wheat (b), and maize (c) datasets. X-axis repre-sents p-value, y-axis indicates number

5 Conclusion

The application of the Fed-ARIMA-OPARBFN (FAOAR) model to agricultural yield forecasting has proven to be effective, with the FAOAR model outperforming the single-source model and the centralized model on several key metrics, as verified through analysis of both the training and validation sets, in terms of prediction accuracy and stability, and a high coefficient of determination. The optimization of the model improves the explanation of data variability and provides more reliable predictions. The ablation experiments further validate the effectiveness of modifying the traditional RBFN kernel function to improve model performance. The excellent performance on the validation set further demonstrates the generalizability of the model and its feasibility in practical applications. The results demonstrate the effectiveness and potential of federated learning in agricultural data analysis, especially in yield time series prediction. The method protects data privacy and facilitates distributed data integration, providing new strategies and insights for future agricultural extensive data analysis.

References

1. Kamilaris, A., Prenafeta-Boldú, F.X.: Deep learning in agriculture: a survey. Comput. Electron. Agric. **147**, 70–90 (2018). https://doi.org/10.1016/j.compag.2018.02.016
2. Alibabaei, K., Gaspar, P.D., Lima, T.M.: Crop yield estimation using deep learning based on climate big data and irrigation scheduling. Energies **14**, 3004 (2021). https://doi.org/10.3390/en14113004
3. Sakamoto, T.: Incorporating environmental variables into a MODIS-based crop yield estimation method for United States corn and soybeans through the use of a random forest regression algorithm. ISPRS J. Photogramm. Remote Sens. **160**, 208–228 (2020). https://doi.org/10.1016/j.isprsjprs.2019.12.012
4. Zhang, Q., Zhao, X., Han, Y., et al.: Maize yield prediction using federated random forest. Comput. Electron. Agric. **210**, 107930 (2023). https://doi.org/10.1016/j.compag.2023.107930
5. McMahan, H.B, Moore, E., Ramage, D., Hampson, S.: Communication-efficient learning of deep networks from decentralized data
6. Chen, Z., Guo, S., Shen, L., et al.: Improving knowledge distillation for federated learning on non-IID Data. In: 2023 IEEE International Conference on Big Data (BigData). IEEE, Sorrento, Italy, pp. 598–607 (2023)
7. Li, L., Fan, Y., Tse, M., Lin, K.-Y.: A review of applications in federated learning. Comput. Ind. Eng. **149**, 106854 (2020). https://doi.org/10.1016/j.cie.2020.106854
8. Kairouz, P., McMahan, H.B., Avent, B., et al.: Advances and open problems in federated learning. FNT Mach. Learn. **14**, 1–210 (2021). https://doi.org/10.1561/2200000083
9. Yang, Q., Liu, Y., Chen, T., Tong, Y.: Federated machine learning: concept and applications. ACM Trans. Intell. Syst. Technol. **10**, 1–19 (2019). https://doi.org/10.1145/3298981
10. Bali, N., Singla, A.: Emerging trends in machine learning to predict crop yield and study its influential factors: a survey. Arch. Comput. Methods Eng. **29**, 95–112 (2022). https://doi.org/10.1007/s11831-021-09569-8
11. Prasad, A.K., Chai, L., Singh, R.P., Kafatos, M.: Crop yield estimation model for Iowa using remote sensing and surface parameters. Int. J. Appl. Earth Obs. Geoinf. **8**, 26–33 (2006). https://doi.org/10.1016/j.jag.2005.06.002
12. Marshall, M., Belgiu, M., Boschetti, M., et al.: Field-level crop yield estimation with PRISMA and Sentinel-2. ISPRS J. Photogramm. Remote Sens. **187**, 191–210 (2022). https://doi.org/10.1016/j.isprsjprs.2022.03.008

13. Mateo-Sanchis, A., Piles, M., Muñoz-Marí, J., et al.: Synergistic integration of optical and microwave satellite data for crop yield estimation. Remote Sens. Environ. **234**, 111460 (2019). https://doi.org/10.1016/j.rse.2019.111460
14. Zhang, G.P.: Time series forecasting using a hybrid ARIMA and neural network model. Neurocomputing **50**, 159–175 (2003). https://doi.org/10.1016/S0925-2312(01)00702-0
15. Tseng, F.-M., Yu, H.-C., Tzeng, G.-H.: Combining neural network model with seasonal time series ARIMA model. Technol. Forecast. Soc. Chang. **69**, 71–87 (2002). https://doi.org/10.1016/S0040-1625(00)00113-X
16. Cadenas, E., Rivera, W.: Wind speed forecasting in three different regions of Mexico, using a hybrid ARIMA–ANN model. Renew. Energy **35**, 2732–2738 (2010). https://doi.org/10.1016/j.renene.2010.04.022
17. Babu, C.N., Reddy, B.E.: A moving-average filter based hybrid ARIMA–ANN model for forecasting time series data. Appl. Soft Comput. **23**, 27–38 (2014). https://doi.org/10.1016/j.asoc.2014.05.028
18. Koutroumanidis, T., Ioannou, K., Arabatzis, G.: Predicting fuelwood prices in Greece with the use of ARIMA models, artificial neural networks and a hybrid ARIMA–ANN model. Energy Policy **37**, 3627–3634 (2009). https://doi.org/10.1016/j.enpol.2009.04.024
19. Box, G.E.P, et al.: Time Series Analysis: Forecasting and Control. Wiley (2015)
20. Lowe, D., Broomhead, D.: Multivariable functional interpolation and adaptive networks. Complex Syst. **2**(3), 321–355 (1988)
21. ICIMOD| RDS --- ICIMOD | RDS. https://rds.icimod.org/Home/DataDetail?metadataId=17200. Accessed 30 Jul 2024
22. The study on the disaster prevention plan for severely affected areas by 1993 disaster in the central development region of Nepal, Japan International Cooperation Agency, Department of Soil Conservation, Ministry of Forest and Soil Conservation, His Majesty's Government of Nepal, Nippon Koei co., Ltd., Tokyo, Japan (1997)
23. Agriculture statistics | Publication category | Ministry of agriculture and livestock development. https://moald.gov.np/publication-types/agriculture-statistics/. Accessed 26 Aug 2024
24. Statistical information on Nepalese agriculture 2079/80[2022/23]. http://nepalindata.com/resource/pdf/STATISTICAL-INFORMATION-ON-NEPALESE-AGRICULTURE-2073-74--2016-17/. Accessed 26 Aug 2024
25. Box, G.E., Pierce, D.A.: Distribution of residual autocorrelations in autoregressive-integrated moving average time series models. J. Am. Stat. Assoc. **65**(332), 1509–1526 (1970)
26. Bali, N., Singla, A.: Deep learning based wheat crop yield prediction model in Punjab region of north India. Appl. Artif. Intell. **35**(15), 1304–1328 (2021)
27. Wang, J., Wang, P., Tian, H., et al.: A deep learning framework combining CNN and GRU for improving wheat yield estimates using time series remotely sensed multi-variables. Comput. Electron. Agric. **206**, 107705 (2023)
28. Sun, J., Di, L., Sun, Z., et al.: County-level soybean yield prediction using deep CNN-LSTM model. Sensors **19**(20), 4363 (2019)
29. Hwang, Y.-S., Bang, S.-Y.: An efficient method to construct a radial basis function neural network classifier. Neural Netw. **10**, 1495–1503 (1997). https://doi.org/10.1016/S0893-6080(97)00002-6

S-CPD: Topological Smoothing-Based Change Point Detection

Harindu Sugathadasa(✉), Sarah Monazam Erfani, and Christopher Leckie

University of Melbourne, Melbourne 3010, Australia
{h.udagekankanange,sarah.erfani}@unimelb.edu.au

Abstract. Change Point Detection in time series aims to identify the time points when the statistical properties of a time series change. A major challenge in this field is how to reduce the effect of noise in the time series on the accuracy of change point detection. This is particularly challenging when the time series exhibits complex patterns of normal behaviour. Recently, Topology Data Analysis has shown the potential to characterise the shape of complex and irregular time series. In this paper, we propose a method called *Smoothing-based Change Point Detector* (S-CPD) (Source code is available here: https://github.com/harinduashan/S-CPD.) which transforms the time signal to a smoothed signal by using the *lower-star* filtration-based Persistent Homology from the field of topological data analysis. We then detect the change points in the transformed signal by using a classification-based score profiling. The benefit of S-CPD is its ability to interpret noisy information as low-persistence topological data and remove it, allowing the refined signal to be effectively used for detecting change points. We demonstrate the effectiveness of our method on two real-world datasets in comparison with several state-of-the-art change point detection methods.

Keywords: Change point detection · Topological data analysis

1 Introduction

Change Point Detection (CPD) is a crucial technique in time series analysis, aimed at identifying points where the statistical properties of the data, such as mean, variance, or other characteristics of the distribution, exhibit significant changes [18]. These *change points* (CPs) often indicate critical transitions or events within the monitored system, with applications ranging from detecting shifts in vital signs in medical diagnostics to identifying structural changes in industrial processes [13]. Despite its importance, CPD methods face challenges such as sensitivity to noise, limitations in capturing complex patterns, and difficulties in detecting subtle changes, which can affect their reliability and adaptability across diverse datasets [2,18].

Topological Data Analysis (TDA) offers a promising approach to these challenges [8]. TDA provides tools for analysing the shape of the data as a manifold, mainly through techniques such as *Persistent Homology* (PH). PH allows the extraction of robust features by capturing the shape and connectivity of

X. Wu et al. (Eds.): PAKDD 2025, LNAI 15875, pp. 147–158, 2025.
https://doi.org/10.1007/978-981-96-8295-9_10

data across multiple scales, called *filtration* [7]. TDA is particularly well-suited for analysing complex and irregular time series data, where traditional statistical methods may fall short. Although TDA has been successfully applied in various fields in time series data [14], its use in CPD frameworks remains an open problem. Previous methods employing TDA in CPD have often relied on computationally intensive approaches, limiting their practicality in real-time or resource-constrained scenarios [19]. While simpler and computationally efficient methods exist within TDA [6], their potential application in CPD has not been extensively investigated, presenting the opportunity to enhance the effectiveness of CPD by leveraging topological information.

A prominent method for CPD is the Classification Score Profile [9] (ClaSP) approach, which provides a framework for detecting CPs based on time series classification and segmentation. ClaSP converts the CPD problem into a time series segmentation (TSS) task and then tries to find the correct segments. Despite its promising results for CPD, its performance can be affected by noise in time series, which can be erroneously flagged as change points. By building on this foundation, our approach, called **S-CPD**, aims to first smooth out the input signal by removing such noisy/unwanted information using TDA as a preprocessing step. Moreover, this step helps to identify both topologically persistent and noisy information which can then be used to filter out such unwanted information and its paired data patterns in the signal. Once the signal is reconstructed by removing noisy patterns, we perform CPD with ClaSP to detect CPs. The key contributions of this work are as follows:

1. Motivated by the potential of computationally efficient TDA methods, we introduce a novel framework, S-CPD, that integrates a TDA-based smoothing technique with CPD to improve CP detection accuracy and robustness.
2. To the best of our knowledge, there is no existing work to detect CPs through the use of a topologically smoothed signal into the CP detectors without introducing significant complexity.
3. We demonstrate the effectiveness of the proposed method on diverse time series datasets, showing improvements over existing methods.
4. We provide insights into the benefits of incorporating topological features to enhance CPD and develop more effective tools for time series analysis.

2 Related Work

CPD aims to detect points when the statistical properties of a time series change, signaling a change in the underlying generative process. It is used in numerous domains, including human activity recognition [4], financial market analysis [11], and sensor monitoring networks [21]. The complexity of CPD ranges from detecting changes in simple mean and variance-based data to more complex simultaneous changes, such as frequency and autocorrelation [2].

Considering non-deep learning methods, BOCD [1] detects CPs via probability distributions using recursive message passing. Although it is effective in online segmentation, it can have difficulty with overlapping changes. PELT [15]

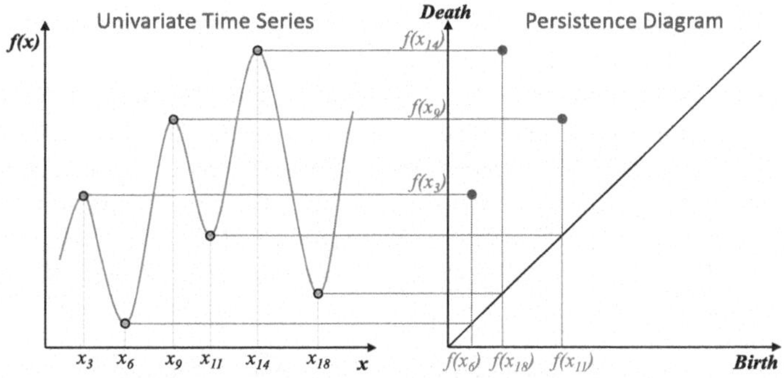

Fig. 1. An example time series on the left, with three local minima and maxima. The birth-death pairs are displayed in the persistence diagram on the right.

models segmentation as an optimisation function, using a pruning method to reduce computational cost. However, it does not directly address the challenge of noisy inputs. Binary segmentation (BinSeg) [17] introduces a greedy recursive approach using a simple and fast mechanism for smaller datasets. FLOSS [10], another heuristic-based method, measures the density of similar subsequences and extracts segments using a greedy heuristic. This works for long TS with a predefined segment count. ESPRESSO [5] extends FLOSS by introducing entropy for segmentation and incorporating positional subsequence information. An open challenge for these methods is to reduce their sensitivity to noise in the time series. In deep learning algorithms, SIMTSeg [3] follows a self-supervised approach to segment TS. It uses an encoder-decoder method with a reverse diffusion process to generate segmentation masks, and Ramanujan subspace projection for capturing intra-period and inter-variate dependencies. However, it is not designed for the case of univariate TS, which are our focus. Classification Score Profile (ClaSP) [9] is another self-supervised classification method, recognizing CPs without true labels. However, it still has difficulty with noisy data.

To address the challenge of CPD in noisy time series, we propose the use of TDA. TDA has been shown to extract useful features from time series [11, 12], but not in the context of CPD. In particular, Rosen et al. [16] introduce topological smoothing for time series data. In the following sections, we propose an approach for topological smoothing in time series CPD.

3 Preliminaries

This section introduces TDA in time series and relevant terminology, referring readers to [6,7,16] for further details.

A key concept in TDA is the simplicial complex, an abstract structure closed under nonempty subsets. TDA commonly assigns a sequence of simplicial complexes to a metric space $(\mathbb{M}, d_{\mathbb{M}})$, parameterized by $\lambda \in \mathbb{R}$, forming a filtration. Persistent homology (PH) analyzes how homological features, such as connected components (0D) merging and cycles (1D) emerging or disappearing, evolve as λ increases, capturing the dynamic topology.

The **Lower-star** filtration is used to analyze the topology of a time series $\mathcal{T}$ (see Fig. 1). Given a continuous function $f : \mathbb{F} \to \mathbb{R}$ on a topological space $\mathbb{F}$, a time step $x \in \mathbb{F}$ is a *critical point* $x^{(c)}$ of f if $f'(x) = 0$. A critical point is *non-degenerate* if $f''(x) \neq 0$, ensuring isolation and local minimum (maximum) behavior (see Definition 1). For any $\lambda \in \mathbb{R}$, the sub-level set is defined as $\mathbb{F}_\lambda = f^{-1}(-\infty, \lambda] = \{x \in \mathbb{F} \text{ s.t. } f(x) \leq \lambda\}$, capturing the function's topology at different thresholds.

Definition 1. *Let $x_0 \in \mathbb{F}$ and $I = (x_0 - \epsilon, x_0 + \epsilon)$ be a time-step and an interval with $\epsilon > 0$, respectively. x_0 is called a local minimum of f if $\exists I$ s.t. $f(x_0) \leq f(x) \quad \forall x \in I$. A local maximum is called if $\exists I$ s.t. $f(x_0) \geq f(x) \quad \forall x \in I$.*

A *birth-death* pair (b, d) is defined based on the changes in the topology of $\mathbb{F}_\lambda$ as λ increases from $-\infty$. Definition 2 provides the connectivity of $\mathbb{F}_\lambda$ changes only when it crosses a $x^{(c)}$.

Definition 2. *A local minimum $x_b^{(c)}$ is considered as a birth point when it introduces a new connected component in $\mathbb{F}_\lambda$ as λ crosses $f(x_b^{(c)})$. At this value, the component is born: $b = f(x_b^{(c)})$. A local maximum $x_d^{(c)}$ is considered as a death point when it causes the merging of two connected components in $\mathbb{F}_\lambda$ as λ crosses $f(x_d^{(c)})$. Specifically, $x_d^{(c)}$ is paired with the higher-valued of the two local minima $x_{b1}^{(c)}, x_{b2}^{(c)}$ such that $f(x_{b1}^{(c)}) > f(x_{b2}^{(c)})$, which represent the merging components: $d = f(x_d^{(c)})$.*

To quantify the importance of the pairs of critical points, the persistence $p \in \mathbb{R}$ of a pair is defined as $p = |d - b|$, which quantifies the lifetime of the topological feature within the function f. Note that the critical points $x_b^{(c)}, x_d^{(c)}$ are not necessarily adjacent. The pairing shows that local minima create new features, while local maxima terminate features through component mergers. All (b, d) pairs are recorded in $\mathcal{C}$ s.t. $\mathcal{C} = \{(b_1, d_1), \cdots, (b_m, d_m)\}$ where m is the total number of birth-and-death pairs.

This concept is represented using a *persistence diagram*, where $(b, d) \in \mathcal{C}$ is assigned to the $x - y$ plot. In the persistence diagram, all points are in the half-space where $x \leq y$, and p of each pair corresponds to the vertical distance from the diagonal $y = x$. This serves as a compact summary of the topological features of f, allowing a direct comparison of time series through their $x^{(c)}$ structures and the p of associated features. Typically, the diagonal is included in the diagram as a reference line to aid in interpreting persistence values.

4 S-CPD: Smoothing-Based Change Point Detector

We now present an overview of our proposed approach for CPD, called S-CPD. Figure 2 visually summarizes the workflow of S-CPD. Section 4.1 explains how

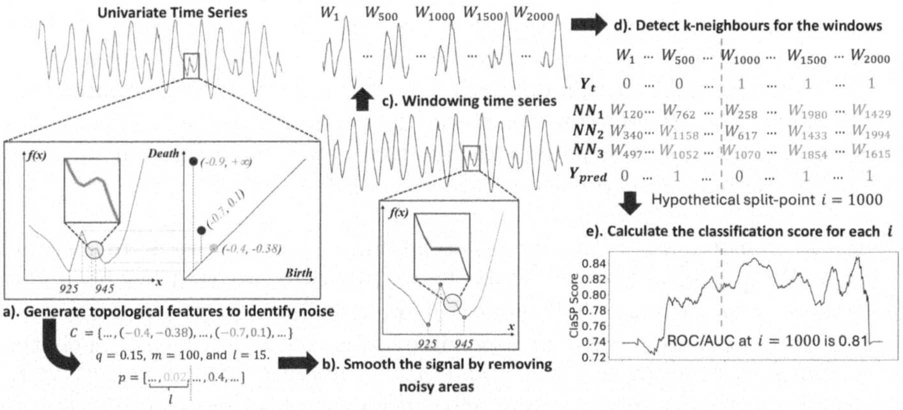

Fig. 2. High-level architecture of *S-CPD*. (*a*) A time series is subjected to *Lower-star* filtration-based PH, and removing l-pairs (yellow-color dot in persistence diagram). The effect is visualized in yellow-color circle and magnified box. (*b*) The reconstruction of smoothed signal by removing l-pairs detected as **noisy**-pairs (both yellow-color circle and magnified box). (*c*) Then overlapping windows are created. (*d*) kNN is used to detect k neighbors to each window, followed by self-supervised-based ground truth labelling (with hypothetical splits at i^{th} input time step). (*e*) Finally, the ClaSP is generated from the cross-validation.

persistent homology is used in time series to extract topological features and identify noisy data (Sub-Fig. 2.*a*), and remove the extracted noisy features to smooth time series (Sub-Fig. 2.*b*). Section 4.2 covers how CPs are detected using a classification-based detector on smoothed signal (Sub-Figs. 2.*c* to 2.*e*).

4.1 Time Series Smoothing

Persistent Homology of Time Series: Given a time series $\mathcal{T} = (f(x_1), \cdots, f(x_n))$ where $f(x_i) \in \mathbb{R}$ over the time steps $[\mathcal{T}] = \{x_1, \cdots, x_n\}$, we construct a 1-dimensional simplicial complex $\mathcal{K} = \{\{x_1\}, \cdots, \{x_n\}, \{x_1, x_2\}, \cdots, \{x_{n-1}, x_n\}\}$. Note that this complex is composed of *0-simplices* representing individual time steps $\{x_1, \cdots, x_n\}$, and *1-simplices* representing edges between consecutive time steps $\{x_i, x_{i+1}\}$. This simplicial complex preserves the sequential structure of $\mathcal{T}$. To analyse the topological structures through persistent homology (PH), we use a *filtration* over $\mathcal{K}$ based on the sorted values of $\mathcal{T}$. Let $\lambda_{min(\mathcal{T})} \leq \cdots \leq \lambda_{f(x_i)} \leq \cdots \leq \lambda_{max(\mathcal{T})}$ represent the sorted values of $\mathcal{T}$. We construct a sequence of subcomplexes $\mathcal{K}_{\mathcal{T}}^{x_j} = \{\sigma \in \mathcal{K} \mid \forall i \in \sigma \mid f(x_i) \leq \lambda_{f(x_j)}\}$ such that $\mathcal{K}_{\mathcal{T}}^{0} = \varnothing \subseteq \mathcal{K}_{\mathcal{T}}^{x_1} \subseteq \cdots \subseteq \mathcal{K}_{\mathcal{T}}^{[\mathcal{T}]} = \mathcal{K}$ where $x_j \in [\mathcal{T}]$. During the filtration process, by using Definition 2, we track the birth and death of connected components as λ increases. PH then tracks the evolution of topological features by using this construction and plots them up in a persistence diagram.

The birth-and-death pairs are extracted from the filtration using PH and stored in a stack. Before processing local extrema, the global minimum is paired with $+\infty$ and the global maximum with $-\infty$. For a given λ, local minima serve as birth values and are added to the stack in ascending order of f. When a local maximum is encountered, it pairs with the most recent birth value in the stack, forming a (b, d) pair. This process continues until the stack is empty.

Topological Smoothing in Time Series: The aim of this process is to maintain the essential topological features while eliminating the noise that has been identified based on persistence. Following [16], we first measure the persistence of all $(b, d) \in \mathcal{C}$ and rank them in ascending order; $p_1 \leq \cdots \leq p_m$ such that $p_1 = min(|d_i - b_i| : (b_i, d_i) \in \mathcal{C})$ and $p_m = max(|d_i - b_i| : (b_i, d_i) \in \mathcal{C})$. The smoothing process involves removing insignificant l-pairs from $\mathcal{C}$ based on the sorted persistence, which is defined as a user-defined parameter q where $0 \leq q \leq 1$, such that $l = \lfloor q \cdot m \rfloor$, where $\lfloor \cdot \rfloor \in \mathbb{N}^+$ is the floor function.

Let $g : \mathcal{F} \rightarrow \mathcal{R}$ be a continuous function. Once the selected l-pair with its time step pairs $(x_{b_i}^{(c)}, x_{d_i}^{(c)})$ are removed, the expectation of new function g remains close to f s.t. $\|f - g\| \leq \epsilon$ where ϵ is a maximum tolerance for acceptable deviation. This function simplification involves eliminating unnecessary oscillations between critical points. Removing such pairs of insignificant critical points results in *monotonic behavior* between neighboring points. This flattening can be expressed mathematically as: $f(x_l) \geq f(x) \geq f(x_r) \quad \forall x \in [x_l, x_b]$ s.t. $min(|f(x_l) - f(x_d^{(c)})|)$ and $min(|f(x_r) - f(x_b^{(c)})|)$ if $x_l < \cdots < x_b^{(c)} < \cdots < x_b^{(c)} < \cdots < x_r)$.

To ensure that the function g remains as close as possible to the original data, we apply *isotonic regression* where this technique finds the monotonic approximation that minimises the sum of squared errors: $min \sum_{i=1}^{n} (f(x_i) - g(x_i))^2$. This ensures minimal residual error while satisfying the monotonic constraint.

4.2 Change Point Detector

We selected *ClaSP* as the change point detector for CPD, as described in [9]. ClaSP works by dividing a univariate time series $\mathcal{T}$ into fixed-length overlapping windows W, where the length of the window ω is chosen to capture similar statistical properties within each window. Next, the k-Nearest-Neighbours (kNNs) are computed for every window based on similarity distances. These kNNs serve as classifiers in the self-supervision process, striking a balance between computational efficiency and effective similarity measurement, helping to distinguish different windows. The algorithm then introduces a split point i to divide $\mathcal{T}$ into left and right segments, with the goal of assessing how well i matches a potential change point. We summarise ClaSP below (see [9] for details).

Algorithm 1 in [9] outlines the binary classification approach for time series segmentation (TSS), which is key to CPD. For each possible split point i, the windows on the left (right) of the split are labelled as zero (one), transforming the segmentation task into a supervised learning problem. The kNNs for each window are classified based on which segment they belong to, and cross-validation is used to measure the dissimilarity between the windows in the left and right segments. A high score for a given i suggests that it could be a change point. These scores are recorded for all split points, creating a classification score profile for $\mathcal{T}$. The local maxima in this profile highlight the points where the difference between segments is strongest, indicating potential change points.

Precomputation of kNNs (kNN-Profiling): We explain the workings of the KNN PROFILE method (Algorithm 2 in [9]). For each window $W \in \mathcal{W}$, we first compute the *z-normalized* window using its mean and standard deviation μ, σ. Next, we calculate the pairwise dot product matrix $\mathcal{D}$ between z-normalised windows, $W_i^{(z)}$ and $W_j^{(z)}$, which then forms the pairwise distance matrix $\text{Dist}(W_i^{(z)}, W_j^{(z)})$ (Eq. 1). To reduce computational overhead, an exclusion zone of length ω is applied around each $W^{(z)}$, excluding comparisons where $|i - j| < \omega/2$. The distances in this zone are set to ∞. Finally, the k windows with the smallest distances from $W^{(z)}$ are selected from $\text{Dist}(W_i^{(z)}, W_j^{(z)})$.

$$\mathcal{D}_{ij} = \sum_{k=1}^{\omega} W_i^{(z)}[k] \cdot W_j^{(z)}[k], \quad \text{Dist}(W_i^{(z)}, W_j^{(z)}) = \sqrt{2\omega \cdot (1 - \frac{\mathcal{D}_{ij}}{\left\|W_i^{(z)}\right\| \left\|W_j^{(z)}\right\|})} \quad (1)$$

Cross-Validation Score: Since $\text{Dist}(W_i^{(z)}, W_j^{(z)})$ and the self-supervised ground truth labels Y_t are independent, cross-validation is used to evaluate different split points (Algorithm 3 in [9]). At each split point during cross-validation, the majority label of kNNs in each window is determined. As the split moves, only Y_t changes, not kNNs, allowing for more accurate classifications. To avoid class imbalance in the evaluation, metrics, such as accuracy, macro F1 or ROC/AUC are preferred, as they are less affected by class imbalance.

Detection of Reoccurring Segments: Binary segmentation can mistakenly group multiple segments into one, increasing variability and complicating CPD, especially with repeating patterns, which can reduce classification accuracy. To address this, a temporal constraint (TC) is introduced to limit the length of $\mathcal{T}$. However, due to the variability of the segment lengths and numbers, static or weighted TCs often fall short. To overcome this, an ensemble method aggregates multiple classification profiles generated using random TCs, combining high-scoring profiles to highlight meaningful splits while ignoring irrelevant ones. Algorithm 4 in [9] describes this process, where ClaSP is first calculated for $\mathcal{T}$, followed by iterations that generate local ClaSPs for random intervals with varying TCs. These are weighted by TC length and merged to maximise scores. After a few iterations, recurring segments are detected with precision, capturing the most predictive splits, making the method efficient and robust.

Segmentation with Unknown Change Points: Detecting multiple or no CPs ($\mathcal{C}$) requires a more advanced method, especially without knowing the available CPs. To handle this, a parameter-free strategy combines recursive segmentation with statistical validation (see ADD_VALID_CPT in Algorithm 5 [9]). Given $\mathcal{T}$ and ω,

ClaSP detects the most prominent peak as the first CP. The Wilcoxon rank sum test [20] then validates this point, distinguishing the left and right segments. This process repeats recursively until no more valid CPs are found (see SEGMENTATION in Algorithm 5 [9]). The benefit of this method is that it measures how significantly predicted CP differs between the left and right segments.

Window Selection: The Summary Statistics Subsequence (SuSS), introduced in Algorithms 6 of [9], is based on the idea that the summary statistics (μ, σ, and range) of Ws with the optimal ω should closely match those of $\mathcal{T}$. The algorithm computes global statistics for $\mathcal{T}$ and rolling statistics for each W over a length range $[1, n]$. Then it calculates the Euclidean distance between the global and rolling statistics, weighted by the square root of the window size. These distances, scaled to avoid bias towards larger windows, produce the SuSS score, which tends to increase with larger windows, as they better align with $\mathcal{T}$. To efficiently find the optimal ω, the algorithm uses an exponential search, followed by a binary search, to pinpoint the best value.

5 Experiments

In this section, we evaluate S-CPD in terms of (*i*) investigating how smoothed signals help to detect CPs through segmentation, and (*ii*) how different smoothing levels impact to change point detection.

Datasets. The Time Series Segmentation Benchmark (TSSB) comprises 75 semi-synthetically generated TS datasets[1] designed by Ermshaus et al. [9]. For our analysis, we used 70 of these datasets, based on assumptions of within-class similarity and moderate periodicity. The design of this dataset include various segments, such as repeating and non-repeating temporal patterns. In addition, the range of each segment varies from simple to complex patterns. Joining these segments marks CPs while ensuring continuity. The selected datasets varied in complexity, containing 2 to 9 unique segments, with additional variations including 6 TS with only one segment and 10 TS with recurring segments. Additionally, we use 30 TS from the UCR Time Series Semantic Segmentation Archive (UTSA) [10], which represent biological, mechanical, or synthetic processes. These datasets are organized into 16 use cases, 8 involving fundamental changes and 7 featuring semi-synthetic changes, such as modified recordings. One TS is entirely synthetic. Some TSs were further processed with techniques such as filtering or resampling, and human experts annotated the CPs and periods.

Baselines. We compare S-CPD with seven state-of-the-art (SOTA) methods for CPD. We use ClaSP itself to compare the smoothing effect. Since FLOSS [10] requires prior knowledge of how many segments are to be detected, we set 2 and 9 segments for UTSA and TSSB, respectively, and use the same optimal

[1] Available in Github: https://github.com/ermshaua/claspy/tree/main,.

threshold of 0.45 for its arc curve, defined in [9], to identify CPs with scores below the threshold. For ESPRESSO [5], we use a chain length of 3 and a maximum of 8 CPs. To evaluate BinSeg [17] and PELT [15], we adopt an autoregressive cost function with a CP penalty of 10. We also compare against Bayesian Online Change Point Detection (BOCD) [1] and a window-based (Window) [18] method with Mahalanobis cost function and a penalty of 20 for CP detection.

Evaluation Metrics. We employ the covering metric, a clustering-based measure to assess how well a segment produced by an algorithm matches the annotated segments provided by human annotators (see Definition 3). It is based on the Jaccard index, which measures the similarity between two segments. This provides a normalised value where a higher value indicates better alignment. This can also compare segments of varying lengths, including missing segments.

Definition 3. *The Jaccard index* $J(s, s')$ *between two segments* s *and* s' *is denoted as* $J(s, s') = \frac{\|s \cap s'\|}{\|s \cup s'\|}$. *The covering metric* $C(S, S')$ *of the true segments* S *by the predicted segments* S' *is calculated as:*

$$C(S, S') = \frac{1}{\|T\|} \sum_{s \in S} \|s\| \cdot \max_{s' \in S'} J(s, s'),$$

where $\|s\|$ *is the length of the segment* s.

Parameter Settings. We adopted the same settings as [9], using $k = 3$ for KNN_PROFILE and the ROC/AUC score for CROSS_VAL to ensure robust performance against outliers. In CAL_CLASP_ENSEMBLE, n_{iter} was set to 30 for stable detection of recurring segments. The threshold $t_{threshold}$ in CAL_SUSS was set to 0.89. A Wilcoxon rank sum test with a p-value of 1e−15 was used to validate change points. For time series smoothing, the smoothing factor q was set to 0.1 for TSSB and 0.15 for UTSA.

Results and Discussion

Segmentation with Smoothed Signals: Table 1 presents the mean and standard deviation of the covering metric for different methods applied to the TSSB and UTSA datasets. The overall wins, ties, and losses for each method are shown in Table 2. The S-CPD method shows significant improvement over the baselines, with a mean covering metric of 87.8% for TSSB, which is 1.1% higher than ClaSP, and a low standard deviation of 17.1%, indicating consistent performance. In comparison, ESPRESSO shows a lower mean of 44.4%. For UTSA, S-CPD achieves 77.1%, ranking second to FLOSS, which requires the number of change points to be predefined. For TSSB, S-CPD records 18 wins, 26 ties, and 26 losses against ClaSP, while for UTSA, it has 12 wins, 8 ties, and 10 losses, highlighting its competitive performance across both datasets.

Table 1. Mean and standard deviation metrics for covering score measured in S-CPD and SOTA methods. The best (second-best) results are highlighted in bold (underline).

Dataset	Covering	S-CPD	ClaSP	FLOSS	BinSeg	PELT	Window	BOCD	ESPRESSO
TSSB	Mean	**87.8%**	86.7%	65.6%	58.2%	59.0%	40.3%	44.0%	44.4%
	STD	17.1%	18.7%	24.3%	26.3%	25.3%	23.4%	20.2%	**15.5%**
UTSA	Mean	77.1%	75.7%	**79.2%**	50.9%	48.4%	53.7%	55.1%	57.7%
	STD	19.5%	21.1%	20.1%	30.4%	29.5%	**13.3%**	14.6%	15.6%

Table 2. Summary of wins/ties/losses of S-CPD against SOTA baselines.

	Dataset	ClaSP	FLOSS	BinSeg	PELT	Window	BOCD	ESPRESSO
S-CPD	TSSB	18/26/26	55/10/5	54/0/16	53/0/17	64/0/6	67/0/3	68/0/2
	UTSA	12/8/10	10/1/19	24/0/6	25/0/5	24/0/6	24/0/6	24/0/6

The superior performance of S-CPD is due to its innovative use of signal smoothing and segmentation. By applying lower-star filtration to time series, S-CPD reduces noise and irrelevant features, improving the accuracy of the ClaSP method to identify optimal ω, recognize patterns and calculate segment distances, thus enhancing change point detection. The results emphasize the role of signal smoothing in improving segmentation accuracy, particularly for noisy datasets such as TSSB. For the UTSA dataset, the presence of variance shifts within segments presents a different challenge. These manually generated signals contain only two CPs, which FLOSS detects using a predefined parameter. In contrast, S-CPD applies ClaSP without prior knowledge of the number of CPs. However, smoothing variance-based signals can disrupt CP detection, suggesting that the lower-star filtration method may require refinement to better handle such shifts.

Effect of Different Smoothing Levels: The results for S-CPD in terms of mean and standard deviation of the covering metric for different q values are shown in Fig. 3, with a focus on the selected threshold values. For TSSB, we choose $q_{threshold} = 0.1$, which achieves the optimal mean covering metric of 87.8%, with a low standard deviation, indicating consistent performance. For UTSA, the chosen $q_{threshold} = 0.15$ yields an optimal mean value of 77.1%, demonstrating that these settings provide reliable results in both datasets.

Comparing the results across different q values, it is clear that $q_{threshold} = 0.1$ for TSSB and $q_{threshold} = 0.15$ for UTSA offer the best balance between mean performance and consistency. While other q values may result in slightly lower mean values, the increased standard deviation shows a loss of consistency, reinforcing the suitability of these choices for both datasets. These thresholds provide a desirable smoothing effect by minimizing the impact of noisy patterns without losing important trends. However, using too high a q value can lead to over-smoothing, potentially missing key change points and lowering mean rates,

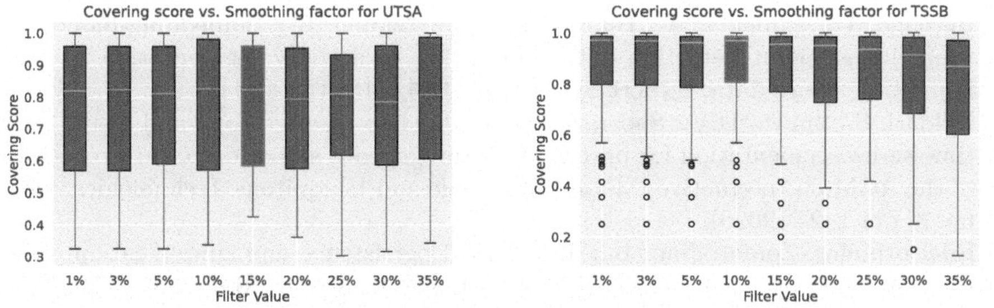

(a) Performance analysis for UTSA. **(b)** Performance analysis for TSSB.

Fig. 3. Box plots of mean and standard deviation metrics for covering score to record performance in different filtering levels q of the S-CPD model.

as shown in Fig. 3. This analysis highlights that while smoothing can improve interpretability, overly aggressive smoothing can obscure critical patterns. The selected thresholds strike a balance between reducing noise and preserving essential structural information, ensuring reliable CP detection across both datasets.

Finally, the complexity of the PH of the signal is $\mathcal{O}(n + m \log m)$, smoothing the signal is $\mathcal{O}(n^2)$, and the overall ClaSP complexity is $\mathcal{O}(\mathcal{C} \cdot n^2)$, resulting in a final complexity of $\mathcal{O}(\mathcal{C} \cdot n^2)$.

6 Conclusion

In this paper, we proposed a method, *S-CPD*, to detect change points in univariate time series data by incorporating a topological smoothing approach. Unlike existing CPD methods, S-CPD incorporates topological smoothing into a classification-based CPD framework to reduce the impact of noise on CPD accuracy. The experiments on real-world data show that compared to other baselines, S-CPD enhanced CP detection noticeably, while on semi-synthetic data, it was the second best without requiring prior knowledge of the number of segments. Future work aims to automatically recognise the appropriate smoothing level.

Acknowledgments. This research was supported by The University of Melbourne's Research Computing Services and the Petascale Campus Initiative.

References

1. Adams, R.P., MacKay, D.J.: Bayesian online changepoint detection. arXiv preprint arXiv:0710.3742 (2007)
2. Aminikhanghahi, S., Cook, D.J.: A survey of methods for time series change point detection. Knowl. Inf. Syst. **51**(2), 339–367 (2017)
3. Bao, X., Zheng, Y., Zhong, J., Chen, L.: SIMTSeg: a self-supervised multivariate time series segmentation method with periodic subspace projection and reverse diffusion for industrial process. Adv. Eng. Inform. **62**, 102859 (2024)

4. Bermejo, U., Almeida, A., Bilbao-Jayo, A., Azkune, G.: Embedding-based real-time change point detection with application to activity segmentation in smart home time series data. Expert Syst. Appl. **185**, 115641 (2021)
5. Deldari, S., Smith, D.V., Sadri, A., Salim, F.: Espresso: entropy and shape aware time-series segmentation for processing heterogeneous sensor data. In: Proceedings of the ACM on Interactive, Mobile, Wearable and Ubiquitous Technologies, vol. 4, no. 3, pp. 1–24 (2020)
6. Edelsbrunner, Zomorodian, L.,: Topological persistence and simplification. Discr. Comput. Geometry **28**, 511–533 (2002). https://doi.org/10.1007/s00454-002-2885-2
7. Edelsbrunner, H., Harer, J., et al.: Persistent homology-a survey. Contemp. Math. **453**(26), 257–282 (2008)
8. El-Yaagoubi, A.B., Chung, M.K., Ombao, H.: Topological data analysis for multivariate time series data. Entropy **25**(11), 1509 (2023)
9. Ermshaus, A., Schäfer, P., Leser, U.: ClaSP: parameter-free time series segmentation. Data Min. Knowl. Disc. **37**(3), 1262–1300 (2023)
10. Gharghabi, S., Ding, Y., Yeh, C.C.M., Kamgar, K., Ulanova, L., Keogh, E.: Matrix profile viii: domain agnostic online semantic segmentation at superhuman performance levels. In: 2017 IEEE International Conference on Data Mining (ICDM), pp. 117–126. IEEE (2017)
11. Gidea, M., Katz, Y.: Topological data analysis of financial time series: landscapes of crashes. Phys. A **491**, 820–834 (2018)
12. Guo, H., Xia, S., An, Q., Zhang, X., Sun, W., Zhao, X.: Empirical study of financial crises based on topological data analysis. Phys. A **558**, 124956 (2020)
13. Jin, M., et al.: A survey on graph neural networks for time series: forecasting, classification, imputation, and anomaly detection. arXiv preprint arXiv:2307.03759 (2023)
14. Karan, A., Kaygun, A.: Time series classification via topological data analysis. Expert Syst. Appl. **183**, 115326 (2021)
15. Killick, R., Fearnhead, P., Eckley, I.A.: Optimal detection of changepoints with a linear computational cost. J. Am. Stat. Assoc. **107**(500), 1590–1598 (2012)
16. Rosen, P., Suh, A., Salgado, C., Hajij, M.: TopoLines: topological smoothing for line charts. arXiv preprint arXiv:1906.09457 (2019)
17. Scott, A.J., Knott, M.: A cluster analysis method for grouping means in the analysis of variance. Biometrics **30**, 507–512 (1974)
18. Truong, C., Oudre, L., Vayatis, N.: Selective review of offline change point detection methods. Signal Process. **167**, 107299 (2020)
19. Truong, P.: An exploration of topological properties of high-frequency one-dimensional financial time series data using TDA (2017)
20. Wilcoxon, F., Katti, S., Wilcox, R.A., et al.: Critical values and probability levels for the Wilcoxon rank sum test and the Wilcoxon signed rank test. Sel. Tables Math. Stat. **1**, 171–259 (1970)
21. Xie, Y., Siegmund, D.: Sequential multi-sensor change-point detection. In: 2013 Information Theory and Applications Workshop (ITA), pp. 1–20. IEEE (2013)

VDASI: VAE-Enhanced Degradation-Aware System Identification Using Constrained Latent Spaces

Saumya Karunadhika[1,3](✉), Ling Luo[1,3], Bastian Oetomo[1,3], Michele Discepola[2,3], and Uwe Aickelin[1,3]

[1] School of Computing and Information Systems, The University of Melbourne, Melbourne, Australia
{ling.luo,bastian.oetomo,uwe.aickelin}@unimelb.edu.au

[2] Department of Chemical Engineering, The University of Melbourne, Melbourne, Australia
michele.discepola@student.unimelb.edu.au

[3] ARC Digital Bioprocess Development Hub, Melbourne, Australia
skarunadhika@student.unimelb.edu.au

Abstract. Degradation processes in industrial systems are notably difficult to monitor due to their hidden, evolving nature and complex nonlinear dynamics, posing significant challenges for accurate system modelling. Reconstruction errors of generative models have been used in literature to capture degradation dynamics; however, these models fail to fully exploit the representations in latent space, often resulting in overgeneralisation and making them ineffective at detecting subtle yet critical degradation patterns. To overcome these limitations, we present a novel system modelling framework, VAE-Enhanced Degradation-Aware System Identification (VDASI), that incorporates constrained latent spaces by integrating Variational Autoencoders (VAE) with Gaussian Process Nonlinear AutoRegressive models with Exogenous Inputs (GP-NARX). The framework operates in two distinct phases: the first captures degradation dynamics by employing reconstruction errors from both the original input space and constrained latent space, generating a reliable Degradation Index, while the second models key process variables. Results demonstrate the capability of VDASI, to outperform conventional approaches across five industrial datasets. By incorporating re-encoding layers and carefully constrained latent space, our approach increases sensitivity to dynamic degradation while avoiding overgeneralisation. This work underscores the critical role of well-constrained latent spaces in effectively capturing evolving degradation phenomena, and the importance of degradation-aware models in system identification.

Keywords: Constrained Latent Space · Degradation Dynamics · Variational Autoencoders · System Modelling · Regularisation

X. Wu et al. (Eds.): PAKDD 2025, LNAI 15875, pp. 159–171, 2025.
https://doi.org/10.1007/978-981-96-8295-9_11

1 Introduction

Degradation in industrial systems poses a significant challenge to maintaining machines and processes in optimal working conditions [5]. These degradation trends can negatively impact performance, leading to unexpected downtime and inefficiencies. Overcoming this challenge requires advanced machine learning (ML) frameworks capable of capturing the hidden, dynamic and nonlinear nature of degradation. This study introduces a novel ML framework, VAE-Enhanced Degradation-Aware System Identification (VDASI), that incorporates constrained latent spaces by integrating Variational Autoencoders (VAEs) with Gaussian Process Nonlinear AutoRegressive models with Exogenous Inputs (GP-NARX) to perform degradation-aware system modelling. The primary motivation for this study stems from bio-processing applications, particularly Single Pass Tangential Flow Filtration (SPTFF), where membrane fouling caused by particle and solute accumulation significantly impacts performance over time and remains unobserved [17]. Degradation trends also pose challenges in broader industrial contexts, causing system models to no longer function as expected in the long term. Incorporating system degradation into modelling efforts is therefore essential for better understanding and optimising these processes.

Degradation detection involves either physics-based methods, which rely on accurate models that may not exist for complex systems, or data-driven methods, which typically require supervised algorithms. While VAEs use unsupervised learning to bypass the need for labelled data, they are prone to "overgeneralisation"—where the model's strong generalisation ability not only better reconstructs normal data, but also degradation data [2]—making them less effective at identifying subtle, yet critical degradation dynamics through reconstruction errors. To address this limitation, our approach introduces constraint networks and re-encoders into the standard VAE —a method inspired by prior work on anomaly detection [20]—and leverages reconstruction errors from both original input and latent spaces. This dual-layered analysis yields a refined Degradation Index (DI), which serves as a pivotal feature in system identification.

Although primarily motivated by bio-processing, we evaluated our approach on five distinct degradation datasets, including robotics, metal coating, and manufacturing, benchmarking its performance against state-of-the-art methods. The results consistently demonstrated superior accuracy and confirmed the framework's adaptability to diverse processes with nonlinear degradation behaviours. By addressing the limitations of overgeneralisation in traditional generative models, our framework provides a more sensitive and accurate tool for degradation monitoring. The *key contributions* of this study are:

- A novel framework utilising VAEs with constrained latent spaces and re-encoders for degradation-aware system modelling.
- Introduction of a unique loss function to derive a DI, aiding in the identification of underlying degradation mechanisms.
- Enhanced prediction accuracy for key process parameters by incorporating hidden degradation dynamics into system models.

The remainder of this paper is organised as follows: Sect. 2 reviews related work; Sect. 3 introduces the followed methodology; Sect. 4 describes the experimental setup; Sect. 5 discusses the results; and finally, Sect. 6 offers concluding remarks.

2 Related Work

Broadly, methodologies for constructing DIs fall into two categories: physics-based and data-driven approaches. Physics-based methods, while valuable, are often limited because accurate physics models are not available for many complex systems. These methods also struggle with uncertainties in real manufacturing environments and difficulties in measuring critical inputs [4]. In contrast, data-driven methods offer greater adaptability to these challenges. Traditional data-driven approaches typically rely on statistical features extracted from time and frequency domains. However, these methods may overlook complex relationships within the data [12].

Machine learning models, particularly deep neural networks (DNNs), have shown promise in improving degradation prediction by automatically learning intricate patterns. However, these models are typically supervised, requiring human intervention to label degradation trends accurately [1,14]. Incorrect or inconsistent labelling can lead to misleading degradation patterns, ultimately reducing model effectiveness.

Unsupervised approaches [13,16], such as deep belief networks (DBNs) and Principal Component Analysis (PCA), address the need for manual labelling by providing robust feature extraction. These methods also mitigate the issue of lacking prior domain knowledge. Autoencoders (AEs), a type of neural network, have been employed in several studies to construct health indices [6,18]. A variant of the AE, the VAE, introduces variational approximations to model the input data. The VAE incorporates an additional loss term, the Kullback-Leibler (KL) divergence, and uses a specialised training method known as stochastic gradient variational Bayes [3]. As shown in [19], features extracted by VAEs tend to contain richer information compared to those obtained by AEs.

In the context of degradation monitoring, VAEs are commonly trained on normal (healthy) data. This allows them to effectively reconstruct normal system dynamics and makes reconstruction errors a useful health indicator [7]. Although VAEs offer the advantage of capturing complex features through generative modelling, this generative capability can also introduce challenges. Specifically, VAEs are prone to "overgeneralisation" [20], where the model becomes so proficient at generating realistic data that it fails to detect subtle degradation dynamics. This overgeneralisation undermines their reliability for degradation identification, as monitoring degradation dynamics requires higher sensitivity to effectively capture gradual changes. This research aims to leverage constrained latent spaces to address the challenges of traditional generative models and enable degradation-aware system modelling.

3 Methodology

The objective of this study is to enhance the predictive accuracy of system models by incorporating DI as an additional feature. The model inputs include lagged multivariate sensor data ($x_{(t-\tau_x):t}$), lagged process parameters ($y_{(t-\tau_y):t}$), and lagged degradation dynamics ($\text{DI}_{(t-\tau_{\text{DI}}):t}$), where τ_x, τ_y, and τ_{DI}, represent the maximum lags considered for sensor data, process parameters, and DI respectively. The system is expressed as:

$$y_{t+k} = f(x_{(t-\tau_x):t}, y_{(t-\tau_y):t}, \text{DI}_{(t-\tau_{\text{DI}}):t}) + e_t \quad (1)$$

where $f : \mathbb{R}^n \to \mathbb{R}$ is a nonlinear mapping that models the system's dynamics by capturing relationships between the inputs and y_{t+k}, the k-step-ahead prediction horizon of process parameter y. Here e_t represents the residual error.

Our methodology follows a two-stage process as shown in Fig. 1. In the first stage, degradation dynamics are estimated using the proposed VAE framework. The VAE's objective is to reconstruct the input through the decoder while minimising reconstruction loss. Since our aim is to capture degradation dynamics, the model is trained on data from the initial cycles of operation, allowing it to produce relatively low reconstruction errors under normal conditions without degradation. As new data is introduced, increasing reconstruction errors over time indicate the onset of degradation. The proposed model integrates a VAE framework with a Constraint Network, which enables better identification of hidden degradation dynamics. In the second stage, we use a GP-NARX model to predict process variables multiple steps ahead. The following subsections outline each component of the framework.

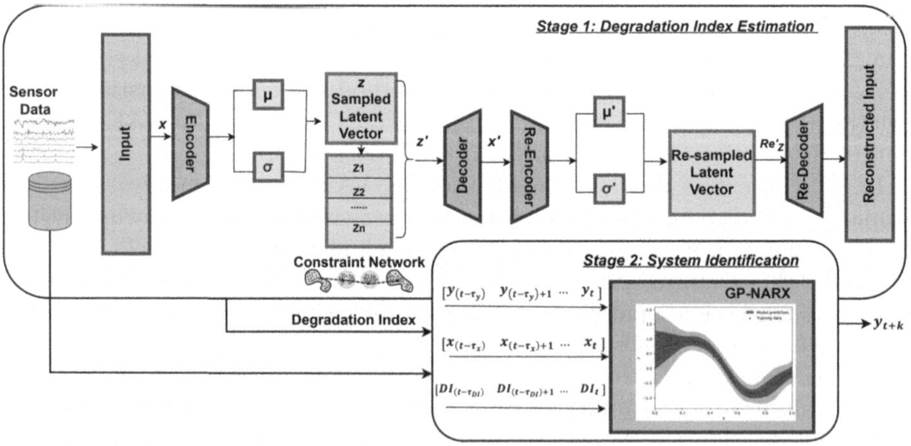

Fig. 1. VDASI Framework

3.1 Degradation Index Estimation

VAE. The Variational Autoencoder (VAE) [3] is a generative deep learning model designed to learn data distributions in an unsupervised manner. VAEs utilise principles from Bayesian inference and consist of an encoder, a latent distribution, and a decoder. Fundamentally, VAEs leverage simple distributions (e.g., Gaussian) which, when combined with neural networks, can approximate complex data distributions. The encoder-decoder pairs utilise bidirectional long short-term memory (BiLSTM) networks. Specifically, the forward process in a VAE involves passing an input sample X through the encoder to estimate the parameters of the latent space. A latent variable $z \in \mathbb{R}^D$ is sampled from the approximate posterior $q_\phi(z \mid x) = \mathcal{N}(z; \mu_\phi(x), \text{diag}(\sigma_\phi(x)^2))$, a multivariate Gaussian with mean $\mu_\phi(x)$ and variance $\sigma_\phi(x)^2$. The prior $p_\theta(z)$ is assumed to be a standard Gaussian $\mathcal{N}(0, I)$. The decoder, parameterised by θ, reconstructs X by modelling $p_\theta(x \mid z)$. This process enables VAEs to define a loss function:

$$\mathcal{L}_{\text{VAE}} = -\mathbb{E}_{q_\phi(z|x)} [\log p_\theta(x \mid z)] + \text{KL}(q_\phi(z \mid x) \| p(z)) \tag{2}$$

where the first term is the reconstruction loss, and the remaining term regularises the latent space. A critical feature of VAEs is the re-parametrisation trick, which makes the sampling operation differentiable. Instead of sampling z directly from $q_\phi(z \mid x)$, the latent variable is re-expressed as $z = \mu_\phi(x) + \epsilon \cdot \sigma_\phi(x)$, where $\epsilon \sim \mathcal{N}(0, I)$ is sampled element-wise from a standard normal distribution. This allows for backpropagation through the entire model. Unlike traditional autoencoders, which are deterministic and lack probabilistic grounding, VAEs provide probabilistic modelling of data, allowing them to generate calibrated distributions rather than mere deterministic reconstructions.

Constraint Network. In degradation detection, standard generative models often struggle to capture degradation dynamics accurately. This is due to two potential issues: (a) the model may not reconstruct normal data with sufficient fidelity, leading to elevated reconstruction errors, or (b) it may incorrectly produce low reconstruction errors for degraded data, which can mask underlying degradation—a phenomenon known as "overgeneralisation" [2]. To address this, a Constraint Network [20] is introduced after the latent sampling step to limit the model's ability to reconstruct degraded data, aligning the reconstruction more closely with normal operating states.

The Constraint Network consists of a matrix $\mathbf{C} \in \mathbb{R}^{\mathbf{z}_{\text{dim}} \times \mathbf{N}}$, where $\mathbf{z}_{\text{dim}}$ is the latent vector dimension, and N defines model complexity. For each latent variable z, cosine similarity with each row of C produces representation coefficients:

$$w_i = \frac{\mathbf{z} \cdot \mathbf{c}_i}{\|\mathbf{z}\| \|\mathbf{c}_i\|}, \quad i = 1, 2, \ldots, N$$

The resulting vector is $\underline{\mathbf{w}} = (w_1, w_2, \ldots, w_N) \in \mathbb{R}^N$. Normalising the entire vector yields $\hat{\mathbf{w}} = (\hat{w}_1, \hat{w}_2, \ldots, \hat{w}_N)$, where $\hat{w}_i = \frac{w_i}{\|\mathbf{w}\|}$ for $i = 1, 2, \ldots, N$.

A sparsity constraint then zeros elements in $\hat{\mathbf{w}}$ below a threshold, reducing the model's expressiveness for degraded states. The final constrained latent vector $\hat{z}$ is computed as:

$$\hat{z} = \sum_i w'_i c_i, \tag{3}$$

where w'_i denotes the thresholded elements of $\hat{\mathbf{w}}$. This approach limits the model's reconstruction of degraded states, enhancing its sensitivity to degradation while maintaining accurate representations of normal conditions.

Re-encoder. In the final step, the reconstructed data is passed through a re-encoder network, where updated latent space parameters μ' and σ', are generated. New latent variables are then sampled from these updated parameters, matching the dimensions of the original latent variables z. The purpose of the re-encoder is to further refine the latent space, allowing the model to better capture features in data both in original and new latent spaces [20]. By using both original and updated latent spaces, the model becomes more effective at distinguishing normal operation from degradation, thereby reducing overgeneralisation.

Loss Function. In the defined VAE, the loss function consists of four components. The first component is derived from the standard VAE loss, which includes reconstruction loss and KL divergence loss. The reconstruction loss minimises the $L2$ distance between the original data X and the reconstructed data X' in the input space, while the KL divergence loss encourages the model's latent distribution to approximate the training data distribution.

$$L_{rec\_x1} = \|X - X'\|_2 \tag{4}$$

$$L_{\mathrm{KL}_1} = \frac{1}{2} \sum_{i=1}^{z_{\mathrm{dim}}} \left[(\mu_i^2 + \sigma_i^2) - 1 - \log(\sigma_i^2)\right] \tag{5}$$

where μ_i and σ_i represent the mean and standard deviation of the latent variable z_i.

The second component stems from the re-encoder, which replicates the KL divergence loss used in the original VAE.

$$L_{\mathrm{KL}_2} = \frac{1}{2} \sum_{i=1}^{z_{\mathrm{dim}}} \left[(\mu_i'^2 + \sigma_i'^2) - 1 - \log\left(\sigma_i'^2\right)\right] \tag{6}$$

The third component is a reconstruction loss applied in the latent space. Here, the error of the latent vector serves as an additional loss function, allowing the model to capture more refined representations in the latent space.

$$L_{\mathrm{lat}} = \left\|Z' - \mathrm{Re}'_Z\right\|_2 \tag{7}$$

where Z' and Re'_Z denote the latent space of the first encoder and re-encoder respectively.

The final loss function is from the final reconstruction of the input:

$$L_{\mathrm{rec\_}x2} = \|X - Re'_X\|_2 \tag{8}$$

where Re'_X denotes the reconstructed input of the re-decoder. The total loss function can be written as:

$$L_{VAE} = L_{rec\_x1} + L_{KL_1} + L_{lat} + L_{KL_2} + L_{rec\_x2} \tag{9}$$

The system's degradation dynamics are assessed using a weighted degradation indicator, which combines loss components from both the original input space and the latent space. To account for variations in scale, we applied dynamic normalisation, adjusting each loss value to a consistent range of [0,1] based on the 5th and 95th percentiles of its distribution. This method reduces the impact of outliers and stabilises the degradation indicator calculation. The **Degradation Index (DI)** is defined as:

$$\mathrm{DI} = \alpha \cdot L^{\mathrm{norm}}_{\mathrm{rec\_x1}} + \beta \cdot L^{\mathrm{norm}}_{\mathrm{lat}} \tag{10}$$

such that $\alpha+\beta = 1$. Here $L^{\mathrm{norm}}_{\mathrm{rec\_x1}}$ and $L^{\mathrm{norm}}_{\mathrm{lat}}$ denote the normalised reconstruction and latent loss functions, respectively.

3.2 System Identification Model GP-NARX

The Nonlinear Autoregressive with Exogenous Inputs (NARX) model is a time series modelling approach that relates the current value of a target series y_t to both its past values and those of an influencing input series x_t. In this study, we use a Gaussian Process (GP) as the nonlinear function within the NARX framework, forming the GP-NARX model. GPs are particularly well-suited for this application as they model complex nonlinear relationships while providing uncertainty estimates, offering both predictions and confidence intervals. This is essential in industrial applications where understanding uncertainty is critical. The GP-NARX model is expressed as:

$$y_t = \mathcal{GP}(y_{t-1}, y_{t-2}, \ldots, x_t, x_{t-1}, \ldots) + \varepsilon_t \tag{11}$$

Here, $\mathcal{GP}$ captures the nonlinear relationships and provides uncertainty estimates, while ε_t accounts for any remaining unexplained variability.

4 Experimental Setup

This section outlines the experimental setup, including datasets, baseline algorithms, and implementation details. The aim of the experimental setup is to benchmark the proposed framework against baseline algorithms and demonstrate its effectiveness across diverse industrial datasets.

4.1 Datasets

SPTFF Dataset. The dataset was collected from a laboratory SPTFF system. Retentate protein concentration (mg/mL) was the key process variable under consideration. Given the system's time delay, we investigated forecasts for the 11th to 15th steps ahead in system modelling for this dataset.

Other Benchmark Datasets. In our study, aside from testing the SPTFF dataset, which was our primary focus, we utilised several publicly available benchmark datasets for degradation process tasks outlined in [8]. The forecast horizon included forecasts at the 1st, 5th, 10th, 15th, and 20th steps ahead for the datasets below.

- **Vega Shrink-wrapper Dataset**: Provides degradation information for a cutting blade component of Vega shrink-wrapper, with motor torque monitored as the key process variable [9].
- **Preventive to Predictive Maintenance Dataset (PHM)**: This dataset captures the observed degradation process, specifically the clogging of filters during the separation of solid particles from gas, with differential pressure serving as the primary process variable [10].
- **E-coating Ultrafiltration Maintenance Dataset (E-Coat)**: This dataset is sourced from an electrophoresis painting process, including degradation progressions of filters, with the temperature in the paint tank tracked as the main process variable [11].
- **Servomotor Degradation Dataset (Servo):** This dataset captures the degradation details of servomotors, highlighting how wear and damage develop over time. Here stator current is the key process variable [15].

4.2 Benchmark Algorithms

To validate our method, we created DIs for the aforementioned datasets using several benchmark algorithms to facilitate performance comparisons with our proposed framework, including **Isolation Forest (IF)**, a state-of-the-art unsupervised algorithm that employs a tree-based approach to isolate anomalies, making it relevant for degradation detection; **Local Outlier Factor (LOF)**, which detects anomalies by comparing the local density of points with that of their neighbours; **Dynamic Time Warping (DTW)**, a technique measuring time-series dissimilarity and is particularly valuable for handling non-linear time distortions, thereby aiding in degradation identification; and **Standard VAE**, a baseline implementation of the VAE without constraint networks or re-encoders.

4.3 Implementations and Configurations

In the first stage, we used the initial cycles of the datasets to train the VDASI, and we estimated the DI for the entire dataset during the inference phase. The process variable of interest was excluded from the dataset at this stage to ensure an unbiased analysis for comparison with other baseline algorithms. For the SPTFF, Vega Shrink-wrapper, PHM, E-Coat and Servo Motor datasets, we used the following configurations as optimal values: (2, 32, 400), (4, 32, 400), (4, 72, 400), (2, 32, 100), (2, 72, 2018) for (z_{dim}, N, training size), respectively. All models were trained for 50 epochs with a learning rate of 1×10^{-5}. The parameters α and β were set to 0.8 and 0.2, respectively. During system identification, a 70:30 train-test split and three-fold cross-validation with grid search were employed. Root Mean Squared Error (RMSE) served as the evaluation metric.

5 Performance Evaluation

We evaluated the model's performance with the incorporation of a degradation indicator during the system identification step. Significant improvements in system models were seen with the inclusion of DI from proposed VDASI. The percentage reductions of test RMSEs for the SPTFF, Vega-Shrink, PHM, E-Coating and Servo datasets using VDASI, compared to the raw degradation-agnostic model, were 15.64%, 34.28%, 18.51%, 5.52% and 52.66%, respectively. For comparison, we also performed system identification using DIs derived from other benchmark algorithms. Figure 2 illustrates estimated DIs across different datasets. As shown in Table 1, for each dataset considered, the lowest average test RMSE was achieved by the proposed VDASI framework, which incorporates a constraint network and re-encoder to capture degradation dynamics.

Table 1. Average normalised RMSE for the considered forecast horizon with ± standard deviation, with bolded values indicating the lowest RMSE

Algorithm	SPTFF	Vega Shrink	PHM	E-Coating	Servo
Without DI	0.3917	1.0267	0.2410	0.8968	0.8856
	± 0.035	± 0.475	± 0.061	± 0.508	± 0.609
VDASI	**0.3304**	**0.6747**	**0.1964**	**0.8474**	**0.4193**
	± 0.031	± 0.128	± 0.043	± 0.521	± 0.230
IF	0.3938	0.8251	0.2612	0.9127	0.6443
	± 0.046	± 0.254	± 0.086	± 0.518	± 0.359
DTW	0.9066	0.9422	0.4042	0.9283	0.5041
	± 0.096	± 0.341	± 0.214	± 0.504	± 0.515
LOF	0.4353	0.8753	0.2661	0.9068	0.4803
	± 0.055	± 0.309	± 0.064	± 0.516	± 0.261
VAE	0.3575	0.7206	0.2011	0.8530	0.8621
	± 0.043	± 0.162	± 0.070	± 0.526	± 0.451

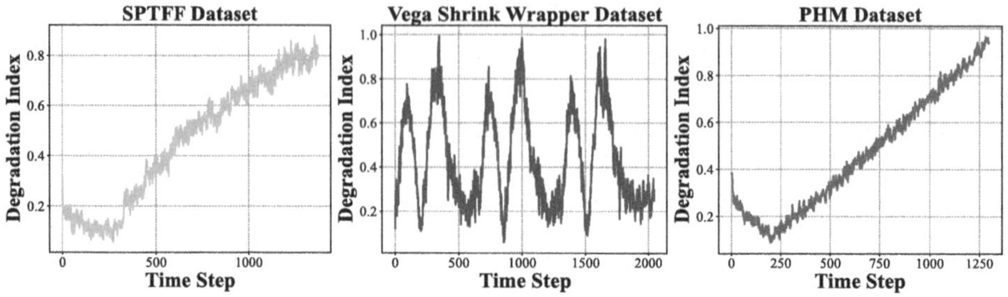

Fig. 2. Estimated smoothed degradation indices from the VDASI framework across different datasets.

5.1 Ablation Study of VDASI Framework Components

We conducted ablation experiments by excluding different components of the VDASI framework to assess the contribution of each element. Comparisons were made between the full VDASI framework, a VAE model with only a re-encoder, and a VAE model with only a constraint network. The results as shown in Table 2 indicated that adding a constraint network or a re-encoder individually did not improve model performance. However, when both components were combined, they contributed to enhanced accuracy, highlighting the synergistic effect of the re-encoder and constraint network within the full VDASI framework. Adding a constraint network only affects the reconstruction loss, while adding a re-encoder alone does not constrain the latent space. This indicates that incorporating latent loss by constraining the latent space into the DI results in more accurate system identification models. Figure 3 shows that with the constraint network (left), the model captures degradation dynamics smoothly and gradually, whereas without it (right), the results are erratic, indicating reduced stability and accuracy. Reconstruction errors in the absence of constraint network struggle to capture degradation dynamics effectively due to overgeneralisation. This highlights how VDASI addresses overgeneralisation and effectively captures degradation trends, thereby improving system modelling accuracy.

Table 2. Average normalised RMSE for the considered forecast horizon with ± standard deviation, with bolded values indicating the lowest RMSE

Algorithm	SPTFF	Vega Shrink	PHM	E-Coating	Servo
VDASI	**0.3304**	**0.6747**	**0.1964**	**0.8474**	**0.4193**
	± 0.031	± 0.128	± 0.043	± 0.521	± 0.230
VAE	0.3799	0.7732	0.2078	0.8704	0.4230
+ ReEncoder	± 0.046	± 0.219	± 0.054	± 0.522	± 0.240
VAE	0.4049	0.7499	0.2235	1.0200	0.9152
+ Constraint Network	± 0.042	± 0.188	± 0.052	± 0.601	± 0.396

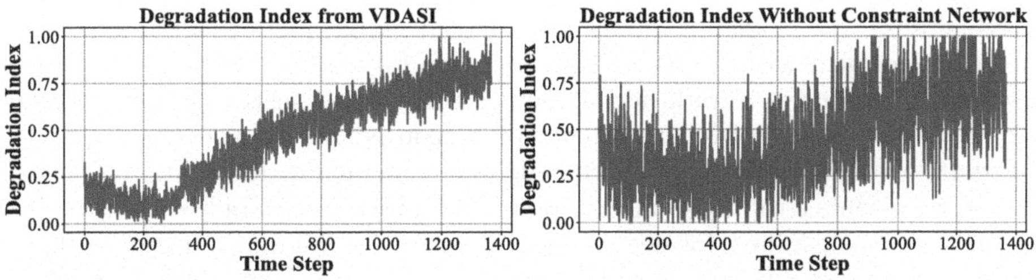

Fig. 3. Comparison of VDASI with and without a constraint network on the SPTFF dataset.

5.2 Sensitivity Analysis

We analysed the effect of the constraint network's parameters and weights of DI on model performance as shown in Table 3. The latent dimension (z_{dim}) was varied from 2 to 8. Additionally, complexity parameter of the constraint network (N), was incrementally varied from 16 to 128. Latent dimensions of 4 and 6 demonstrated the best results on average. This can be attributed to smaller dimensions failing to capture the full representation of the data, while larger dimensions tend to overfit. Thus, a middle range is favourable for z_{dim}, balancing the trade-off between underfitting and overfitting. Furthermore, lower values of N outperformed higher values on average. This trend can be attributed to the fact that smaller N results in a simpler representation of constraints, reducing the risk of overfitting. The DI weights (α, β) were varied from $(0.6, 0.4)$ to $(0.9, 0.1)$, with weights $(0.8, 0.2)$ yielding the best results on average.

Table 3. Sensitivity Analysis: Mean normalised RMSE, with bolded values indicating the lowest average RMSE for z_{dim}, N, and (α, β) values

z_{dim}				N			
2	4	6	8	16	32	64	128
0.5322	0.4921	**0.4910**	0.5164	**0.5122**	0.5168	0.5236	0.5171
(α, β) weights							
(α = **0.9**, β = **0.1**)		(α = **0.8**, β = **0.2**)		(α = **0.7**, β = **0.3**)		(α = **0.6**, β = **0.4**)	
0.4974		**0.4970**		0.5232		0.5448	

6 Conclusion

We propose VDASI, a degradation-aware system modelling framework that enhances accuracy by exploring the latent spaces of a VAE. Proposed method integrates a VAE with a constraint network and a re-encoder, to address the overgeneralisation challenge in traditional generative models for degradation detection, coupled with a GP-NARX. The results validate that our framework successfully captures degradation dynamics through reconstruction errors from original input space as well as latent space, consistently outperforming state-of-the-art algorithms and models that do not account for degradation. Overall, VDASI represents a robust solution, applicable across diverse industrial domains, setting a new benchmark for degradation-sensitive modelling. In future work, we aim to experiment with constrained latent spaces of other generative models.

Acknowledgement. This research was supported under the Australian Research Council's Industrial Transformation Research Program (ITRP) funding scheme (project number IH210100051). The ARC Digital Bioprocess Development Hub is a collaboration between The University of Melbourne, University of Technology Sydney, RMIT University, CSL Innovation Pty Ltd, Global Life Sciences Solutions Australia Pty Ltd. (Cytiva) and Patheon Biologics Australia Pty Ltd. We acknowledge Prof. Sandra Kentish and Prof. Sally Gras for their valuable insights into the bioprocessing application in this work.

References

1. Chen, L., et al.: Health indicator construction of machinery based on end-to-end trainable convolution recurrent neural networks. J. Manuf. Syst. **54**, 1–11 (2020)
2. Kim, C., et al.: DeepNAP: deep neural anomaly pre-detection in a semiconductor fab. Inf. Sci. **457**, 1–11 (2018)
3. Kingma, D.P., Welling, M., et al.: Auto-encoding variational bayes (2013)
4. Li, P., et al.: A novel scalable method for machine degradation assessment using deep convolutional neural network. Measurement **151**, 107106 (2020)
5. de Lima, M.J., et al.: HealthMon: an approach for monitoring machines degradation using time-series decomposition, clustering, and metaheuristics. Comput. Ind. Eng. **162**, 107709 (2021)
6. Lin, X., et al.: Fault diagnosis of aero-engine bearing using a stacked auto-encoder network. In: ITOEC, pp. 545–548. IEEE (2018)
7. Malhotra, P., et al.: Multi-sensor prognostics using an unsupervised health index based on lstm encoder-decoder. arXiv preprint arXiv:1608.06154 (2016)
8. Mauthe, F., et al.: Overview of publicly available degradation data sets for tasks within prognostics and health management. arXiv:2403.13694 (2024)
9. One year industrial component degradation dataset. https://www.kaggle.com/datasets/inIT-OWL/one-year-industrial-component-degradation
10. Preventive to predictive maintenance dataset. https://www.kaggle.com/datasets/prognosticshse/preventive-to-predicitve-maintenance/data
11. Process data for predictive maintenance dataset. https://www.kaggle.com/datasets/boyangs444/process-data-for-predictive-maintenance
12. Qin, Y., et al.: Gated dual attention unit neural networks for remaining useful life prediction of rolling bearings. IEEE Trans. Ind. Inf. **17**(9), 6438–6447 (2020)
13. Rai, A., Upadhyay, S.H.: Bearing performance degradation assessment based on a combination of empirical mode decomposition and k-medoids clustering. Mech. Syst. Signal Process. **93**, 16–29 (2017)

14. She, D., Jia, M.: Health indicator construction of rolling bearings based on deep convolutional neural network considering phase degradation. In: PHM-Paris, pp. 373–378. IEEE (2019)
15. Subramanian, A., et al.: Servomotor dataset: modeling health in mechanisms with typically intermittent operation. In: PHM Society Conference, vol. 15, no. 1 (2023)
16. Xu, F., et al.: Health indicator construction for roller bearing based on an unsupervised deep belief network with a novel sigmoid zero local minimum point model. Struct. Health Monit. **20**(4), 2110–2123 (2021)
17. Xu, H., et al.: A simple method to identify the dominant fouling mechanisms during membrane filtration. Membranes **10**(8), 171 (2020)
18. Yang, Z., et al.: Automatic extraction of a health indicator from vibrational data by sparse autoencoders. In: ICSRS, pp. 328–332. IEEE (2018)
19. Yao, R., et al.: Unsupervised anomaly detection using variational auto-encoder based feature extraction. In: ICPHM, pp. 1–7. IEEE (2019)
20. Zhang, C., et al.: VESC: a new variational autoencoder based model for anomaly detection. Int. J. Mach. Learn. Cybern. **14**(3), 683–696 (2023). https://doi.org/10.1007/s13042-022-01657-w

Disentangled Mode-Specific Representations for Tensor Time Series via Contrastive Learning

Kohei Obata[1(✉)], Taichi Murayama[2], Zheng Chen[1], Yasuko Matsubara[1], and Yasushi Sakurai[1]

[1] SANKEN, Osaka University, Osaka, Japan
{obata88,chenz,yasuko,yasushi}@sanken.osaka-u.ac.jp
[2] Yokohama National University, Kanagawa, Japan
murayama-taichi-bs@ynu.ac.jp

Abstract. Multi-mode tensor time series (TTS) can be found in many domains, such as search engines and environmental monitoring systems. Learning representations of a TTS benefits various applications, but it is also challenging since the complexities inherent in the tensor hinder the realization of rich representations. In this paper, we propose a novel representation learning method designed specifically for TTS, namely MoST. Specifically, MoST uses a tensor slicing approach to reduce the complexity of the TTS structure and learns representations that can be disentangled into individual non-temporal modes. Each representation captures mode-specific features, which are the relationship between variables within the same mode, and mode-invariant features, which are in common in representations of different modes. We employ a contrastive learning framework to learn parameters; the loss function comprises two parts intended to learn representation in a mode-specific way and mode-invariant way, effectively exploiting disentangled representations as augmentations. Extensive experiments on real-world datasets show that MoST consistently outperforms the state-of-the-art methods in terms of classification and forecasting accuracy. Code is available at https://github.com/KoheiObata/MoST.

Keywords: Tensor time series · Contrastive learning

1 Introduction

With the rapid growth of digital innovation for the collection of data, a time series sequence is generated from multiple attributes or modes, forming a tensor time series (TTS).[1] This complex data is increasingly prevalent across divers domains, ranging from environment monitoring [28] to financial analysis [26]. An instance of a TTS is online activity data [17], which records the search volumes

[1] Here, we mainly assume that the TTS has three or more modes.

X. Wu et al. (Eds.): PAKDD 2025, LNAI 15875, pp. 172–184, 2025.
https://doi.org/10.1007/978-981-96-8295-9_12

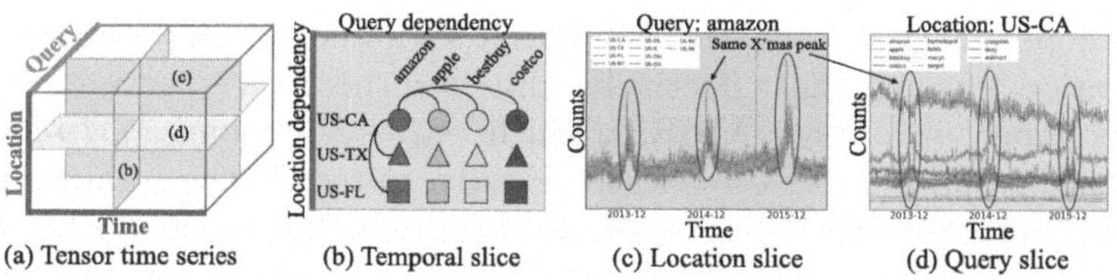

Fig. 1. Illustrations of a tensor time series and three slices along different modes. (a) A tensor time series with three modes: location, query, and time. (b) Location and query dependencies of a temporal slice. (c), (d) Location and query slices have their own intra-mode dependencies, but temporal dependencies are common.

of various queries at various locations for each timestamp, in the form {Location, Query, Time} (Fig. 1(a)). Each temporal slice of the 3rd-order TTS (Fig. 1(b)) is a matrix where each non-temporal mode (i.e., location and query) has several variables. As depicted in Fig. 1(c) and (d), each non-temporal mode slice consists of a multivariate time series (MTS) that contains the interactions among corresponding variables and the dependencies between its past and future [21]. Since TTS has a multifaceted structure with intricate interactions, effectively learning meaningful representations from such data is a long-standing challenge.

Modeling TTS requires us to consider the multiple non-temporal modes since tensors arise from the rich information of each mode. Thus, tensor studies normally assume that different modes are associated with different types of dependencies [2,10,20,31], referred to as *intra-mode dependencies*. As shown in Fig. 1(b), location and query dependencies are examples of intra-mode dependencies. Moreover, while different mode slices in the TTS have different intra-mode dependencies, there are common features invariant in modes as they originate from the same TTS; that is, they have similar *temporal dependencies*. As shown in Fig. 1(c) and (d), the search volumes peak before and after every Christmas, meaning they have similar seasonality regardless of locations and queries. Designing a framework that accommodates these intra-mode dependencies and temporal dependencies inherent in the TTS is expected to facilitate the learning of rich representations.

Despite the number of studies in representation learning, existing methods take little account of the TTS structure [34]. Tensor decomposition is a conventional method for handling a tensor that can address intra-mode dependencies, which aims to provide a low-rank representation that minimizes a reconstruction error and additional penalty terms (e.g., Frobenius norm or KL-divergence) [8]. Although a compact representation is an essential principle for feature reduction, such objective functions mainly aim for accurate reconstruction and are unsuitable for obtaining rich representation that can be applied to downstream tasks. Therefore, recent works incorporate deep learning techniques to further facilitate the representation learning from tensor decomposition. For instance, ATD [31], a self-supervised representation learning method for tensors, combines tensor decomposition and contrastive learning (CL) and succeeds in learning representations for the downstream classification task. However, such tensor decomposition methods usually treat all modes equally, resulting in overlooking temporal dependencies. Thus, there is a need to develop a new approach to TTS representation learning that effectively addresses the TTS structure.

In this paper, we present **MoST**, which learns *disentangled **Mo**de-**S**pecific representations for **T**ensor time series.* Firstly, to reduce the complexity of the TTS structure, we utilize a tensor slicing approach that slices a TTS into each non-temporal mode to make sets of mode slices, where each set eventually converts into corresponding mode-specific representations. After the slicing, slices are independently mapped into latent spaces via a slice feature encoder so that the model can capture intra-mode dependencies, and then are aggregated to generate mode-specific representations. The parameters of the model are learned via a CL framework, where we effectively exploit disentangled representations as augmentations to learn mode-specific and mode-invariant features inherent in the TTS. The contributions of this paper can be summarized as follows.

- We propose MoST, a new representation learning method specifically designed to leverage the TTS structure. To the best of our knowledge, this is the first work to provide representations of TTS via CL.
- To learn mode-specific representations, we introduce tensor slicing and propose corresponding contrastive losses.
- We conduct extensive experiments on 11 real-world datasets to verify the effectiveness and efficiency of our proposed model in downstream tasks.

2 Related Work

Tensor Time Series Analysis. Analyzing TTS benefits a variety of applications, such as environmental monitoring [5,28], biomedical analysis [16], and online activity data [19]. Traditional approaches to tensor analysis typically involve tensor/matrix decomposition: Tucker/CP decomposition and SVD [1,12]. They obtain a lower-dimensional representation that captures a good summarization of a tensor/matrix, but they do not consider the temporal dependencies [31]. To complement the temporal information, tensor/matrix decomposition is further combined with a dynamical system [2,9,24]. For example, SSMF [11] is an online forecasting method that combines a dynamical system with non-negative matrix factorization to capture seasonal trends in TTS. Recent work integrates deep learning models with tensor decomposition and is intended to learn invariant tensor features for forecasting [10]. However, since the dynamical system conveys the information recurrently from the one previous timestamp, they do not sufficiently capture long-term temporal dependencies. Some models are designed to analyze a tensor from a specific modality by utilizing their intrinsic characteristics. These include web data models that utilize the Lotka-Volterra equations [17,19] and spatio-temporal models that utilize a pre-defined graph [27]. Despite their success, these studies are data and task-specific; how to construct a general-purpose model that can extend to various TTS remains a challenge.

Time Series Contrastive Learning. Self-supervised learning enables transferable representations for diverse downstream tasks [34]. Inspired by the success

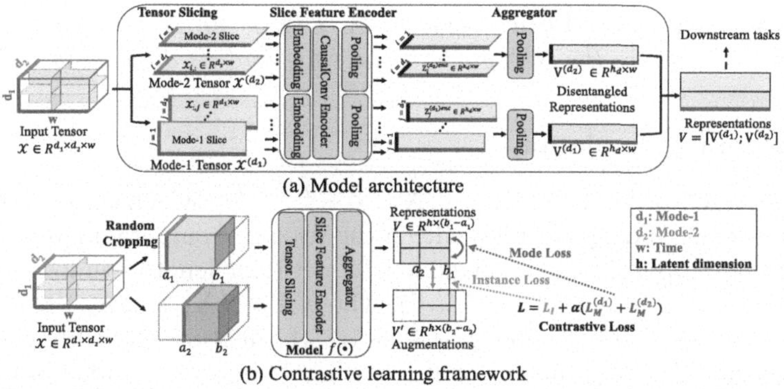

Fig. 2. (a) MoST slices the TTS along each non-temporal mode and independently feeds a slice into the slice feature encoder to learn a representation of the slice. Then, the aggregator is applied to summarize mode information. (b) The parameter of the model is learned via contrastive loss, which is composed of mode loss and instance loss. Mode loss utilizes the representations from different sliced tensors, while instance loss utilizes the representations generated by random cropping as contrastive augmentations.

in computer vision and NLP [23,36], CL has recently emerged as a prominent technique for learning invariant representations in time series data [6,13]. For example, CPC [22], a notable CL method, learns representations by predicting future sequences in latent space, and TNC [25] captures temporal smoothness by defining the neighborhood distribution. However, they generate positive samples from neighboring timestamps and fail to capture global features in time series. A recent successful method is TS2Vec [33], which makes it possible to learn global features for time series by introducing hierarchical and instance losses. The learned representations are generic to downstream tasks, but some methods focus particularly on learning task-specific representations, such as classification [7,18], imputation [14], and change point detection [4]. There is a recent promising trend that learns disentangled representations of time series through CL. As time series can be decomposed into seasonal and trend or spectral and temporal domains, CoST [30] learns disentangled seasonal-trend representations for forecasting, and TF-C [35] and BTSF [32] exploit representations from spectral and temporal domains, to obtain global features in time series.

Our work distinguishes itself from these methods in the following aspects: i) MoST learns representations for TTS that can be applied to any downstream tasks. ii) MoST learns disentangled representations capturing different *mode-specific features* through a tensor slicing approach and contrasts the disentangled representations to learn *mode-invariant features.*

3 Proposed MoST

In this section, we formally define the TTS representation learning problem and then introduce our proposed approach, MoST.

Definition 1. (Tensor Time Series). *A TTS is an Nth-order tensor $\mathcal{X} \in \mathbb{R}^{d_1 \times \cdots \times d_{N-1} \times T}$, where the Nth mode is the time and its dimension is T.*

3.1 Problem Formulation

We consider a 3rd-order TTS$(X_1, \ldots, X_T) \in \mathbb{R}^{d_1 \times d_2 \times T}$, which consists of a set of observations of d_1 and d_2 variables within the two non-temporal modes at T time points for the temporal mode.[2] Given a TTS with a w length window, we aim to learn feature representations $V = f(\mathcal{X})$, where $\mathcal{X} = (X_1, \ldots, X_w) \in \mathbb{R}^{d_1 \times d_2 \times w}$, $V = (v_1, \ldots, v_w) \in \mathbb{R}^{h \times w}$, and $f(\cdot)$ denotes the feature embedding function, which projects $(d_1 \times d_2)$-dimensional raw signals into an h-dimensional latent space for each time point.

Our MoST aims to learn the disentangled mode-specific representations, i.e., $V = [V^{(d_1)}; V^{(d_2)}] \in \mathbb{R}^{h \times w}$, where $V^{(d_1)} \in \mathbb{R}^{h_d \times w}$ is a mode-1-aware representation and $V^{(d_2)} \in \mathbb{R}^{h_d \times w}$ is a mode-2-aware representation, such that $h_d = h/2$.

3.2 Model Architecture

Our proposed model, illustrated in Fig. 2(a), consists of three components: (i) **Tensor slicing** transforms the tensor for learning disentangled mode-specific representations. (ii) **Slice feature encoder** captures intra-mode and temporal dependencies and learns a representation of a slice individually. (iii) **Aggregator** summarizes the mode information to form a representation for a particular mode.

Tensor Slicing. We describe a method that effectively leverages the tensor structure, which helps learn rich representations from a TTS. Concretely, the tensor is processed into a set of mode-1,2 tensors, each of which represents a set of mode-1,2 slices. The tensor slicing is written as follows:

$$\{\mathcal{X}^{(d_1)}, \mathcal{X}^{(d_2)}\} = \textbf{Slice}(\mathcal{X}), \tag{1}$$

where $\mathcal{X}^{(d_1)}$, is a mode-1 tensor, and $\mathcal{X}^{(d_2)}$ is a mode-2 tensor. $\mathcal{X}^{(d_1)}$ is a set of mode-1 slices, i.e., $\mathcal{X}^{(d_1)} = \{\mathcal{X}_{:,j}\}_{j=1}^{d_2}$, where $\mathcal{X}_{:,j} \in \mathbb{R}^{d_1 \times w}$ is a mode-1 slice at the j-th mode-2 of length w, s.t., $j = 1, \ldots, d_2$ is an index of slices. Likewise, $\mathcal{X}^{(d_2)}$ is a set of mode-2 slices, i.e., $\mathcal{X}^{(d_2)} = \{\mathcal{X}_{i,:}\}_{i=1}^{d_1}$, where $\mathcal{X}_{i,:} \in \mathbb{R}^{d_2 \times w}$ is a mode-2 slice at the i-th mode-1, s.t., $i = 1, \ldots, d_1$.

Slice Feature Encoder. After slicing the TTS, each mode-1,2 slice is fed independently into a slice feature encoder, which we call a mode-independence (MI) approach. Our slice feature encoder maps the observed signals to the latent representations and learns intra-mode and temporal dependencies. It consists of an embedding layer and a causal convolutional encoder. Given that the procedure

[2] In this work, we focus on a 3rd-order tensor for simplicity; our method, however, can be generalized to an Nth-order tensor.

is identical for mode-1,2 tensors, $\mathcal{X}^{(d_1)}$ and $\mathcal{X}^{(d_2)}$, we only describe the procedure for mode-1 tensor, $\mathcal{X}^{(d_1)}$.

Each mode-1 slice $\mathcal{X}_{:,j}$ of $\mathcal{X}^{(d_1)}$ is mapped to the latent space of dimension h via a trainable linear projection $W^{(d_1)pro} \in \mathbb{R}^{h \times d_1}$. Then, a determinative additive temporal embedding [37] $W^{tem} \in \mathbb{R}^{h \times w}$ is used to monitor the temporal order of a series: $Z_j^{(d_1)emb} = W^{(d_1)pro}\mathcal{X}_{:,j} + W^{tem}$, where $Z_j^{(d_1)emb} \in \mathbb{R}^{h \times w}$ denotes the input to be fed into a causal convolutional encoder.

To supply long-range temporal information, we employ a causal convolutional encoder and capture information from different time scales [30]. The causal convolutional encoder consists of a stack of $L+1$ one-dimensional causal convolution blocks and an average-pooling layer. We use $L = 7$ in this paper. Each block has h input channels and h_d output channels, where the kernel size of the k-th block is 2^k. The parameters of the causal convolutional encoder are shared in different modes. Then, an average-pooling layer is used to summarize the latent representation across different time scales.

$$Z_{j,k}^{(d_1)cau} = \mathbf{CausalConvBlock}(Z_j^{(d_1)emb}, 2^k), \tag{2}$$

$$Z_j^{(d_1)enc} = \mathbf{AveragePooling}(Z_{j,0}^{(d_1)cau}, \ldots, Z_{j,L}^{(d_1)cau}). \tag{3}$$

Aggregator. We have embedded representations of each slice $Z_j^{(d_1)enc} \in \mathbb{R}^{h_d \times w}$, where $j = 1, \ldots, d_2$ and, $Z_i^{(d_2)enc} \in \mathbb{R}^{h_d \times w}$, where $i = 1, \ldots, d_1$, respectively. An average-pooling layer or a max-pooling layer is applied depending on the dataset to summarize the mode information. Through this process, we acquire mode-1-aware representations $V^{(d_1)}$ and mode-2-aware representations $V^{(d_2)}$.

$$V^{(d_1)} = \mathbf{Pooling}(Z_1^{(d_1)enc}, \ldots, Z_{d_2}^{(d_1)enc}), \tag{4}$$

$$V^{(d_2)} = \mathbf{Pooling}(Z_1^{(d_2)enc}, \ldots, Z_{d_1}^{(d_2)enc}). \tag{5}$$

We concatenate them and obtain the final representations $V = [V^{(d_1)}; V^{(d_2)}]$.

3.3 Contrastive Learning

We utilize CL, a self-supervised learning method, to optimize the entire network, resulting in more comprehensive representations of TTS [3]. We define a contrastive loss Eq. (8) that consists of **instance loss** Eq. (6) and **mode loss** Eq. (7), each of which is a variant of InfoNCE [22] that aims to capture the *mode-specific features* and *mode-invariant features*, respectively.

Instance Loss. Instance loss aims to learn the mode-specific features. As Fig. 2(b) shows, we apply random cropping to the input tensor before feeding it to the model and create two different augmentations. Random cropping treats the representations at the same timestamp in two different augmentations as positive samples, whereas it treats those of other time series samples at the same timestamp as negative samples [33]. Concretely, we randomly sample two overlapping sequences from $\mathcal{X}$, hence $(X_{a_1}, \ldots, X_{b_1})$, $(X_{a_2}, \ldots, X_{b_2})$, s.t.

$0 < a_1 \leq a_2 \leq b_1 \leq b_2 \leq w$. The overlapped sequence $(X_{a_2}, \ldots, X_{b_1})$ is used as a positive sample to learn position-agnostic representations. This is only applied in the training.

Let $V_{i,t}$ denote the representations for the i-th index of the input time series sample at the timestamp t. Then, $V'_{i,t}$ is a positive sample of $V_{i,t}$, which denotes the representations for the i-th sample at the same timestamp t but from a different augmentation. We calculate instance loss as:

$$\mathcal{L}_I = \frac{1}{BN} \sum_i^B \sum_t^N -log \frac{exp(V_{i,t} \cdot V'_{i,t})}{\sum_{j=1}^B (exp(V_{i,t} \cdot V'_{j,t}) + \mathbb{1}_{[i \neq j]} exp(V_{i,t} \cdot V_{j,t}))}, \tag{6}$$

where B is the batch size, $N = b_1 - a_2$ is the length of the overlapped sequence. We use the representations of other samples at timestamp t as negative samples.

Mode Loss. To learn the mode-invariant features in different mode-specific representations, we treat representations of different modes at the same timestamp as a positive sample, while we treat those of other time series samples at the same timestamp of different modes as negative samples. The approach attempts to minimize the similarity among different mode-specific representations of the same timestamp.

We describe the mode loss for mode-1 $\mathcal{L}_M^{(d_1)}$. However, the mode loss for mode-2 $\mathcal{L}_M^{(d_2)}$ is defined as being the same as $\mathcal{L}_M^{(d_1)}$ but in a different mode. Let $V_{i,t}^{(d_1)}$ denote the mode-1-aware representations for the i-th index of the input time series sample at the timestamp t. $V_{i,t}^{(d_2)}$ is a positive sample of $V_{i,t}^{(d_1)}$, which denotes the mode-2-aware representations for the i-th sample at the same timestamp t. The mode loss for mode-1 is defined as:

$$\mathcal{L}_M^{(d_1)} = \frac{1}{BN} \sum_i^B \sum_t^N -log \frac{exp(V_{i,t}^{(d_1)} \cdot V_{i,t}^{(d_2)})}{\sum_{j=1}^B exp(V_{i,t}^{(d_1)} \cdot V_{j,t}^{(d_2)})}. \tag{7}$$

We use the representations of other samples at timestamp t from a different mode as negative samples.

Contrastive Loss. The overall loss function is the weighted sum of the instance loss and the mode loss of each non-temporal mode.

$$\mathcal{L} = \mathcal{L}_I + \alpha(\mathcal{L}_M^{(d_1)} + \mathcal{L}_M^{(d_2)}), \tag{8}$$

where α is a hyper-parameter that controls the importance of the mode loss.

4 Experiments

4.1 Classification

Settings. We follow TS2Vec [33], where max-pooling is applied over all timestamps to obtain the representations for a sequence. We then train a logistic regressor model with an L2 norm penalty. We use two real-world 3rd-order TTS

Table 1. Classification results with Acc. Best results are in **bold**, and second best results are underlined (higher is better). The average of five runs is reported.

Dataset	MoST	CoST	TS2Vec	TS-TCC	ATD
Daily	**0.726**	0.164	0.688	0.478	0.249
Realdisp	**0.766**	0.126	0.663	0.190	0.151

motion sensor datasets. (1) **Daily** consists of five units and nine statistical features, with 19 activity labels.[3] (2) **Realdisp** comprises nine units and 13 features, with 33 activity labels.[4] We compare our method with the following four state-of-the-art self-supervised representation learning methods for time series (CoST [30], TS2Vec [33], and TS-TCC [7]) and for tensor (ATD [31]). We use the classification accuracy (Acc) as an evaluation metric.

Results. The evaluation results are shown in Table 1. In general, MoST outperforms other baselines. This is because MoST exploits the TTS structure, including intra-mode and temporal dependencies, resulting in learning rich representations. CoST learns disentangled representations of trend and seasonality, but the representations are not suitable for classification. ATD jointly learns tensor decomposition and representations through CL, but fails to capture temporal information in TTS. TS-TCC learns representations for time series classification and shows better results than CoST and ATD. TS2Vec is less accurate than MoST because it does not exploit the TTS structure.

4.2 Forecasting

Settings. We use the representations of the last timestamp to predict future T_f observations. Then, following [30,33], we train a linear regression model with an L2 norm penalty to directly predict future observations. We use nine real-world datasets of 3rd-order TTS. (1) **Google Trends** consists of daily web-search counts, including seven query sets from [19], resulting in seven datasets: four from the 50 US states (US#1∼#3) and three from the top 50 countries by GDP (World#1∼#4). (2) **KnowAir** consists of PM2.5 concentrations and weather forecasting data, 12 sensor values in total, taken every three hours covering 184 cities [28]. (3) **NYC-CB** contains NYC CitiBike[5] ride trips obtained from the 25 starting and ending positions taken every three hours. We compare our method with the following six state-of-the-art methods for CL (CoST [30] and TS2Vec [33]), end-to-end forecasting (Informer [37] and LaST [29]), and tensor/matrix decomposition (NET^3 [10] and SSMF [11]).

[3] https://archive.ics.uci.edu/dataset/256.

[4] https://archive.ics.uci.edu/dataset/305.

[5] https://s3.amazonaws.com/tripdata/index.html.

Table 2. Forecasting performance (lower is better). The average of five runs is reported.

	Methods	Self-supervised representation learning						End-to-end forecasting				Tensor forecasting			
		MoST		CoST		TS2Vec		Informer		LaST		NET^3		SSMF	
	T_f	MSE	MAE	MSE	MAE	MSE	MAE	MSE	MAE	MSE	MAE	MSE	MAE	MSE	MAE
US#1	2 week	**0.752**	**0.590**	0.900	0.668	1.153	0.806	1.364	0.818	1.003	0.694	1.501	0.915	1.380	0.824
	8 week	**0.811**	**0.609**	1.002	0.685	1.133	0.751	1.125	0.735	1.060	0.731	2.143	1.058	1.353	0.815
	32 week	**0.949**	**0.661**	1.132	0.714	1.355	0.822	1.242	0.770	1.449	0.839	1.599	0.866	1.355	0.811
US#2	2 week	**0.653**	**0.566**	0.676	0.575	0.725	0.600	0.855	0.655	0.709	0.588	0.805	0.630	1.385	0.862
	8 week	**0.695**	**0.580**	0.729	0.595	0.766	0.613	0.891	0.670	0.726	0.601	0.891	0.680	1.396	0.860
	32 week	**0.731**	**0.597**	0.786	0.627	0.832	0.646	0.864	0.647	0.793	0.621	1.013	0.733	1.562	0.913
US#3	2 week	**0.591**	**0.511**	0.628	0.525	0.665	0.549	0.760	0.576	0.674	0.559	0.682	0.563	1.281	0.816
	8 week	0.580	**0.506**	0.610	0.519	0.625	0.530	0.597	0.535	**0.563**	0.516	0.697	0.576	1.268	0.815
	32 week	**0.521**	**0.500**	0.541	0.509	0.563	0.524	0.578	0.548	0.528	0.515	1.268	0.732	1.290	0.841
US#4	2 week	**0.439**	**0.500**	0.465	0.517	0.497	0.541	0.723	0.656	0.517	0.555	0.547	0.567	1.841	0.998
	8 week	**0.448**	**0.506**	0.477	0.526	0.515	0.552	0.683	0.630	0.527	0.553	0.647	0.625	1.954	1.030
	32 week	**0.463**	**0.513**	0.486	0.528	0.525	0.554	0.708	0.637	0.533	0.557	0.709	0.657	1.996	1.044
World#1	2 week	**0.705**	**0.402**	0.716	0.420	0.796	0.470	0.761	0.457	0.802	0.464	0.716	0.438	1.985	1.035
	8 week	**0.706**	**0.418**	0.720	0.441	0.817	0.500	0.773	0.468	0.798	0.476	0.800	0.515	1.979	1.037
	32 week	**0.681**	**0.435**	0.734	0.484	0.877	0.548	0.700	0.446	0.822	0.508	0.966	0.681	1.764	0.968
World#2	2 week	0.949	**0.573**	0.953	0.591	1.172	0.687	1.129	0.652	**0.942**	0.584	1.449	0.792	1.663	0.848
	8 week	0.982	**0.590**	0.984	0.611	1.257	0.732	1.042	0.654	**0.951**	0.632	1.505	0.822	1.608	0.836
	32 week	**1.104**	**0.674**	1.147	0.692	1.670	0.891	1.124	0.683	1.176	0.702	1.499	0.824	1.545	0.823
World#3	2 week	**1.566**	0.823	1.593	**0.811**	1.679	0.818	1.670	0.844	1.634	0.838	1.599	0.818	2.125	0.949
	8 week	1.583	0.825	1.600	**0.811**	1.697	0.819	**1.530**	0.813	1.583	0.819	1.640	0.835	2.099	0.946
	32 week	1.638	0.823	1.665	0.814	1.765	0.833	**1.616**	**0.813**	1.752	0.857	1.678	0.830	2.116	0.949
KnowAir	2 week	**0.635**	**0.537**	0.637	0.552	0.873	0.703	0.819	0.667	0.710	0.595	1.024	0.796	7.006	2.163
	4 week	**0.651**	**0.545**	0.660	0.564	0.881	0.706	0.861	0.698	0.725	0.608	1.007	0.785	7.042	2.170
	8 week	**0.659**	**0.550**	0.677	0.574	0.867	0.697	0.862	0.697	0.691	0.586	1.023	0.793	7.116	2.185
NYC-CB	1 day	**0.905**	**0.485**	0.926	0.489	1.172	0.556	1.507	0.722	0.989	0.545	1.355	0.663	1.817	0.652
	1 week	**0.950**	**0.497**	0.958	0.501	1.203	0.564	1.568	0.753	0.957	0.539	1.310	0.639	1.844	0.660
	2 week	0.981	**0.506**	0.984	0.510	1.228	0.572	1.506	0.742	**0.959**	0.540	1.271	0.612	1.874	0.669

Results. The experimental results are presented in Table 2. Overall, our model outperforms most of the comparative methods, confirming the benefits of our model architecture. CoST and TS2Vec, which are CL methods for time series, are less accurate than our method. The most significant difference between our method and the others is that MoST uses a TTS structure that significantly contributes to improved accuracy.

4.3 Ablation Study

We describe ablation studies concerning the architecture of our method. Table 3 presents the mean results of the classification datasets and the US datasets on all forecast horizon settings with different architectures.

We first perform an ablation study to understand the effects of the tensor slicing and MI approaches. We conduct a comparison with the five alternative pre-

Table 3. Performance of forecasting and classification for the different (a) tensor slicing and MI approaches, (b) architectures, and (c) loss functions.

		Forecasting		*Classification*
		Avg. MSE	Avg. MAE	Avg. Acc
MoST		**0.636**	**0.553**	**0.746**
(a)	w/o Mode-2 dependency (M1D)	0.811 (−27.5%)	0.630 (−13.9%)	0.714 (−4.4%)
	w/o Mode-1 dependency (M2D)	0.650 (−2.2%)	0.560 (−1.1%)	0.599 (−19.8%)
	Random	0.656 (−3.2%)	0.556 (−2.3%)	0.626 (−16.2%)
	Channel-independence (CI)	0.651 (−2.3%)	0.559 (v1.0%)	0.667 (−10.6%)
	Channel-dependence (CD)	0.912 (−43.3%)	0.666 (−20.0%)	0.405 (−45.7%)
(b)	w/o Temporal Embedding	0.693 (−8.9%)	0.561 (−4.4%)	-
	w/o Causal Conv Encoder	0.726 (−14.1%)	0.593 (−8.6%)	0.536 (−28.1%)
(c)	w/o Instance Loss	0.644 (−1.2%)	0.557 (−0.7%)	0.730 (−2.2%)
	w/o Mode Loss	0.636 (−0.0%)	0.553 (−0.0%)	0.732 (−1.9%)
	MSE Loss	0.823 (−29.1%)	0.643 (−16.2%)	-
	MSE and Contrastive Loss	1.342 (−111.0%)	0.852 (−54.2%)	-

processing methods. **Mode-1/2 dependency (M1D/M2D)**: Feeds only the mode-1/2 tensor $\mathcal{X}^{(d_1)}/\mathcal{X}^{(d_2)}$ into the slice feature encoder. **Random**: Creates a randomly organized tensor by flattening, randomizing variables, and restoring the original shape, applied once before learning. **Channel-independence (CI)**: Feeds $\{\mathcal{X}_{1,1}, \mathcal{X}_{1,2}, \ldots, \mathcal{X}_{d_1,d_2}\}$ independently into the slice feature encoder. **Channel-dependence (CD)**: Feeds a flattened tensor $X^{flat} \in \mathbb{R}^{d_1 d_2 \times w}$ into the slice feature encoder. Table. 3(a) demonstrates the significant impact of our approaches on improving prediction and classification accuracy. Comparison with M1D and M2D underscores the importance of both mode-1 and mode-2 dependencies for rich representations. Additionally, comparisons with Random, CI, and CD demonstrate the superiority of our MI approach, which leverages the TTS structure to learn intra-mode dependencies effectively.

We evaluate the impact of our architecture and loss functions. Table. 3(b) reveals that both temporal embedding and the causal convolutional encoder enhance performance. Table. 3(c) presents results w/o instance and mode losses and end-to-end training using forecasting loss functions. Instance loss proves crucial for forecasting, while both losses improve classification accuracy. Furthermore, our model, which optimizes representations and forecasting separately, outperforms end-to-end training with supervised MSE forecasting loss.

4.4 Case Study

We use synthetic data to compare the representation ability of MoST, CoST and TS2Vec to learn intra-mode dependencies We generate a synthetic dataset by defining three different intra-mode dependencies for each non-temporal mode, and taking the cross product to form nine time series.

We visualize the learned representations with t-SNE [15] in Fig. 3. CoST and TS2Vec are unable to distinguish between different intra-mode dependencies. In contrast, our approach is able to discriminate between various intra-mode dependencies in the latent space thanks to its disentangled representations.

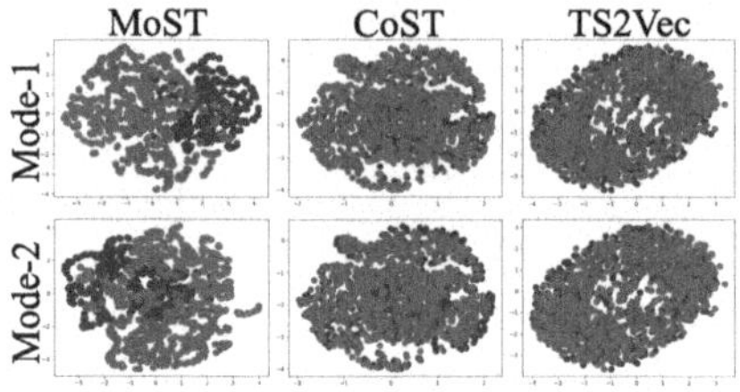

Fig. 3. t-SNE visualizations of the learned representations. The colors represent the three distinct intra-mode dependencies for (top) mode-1 and (bottom) mode-2. For MoST, representations of mode-1 $V^{(d_1)}$ and mode-2 $V^{(d_2)}$ are shown. (Color figure online)

5 Conclusion

In this study, we proposed MoST, a representation learning method that learns disentangled mode-specific representations for TTS. MoST consists of the tensor slicing approach and a newly defined contrastive loss to effectively model the TTS. Thanks to our proposed architecture, MoST captures intra-mode and temporal dependencies. The learned representations can be applied to any downstream tasks, including classification and forecasting, and extensive experiments show that our method outperforms the state-of-the-art methods. MoST has the potential to contribute to enhancing analysis and our understanding of TTS.

Acknowledgments. The authors would like to thank the anonymous referees for their valuable comments and helpful suggestions. This work was supported by JSPS KAKENHI Grant-in-Aid for Scientific Research Number JP23K16889, JP24K20778, JST CREST JPMJCR23M3.

References

1. Bahadori, M.T., et al.: Fast multivariate spatio-temporal analysis via low rank tensor learning. In: NeurIPS, vol. 27 (2014)
2. Cai, Y., et al.: Facets: Fast comprehensive mining of coevolving high-order time series. In: KDD, pp. 79–88 (2015)

3. Chen, T., et al.: A simple framework for contrastive learning of visual representations. In: ICML, pp. 1597–1607. PMLR (2020)
4. Deldari, S., et al.: Time series change point detection with self-supervised contrastive predictive coding. In: WWW (2021)
5. Deng, J., et al.: Learning gaussian mixture representations for tensor time series forecasting. In: IJCAI, pp. 2077–2085 (2023)
6. Duan, J., et al.: MF-CLR: Multi-frequency contrastive learning representation for time series. In: ICML Proceedings of Machine Learning Research, vol. 235, pp. 11918–11939. PMLR (2024)
7. Eldele, E., et al.: Time-series representation learning via temporal and contextual contrasting. In: IJCAI, pp. 2352–2359 (2021)
8. Hong, D., et al.: Generalized canonical polyadic tensor decomposition. SIAM Rev. **62**(1), 133–163 (2020)
9. Hooi, B., et al.: SMF: drift-aware matrix factorization with seasonal patterns. In: SIAM International Conference on Data Mining, pp. 621–629 (2019)
10. Jing, B., et al.: Network of tensor time series. In: WWW, pp. 2425–2437 (2021)
11. Kawabata, K., et al.: SSMF: shifting seasonal matrix factorization. NeurIPS **34**, 3863–3873 (2021)
12. Lee, D., et al.: Robust factorization of real-world tensor streams with patterns, missing values, and outliers. In: ICDE, pp. 840–851. IEEE (2021)
13. Lee, S., et al.: Soft contrastive learning for time series. In: ICLR (2024)
14. LIU, S., et al.: Multivariate time-series imputation with disentangled temporal representations. In: ICLR (2023)
15. Van der Maaten, L., et al.: Visualizing data using t-SNE. J. Mach. Learn. Res. **9**(11), 2579–2605 (2008)
16. Madabhushi, A., et al.: Image analysis and machine learning in digital pathology: challenges and opportunities. Med. Image Anal. **33**, 170–175 (2016)
17. Matsubara, Y., et al.: Non-linear mining of competing local activities. In: WWW, pp. 737–747 (2016)
18. Meng, Q., et al.: MHCCL: masked hierarchical cluster-wise contrastive learning for multivariate time series. In: AAAI, pp. 9153–9161 (2023)
19. Murayama, T., et al.: Mining reaction and diffusion dynamics in social activities. In: CIKM, pp. 1521–1531 (2022)
20. Obata, K., et al.: Dynamic multi-network mining of tensor time series. In: WWW, pp. 4117–4127 (2024)
21. Obata, K., et al.: Mining of switching sparse networks for missing value imputation in multivariate time series. In: KDD, pp. 2296–2306 (2024)
22. van den Oord, A., et al.: Representation learning with contrastive predictive coding. CoRR abs/1807.03748 (2018). http://arxiv.org/abs/1807.03748
23. Rethmeier, N., et al.: A primer on contrastive pretraining in language processing: methods, lessons learned, and perspectives. ACM Comput. Surv. **55**(10), 1–17 (2023)
24. Rogers, M., et al.: Multilinear dynamical systems for tensor time series. In: NeurIPS, vol. 26 (2013)
25. Tonekaboni, S., et al.: Unsupervised representation learning for time series with temporal neighborhood coding. In: ICLR (2021)
26. Tsay, R.S.: Analysis of Financial Time Series. John Wiley & Sons (2005)
27. Wang, S., et al.: Deep learning for spatio-temporal data mining: a survey. TKDE **34**, 3681–3700 (2020)
28. Wang, S., et al.: PM2. 5-GNN: a domain knowledge enhanced graph neural network for PM2. 5 forecasting. In: SIGSPATIAL, pp. 163–166 (2020)

29. Wang, Z., et al.: Learning latent seasonal-trend representations for time series forecasting. In: NeurIPS (2022)
30. Woo, G., et al.: CoST: contrastive learning of disentangled seasonal-trend representations for time series forecasting. In: ICLR (2022)
31. Yang, C., et al.: ATD: augmenting CP tensor decomposition by self supervision. NeurIPS **35**, 32039–32052 (2022)
32. Yang, L., et al.: Unsupervised time-series representation learning with iterative bilinear temporal-spectral fusion. In: ICML, pp. 25038–25054. PMLR (2022)
33. Yue, Z., et al.: TS2VEC: towards universal representation of time series. In: AAAI, pp. 8980–8987 (2022)
34. Zhang, K., et al.: Self-supervised learning for time series analysis: Taxonomy, progress, and prospects. arXiv preprint arXiv:2306.10125 (2023)
35. Zhang, X., et al.: Self-supervised contrastive pre-training for time series via time-frequency consistency. In: NeurIPS (2022)
36. Zhang, Y., et al.: Contrastive spatio-temporal pretext learning for self-supervised video representation. In: AAAI, pp. 3380–3389 (2022)
37. Zhou, H., et al.: Informer: beyond efficient transformer for long sequence time-series forecasting. In: AAAI, pp. 11106–11115 (2021)

PFformer: A Position-Free Transformer Variant for Extreme-Adaptive Multivariate Time Series Forecasting

Yanhong Li and David C. Anastasiu(✉)

Santa Clara University, Santa Clara 95094, CA, USA
danastasiu@scu.edu

Abstract. Multivariate time series (MTS) forecasting is vital in fields like weather, energy, and finance. However, despite deep learning advancements, traditional Transformer-based models often diminish the effect of crucial inter-variable relationships by singular token embedding and struggle to effectively capture complex dependencies among variables, especially in datasets with rare or extreme events. These events create significant imbalances and lead to high skewness, complicating accurate prediction efforts. This study introduces PFformer, a position-free Transformer-based model designed for single-target MTS forecasting, specifically for challenging datasets characterized by extreme variability. PFformer integrates two novel embedding strategies: Enhanced Feature-based Embedding (EFE) and Auto-Encoder-based Embedding (AEE). EFE effectively encodes inter-variable dependencies by mapping related sequence subsets to high-dimensional spaces without positional constraints, enhancing the encoder's functionality. PFformer shows superior forecasting accuracy without the traditional limitations of positional encoding in MTS modeling. We evaluated PFformer across four challenging datasets, focusing on two key forecasting scenarios: long sequence prediction for 3 days ahead and rolling predictions every four hours to reflect real-time decision-making processes in water management. PFformer demonstrated remarkable improvements, from 20% to 60%, compared with state-of-the-art models.

Keywords: Deep Learning · Time Series Forecasting · Hydrology · Transformer

1 Introduction

Effective time series forecasting is critical for strategic decision-making in many industries, including meteorology [5,24], financial markets [12,14], and energy management [8]. Advancements in deep learning have resulted in a wave of complex multivariate time series (MTS) models that have greatly improved forecasting accuracy by utilizing intricate data relationships, especially where sporadic but major extreme events [20,21] such as flash floods and droughts exist.

This research was made possible through resources provided by Supermicro and NVIDIA.

X. Wu et al. (Eds.): PAKDD 2025, LNAI 15875, pp. 185–197, 2025.
https://doi.org/10.1007/978-981-96-8295-9_13

After its immense success in language processing and computer vision, the Transformer model [15], known for its strong capabilities in depicting pairwise dependencies in sequences, is now making significant inroads into time series forecasting [19,23,25]. However, these models typically compress multiple variables of the same timestamp into a single token, potentially obscuring vital multivariate correlations. As highlighted by Liu et al. [9] and Zhang and Yan [22], data points within the same timestamp often represent distinct physical phenomena and are measured inconsistently. This embedding into a singular token can erase crucial inter-variable relationships, and the localized receptive field of a single timestamp may not capture beneficial information effectively, especially when dealing with events that are not aligned in time.

Moreover, the application of permutation-invariant attention mechanisms, which fail to consider the importance of sequence order, may not be suitable for analyzing temporal data where order significantly impacts series variations. Consequently, these flaws can limit the Transformer's capacity to represent and generalize across various multivariate time series effectively. Even with closely connected multivariate inputs available, like the streamflow and rain time series, the Transformer struggles to use these correlations to accurately predict time series with extreme events [6].

In response to these challenges, we propose PFformer, a position-free Transformer variant for adaptive multivariate time series forecasting. We evaluated PFformer on four separate datasets and found that PFformer significantly outperforms state-of-the-art baselines by 20% to 60%. Additionally, we carried out several ablation studies to understand the effects of specific design decisions.

2 Related Work

To fully leverage auxiliary variables in prediction, multivariate time series studies have utilized various algorithms, from traditional methods like vector autoregression and multivariate exponential smoothing to advanced deep learning methods. Vector autoregressive (VAR) models [10] are statistical models that assume linear dependencies both across different dimensions and over time. Wang et al. [16] developed a hybrid model that combines Empirical Mode Decomposition (EMD), Ensemble EMD (EEMD), and ARIMA for long-term streamflow forecasting. Additionally, graph neural networks (GNNs) [18] explicitly capture cross-dimensional dependencies by combining temporal and graph convolutional layers.

Recently, transformer-based techniques such as Autoformer [17] and FEDFormer [25] have been introduced to solve this problem. PatchTST [13] enhances time series modeling by using patching techniques to extract local semantics and maintain channel independence. Crossformer [22] features a cross-scale embedding layer and Long Short Distance Attention (LSDA) to effectively capture cross-time and cross-variable dependencies in MTS, and iTransformer [9] optimizes Transformer inputs to improve time-series modeling, focusing on more accurate data interpretation and prediction.

Despite extensive research on time series prediction, deep learning models face difficulties when dealing with time series data that contain rare or extreme occurrences because of the obvious imbalance in the dataset. Predicting these types of data is notably difficult as their distribution is heavily influenced by extreme values, leading to high skewness. This calls for the creation of specialist models intended for precise forecasting of extreme events. For example, an earlier model we designed, NEC+ [7] is specifically designed to accommodate extreme values in hydrologic flow prediction by employing the Gaussian mixture model (GMM) [1] and training three predictors in parallel. Another model we previously proposed, DAN [6], learns and merges rich representations to adaptively predict streamflow.

However, few of these prior works have addressed both long-term predictions in prolonged sequences and used auxiliary variables effectively to enhance extreme value prediction. To bridge this gap, we propose PFformer to address the challenges of highly skewed single-target MTS forecasting in the hydrology domain. Experiments on four real-life hydrologic streamflow datasets show that PFformer significantly outperforms state-of-the-art methods for hydrologic and long-term time series prediction.

3 Preliminaries

3.1 Problem Statement

Suppose we have a collection of m ($m >= 1$) related univariate time series, with each row in the input matrix corresponding to a different time series. We are going to predict the next h time steps for the first time series x_1, given historical data from multiple length-t observed series. The problem can be described as,

$$\begin{bmatrix} x_{1,1} & \cdots & x_{1,t} \\ x_{2,1} & \cdots & x_{2,t} \\ \vdots & \ddots & \vdots \\ x_{m,1} & \cdots & x_{m,t} \end{bmatrix} \in \mathbb{R}^{m \times t} \rightarrow [x_{1,t+1}, \ldots, x_{1,t+h}] \in \mathbb{R}^{h},$$

where $x_{i,j}$ denotes the value of time series i at time j. The matrix on the left are the inputs, and $x_{1,t+1}$ to $x_{1,t+h}$ are the outputs of our method. We first define this task by modeling the objective time series x_1 as the *ordinary series* and the group of related time series x_2 to x_m as *auxiliary series*.

3.2 Data Descriptions

Our study uses a hydrologic dataset first introduced in [6] that captures streamflow from four California streams: Ross, Saratoga, UpperPen, and SFC. Given California's lack of rainfall during the summer, we follow the same problem design as in [6] and focus on forecasting the months from September to May, deliberately excluding the summer period. Data for training and validation was

Table 1. Input Stream Data Statistics

Statistic / Stream	Ross	Saratoga	UpperPen	SFC
min	0.00	0.00	0.00	0.00
max	1440.00	2210.00	830.00	7200.00
mean	2.91	5.77	6.66	20.25
std. deviation	24.43	26.66	21.28	110.03
skewness	19.84	19.50	13.42	18.05
kurtosis	523.16	697.78	262.18	555.18

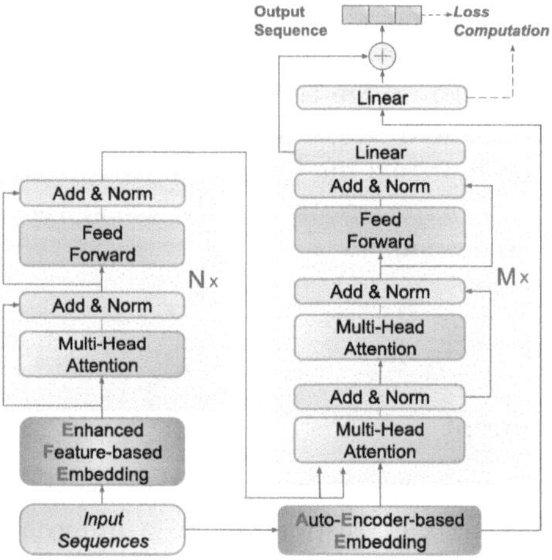

Fig. 1. *The PFformer framework*; $\oplus$ denotes the element-wise addition. The PFformer framework is a transformer-based variant optimized for multivariate time series forecasting. It replaces the encoder's positional encoding layer with Enhanced Feature-based Embedding (EFE) to capture complex inter-variable relationships. In the decoder, Auto-Encoder-based Embedding (AEE) substitutes the positional embedding, enabling direct, fixed-length predictions without error propagation or masking.

drawn from January 1988 to August 2021 and we aim to accurately project the streamflow for the subsequent year (September 2021 to May 2022), with predictions made every four hours. Each prediction estimates the upcoming 3 days based on the preceding 15 days of data. The performance metrics we employed are Root Mean Square Error (RMSE) and Mean Absolute Percentage Error (MAPE). We add 1 to both the true value and the predicted value to avoid instability in the calculation when the true values are close to zero. Since the sensors measure the streamflow and precipitation every 15 min, we are attempting a lengthy forecasting horizon ($h = 288$). Table 1 shows several statistics of our input time series, which provide valuable insights into the shape and distribution of the data, including min, max, mean, median, variance, skewness, and kurtosis. The presence of high skewness and kurtosis values suggests that our data exhibit significant asymmetry and departure from the symmetric bell-shaped curve of a Normal distribution. This implies that there are more extreme values or outliers on the higher end of the distribution.

4 Methods

4.1 PFformer Architecture

Figure 1 illustrates the PFformer architecture, a transformer-based framework optimized for multivariate time series forecasting. It consists of two main components: the encoder and the decoder, both structured to enhance forecasting accuracy through advanced embedding techniques.

Encoder. As shown in the left part of Fig. 1, the encoder section of our model closely follows the standard Transformer encoder design aside from the embedding layer, which uses Enhanced Feature-based Embedding (EFE). This strategy maps input sequences to high-dimensional spaces without positional encoding and effectively captures complex inter-variable relationships. These are then processed through layers of multi-head attention and feed-forward networks, each featuring an "Add and Norm" step for output integration and normalization, crucial for stabilizing the learning process.

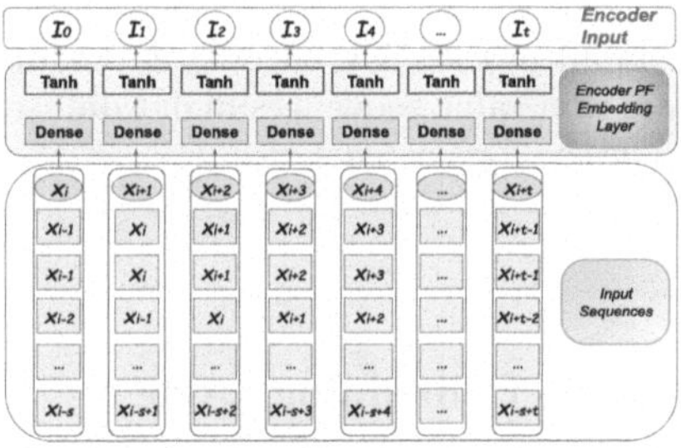

Fig. 2. *PFformer encoder embedding module.* X and A are the stream and rain value sequences in our dataset. The output of EFE is the input to the transformer encoder.

Decoder. PFformer bypasses the use of target sequences as inputs to avoid error propagation common with recursive prediction methods. Instead, it predicts all outcomes directly, eliminating the need for both masking mechanisms and padding due to the fixed prediction length. The AEE layer is used to create rich representations that enhance input data for the attention mechanisms. Additionally, we simplified the cross-attention layer by applying attention computing directly on the output of AEE in the transformer decoder, enabling effective learning from the richly represented embedding layers.

Output. To enhance the model's adaptability to extreme values, the AEE module also plays a critical role in the loss calculation. Specifically, the output from

AEE passes through a linear layer, which is combined with the output from the decoder's linear layer to form the final output, as further described in Sect. 4.5.

4.2 Enhanced Feature-Based Embedding (EFE)

Figure 2 shows the EFE module in PFformer. For an input sequence of length t, taking the first time point as an example, EFE combines the predicted sequence value at this time point, corresponding values from auxiliary sequences, and s preceding values from each of these auxiliary sequences. These elements are concatenated to form a multi-spatial subsequence unique to that time point. This subsequence is then processed through a dense layer coupled with a nonlinear activation function. Then, the output of the EFE module serves as the input vector for the encoder at this fixed time point. Across all time points, this forms a high-dimensional input sequence for the encoder layer. The EFE mechanism focuses on capturing the intricate relationships between different time series variables across time.

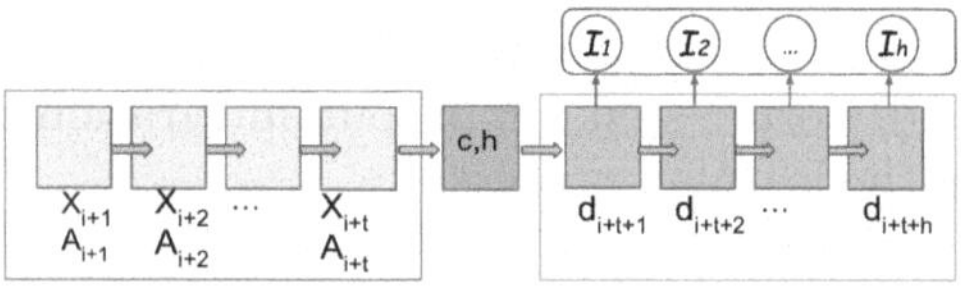

Fig. 3. *AEE module.* The input of the AEE encoder is the aligned multivariate series. X and A are the stream and rain value sequences in our dataset. The input of the AEE decoder is the time stamp of the forecasted series.

4.3 Auto-Encoder-Based Embedding (AEE)

Figure 3 illustrates the Auto-Encoder-Based Embedding (AEE) layer, which is designed to handle aligned multivariate time series data, such as streamflow and rainfall in our dataset. The encoder module takes as input the aligned multivariate series, where X represents the streamflow sequence, and A corresponds to the rainfall sequence. These sequences are processed step-by-step to generate the latent states c and h, which summarize the temporal patterns and dependencies in the input data. The decoder module then uses the latent states, along with the time stamps of the forecasted series, to predict future values.

The output of the AEE fulfills two key roles. First, it is integrated with the outputs from the PFformer decoder, enhancing the overall prediction accuracy. Second, its capacity for short-term prediction is measured as a component of the loss function. This inclusion acts as a penalty term, underscoring the critical role of auxiliary variables in refining short-term forecast accuracy.

4.4 Clustering-Based Oversampling Policy

Since the total length of each time series in our dataset is approximately 1.4 million, the sampling strategy is crucial during model training. However, simply oversampling extreme values can degrade overall prediction quality for the rest of the time series, as previously demonstrated in research. To address this, we propose a Clustering-Based Oversampling Policy which aims to capture significant data points based on statistic distributions. We employ a Gaussian Mixture Model (GMM) to cluster the data into M clusters with mean values $\mu_1, \mu_2, \ldots, \mu_M$. Since the cluster with the highest mean value z represents those extreme values, we marked a data point as important if its value exceeds $\eta \times z$. For each peak point, we treat step size s and scope ν as hyperparameters. The sampling starts $\nu/2$ points to the left of an identified peak, and samples are collected every s points, effectively generating ν/s samples around each peak.

To manage the volume of oversampled data, we set a ratio, os, which caps the oversampled data at $os\%$ of the total training set volume. We employ grid search to determine the optimal value of os to ensure our sampling method remains targeted and efficient without overwhelming the model with excessive data.

Table 2. 3-day/4-hour Long-Term ($h = 288$) Series Forecasting Results

Methods	RMSE								MAPE							
	Ross		Saratoga		UpperPen		SFC		Ross		Saratoga		UpperPen		SFC	
	3 d	4 h	3 d	4 h	3 d	4 h	3 d	4 h	3 d	4 h	3 d	4 h	3 d	4 h	3 d	4 h
FEDformer	6.01	3.95	6.01	4.82	3.05	2.55	23.54	17.11	2.10	2.05	1.55	1.54	1.87	1.75	2.35	2.16
Informer	7.84	6.76	5.04	3.78	5.88	5.00	39.89	23.21	4.05	4.71	1.43	1.54	4.10	3.99	8.64	3.61
Nlinear	6.10	2.76	5.23	4.13	1.57	0.51	18.47	5.08	1.99	0.52	0.83	0.82	0.45	0.16	0.92	0.52
Dlinear	7.16	3.31	4.33	1.79	3.53	1.35	21.62	8.75	3.10	1.15	1.40	0.65	2.35	0.69	2.74	1.45
LSTM-Atten	7.35	6.84	6.49	5.59	6.35	4.75	34.17	23.09	3.74	4.10	1.80	1.79	4.76	3.67	9.90	6.25
NEC+	9.44	<u>2.07</u>	1.88	<u>0.26</u>	2.22	<u>0.33</u>	17.00	**2.36**	4.80	<u>0.45</u>	0.17	<u>0.07</u>	0.95	<u>0.06</u>	1.07	<u>0.07</u>
iTransformer	4.56	2.14	2.37	0.94	1.12	0.58	17.04	11.00	0.57	0.43	0.27	0.18	<u>0.11</u>	<u>0.06</u>	0.47	0.54
DAN	<u>4.25</u>	2.61	<u>1.80</u>	0.62	<u>1.10</u>	0.43	<u>15.23</u>	3.73	**0.07**	0.46	<u>0.14</u>	0.22	0.15	0.07	<u>0.26</u>	0.22
PFformer	**4.21**	**1.52**	**1.69**	**0.22**	**1.01**	**0.24**	**14.98**	<u>2.86</u>	<u>0.10</u>	**0.03**	**0.10**	**0.04**	**0.06**	**0.01**	**0.18**	**0.06**

4.5 Multiple-Objective Loss Function

To force the AEE to learn rich representations, we use multiple loss functions [2, 3,11] when training the PFformer model. We build our loss items as follows,

$$\mathcal{L}_1 = RMSE(\hat{y}_{aux}[:s], y[:s]), \quad \mathcal{L}_2 = RMSE(\hat{y}, y).$$

where $\hat{y}$ is the output of PFformer, $\hat{y}_{aux}$ is the output of the AEE module after a linear transformation, and s represents the length of the short-term interval, which is set to 16 (4 h) in our experiments. The $\mathcal{L}_1$ regularization component is responsible for short-term prediction accuracy, while $\mathcal{L}_2$ focuses on the overall model accuracy. Then, the overall loss is composed as,

$$\mathcal{L} = \lambda \times \mathcal{L}_1 + \mathcal{L}_2, \qquad \lambda = max(-1 \cdot e^{\frac{epoch}{45}} + \alpha, \beta).$$

λ acts as a scaling factor for regularization, initially set high to steer the AEE to be more responsible for short-term prediction accuracy. This 'teacher mode' diminishes over time; λ starts at α and decreases to β over epochs.

5 Evaluation

5.1 Experimental Settings

This study utilized the same time series data as in [7]. In our experiments, we consistently set $N = 3$, $M = 1$, and $\eta = 1.2$ across all datasets (Ross, Saratoga, UpperPen, and SFC). The oversampling percentages $os\%$ were adjusted to 20% for Ross, Saratoga, and UpperPen, and to 15% for SFC, reflecting the unique distributions of each dataset. The oversampling policies, denoted as (s, v), and the α and β values for the regularization parameters were tailored to the distinct characteristics of each dataset. We performed a grid search with $s \in \{1, 2, 4\}$, $v \in \{4, 8, 16\}$, $\alpha \in \{1, 1.2, 1.5, 1.8, 2, 2.2\}$, and $\beta \in \{0.5, 0.6, \ldots, 1\}$ and obtained the best results at (s, v, α, β) of (1,8,1.5,0.9), (2,16,1.8,0.8), (4,16,2.0,0.8), and (2,16,2.0,0.9) for the Ross, Saratoga, UpperPen, and SFC datasets, respectively. The hidden dimensions for the attention and linear layers were set to [384, 268, 288, 300], and for the AEE LSTM layer to [384, 268, 320, 256]. Furthermore, the AEE featured one LSTM layer for SFC and two for the other sensors. All models were trained for a maximum of 40 epochs with early stopping triggered after four consecutive epochs without improvement. We used the Adam gradient descent algorithm with a starting learning rate of 0.0005, decaying by 0.9 after every epoch, and set the batch size to 24 for SFC, 48 for Saratoga and UpperPen, and 96 for Ross. All models were trained using PyTorch 1.11.0+cu102 on a Supermicro SYS-420GP-TNAR+ system equipped with NVIDIA HGX A100 8-way GPUs (80 GB RAM each) running Rocky Linux 9.4 (Blue Onyx), but only used one GPU for all training and inference. We also verified training and inference of our model on a system with one NVIDIA V100 GPU (32 GB RAM).

5.2 Baseline Methods

We compared our proposed method, PFformer[1], against a wide array of state-of-the-art time series and hydrologic prediction methods we discussed in Sect. 2. FEDFormer [25] enhances the standard Transformer by incorporating seasonal-trend decomposition. Informer [23] introduces a prob-sparse self-attention mechanism tailored for long-term time series prediction. NEC+ [7] utilizes LSTM-based models optimized for hydrologic time series prediction in series with

[1] Code for our method is at https://github.com/davidanastasiu/pfformer.

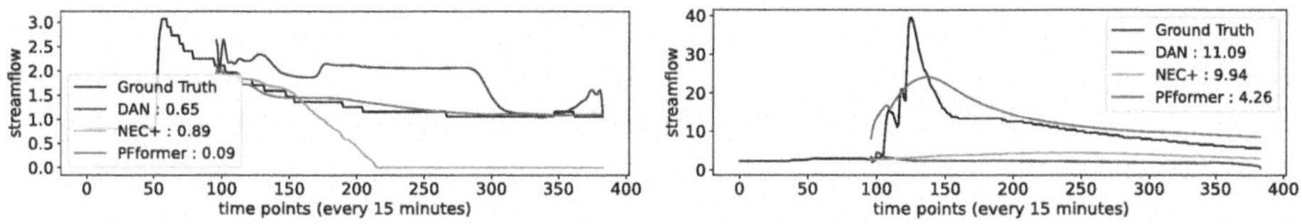

Fig. 4. Comparative examples with the best baselines reveal that PFformer excels in predicting streamflow values with greater accuracy, both in the short and long terms.

extreme events. NLinear [19] is an effective linear model with one order difference preprocessing for long-term time series, while DLinear [19] focuses on decomposing trends for similar applications. Attention-LSTM [4] serves as a state-of-the-art multivariate model in hydrology. Finally, iTransformer [9] achieved state-of-the-art results on challenging multivariate time series prediction challenges, and DAN [6] learns and merges rich representations to adaptively predict streamflow.

5.3 Main Results

The experimental results, detailed in Table 2, highlight the best and second-best values for each metric, marked in bold and underline, respectively. Overall, PFformer consistently outperforms other models. While models tailored for hydrologic data forecasting, like NEC+ and DAN, perform better than more generic time series prediction models, they do not match the effectiveness of PFformer. Specifically, PFformer not only surpasses DAN across all RMSE metrics but also achieves remarkable improvements of 23% to 64% in 4-hour rolling prediction RMSE. Compared to NEC+ and iTransformer, PFformer shows an average increase of 22% and 37% across all RMSE metrics respectively, showing its extreme-adaptive ability.

Transformer-based methods such as FEDFormer and Informer struggle significantly with datasets that exhibit large variance and fail to adapt to extreme values effectively, not fully leveraging the attention mechanism's potential. While iTransformer effectively utilizes variate tokens to capture multivariate correlations through attention, it risks losing intra-variable temporal relationships by focusing solely on inverted dimensions. This limitation is significant in forecasting time series with extreme values, where both inter-variable relationships and intra-variable temporal sequences are crucial for accuracy. Our EFE and AEE address this by preserving both. In contrast, fully connected networks like NLinear and DLinear, which decompose data into main trends and residuals, perform better than some Transformer-based methods, particularly in rolling prediction scenarios. However, they still fall short of achieving high accuracy.

3-Days Prediction. In the realm of single-shot 3-day predictions, PFformer consistently outperforms baselines in RMSE across all four sensors. Notably, when compared to DAN, PFformer achieves a 60% improvement in MAPE for the UpperPen dataset. Furthermore, against NEC+, PFformer demonstrates substantial enhancements on the Ross and UpperPen datasets, with over 50% improvement in RMSE and over 90% in MAPE. For the Saratoga dataset, PFformer outperforms iTransformer by 28% and 62% in RMSE and MAPE, respectively.

4-Hour Rolling Prediction. In the scenario of 4-hour rolling predictions, which is more common in practical applications, the PFformer model showcases exceptional performance. Compared to DAN, it achieves an average RMSE improvement of 43% and a MAPE improvement of 83%. Against NEC+, PFformer demonstrates a 22% improvement in RMSE and a 55% increase in MAPE. Compared with iTransformer, PFformer is superior by 59% and 85% in RMSE and MAPE, respectively. This scenario holds significant practical value, reflecting the model's strong applicability in real-world settings.

Visual Analysis. Figure 4 shows some example predictions for PFformer and the next two best performing algorithms. As seen in the figure, PFformer's predictions closely match the ground truth, particularly in datasets with considerable oscillations. Notably, PFformer more accurately captures short-term oscillations and severe values than DAN, as evidenced by lower RMSE values. This improved performance is primarily due to the rich expressive capabilities of the embeddings mixed with the attention mechanism.

6 Ablation Study

In this ablation study, we aim to explore several key research questions that address the prediction effectiveness of our model, including (1) the impact of our clustering-based oversampling policy on model performance; (2) the significance of subsequence length s in our model's EFE module; (3) the influence of loss function parameters on its predictive accuracy; and (4) the overall effect of both the EFE and AEE modules.

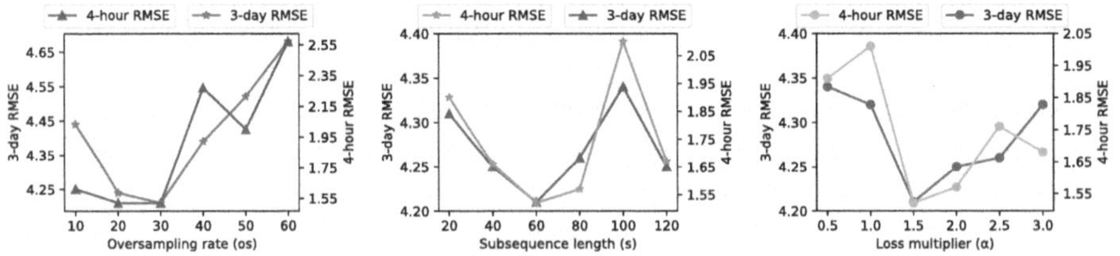

Fig. 5. PFformer RMSE on Ross given different oversampling percentage os (left), EFE subsequence length s (center), and loss multiplier α (right) values, under both 3-day prediction (y1-axis) and 4-hour prediction (y2-axis) scenarios.

6.1 Effect of Parameters

To address the first three research questions, we first established the optimal parameter combination for the Ross dataset as our baseline. Then, fixing all other parameters, adjusted only one parameter at a time to analyze its influence on model performance. Figure 5 shows our results for oversampling percentages $os \in \{10, 20, 30, 40, 50, 60\}$ (left), the EFE subsequence length $s \in \{20, 40, 60, 80, 100, 120\}$ (center), and the loss multiplier parameter $\alpha \in \{0, 1, 1.5, 2.0, 2.5, 3.0\}$ (right), for both 3-day prediction (y1-axis) and 4-hour prediction (y2-axis) scenarios.

Oversampling. For the 3-day RMSE, the model performs best at a 20% rate, while short-term predictions are more sensitive to extreme values in the training set, but a 20% rate is the best overall.

EFE Subsequence Length. Regarding the EFE parameter s, theoretically, a larger s contains more information. By adjusting the accompanying hidden size and number of layers, a better model might be found, but this also increases redundancy and computation load. In our experiments, $s = 60$ was best.

Table 3. PFformer *RMSE* With Different Embeddings

Dataset	Type	EFE/AEE	Position	Token
Ross	3-day	4.21	4.41	4.55
	rolling	1.52	2.24	2.61
SFC	3-day	14.98	21.68	21.38
	rolling	2.86	17.13	16.72
UpperPen	3-day	1.01	3.91	2.44
	rolling	0.24	3.92	1.47
Saratoga	3-day	1.69	2.59	3.88
	rolling	0.29	1.71	3.88

Loss Regularization. Higher values of the loss regularization parameter α force PFformer to focus more on the accuracy of short-term predictions, but should be balanced with the need for overall good long-term predictions. In our experiment, $\alpha = 1.5$ produced the best results in both 3-day ahead and 4-hour predictions.

6.2 Effect of Architecture

In Table 3, the first column shows the RMSE values obtained when using both the EFE and AEE layers. In the second column, we removed EFE and AEE and reverted to using the original Transformer's combination of position embedding and token embedding. In the third column, we eliminated position embedding altogether and solely utilized token embedding. The results show that the EFE and AEE introduced by PFformer significantly enhance overall predictive performance by enabling the attention layer to focus on short-term performance in a rolling prediction mode. Moreover, the position embedding inherent in traditional Transformers has an inconsistent effect on time series forecasting, failing to enhance performance on datasets like SFC and UpperPen.

7 Conclusion

This study has successfully demonstrated the efficacy of position-free trainable embedding techniques—Enhanced Feature-based Embedding (EFE) and Auto-

Encoder-based Embedding (AEE)—in improving hydrologic flow prediction. By comparing these techniques against traditional Transformer embeddings and other state-of-the-art methods, we have shown that our method significantly enhances forecasting accuracy, especially in scenarios with extreme values. Our findings suggest that traditional positional information may be less crucial for time series than previously thought, indicating a promising direction for future research to further refine these models for broader applications in complex time series forecasting scenarios.

References

1. Day, N.E.: Estimating the components of a mixture of normal distributions. Biometrika **56**(3), 463–474 (1969). https://doi.org/10.1093/biomet/56.3.463
2. Farfade, S.S., Saberian, M.J., Li, L.J.: Multi-view face detection using deep convolutional neural networks. In: Proceedings of the 5th ACM on International Conference on Multimedia Retrieval, pp. 643–650 (2015)
3. Kendall, A., Gal, Y., Cipolla, R.: Multi-task learning using uncertainty to weigh losses for scene geometry and semantics. In: Proceedings of the IEEE Conference on Computer Vision and Pattern Recognition, pp. 7482–7491 (2018)
4. Yan, L., Chen, C., Hang, T., Hu, Y.: A stream prediction model based on attention-LSTM. Earth Sci. Inf. **14**(2), 723–733 (2021). https://doi.org/10.1007/s12145-021-00571-z
5. Li, Y., Anastasiu, D.C.: Multivariate segment expandable encoder-decoder model for time series forecasting. IEEE Access **12**, 185012–185026 (2024). https://doi.org/10.1109/ACCESS.2024.3513256
6. Li, Y., Xu, J., Anastasiu, D.: Learning from polar representation: an extreme-adaptive model for long-term time series forecasting. In: Proceedings of the AAAI Conference on Artificial Intelligence, vol. 38, pp. 171–179 (2024)
7. Li, Y., Xu, J., Anastasiu, D.C.: An extreme adaptive time series prediction model based on probability-enhanced LSTM neural networks. In: Proceedings of the AAAI Conference on Artificial Intelligence, vol. 37, pp. 8684–8691 (2023)
8. Lin, Y.H., Tang, H.S., Shen, T.Y., Hsia, C.H.: A smart home energy management system utilizing neurocomputing-based time-series load modeling and forecasting facilitated by energy decomposition for smart home automation. IEEE Access **10**, 116747–116765 (2022)
9. Liu, Y., et al.: iTransformer: inverted transformers are effective for time series forecasting. arXiv preprint arXiv:2310.06625 (2023)
10. Lütkepohl, H.: Vector autoregressive models. In: Handbook of Research Methods and Applications in Empirical Macroeconomics, pp. 139–164. Edward Elgar Publishing (2013)
11. Ma, J., Zhao, Z., Yi, X., Chen, J., Hong, L., Chi, E.H.: Modeling task relationships in multi-task learning with multi-gate mixture-of-experts. In: Proceedings of the 24th ACM SIGKDD International Conference on Knowledge Discovery & Data Mining, pp. 1930–1939 (2018)
12. Mohan, S., Mullapudi, S., Sammeta, S., Vijayvergia, P., Anastasiu, D.C.: Stock price prediction using news sentiment analysis. In: 2019 IEEE Fourth International Conference on Big Data Computing Service and Applications (BigDataService). BDS 2019. IEEE (2019)

13. Nie, Y., Nguyen, N.H., Sinthong, P., Kalagnanam, J.: A time series is worth 64 words: long-term forecasting with transformers. arXiv preprint arXiv:2211.14730 (2022)
14. Sako, K., Mpinda, B.N., Rodrigues, P.C.: Neural networks for financial time series forecasting. Entropy **24**(5), 657 (2022)
15. Vaswani, A., et al.: Attention is all you need. Adv. Neural Inf. Process. Syst. **30** (2017)
16. Wang, Z.Y., Qiu, J., Li, F.F.: Hybrid models combining EMD/EEMD and ARIMA for long-term streamflow forecasting. Water **10**(7), 853 (2018). https://doi.org/10.3390/w10070853
17. Wu, H., Xu, J., Wang, J., Long, M.: Autoformer: decomposition transformers with auto-correlation for long-term series forecasting. Adv. Neural. Inf. Process. Syst. **34**, 22419–22430 (2021)
18. Wu, Z., Pan, S., Long, G., Jiang, J., Chang, X., Zhang, C.: Connecting the dots: multivariate time series forecasting with graph neural networks. In: The 26th ACM SIGKDD International Conference on Knowledge Discovery & Data Mining, pp. 753–763 (2020)
19. Zeng, A., Chen, M., Zhang, L., Xu, Q.: Are transformers effective for time series forecasting? arXiv preprint arXiv:2205.13504 (2022)
20. Zhang, M., Ding, D., Pan, X., Yang, M.: Enhancing time series predictors with generalized extreme value loss. IEEE Trans. Knowl. Data Eng. **35**(2), 1473–1487 (2021)
21. Zhang, Y., et al.: Data regression framework for time series data with extreme events. In: 2021 IEEE International Conference on Big Data (Big Data), pp. 5327–5336 (2021). https://doi.org/10.1109/BigData52589.2021.9671387
22. Zhang, Y., Yan, J.: Crossformer: transformer utilizing cross-dimension dependency for multivariate time series forecasting. In: The Eleventh International Conference on Learning Representations (2023)
23. Zhou, H., et al.: Informer: beyond efficient transformer for long sequence time-series forecasting. In: Proceedings of the AAAI Conference on Artificial Intelligence, vol. 35, pp. 11106–11115 (2021)
24. Zhou, S., Anastasiu, D.C.: Long-term hydrologic time series prediction with LSPM. In: Proceedings of the 33rd ACM International Conference on Information and Knowledge Management, pp. 4308–4312. CIKM '24, Association for Computing Machinery, New York, NY, USA (2024). https://doi.org/10.1145/3627673.3679957
25. Zhou, T., Ma, Z., Wen, Q., Wang, X., Sun, L., Jin, R.: FEDformer: frequency enhanced decomposed transformer for long-term series forecasting. In: International Conference on Machine Learning, pp. 27268–27286. PMLR (2022)

Advancing Long-Term High-Frequency Dissolved Oxygen Forecasting for Australian Rivers

Amelia Genova[1,2], Huidong Jin[1(✉)], John Taylor[2], Joel Janek Dabrowski[4], Peter Toscas[5], and Klaus Joehnk[3]

[1] Data61, CSIRO, Canberra, Australia
warren.jin@csiro.au
[2] School of Computing, College of Engineering, Computing and Cybernetics, The Australian National University, Canberra, Australia
john.taylor@anu.edu.au
[3] CSIRO Environment, Canberra, Australia
klaus.joehnk@csiro.au
[4] Data61, CSIRO, Brisbane, Australia
joel.dabrowski@csiro.au
[5] Data61, CSIRO, Melbourne, Australia
peter.toscas@csiro.au

Abstract. Dissolved Oxygen (DO) is crucial for sustaining healthy river ecosystems. Recent low DO events in Australia's Murray-Darling basin have highlighted the need for accurate, long-term DO forecasts. Process-based models are traditionally used but often fail to outperform simple persistence models beyond a few days. Most data-driven DO models have concentrated on high-frequency DO predictions a few days ahead, or on daily averages that may not capture daily DO minima. Building on the success of transformer-based models for high-frequency time series in long-term forecasting, this paper introduces a novel model, ARPatchTST (AutoRegressive Patch Time Series Transformer). It employs a block autoregressive technique, enabling the State-Of-The-Art (SOTA) PatchTST to utilise forecasted block DO concentrations. By leveraging both blocks and patches, ARPatchTST effectively captures intricate local and cyclic patterns from water quality data to improve forecast accuracy of a transformer. Extensive experimental results from three representative monitoring sites in the Murray-Darling basin show that the proposed model outperforms five benchmark models, including the SOTA PatchTST and a persistence model, for 4- or 7-day forecast horizons at 15 min frequency, or daily DO minima. None of the other models can clearly outperform the simple baseline model, persistence, in forecasting daily minima at site 425012, which has experienced several recent fish kill events. Additionally, paired t-tests over 10 independent runs indicate that its improvements over PatchTST are often statistically significant. This work opens several avenues for future research, which are also discussed.

X. Wu et al. (Eds.): PAKDD 2025, LNAI 15875, pp. 198–209, 2025.
https://doi.org/10.1007/978-981-96-8295-9_14

Keywords: Dissolved oxygen · long-term forecasts · high-frequent time series · AutoRegression · Transformer

1 Introduction

Dissolved Oxygen (DO) is a critical indicator of aquatic ecosystem health, essential for the survival of fish and other aquatic organisms [11]. However, DO levels can fluctuate significantly over time, threatening ecosystem stability. At several locations in the Murray-Darling Basin in Australia, DO levels on several occasions over the last decade dropped sharply, plunging from around 7 mg/L to below 2 mg/L within a matter of hours, resulting in hypoxic conditions and widespread fish deaths [13,16]. These events highlight low DO concentrations as a pressing environmental issue and underscore the need for accurate long-term forecasting and effective mitigation strategies, such as increasing water mixing or introducing oxygen-rich water from upstream reservoirs if available, or using artificial oxygenisation [13].

Increasing interest in DO forecasting has driven extensive research in recent decades, yet many factors influencing DO variability remain poorly understood and dynamic [21]. Although DO exhibits predictable daily and seasonal cycles, its sensitivity to external factors complicates pattern extraction and hinders accurate long-term forecasting [10]. Traditionally, DO forecasting has relied on process-based models to simulate complex physical, chemical, and biological interactions [11,16,17,21]. However, these models are constrained by data limitations, environmental assumptions, and the need for expert recalibration [17,19], often delivering unsatisfactory performance beyond three days and failing to outperform simple persistence methods [17].

By leveraging advanced computational power and large volumes of observations collected by in-situ sensors, data-driven models offer significant flexibility and adaptability. These models extract patterns directly from data, allowing them to adapt to complex systems without requiring extensive domain-specific knowledge, making them particularly valuable where process-based models are limited or unavailable [19]. Most long-term DO forecasts focus on daily or weekly averages [19,21], which smooth the data and may overlook critical daily minima that are essential for mitigating fish kill events. Some studies have targeted daily DO minima but are typically limited to forecasts for one or two days or rely on forecasted weather data, e.g., [4]. Others have attempted high-frequency DO forecasting due to the immediate impacts of DO fluctuations, but these efforts often span only 24 to 48 h due to challenges in capturing long-term patterns in noisy data [1,10]. As a result, long-term high-frequency DO forecasting remains underexplored, despite its critical role in environmental planning and hypoxia mitigation.

Recent advances in time series forecasting, particularly deep learning models based on transformers, have shown promise in addressing the challenges of long-term time series forecasting [2,12,18,20], with the Patch Time Series Transformer (PatchTST) being especially notable [12]. This study is the first

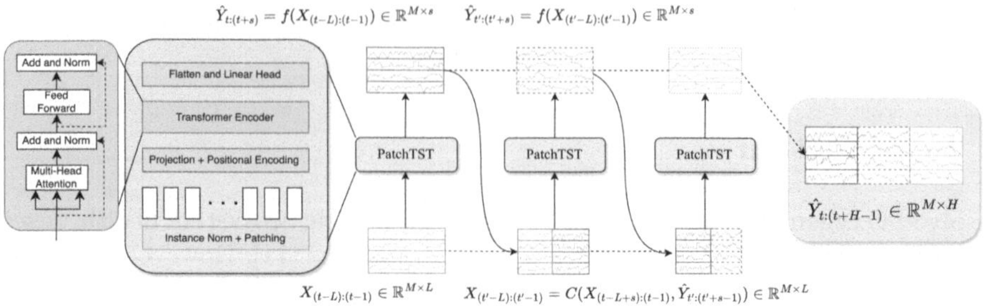

Fig. 1. ARPatchTST architecture illustrated on the block size s chosen as one-third of the forecast horizon H and PatchTST is illustrated in the left two panels [12]

to utilise state-of-the-art (SOTA) models for long-term dissolved oxygen (DO) forecasting based on observations recorded at a 15 min frequency. Building on the success of PatchTST, this paper proposes a novel model, AutoRegressive PatchTST (ARPatchTST), which integrates block autoregressive techniques into data patches used as input tokens for a transformer. ARPatchTST outperforms the persistence model and several SOTA time series forecasting models, including PatchTST [12] for 15 min intervals and daily minima for 4 or 7 days forecasts. Notably, none of the other models clearly surpass the simple persistence baseline in forecasting daily minima at site 425012 which recently experienced hypoxic conditions. Statistical tests confirm the significant improvements of ARPatchTST over PatchTST.

2 An Advanced Model: ARPatchTST

Motivated by the real-world challenges of detailed long-term DO forecasting, we apply SOTA time series models at a 15 min interval [2,12,18]. PatchTST [12] often demonstrates superior performance, as shown in Table 2. By processing segments of time series as patches and feeding them into a transformer, PatchTST efficiently captures long-term temporal dependencies. To further leverage cyclic or more intricate patterns, we introduce ARPatchTST, which incorporates block autoregression into the PatchTST framework (Fig. 1). This allows ARPatchTST to utilise its own strong short-term forecasts to improve long-term predictions. This approach results in statistically significant improvements over PatchTST.

To facilitate understanding of ARPatchTST, we briefly outline PatchTST [12], a SOTA time series forecast model. PatchTST achieves efficient forecasts by feeding sub-series patches as tokens to a transformer encoder for high-frequency time series. By using patches and attention in the transformer, the model can more efficiently capture long-term temporal dependencies in the data. Given an M-dimensional time series before time t, $X_{(t-L):(t-1)} \in \mathbb{R}^{M\times L}$ with a look-back window of length L, each series i is divided into a sequence of patches $x_p^{(i)} \in \mathbb{R}^{P\times N}$ where P is the patch size and N is the number of patches with $P \times N \geq L$. As depicted in the second panel on the left of Fig. 1, each patch is mapped into the latent space of dimension D via a linear projection, $x_d^{(i)} = W_p x_p^{(i)} + W_{\text{pos}}$ where $W_p \in \mathbb{R}^{D \times P}$ is the projection matrix and $W_{\text{pos}} \in \mathbb{R}^{D \times N}$ is the position encoding that preserves temporal order. Multi-head attention is applied in the encoder to transform these projected patches into the query $Q_h^{(i)} = (x_d^{(i)})^\top W_Q^h$, key $K_h^{(i)} = (x_d^{(i)})^\top W_K^h$, and value $V_h^{(i)} = (x_d^{(i)})^\top W_V^h$ matrices for each head $h \in 1 : H$. Then the output of the attention mechanism is computed from a scaled dot-product between the query and key matrices followed by a softmax operation to generate the attention weights which is then applied to the value matrix as $(O_h^{(i)})^T = \text{Attention}(Q_h^{(i)}, K_h^{(i)}, V_h^{(i)}) = \text{Softmax}\left(\frac{Q_h^{(i)} K_h^{(i)^T}}{\sqrt{d_k}}\right) V_h^{(i)}$. The attention output, O, is processed through a feed-forward network with residual connections to produce the latent representation $z^{(i)} \in \mathbb{R}^{D \times N}$. After processing the patches, a flattening layer followed by a linear head is used to generate the H forecast values $\hat{x}^{(i)} = (\hat{x}_t^{(i)}, \hat{x}_{t+1}^{(i)}, \ldots, \hat{x}_{t+H-1}^{(i)}) \in \mathbb{R}^{1 \times H}$.

ARPatchTST integrates block autoregression into PatchTST in order to improve short-term dependency modelling. Autocorrelation analysis of DO observations at several water quality monitoring sites revealed a strong influence of recent historical values on current DO levels. By integrating a block AR mechanism into the PatchTST framework, we enable the model to leverage these temporal dependencies, leading to more accurate and reliable long-term predictions. This mechanism operates on sub-series, capturing local or cyclic patterns. This is achieved by dividing the target horizon H into smaller sub-horizons of s steps, or blocks. The order of the block AR mechanism, p, is given as $\lceil \frac{H}{s} \rceil$ ($p = 3$ is illustrated in Fig. 1).

As illustrated in the third panel of Fig. 1, ARPatchTST takes an input sequence $X_{(t-L):(t-1)} \in \mathbb{R}^{M \times L}$ consisting of L past time steps to predict the next s steps:

$$\hat{Y}_{t:(t+s-1)} = \text{PatchTST}\left(\text{InstanceNorm}\left(X_{(t-L):(t-1)}\right)\right) \in \mathbb{R}^{M \times s} \quad (1)$$

To address distribution shifts between input and output of the model, instance normalisation is applied to each series in the input sequence before applying the PatchTST model. As such each series is normalised to have zero mean and unit variance, and then the original mean and variance are incorporated back into the output $\hat{Y}$ [12].

The model then updates the input sequence by concatenating the forecasted block with the original input sequence to form a new input matrix:

$$X_{(t'-L):(t'-1)} = C\left(X_{(t-L+s):(t-1)}, \hat{Y}_{t':(t'+s-1)}\right) \in \mathbb{R}^{M \times L} \quad (2)$$

where $t' \equiv t$ and the C function concatenates inputs together. This input matrix is then input to PatchTST to forecast the next block as in Eq. (1).

This process is recursively applied until the target horizon H is reached, then we generate the final forecast horizon by concatenating all the predicted sub series values:

$$\hat{Y}_{t:(t+H-1)} = \text{C}\left(\hat{Y}_{t:(t+s-1)}, \hat{Y}_{(t+s):(t+2s-1)}, \ldots, \hat{Y}_{(t+(\lceil \frac{H}{s} \rceil-1)s):(t+H-1)}\right). \quad (3)$$

Table 1. Metadata of water quality data from three sites in the Murray-Darling Basin, Australia

Site ID	Site Name	Start Time	End Time	#Vari	Long	Lat
409215	Murray River at Barmah	2022-06-10 10:00	2024-03-18 15:00	6	144.961	−36.018
425012	Darling River upstream Weir 32	2022-01-28 14:15	2024-03-05 09:30	5	142.340	−32.360
409397	Little Budgee Creek	2019-07-03 15:00	2024-03-18 15:00	3	145.006	−35.895

Thus, ARPatchTST processes input data in patches and outputs data in blocks.

In our model, training is preformed using mean squared error (MSE) loss, where the model learns to minimise the expected loss over the target horizon H across all time steps and channels, as in [12]. By minimising the expected MSE, the model effectively reduces potential error propagation effects that may arise from the block AR mechanism. The loss is calculated from each time series channel and then averaged across all M channels across H time steps to obtain the objective loss as $\mathcal{L}_{MSE} = \mathbb{E}_x \left[\frac{1}{M} \sum_{i=1}^{M} \left(\frac{1}{H} \sum_{t=1}^{H} \left(\hat{y}_t^{(i)} - y_t^{(i)} \right)^2 \right) \right]$.

3 Experimental Setup and Experiments

This study conducts extensive experiments across three water quality monitoring sites in the Murray-Darling Basin, Australia. This region is a significant hydrological and ecological area in southeastern Australia, where hypoxic conditions have led to major fish kill events [16], particularly recently at Menindee Town on the Darling River Lake, near site 425012 [13]. At each monitoring site, in-situ sensors continuously measure various water quality variables at 15 min intervals, including DO, pH, water temperature, water level, water flow, turbidity, and electrical conductivity. Site 409215 included all seven variables, but electrical conductivity was found to have little correlation with DO and was therefore excluded from our experiments. These high-frequency data were downloaded from government websites[1] which are publicly accessible. Metadata including site IDs, names, locations, data start and end times, the number of variables and geographic locations are listed in Table 1. Of the three sites selected, two are associated with the Murray River: site 409215 directly on the main stem, and site 409397 on its tributary, Little Budgee Creek. The third site (425012) is located on the Darling River.

During data preprocessing, water quality variables with a large proportion of missing values (>50%) were removed from a dataset. To enable a fair comparison among forecast models, we used another SOTA time series model DLinear [20] for imputing missing values for each variable. The dataset for each site was divided into training, validation, and test sets in a 60:10:30 ratio, respectively. The last 30% of the time series was used for testing, with test cases generated

[1] data.water.vic.gov.au/water-quality; www.waternsw.com.au/water-services/water-data.

at 15 min intervals over time. There were, e.g., about 18,285 test cases for the forecast horizon of 384 time steps for site 409215.

Our experiments focus on long-term forecasting with forecast horizons H of 384 and 672 15 min intervals, corresponding to 4 and 7 days, respectively. To compare the forecast performance, we use four metrics for both, all the observations at 15 min and daily minima. Two commonly-used metrics, Mean Absolute Error (MAE, $\frac{1}{H}\sum_{t=1}^{H}|y_t - \hat{y}_t|$ with y_t for DO observation at time t or daily minimum on day t) and Root Mean Square Error (RMSE, $\sqrt{\frac{1}{H}\sum_{t=1}^{H}(y_t - \hat{y}_t)^2}$) provide quantitative measures of the discrepancy between predicted and observed DO values in the same unit as DO, i.e., mg/L. Both are lower-is-better metrics, meaning smaller values indicate better model performance. Two higher-is-better metrics are commonly used to assess the accuracy of hydrological and water quality models, such as DO [17,19]: Nash-Sutcliffe Efficiency (NSE, $1 - \frac{\sum(y_t - \hat{y}_t)^2}{\sum(y_t - \bar{y})^2}$), and Willmott's Index of Agreement (d, $1 - \frac{\sum(y_t - \hat{y}_t)^2}{\sum(|y_t - \bar{y}| + |\hat{y}_t - \bar{y}|)^2}$). Both reach a maximum of 1. The results reported in Sect. 4 are averaged over all testing cases and 10 independent runs with 10 random seeds for each site.

To assess the effectiveness of ARPatchTST in forecasting DO levels, several SOTA deep learning models, including PatchTST [12], TimesNet [18], and TiDE [2], were selected as benchmarks. TimesNet transforms 1D time series into a 2D representation based on multiple periods and uses 2D convolution to capture both intraperiod and interperiod variations. TiDE [2] leverages an MLP model with linear layers to effectively handle covariates in relation to temporal elements. LightGBM is also chosen as a benchmark because our initial results and previous research indicated its superiority over traditional machine learning techniques [8]. We also chose the persistence model, a widely recognised baseline in the environmental domain [17]. It uses the corresponding value for DO from the current time as the prediction for the forecast time step, $\hat{DO}_t = DO_{t-H}$ where t indicates the target time point and H is the forecast horizon, such as 4- or 7- days.

All models used a fixed-length look-back window of L=336 time steps (equivalent to 3.5 days), according to some initial test results. The training of all deep learning models, including the proposed ARPatchTST, used the ADAM optimiser with exponential learning rate decay [12,18]. An early stopping mechanism was implemented, whereby training ceased if the validation loss did not improve for five consecutive epochs [12]. The model version with the lowest validation loss was selected for the testing phase to minimise overfitting. For all testing sites, certain hyperparameters were standardized to ensure consistency across experiments. For example, dimensions of embeddings and hidden representations were set to 128, hidden layer dimension in the feed forward network within each encoder layer was 256, the number of encoder layers was 3, and learning rate was initialised at 0.0001. The new hyperparameter of ARPatchTST, the autoregressive block size s, was selected for each site based on validation results, with values ranging from 48, 96, and 192. All models were implemented in Python 3.8.5 using pytorch, scikit-learn, DARTs, and Time Series Library, shared on

Table 2. Forecasting results for six models across three sites and four metrics, averaged over 384 future time steps for DO. **Bold** and underlined cells indicate the best and second-best performances, respectively, for each site-metric combination or row.

Site	Metric	ARPatchTST	PatchTST	TimesNet	TiDE	LightGBM	persistence
409215	MAE	**0.401**	0.402	0.421	0.484	0.509	0.498
	RMSE	0.613	0.627	**0.606**	0.691	0.734	0.701
	NSE	**0.690**	0.676	0.687	0.607	0.556	0.594
	d	0.918	0.916	**0.919**	0.889	0.859	0.892
425012	MAE	**0.656**	0.673	0.695	0.872	1.368	0.751
	RMSE	**0.867**	0.888	0.898	1.146	1.702	0.982
	NSE	**0.757**	0.745	0.740	0.575	0.064	0.688
	d	**0.935**	0.932	0.927	0.864	0.727	0.918
409397	MAE	0.300	**0.299**	0.316	0.376	0.521	0.450
	RMSE	**0.479**	**0.479**	0.491	0.508	0.699	0.680
	NSE	**0.930**	**0.930**	0.927	0.922	0.852	0.860
	d	**0.982**	**0.982**	0.981	0.980	0.958	0.964

github. Deep learning models and their optimised variants were trained on an NVIDIA V100 GPU node equipped with two 24-core Intel Xeon Platinum 8268 processors.

4 Forecasting Results and Comparisons

Table 2 lists the average performance metrics of six models across three sites, with forecasts averaged over 384 time steps (i.e., up to 4 days ahead). At site 409215, ARPatchTST achieves the lowest MAE (0.401), the second-lowest RMSE (0.613), the highest NSE (0.690), and the second-highest *d* (0.918). It outperforms other models across the four metrics, except when compared to TimesNet on 409215, where it outperforms other models on two of the four metrics, demonstrating comparable performance. However, PatchTST underperforms compared to TimesNet, falling short on three out of four metrics. For MAE, PatchTST ranks second with a value of 0.402. Both ARPatchTST and PatchTST achieve MAE values that are at least 4.6% lower than those of the other four models. While the improvements of ARPatchTST over PatchTST range from 0.19% to 2.10%, paired t-tests over 10 runs indicate that these improvements are statistically significant at the 0.05 level, except for MAE (as shown in Table 3). At site 425012, ARPatchTST achieves the best results across all four metrics, while PatchTST consistently ranks second, as shown in Table 2. For these two sites, PatchTST also consistently outperforms TimesNet, which, in turn, surpasses the other three models: TiDE, LightGBM, and persistence. The relative improvements of ARPatchTST over PatchTST range from 0.07% to 2.52%, with all improvements being statistically significant, as indicated by

Table 3. Improvement (%) and p-values of ARPatchTST versus PatchTST across four metrics for forecasts over 384 time steps, based on 10 independent runs per site. **Bold** cells indicate p-values less than 0.05.

Site	Metrics	MAE	RMSE	NSE	d
409215	**Improvement(%)**	0.25	2.10	1.99	0.19
	p-value	1.91E-01	**3.13E-05**	**3.06E-05**	**9.19E-04**
425012	**Improvement(%)**	2.52	2.30	1.55	0.38
	p-value	**1.30E-07**	**2.41E-08**	**2.64E-08**	**7.77E-08**
409397	**Improvement(%)**	−0.32	−0.07	−0.01	0.00
	p-value	2.23E-01	4.22E-01	4.18E-01	4.24E-01

p-values from paired t-tests ranging from 2.41×10^{-8} to 1.30×10^{-7} (details in Table 3). At site 409397, both ARPatchTST and PatchTST consistently outperform the other four models across all four metrics. ARPatchTST shows nearly identical performance to PatchTST, except for a slight increase in MAE.

Considering these three sites together, ARPatchTST achieves the best results in 9 out of 12 site-metric combinations and ranks second in the remaining three. The relative improvement of ARPatchTST over PatchTST across the tree sites ranges from 0.18% on d to 1.67% on $RMSE$, with all improvements being statistically significant, as indicated by p-values less than 8.00×10^{-4}. ARPatchTST, PatchTST, and TimesNet outperform the baseline model, persistence, at all three sites in all four metrics. TiDE outperforms it at two sites (409215 and 409397).

As listed in Table 4 for up to a 672-step horizon, ARPatchTST continues to demonstrate improved performance. It achieves the best results in 11 of 12 site-metric combinations. For the only exception, namely MAE at site 409215, it ranks second, slightly worse than PatchTST. For the other three metrics at these sites, ARPatchTST outperforms PatchTST. Only ARPatchTST consistently outperforms the other four models – TimesNet, TiDE, LightGBM, and persistence. For example, TimesNet outperforms PatchTST on two metrics at sites 409215 and 425012, making it only comparable to TimesNet at these two sites. The average improvements across the three sites on these metrics of ARPatchTST over PatchTST range from 0.24% at d to 2.45% at NSE, and these improvements, though small, are statistically significant with p-values as low as 2.94×10^{-4}. Compared to the baseline persistence model, ARPatchTST, PatchTST, and TimesNet outperform it at all three sites, while TiDE outperforms it at only one site. TiDE performs relatively worse with longer forecast horizons.

From Tables 2 and 4, it is observed that the more complex deep learning models, ARPatchTST, PatchTST, and TimesNet, consistently outperform the baseline persistence model in all instances. In contrast, the relatively simpler deep learning model TiDE does not consistently outperform the baseline, particularly at site 425012. This site seems more complex, with higher variability,

Table 4. Forecast testing results for six models across three sites and four metrics, averaged over 672 future time steps for DO. **Bold** and <u>underlined</u> cells highlight the best and second-best performances, respectively, in each row.

Site	Metric	ARPatchTST	PatchTST	TimesNet	TiDE	LightGBM	persistence
409215	MAE	<u>0.521</u>	**0.520**	0.545	0.618	0.661	0.605
	RMSE	**0.775**	0.799	<u>0.794</u>	0.860	0.920	0.853
	NSE	**0.493**	0.461	<u>0.468</u>	0.376	0.285	0.386
	d	**0.862**	<u>0.861</u>	0.853	0.808	0.770	0.834
425012	MAE	**0.739**	<u>0.758</u>	0.772	1.681	1.358	0.830
	RMSE	**0.968**	0.991	<u>0.987</u>	2.041	1.684	1.076
	NSE	**0.690**	0.675	<u>0.677</u>	−0.380	0.060	0.616
	d	**0.917**	<u>0.912</u>	0.904	0.677	0.714	0.898
409397	MAE	**0.389**	<u>0.395</u>	0.414	0.535	0.623	0.523
	RMSE	**0.621**	<u>0.630</u>	0.664	0.705	0.821	0.788
	NSE	**0.883**	<u>0.879</u>	0.875	0.849	0.795	0.811
	d	**0.970**	<u>0.969</u>	0.968	0.962	0.939	0.952

noise, and larger fluctuations. LightGBM does not outperform the persistence model at the three sites. Based on these results, it can be concluded that long-term, high-frequency DO forecasts generally require more complex deep learning models that can capture temporal intricacies and handle complex multivariate relationships.

We now discuss the test performance on predicting daily DO minima. Table 5 lists the testing results of six models for forecasts up to 4 days ahead. In all 12 site-metric combinations, ARPatchTST demonstrates the best performance. PatchTST secures second place in 10 of those combinations. The clear advantage can be observed at site 425012, where PatchTST's performance is marginally lower than TimesNet's for MAE and the index of agreement d, whereas ARPatchTST clearly outperforms TimesNet. Averaging over the three sites, ARPatchTST shows improvements of 0.64% on d to 6.48% on MAE over PatchTST. Furthermore, the improvements are statistically significant for four metrics at the 0.05 level. Note that ARPatchTST clearly outperforms the baseline persistence model at all three sites. PatchTST and TimesNet outperform the baseline at two sites but perform comparably at site 425012. The test data window for this site includes several recent fish kill events, where DO drops rapidly. Similar comparison results are observed for 7-day ahead forecasts. However, due to space limitations, these detailed findings are not included in this paper.

Model performance for DO or water quality forecasts can be considered satisfactory if the NSE exceeds 0.5 and Willmott's Index of Agreement (d) is greater than 0.9. Based on the results presented in this section, it is evident that ARPatchTST can generate satisfactory DO forecasts, unlike other models, including PatchTST. The incorporation of an autoregressive (AR) compo-

Table 5. Forecast testing results for six models across three sites and four metrics, for daily minima up to 4 days. **Bold** and underlined cells highlight the best and second-best performances, respectively, in each row.

Site	Metric	ARPatchTST	PatchTST	TimesNet	TiDE	LightGBM	persistence
409215	MAE	**0.532**	0.543	0.594	0.662	0.742	0.687
	RMSE	**0.790**	0.792	0.853	0.938	1.051	0.914
	NSE	**0.610**	0.605	0.543	0.448	0.307	0.476
	d	**0.885**	0.883	0.866	0.831	0.775	0.855
425012	MAE	**0.885**	0.991	0.988	1.319	1.935	0.984
	RMSE	**1.103**	1.139	1.207	1.629	2.298	1.241
	NSE	**0.632**	0.607	0.559	0.196	-0.598	0.534
	d	**0.899**	0.884	0.885	0.779	0.652	0.885
409397	MAE	**0.337**	0.341	0.358	0.405	0.625	0.504
	RMSE	**0.498**	0.502	0.527	0.560	0.865	0.750
	NSE	**0.928**	0.925	0.918	0.907	0.779	0.834
	d	**0.981**	0.980	0.979	0.975	0.934	0.958

nent in ARPatchTST is crucial for this enhanced performance. By capturing short-term dependencies within the data, the model can identify stable patterns, reducing sensitivity to random noise or one-off fluctuations. This characteristic enables ARPatchTST to produce smoother predictions, aligning with consistent DO behaviours over short periods. Leveraging the strengths of PatchTST, ARPatchTST can effectively project these short-term trends into the long term, resulting in superior performance compared to PatchTST alone.

5 Conclusion and Future Work

This paper has proposed ARPatchTST (AutoRegressive Patch Time Series Transformer), a new model for forecasting long-term high-frequency dissolved oxygen (DO) levels and daily minimum DO levels. By incorporating a block autoregressive (AR) component, ARPatchTST builds upon the state-of-the-art PatchTST model [12]. While PatchTST effectively captures long-term and local trends using transformers and patches, the added AR component enhances its sensitivity to short-term temporal dependencies, which is critical for handling noisy, high-frequency 15 min real-world water quality data. The model has comprehensively been tested across three representative sites in the Murray-Darling basin, Australia where recent hypoxic events have been reported, demonstrating superior performance over other benchmark models including PatchTST and the persistence model across four metrics. None of the other models clearly outperformed the simple baseline model, persistence, in forecasting daily minima at site 425012 whose test data window includes a massive fish kill event. Paired t-tests on 10 independent runs have indicated that the improvements of ARPatchTST over PatchTST are statistically significant.

Future research could build on this work by extending the model to forecast a broader range of water quality variables, such as temperature, and nutrient levels, thereby enhancing its applicability. Improving the ARPatchTST model's adaptability and accuracy for long-term DO forecasting remains a key priority, which may involve refining its architecture to better capture temporal patterns, optimising training techniques, or integrating richer datasets. Another promising direction is to quantify forecast uncertainty, potentially increasing trust in the model's predictions, particularly for extended horizons [7]. These advancements aim to support more informed decision-making, improve economic outcomes in water resource management, and promote sustainable aquatic ecosystems. Addressing these challenges could significantly advance water quality forecasting and its practical applications [14]. Additionally, extending long-term time series forecasting techniques like ARPatchTST to other domains - such as daily weather, crop water stress and yield forecasts [3,5,6,9,15] - could further demonstrate the model's versatility and impact.

Acknowledgements. We appreciate the constructive comments from the four reviewers. This work was supported by Digital Water and Landscapes and AquaWatch projects from CSIRO Environment Research Unit, and the Queensland government DCAP program. This research was undertaken with the assistance of resources from the National Computational Infrastructure (NCI Australia), an NCRIS enabled capability supported by the Australian Government.

References

1. Dabrowski, J.J., Rahman, A., Hellicar, A., Rana, M., Arnold, S.: Deep learning for prawn farming forecasting and anomaly detection. In: PAKDD 2022, PT III, vol. 13282, pp. 29–41 (2022)
2. Das, A., Kong, W., Leach, A., Mathur, S.K., Sen, R., Yu, R.: Long-term forecasting with TiDE: time-series dense encoder. arXiv preprint arXiv:2304.08424 (2023)
3. Gobbett, D.L., et al.: Minimum temperature mapping augments Australian grain farmers' knowledge of frost. Agric. For. Meteorol. **304**, 108422 (2021)
4. Hu, Y., Liu, C., Wollheim, W.M.: Prediction of riverine daily minimum dissolved oxygen concentrations using hybrid deep learning and routine hydrometeorological data. Sci. Total Environ. **918**, 170383 (2024)
5. Jin, H., Jiang, W., Chen, M., Li, M., Bakar, K.S., Shao, Q.: Downscaling long lead time daily rainfall ensemble forecasts through deep learning. Stoch. Env. Res. Risk Assess. **37**(8), 3185–3203 (2023)
6. Jin, H., Li, M., Hopwood, G., Hochman, Z., Bakar, K.S.: Improving early-season wheat yield forecasts driven by probabilistic seasonal climate forecasts. Agric. For. Meteorol. **315**, 108832 (2022)
7. Jin, H., Mahani, M.E., Li, M., Shao, Q., Crimp, S.: Probabilistic seasonal precipitation forecasts using quantiles of ensemble forecasts. Stoch. Env. Res. Risk Assess. **38**(5), 2041–2063 (2024)

8. Ke, G., et al.: LightGBM: a highly efficient gradient boosting decision tree. In: NIPS, vol. 30 (2017)
9. Li, M., et al.: Improvements in subseasonal forecasts of rainfall extremes by statistical postprocessing methods. Weather Clim. Extremes **34**, 100384 (2021)
10. Liu, Y., Zhang, Q., Song, L., Chen, Y.: Attention-based recurrent neural networks for accurate short-term and long-term dissolved oxygen prediction. Comput. Electron. Agric. **165**, 104964 (2019)
11. Mosley, L.M., Wallace, T., Rahman, J., Roberts, T., Gibbs, M.: An integrated model to predict and prevent hypoxia in floodplain-river systems. J. Environ. Manage. **286**, 112213 (2021)
12. Nie, Y., Nguyen, N.H., Sinthong, P., Kalagnanam, J.: A time series is worth 64 words: long-term forecasting with transformers. In: The Eleventh International Conference on Learning Representations (2023)
13. Office of the NSW Chief Scientist & Engineer: Independent review into the 2023 fish deaths in the Darling-Baaka river at Menindee, Technical report (2023). www.chiefscientist.nsw.gov.au/independent-reports/menindee-fish-deaths
14. Roman, K.: The case for probabilistic forecasting in hydrology. J. Hydrol. **249**(1–4), 2–9 (2001)
15. Shao, Q., et al.: A new probabilistic forecasting model for canopy temperature with consideration of periodicity and parameter variation. Agric. For. Meteorol. **265**, 88–98 (2019)
16. Whitworth, K.L., Baldwin, D.S., Kerr, J.L.: Drought, floods and water quality: drivers of a severe hypoxic blackwater event in a major river system (the Southern Murray-Darling Basin, Australia). J. Hydrol. **450**, 190–198 (2012)
17. Woelmer, W.M., Thomas, R.Q., Olsson, F., Steele, B.G., Weathers, K.C., Carey, C.C.: Process-based forecasts of lake water temperature and dissolved oxygen outperform null models, with variability over time and depth. Ecol. Inf. **83**, 102825 (2024)
18. Wu, H., Hu, T., Liu, Y., Zhou, H., Wang, J., Long, M.: TimesNet: temporal 2D-variation modeling for general time series analysis. In: The Eleventh International Conference on Learning Representations (2023)
19. Tungand, T.M., Yaseen, Z.M.: A survey on river water quality modelling using artificial intelligence models: 2000–2020. J. Hydrol. **585**, 124670 (2020)
20. Zeng, A., Chen, M., Zhang, L., Xu, Q.: Are transformers effective for time series forecasting? In: Proceedings of the AAAI Conference on Artificial Intelligence, vol. 37, pp. 11121–11128 (2023)
21. Zhi, W., et al.: From hydrometeorology to river water quality: can a deep learning model predict dissolved oxygen at the continental scale? Environ. Sci. Technol. **55**(4), 2357–2368 (2021)

CNO-Former: Chaotic Neural Oscillatory Transformer for Social Media Text Generation

Nuobei Shi[1,4], Jiapeng Yu[2], Hengyi Luo[3], Jin Ma[4], and Raymond Lee[1](✉)

[1] Beijing Normal University-Hong Kong Baptist University United International College, Zhuhai 519000, China
{clarissanbshi,raymondshtlee}@uic.edu.cn

[2] University of Warwick, Coventry CV4 7AL, UK
Jiapeng.Yu@warwick.ac.uk

[3] University of Illinois Urbana-Champaign, Urbana, IL 61801, USA
hengyil2@illinois.edu

[4] Hong Kong Baptist University, Hong Kong SAR 999077, China
majing@comp.hkbu.edu.hk

Abstract. Transformers have become foundational in NLP applications, including social media dialogue learning and text generation. Large Language Models (LLMs), pre-trained on extensive data, progressively acquire knowledge for diverse NLP tasks. However, vanilla self-attention requires quadratic time and space, posing challenges in balancing computational efficiency and information representation. Models like Linformer, Performer, Reformer, Longformer, and Synthesizer introduce sparse attention and low-rank projections to address this but suffer from significant information loss and low-resolution representations. This paper proposes the Chaotic Neural Oscillatory Transformer (CNO-former), which uses chaotic dynamics in dot-product attention within multi-head and masked multi-head self-attention. This improves computational efficiency, convergence speed, and reduces information loss. Comprehensive testing against traditional and state-of-the-art models demonstrates that CNO-former achieves superior performance in preserving information and minimizing loss. It outperforms benchmarks on BLEU, F1, and PPL metrics for sentence evaluation, underscoring its effectiveness in text generation tasks.

Keywords: Chaotic Neural Oscillatory Transformer (CNO-former) · Social Media Text Generation · Efficient Transformers · Lee-Oscillator · Dialogue Learning in NLP · Chaotic Dynamics in Self-Attention

1 Introduction

Transformers play a crucial role in NLP applications, utilizing self-attention mechanisms in encoders and decoders to extract information and generate tokens. Pre-trained models fine-tuned on large datasets enhance computational efficiency and improve language model performance. However, Vanilla Transformers require significant time and space, posing challenges in processing complex input sequences.

X. Wu et al. (Eds.): PAKDD 2025, LNAI 15875, pp. 210–222, 2025.
https://doi.org/10.1007/978-981-96-8295-9_15

To address these limitations, a series of "X-former" models, such as Linformer [1], Performer [2], Reformer [3], and Longformer [4], focus on sparse attention and low-rank approximations to reduce computational costs. Sparse attention improves efficiency by limiting token capacity, while low-rank projection simplifies input sequences but risks information loss, limiting their effectiveness for NLP tasks.

This paper presents the Chaotic Neural Oscillatory Transformer (CNO-former), the architecture shown on Fig. 1, an enhanced version of the Vanilla Transformer that integrates chaotic dynamics through the Lee-oscillator [5], based on chaos theory and neural dynamics. By modifying neuron states, CNO-former improves computational efficiency, accelerates convergence, and reduces information loss. Its chaotic dynamic projection strengthens attention scores by expanding correlations across input sequences, enhancing global representation while preserving data magnitude. Unlike other Transformer variants, CNO-former achieves a balance between optimizing computational processes and maintaining representation quality.

The study assesses CNO-former's encoder-decoder performance in social media text generation using datasets of varying input lengths, focusing on contextual feature extraction and chatbot text generation [6]. Results show that CNO-former surpasses Vanilla Transformers, traditional RNNs, Linformer, and Synthesizer in BLEU [7], F1, and PPL [8] metrics, demonstrating superior efficiency and effectiveness in NLP tasks, especially dialogue generation and social media applications. Figure 1 illustrates the CNO-former system architecture for dialogue generation.

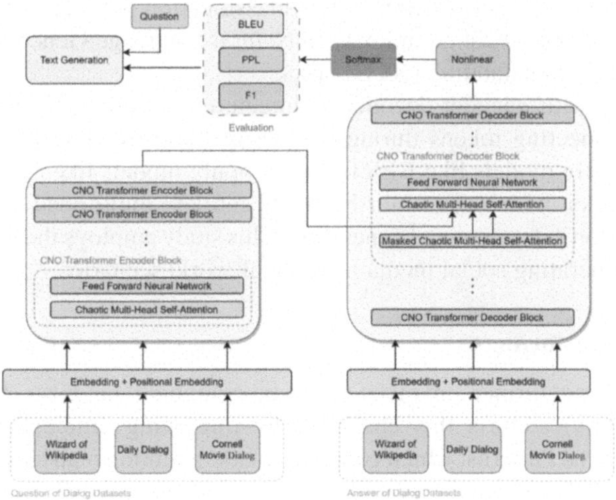

Fig. 1. System architecture for dialogue generation

The contributions of this paper are to:

1. Propose the Chaotic Neural Oscillatory-Transformer (CNO-former), an advanced layer enhancing scaled dot-product attention through chaotic dynamic projection, facilitating automatic association of input sequences.

2. Implement chaotic dynamic projection by transforming linear input into a nonlinear oscillator matrix using a sigmoid-based Lee-Oscillator bifurcation diagram. This technique extracts global information and enhances sequence correlation through gradients.
3. Modify the attention matrix distribution by transitioning neuron states in the hidden layer to a chaotic state, significantly improving computational efficiency and convergence speed.
4. Apply CNO-former to social media text generation tasks, particularly for real-world conversational commands such as English chatbot applications.

2 Related Work

RNN-based models have transitioned into Transformer-based models, advancing from Language Models (LM) to Large Language Models (LLMs) [9]. This section reviews (1) traditional RNN-based models and Transformers and (2) Efficient Transformers addressing computational efficiency in processing large-scale data.

2.1 Traditional RNN-Based Models and Transformers

RNNs, including GRU [10] and LSTM [11], have been widely applied in NLP tasks like text generation, utilizing sequential memory from past and current inputs. While LSTM addressed gradient issues in vanilla RNNs with hidden states similar to attention mechanisms, it struggled to generate complete sentences consistently under uniform training [12]. RNNs performed better in text classification but were heavily reliant on input quality and grammatical complexity in Natural Language Generation (NLG) and Natural Language Understanding (NLU) tasks.

Transformers, with encoder-decoder architectures, introduced scaled dot-product self-attention, connecting tokens through relevance-based pooling [13]. Key variants include encoder-only models like BERT, decoder-only models like GPT, and encoder-decoder frameworks, which use multi-head and masked multi-head self-attention for efficient information extraction and generation. This study employs the encoder-decoder framework for generating social media text in real-world scenarios.

2.2 Efficient Transformers

Self-attention in Transformers involves quadratic computational costs. Efficient Transformers (X-formers) address this challenge by improving efficiency through three approaches: (1) Fixed Patterns, which constrain sparse attention matrices [1]; (2) Combination of Patterns (CP), which merge multiple access patterns; and (3) Learnable Patterns (LP). Notable examples include Longformer [4], Big Bird [14], Axial Transformer [15], Linformer [1], Synthesizer [16], Compressive Transformer [17], and Set Transformer [18]. This paper introduces the CNO-former, which enhances efficiency with a chaotic oscillation layer incorporating normalization and ReLU activation at the residual unit. This design improves computational performance while reducing resource demands. Experimental results demonstrate significant advancements over baseline models such as Linformer and Synthesizer, achieving superior efficiency without compromising effectiveness.

3 Methodology

The CNO-former builds upon the Vanilla Transformer model by integrating an oscillator layer into the scaled dot-product attention mechanism, impacting both multi-head and masked multi-head self-attention within the encoder and decoder. The Lee-oscillator layer transforms neuron states from linear projection or classification into nonlinear dynamic projection using a set of parameters with a sigmoid transfer function for Q and K. Unlike sparse attention mechanisms, the chaotic oscillator accelerates computation, enhances convergence efficiency, and reduces loss, resulting in improved model performance. The structure of the CNO-former model is depicted in Fig. 2.

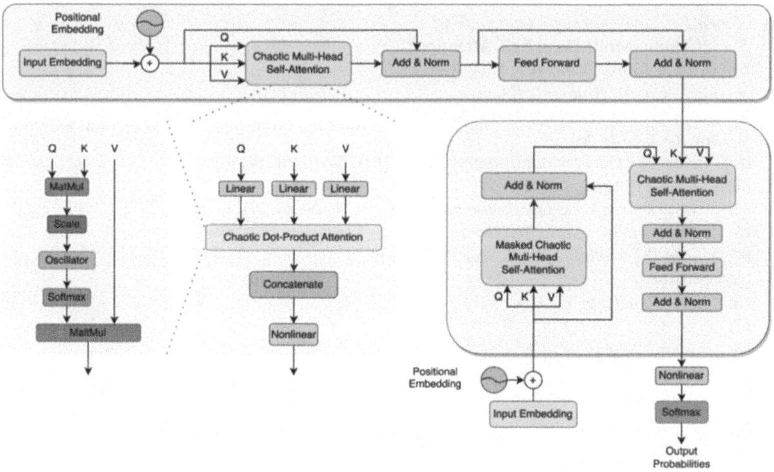

Fig. 2. Structure of proposed CNO-former in social media text generation

3.1 CNO-Former

The CNO-former structure includes an encoder and decoder designed for large-scale social media text generation tasks. Based on the Vanilla Transformer, it integrates chaos theory and oscillatory dynamics into the scaled dot-product attention mechanism, a key component for multi-head and masked head attention in auto-regression tasks [19]. An oscillator layer is introduced before the SoftMax function, acting as a nonlinear transfer function to dynamically alter neuron states in the network.

Algorithms 1 and 2 detail the encoder and decoder workflows. Chaotic behaviour within dot-product attention appears in steps 9–15 of Algorithm 1 and steps 11–17 of Algorithm 2 together with Eqs. (1)-(5) for Lee-Oscillator nonlinear projection, where SoftMax functions as the self-attention activation mechanism. In text generation tasks, SoftMax calculates probabilities for each word, selecting the next word based on these probabilities.

By replacing the multi-head self-attention pattern with a chaotic counterpart, the vector representation scores are adjusted to enhance information extraction and association. The SoftMax function projects input sequences into the range [0,1], ensuring the output probabilities sum to 1. During self-attention in the encoder, SoftMax generates output values corresponding to the probability of each word in the sequence.

Algorithm 1 CNO-former Encoder Layer

```
Input: in_emb, input embedding vector
Output: out_enc, output Encoder vector
Parameters: H, number of attention heads
            N, number of layers of CNO-former
/*Encoder Layer*/
for n = 0 to N do
  /*Layer Norm*/
  lnorm1 = LayerNorm(in_emb,γ_l1,β_l1)
  /*Chaotic Multi-Head Self-Attention*/
  value = Conv1D(lnorm1, W_v, b_v)
  key = Conv1D(lnorm1, Wv, bv)
  query = Conv1D(lnorm1, Wv, bv)
  for h = 0 to H do
    scores = Devide(Matul(query[h], key^T[h]), Sqrt(len(query))
    scores = Mask_fill(scores)
    Osc_scores = LeeOscillator(scores, a_1, a_2, θ_a, b_1, b_2, θ_b)
    Osc_scores = Softmax(Osc_scores)
    attn'[h] = Matul(Osc_scores , value[h])
  end for
  attn = Sync(attn')
  c_attn' = Conv1D(attn, W_a, b_a)
  c_attn = Sync(c_attn')
  /*Residual*/
  c_attn = c_attn + in_emb
  /*Layer Norm*/
  lnorm2 = LayerNorm(c_attn ,γ_l2,β_l2)
  /*Feed-Forward*/
  ffn1' = ReLU(Conv1D(lnorm2, W_f1, b_f1))
  ffn1 = Sync(ffn1')
  ffn2' = Conv1D(ffn1, W_f2, b_f2)
  ffn2 = Sync(ffn2')
  /*Residual*/
  in_emb = ffn2 + c_attn
end for
out_enc = in_emb
```

Algorithm 2 CNO-former Decoder Layer

```
Input: out_enc, Encoder layer output vector
    in_emb, input embedding vector
Output: out_dec, output Decoder vector
Parameters: H, number of attention heads
    N, number of layers of CNO-former
/*Decoder Layer*/
for n = 0 to N do
  /*Input Encoder*/
  lnorm1 = Encoder(in_emb)
  /*Residual*/
  lnorm1= lnorm1 + in_emb
  /*Masked Chaotic Multi-Head Self-Attention*/
  value = Conv1D(out_enc, W_v, b_v)
  key = Conv1D(out_enc, Wv, bv)
  query = Conv1D(lnorm1, Wv, bv)
  for h = 0 to H do
    scores = Devide(Matul(query[h], key^T[h]), Sqrt(len(query)))
    scores = Mask_fill(scores)
    Osc_scores = LeeOscillator(scores, a_1, a_2, θ_a, b_1, b_2, θ_b)
    Osc_scores = Softmax(Osc_scores)
    attn'[h] = Matul(Osc_scores , value[h])
  end for
  attn = Sync(attn')
  c_attn' = Conv1D(attn, W_a, b_a)
  c_attn = Sync(c_attn')
  /*Residual*/
  c_attn = c_attn + lnorm1
  /*Layer Norm*/
  lnorm2 = LayerNorm(c_attn ,γ_l2,β_l2)
  /*Feed-Forward*/
  ffn1'= ReLU(Conv1D(lnorm2, W_f1, b_f1))
  ffn1 = Sync(ffn1')
  ffn2' = Conv1D(ffn1, W_f2, b_f2)
  ffn2 = Sync(ffn2')
  /*Residual*/
  in_emb = ffn2 + c_attn
end for
out_dec = in_emb
```

3.2 Lee-Oscillator

The parameters of the Lee-Oscillator, as applied in TCAN [20] for auto-associator tasks, enhance token correlations across the entire input sequence through nonlinear projection. This approach addresses long-sequence processing challenges by improving computational efficiency, accelerating convergence, reducing loss, and preserving information. With its chaotic transient progressive transfer in neural dynamics, the Lee-Oscillator equations from Eqs. (1)–(5) applied into CNO-former in Algorithm 1 and 2 for nonlinear projection effectively transforms the distribution of input sequence vector representations. The corresponding equations is as follows:

$$E(t+1) = Sig[e_1 E(t) - e_2 I(t) + S(t) - \xi_E] \quad (1)$$

$$I(t+1) = Sig[i_1 E(t) - i_2 I(t) - \xi_I] \tag{2}$$

$$\Omega(t) = Sig[S(t)] \tag{3}$$

$$Lee(t) = [E(t) - I(t)]e^{-kS^2(t)} + \Omega(t) \tag{4}$$

Where E, I, Ω and Lee are excitatory, inhibitory, input and output neurons respectively, e_1, e_2, i_1 and i_2 are weights, ξ_E and ξ_I are thresholds, $S(t)$ is input value. Lee-Oscillator output is the combining result of $Lee(t) = (1, 2, \ldots, T)$.

$$Sig(x) = \frac{1}{1 + e^{-x}} \tag{5}$$

3.3 Chaotic Scaled Dot-Product Attention

Chaotic dot-product attention is applied in chaotic multi-head and masked chaotic multi-head attention. Scaled dot-product attention serves as the self-attention kernel for training the encoder and decoder, proving superior to other methods. Here, Q, K and V represent input sequences that undergo scaled dot-product attention via linear projection to calculate attention scores, facilitating encoder representation extraction in multi-head self-attention. To enhance this process, a Lee-oscillator layer is introduced before the SoftMax function, transitioning the projection from static to dynamic. This dynamic projection calculates scaled dot-product attention scores while extracting token relationships for generating model representations in the decoder.

The chaotic dynamic projection procedure, illustrated in Fig. 3, begins with Q and K diagonals as input sequences in a dialogue generation model. Initially, tokens are equal without attention scores. Chaotic dynamic projection, operating at high frequencies, computes attention through oscillatory transformations, projecting input sequences into an oscillator projection matrix. This enhances token correlations in the attention matrix by leveraging gradients. The number of chaotic dynamic projections corresponds to d, the scaled dot-product attention dimension, maximizing the oscillator's chaotic capacity within neurons.

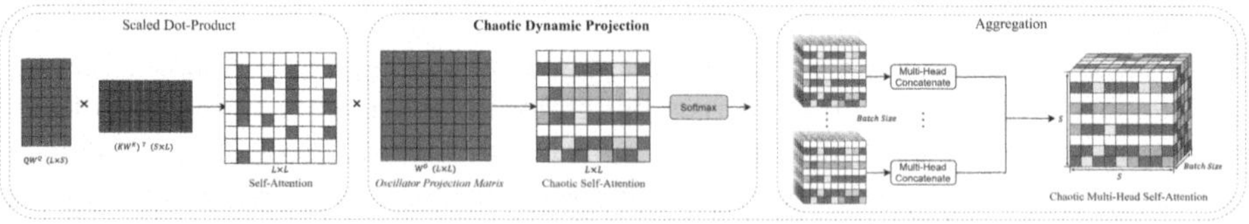

Fig. 3. Calculation procedure of Chaotic multi-head self-attention

4 Experiment

4.1 Experiment Setup

The CNO-former model is trained on three social media dialogue datasets designed to align with English language learning at the university level:

1. Daily Dialog [21]: This dataset compiles QA pairs from English language learning platforms, similar to the 2011 Twitter dialog dataset. It spans 10 cross-cultural topics, including School Life, Culture, Education, Attitude and Emotion, Relationships, Tourism, Health, Work, Politics, and Finance. The dataset avoids idiomatic expressions common in social media to enhance learning quality and establish a pre-trained knowledge base for an English chatbot with correct syntax QA pairs. It is less noisy, making it suitable for new language learners.
2. Wizard of Wikipedia (WoW) [22]: An open-domain dialog dataset based on Wikipedia content from 2019, featuring academic properties. Conversations are role-based, with a knowledge expert ("wizard") and a learner ("apprentice"), distinguishing it from casual Daily Dialog content.
3. Cornell Movie Dialog Corpus [23]: This dataset includes 220,579 conversations between 9,035 characters across 617 movies, derived from IMDB. QA pairs vary based on individual character traits. For training, irrelevant information like genres and release years is excluded, focusing solely on dialogue pairs.

During data preprocessing, text is converted into vectors for model training. A word map is generated based on dataset word counts, assigning an index to each word in a sentence. Special tokens < SOS > (start of sentence) and < EOS > (end of sentence) are added. Sentences are padded to a maximum length, embedded as vectors with positional encoding, and converted into tensor format for training.

Model evaluation differs from classification tasks, which typically use accuracy as a benchmark. Due to the nature of text generation, testing data is limited to 0.01 of the datasets to assess performance on unbroken text generation. This evaluation excludes pre-trained weights from Large Language Models, focusing on BLEU, PPL, and F1 metrics for NLG analysis.

The hardware environment includes an Intel(R) Xeon(R) Platinum 8338C CPU, 80GB memory, and an NVIDIA GeForce RTX 3090 GPU. The implementation is built using PyTorch 1.11.0, based on CUDA 11.3, on Ubuntu 20.04.

4.2 Model Performance

BLEU, PPL, and F1 metrics are used for NLG analysis to evaluate sentence generation in dialogue contexts. Experimental results demonstrate that the CNO-former model outperforms the Vanilla Transformer in computational efficiency, convergence speed, and loss reduction. Specifically, the CNO-former, utilizing the Lee-Oscillator, was tested for loss comparison, as loss reflects information preservation across the entire model, particularly highlighting the layer differences between the CNO-former and the Vanilla Transformer.

Table 1. Model Layers Comparison in WoW

Layer	Method	Loss	BLEU	F1	PPL
1 layer	Transformer	0.7013	0.3980	0.7543	2.0164
	CNO-former	0.3488	0.6542	0.8737	1.4174
2layers	Transformer	0.2107	0.7748	0.9040	1.2424
	CNO-former	0.1109	0.8864	0.9495	1.1173
3layers	Transformer	0.1185	0.7565	0.9263	1.1258
	CNO-former	**0.0483**	**0.9651**	**0.9843**	**1.0495**

The WoW dataset was utilized for loss comparison using a medium sequence length of 50 over 100 epochs. As shown in Table 1, the CNO-former outperformed the Vanilla Transformer, achieving an overall loss reduction of approximately 50%. Notably, in a 2-layer configuration, the CNO-former demonstrated greater loss reduction than the Vanilla Transformer with a deeper 3-layer setup. Consistently across all configurations, the CNO-former reduced loss by about 50% compared to the Vanilla Transformer, highlighting its efficiency and effectiveness.

Additionally, the CNO-former demonstrated significantly higher convergence speed, addressing the quadratic time and space cost challenges of the Vanilla Transformer. Results for three datasets with varying input sequence lengths (short, medium, and long) are listed in Table 2. With the same parameter settings for training, testing, and optimization, the CNO-former achieved faster convergence.

For the Daily Dialog and Cornell Movie Dialog datasets, representing the shortest and longest sequences, the CNO-former required only 1/5 of the epochs needed by the Vanilla Transformer (e.g., 20–21 epochs out of 100). For the WoW dataset, the CNO-former achieved convergence in less than 1/3 of the epochs required by the Vanilla Transformer, as shown in Table 2. In the meantime, the computational efficiency for whole model training improved and saves computational cost.

Table 2. Model Convergence Comparison by Epoch

Datasets	Method	Epoch	Loss	BLEU	F1	PPL
Daily Dialog	Transformer	100	**0.1035**	0.6751	0.8794	**1.1090**
	CNO-former	**21**	0.1848	**0.6806**	**0.8848**	1.2029
Wizard of Wikipedia	Transformer	100	**0.1185**	0.7565	**0.9263**	1.1258
	CNO-former	**29**	0.2094	**0.7617**	0.9083	**1.2329**
Cornell Movie Dialog	Transformer	100	0.2963	0.6756	0.9154	1.3449
	CNO-former	**20**	**0.2638**	**0.7120**	**0.9320**	**1.3019**

5 Comparison and Discussion

This section presents a comparison of the CNO-former model with three baseline models, excluding the Vanilla Transformer, for social media text generation. The models compared are the traditional RNN, Synthesizer, Linformer, and Vanilla Transformer, transitioning from traditional to state-of-the-art approaches. The datasets used for evaluation include Daily Dialog, Wizard of Wikipedia, and Cornell Movie Dialog. Four performance metrics: Loss, BLEU, F1, and PPL were used to assess model performance. The experiment aimed to verify the CNO-former's performance relative to other models, extending the baselines to include four models ranging from RNN to the state-of-the-art Synthesizer. All experiments used the same architecture and parameters for consistency:

1. Transformer [19]: The baseline model for CNO-former, with the chaotic oscillator layer added at the dot-product attention, a critical component of self-attention.
2. RNN [23]: Representing the foundational encoder-decoder framework in NLP tasks before pre-trained LLMs.
3. Synthesizer [16]: A model replacing dot-product self-attention by synthesizing attention weights without token-token interaction, substituting query-key-value computations with a synthetic alignment matrix.
4. Linformer [1]: Focused on reducing time and space complexity in self-attention by approximating it with a low-rank matrix, transforming quadratic complexity into linear for improved efficiency and memory usage.

Table 3 summarizes model performance after 100 epochs. CNO-former consistently outperformed all models, achieving convergence in fewer than 30 epochs. As BLEU and F1 scores increased, PPL decreased, improving the clarity of generated text.

RNN and Synthesizer showed the highest loss values, exceeding other models by more than 0.2, indicating significant information loss. In contrast, CNO-former preserved vector representations more effectively. While Synthesizer was designed to challenge the centrality of dot-product self-attention, its performance lagged significantly. For example, in the Cornell Movie Dialog dataset, Synthesizer had a loss of 1.2830 compared to CNO-former's 0.0912—a 92.89% improvement. On the WoW dataset, the improvement was 83.17%.

For PPL, which evaluates word-level predictions, CNO-former, Linformer, and Vanilla Transformer performed similarly due to their shared self-attention structures, as shown in Fig. 4. RNN and Synthesizer, with different architectures, had higher PPL values, reflecting their limited ability to reduce conversational complexity. For BLEU and F1 metrics, which measure text quality, CNO-former consistently outperformed Linformer. On the Daily Dialog dataset, CNO-former achieved a BLEU score of 0.8778, 19.73% higher than Linformer's 0.7331. For F1, in the Cornell Movie Dialog dataset, CNO-former scored 0.9801 compared to Linformer's 0.9122, a 7.44% improvement.

In summary, traditional RNNs with encoder-decoder architectures underperformed compared to Transformer-based models. Synthesizer, which replaced dot-product self-attention with a synthetic alignment matrix, demonstrated the weakest performance.

Table 3. Baseline Models Comparison

Method	Loss	BLEU	F1	PPL
Daily Dialog				
RNN	1.0777	0.1425	0.3443	2.9380
Transformer	0.1035	0.6751	0.8794	1.1090
Synthesizer	0.2822	0.4033	0.6814	1.3260
Linformer	0.2190	0.7331	0.9208	1.2448
CNO-former	**0.0475**	**0.8778**	**0.9691**	**1.0487**
Wizard of Wikipedia				
RNN	0.2829	0.4741	0.6861	1.3271
Transformer	0.1185	0.7565	0.9263	1.1258
Synthesizer	0.5338	0.5380	0.7379	1.7053
Linformer	0.0988	0.8813	0.9465	1.1038
CNO-former	**0.0483**	**0.9651**	**0.9843**	**1.0495**
Cornell Movie Dialog				
RNN	0.6590	0.4067	0.6042	1.9329
Transformer	0.2963	0.6756	0.9154	1.3449
Synthesizer	1.2830	0.2919	0.5750	3.6037
Linformer	0.2511	0.7686	0.9122	1.2855
CNO-former	**0.0912**	**0.8226**	**0.9801**	**1.0955**

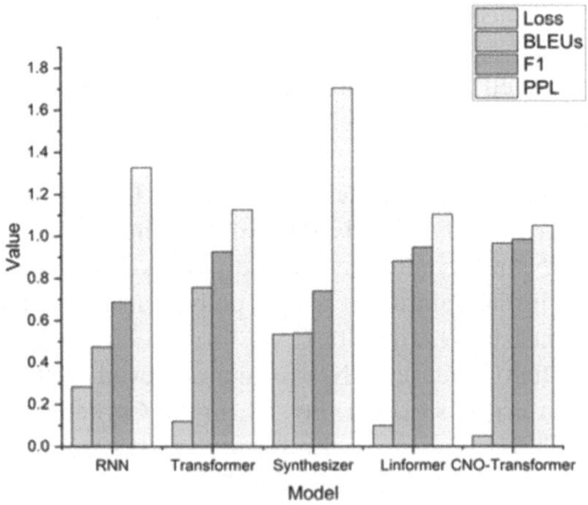

Fig. 4. Baseline model comparison on Wizard of Wikipedia

Among Transformer variants, Linformer and CNO-former outperformed Vanilla Transformer, with CNO-former achieving the best overall results. This chaotic dynamic projection improves vanilla attention score patterns and strengthens token correlations, as evaluated by BLEU, PPL, and F1 metrics for sequence generation.

6 Conclusion

This paper introduces the CNO-former, a model based on the Vanilla Transformer, which modifies neuron states in dot-product self-attention, the core of self-attention, to reduce computation costs and information loss. By integrating the Lee-Oscillator, the chaotic multi-head and masked multi-head self-attention mechanisms in the encoder and decoder improve computational efficiency, accelerate convergence, and enhance vector representation preservation.

The CNO-former was evaluated against traditional and state-of-the-art models, demonstrating superior information preservation and loss reduction. It outperformed benchmarks in BLEU, F1, and PPL metrics for sentence evaluation, highlighting its effectiveness in text generation tasks.

As a social media text generation model for English Language Learning chatbots, CNO-former provides a robust alternative to the Vanilla Transformer for pre-training on large datasets and extending knowledge bases. It was tested on three datasets—Daily Dialog, Wizard of Wikipedia, and Cornell Movie Dialog—focused on daily conversations in social platforms for university-level language learning.

Future work could further optimize computation, convergence, and information correlation. Additionally, the CNO-former's auto-associative capabilities could be extended to long-sequence tasks, such as text summarization, using Long Range Arena benchmarks for document-level analysis. Its adaptability positions it as a valuable variant of the Vanilla Transformer for a wide range of NLP and non-NLP applications.

References

1. Wang, S., Li, B.Z., Khabsa, M., Fang, H., Ma, H.: Linformer: self-attention with linear complexity. arXiv preprint arXiv:2006.04768 (2020)
2. Choromanski, K., et al.: Rethinking attention with performers. arXiv preprint arXiv:2009.14794 (2020)
3. Kitaev, N., Kaiser, Ł., Levskaya, A.: Reformer: the efficient transformer. arXiv preprint arXiv:2001.04451 (2020)
4. Beltagy, I., Peters, M.E., Cohan, A.: Longformer: the long-document transformer. arXiv preprint arXiv:2004.05150 (2020)
5. Lee, R.S.T.: Chaotic type-2 transient-fuzzy deep neuro-oscillatory network for worldwide financial prediction. IEEE Trans. Fuzzy Syst. **28**, 731–745 (2020)
6. Li, J., Song, H., Li, J.: Transformer-based question text generation in the learning system. In: Proceedings of the 6th International Conference on Innovation in Artificial Intelligence, pp. 50–56. ACM, New York (2022)
7. Papineni, K., Roukos, S., Ward, T., Zhu, W.-J.: BLEU: a method for automatic evaluation of machine translation. In: Proceedings of the 40th Annual Meeting of the Association for Computational Linguistics, pp. 311–318. ACL, Morristown (2001)

8. Zhang, H., Song, H., Li, S., Zhou, M., Song, D.: A survey of controllable text generation using transformer-based pre-trained language models. arXiv preprint arXiv:2201.05337 (2022)
9. Li, J., Tang, T., Zhao, W.X., Nie, J., Wen, J.: Pretrained language models for text generation: a survey. arXiv preprint arXiv:2201.05273 (2022)
10. Chung, J., Gulcehre, C., Cho, K., Bengio, Y.: Empirical evaluation of gated recurrent neural networks on sequence modeling. arXiv preprint arXiv:1412.3555 (2014)
11. Hochreiter, S., Schmidhuber, J.: Long short-term memory. Neural Comput. **9**, 1735–1780 (1997)
12. Iqbal, T., Qureshi, S.: The survey: text generation models in deep learning. J. King Saud Univ. Comput. Inf. Sci. **34**, 2515–2528 (2022)
13. Tay, Y., Dehghani, M., Bahri, D., Metzler, D.: Efficient transformers: a survey. ACM Comput. Surv. **55**, 1–28 (2023)
14. Zaheer, M., et al.: Big bird: transformers for longer sequences. In: Advances in Neural Information Processing Systems (2020). arXiv:2007.14062
15. Ho, J., Kalchbrenner, N., Weissenborn, D., Salimans, T.: Axial attention in multidimensional transformers. arXiv preprint arXiv:1912.12180 (2019)
16. Tay, Y., Bahri, D., Metzler, D., Juan, D.-C., Zhao, Z., Zheng, C.: Synthesizer: rethinking self-attention in transformer models. arXiv preprint arXiv:2005.00743 (2020)
17. Rae, J.W., Potapenko, A., Jayakumar, S.M., Lillicrap, T.P.: Compressive transformers for long-range sequence modelling. arXiv preprint arXiv:1911.05507 (2019)
18. Lee, J., Lee, Y., Kim, J., Kosiorek, A.R., Choi, S., Teh, Y.W.: Set transformer: a framework for attention-based permutation-invariant neural networks. arXiv:1810.00825 (2018)
19. Vaswani, A., et al.: Attention is all you need. In: Proceedings of the 31st International Conference on Neural Information Processing Systems, pp. 6000–6010 (2017)
20. Lee, R.S.T.: A transient-chaotic autoassociative network (TCAN) based on lee oscillators. IEEE Trans. Neural Netw. **15**, 1228–1243 (2004)
21. Li, Y., Su, H., Shen, X., Li, W., Cao, Z., Niu, S.: DailyDialog: a manually labelled multi-turn dialogue dataset. arXiv preprint arXiv:1710.03957 (2017)
22. Dinan, E., Roller, S., Shuster, K., Fan, A., Auli, M., Weston, J.: Wizard of Wikipedia: knowledge-powered conversational agents. In: 7th International Conference on Learning Representations (ICLR 2019) (2018). arXiv:1811.01241
23. Ren, D., Cai, Y., Chan, W.H., Li, Z.: A Clustering Based Adaptive Sequence-to-Sequence Model for Dialogue Systems. In: 2018 IEEE International Conference on Big Data and Smart Computing, pp. 775–781 (2018)

Multilingual Non-factoid Question Answering with Answer Paragraph Selection

Ritwik Mishra[1(✉)], Sreeram Vennam[2], Rajiv Ratn Shah[1], and Ponnurangam Kumaraguru[2]

[1] Indraprastha Institute of Information Technology Delhi, Delhi, India
ritwikm@iiitd.ac.in

[2] International Institute of Information Technology Hyderabad, Hyderabad, India

Abstract. Most existing Question Answering Datasets (QuADs) primarily focus on factoid-based short-context Question Answering (QA) in high-resource languages. However, the scope of such datasets for low-resource languages remains limited, with only a few works centered on factoid-based QuADs and none on non-factoid QuADs. Therefore, this work presents MuNfQuAD, a multilingual QuAD with *non-factoid* questions. It utilizes interrogative sub-headings from BBC news articles as questions and the corresponding paragraphs as silver answers. The dataset comprises over 578K QA pairs across 38 languages, encompassing several low-resource languages, and stands as the largest multilingual QA dataset to date. Based on the manual annotations of 790 QA-pairs from MuNfQuAD (golden set), we observe that 98% of questions can be answered using their corresponding silver answer. Our fine-tuned Answer Paragraph Selection (APS) model outperforms the baselines. The APS model attained an accuracy of 80% and 72%, as well as a macro F1 of 72% and 66%, on the MuNfQuAD testset and the golden set, respectively. Furthermore, the APS model effectively generalizes a certain language within the golden set, even after being fine-tuned on silver labels. We also observe that the fine-tuned APS model is beneficial for reducing the context of a question verifying the utility of the proposed dataset.

Keywords: multilingual corpora · datasets for low resource languages · corpus creation · benchmarking · multilingual QA · open-domain QA

1 Introduction

A typical Question Answering Dataset (QuAD) conventionally comprises question-answer pairs [5]. However, certain QuADs are characterized by an additional component called *evidence* or *context* accompanying each question [28]. This contextual information is expected to provide sufficient details to address the corresponding question, leading to these QuADs being referred to as

X. Wu et al. (Eds.): PAKDD 2025, LNAI 15875, pp. 223–236, 2025.
https://doi.org/10.1007/978-981-96-8295-9_16

Reading-Comprehension (RC) datasets as well. A majority of RC datasets focus on factoid answers, typically short phrases or named entities [32]. For example, using the Wikipedia page of India as the *context*, a factoid question such as *Who was the first Prime Minister of India?* would have the corresponding factoid answer *Jawaharlal Nehru*.

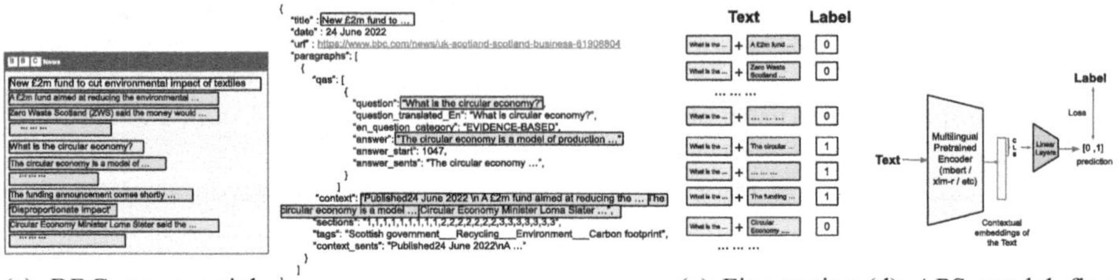

(a) BBC news article on web. (b) Extracted data dictionary. (c) Fine-tuning data. (d) APS model fine-tuning.

Fig. 1. An illustration depicting the data collection process and fine-tuning of the Answer Paragraph Selection (APS) model. (a) Presents a BBC news article accessed via the URL: https://www.bbc.com/news/uk-scotland-scotland-business-61908804. The scraped data dictionary is shown in (b). In subfigure (c), Label-0 indicates that the given paragraph is not included in the silver answer, while Label-1 signifies that it is. APS model architecture and the fine-tuning process are illustrated in (d).

As compared to factoid questions, non-factoid questions have long descriptive answers consisting of multiple sentences or paragraphs. A non-factoid question could be framed as *How did Jawaharlal Nehru become the first Prime Minister of India?* Evidence suggests that modern search engines are unable to answer non-factoid questions effectively [7]. Moreover, even humans find it difficult to answer non-factoid questions [6]. In order to automatically answer non-factoid questions, large non-factoid QuADs are needed to fine-tune Question-Answering (QA) models. Multilingual QA models face additional challenges due to the lack of such resources, motivating the development of a multilingual QA dataset specifically designed for non-factoid questions.

In this study, we automatically extract Question-Answer pairs and their corresponding news articles from the British Broadcasting Corporation (BBC) website in multiple languages[1]. Except for the golden set, the dataset is not manually annotated since it relies on the hypothesis put forth by Soleimani et al. [32] that *all the paragraphs succeeding an interrogative subheading contains its answer.* Therefore, we refer to this dataset as having silver labels/answers. Previous studies have indicated that silver labels have proven beneficial for constructing text classifiers in domains with limited availability of gold labels, such as legal [25], medical [26], and news [11] domains. Our evaluation on silver labels against the gold labels reveals that 98% of the questions were effectively answered by their silver answers. Our main contributions are outlined as follows:

[1] https://www.bbc.co.uk/ws/languages.

1. We release MuNfQuAD[2], a multilingual question-answering dataset partitioned into train, validation, and test splits. It contains more than 578K Question-Answer pairs in 38 different languages.
2. We also release multilingual Answer Paragraph Selection (APS) models fine-tuned on MuNfQuAD, leveraging base variants of different pretrained encoders.

Table 1. Attributes of different multilingual QuADs. For languages like Chinese and Japanese, which lack whitespaces as word boundaries, we used *jieba* and *MeCab* python libraries for tokenization, respectively. Conversely, for other languages, we adopted whitespace-based tokenization. $^{\alpha}$Google Research India (2021). $^{\beta}$Question to context mapping was not given in MMQA, so we greedily built a mapping. $^{\gamma}$MQA is a subset of SQuAD automatically translated into Hindi, but ∼92% of its (start, end) indices are incorrect. $^{\delta}$We concatenate all (ten) contexts of a question to form a single context in XQA. $^{\delta}$In the bAbI dataset, we take the text above a question as its context.

Dataset Name	Avg #tokens			Type	#Languages	#Samples ↓
	Q	C	A			
$^{\alpha}$Chaii	7	1694	2	Factoid Span Detection	2	1.1K
$^{\beta}$MMQA [14]	9	314	7	Factoid Span Detection	2	2.7K
XQuAD [2]	12	155	4	Factoid Span Detection	11	13K
$^{\gamma}$MQA [13]	10	126	2	Factoid Span Detection	2	36K
XOR QA [3]	6	–	4	Open Retreival QA	7	40K
MLQA [17]	8	117	3	Factoid Span Detection	7	46K
TyDiQA-GoldP [9]	5	76	4	Factoid Span Detection	11	54K
TyDiQA-MinSpan [9]	5	2825	4	Factoid Span Detection	11	78K
TyDiQA-SelectP [9]	6	2891	80	Answer Paragraph Selection	11	90K
$^{\delta}$XQA [20]	17	5326	2	Factoid Span Detection	9	90K
MKQA [21]	8	–	2	Open Domain QA	26	260K
$^{\delta}$bAbI [36]	5	21	1	Factoid Span Detection	2	330K
MuNfQuAD	6	875	165	Answer Paragraph Selection	38	578K

2 Related Works

Our work closely resembles NLQuAD [32], but it was designed exclusively for the English language. For a comprehensive review of English QA datasets, readers can refer to [31]. The main focus of our work is on multilingual QuADs (mQuADs). Table 1 contains the statistics of different mQuADs. While multilingual datasets like BOLT [33] and ResPubliQA [27] have associated publications [8,24], we were unable to locate the datasets on the open web. We observed that the field of multilingual non-factoid question answering has received limited research attention.

[2] https://github.com/ritwikmishra/MuNfQuAD.

3 Data Curation

We aimed to create a multilingual QA dataset with non-factoid questions. To achieve this, we utilized automated scraping of the BBC news website, gathering news articles and corresponding question-answer pairs based on the work of Soleimani et al. [32]. In our work, an interrogative subheading is identified by the trailing question mark (or equivalent symbol of that language) in the subheading text. In Fig. 1(a), the web interface of a BBC news article is depicted, while Fig. 1(b) illustrates the scraped data in dictionary format. The construction of non-factoid QA datasets through web scraping is a well-established practice in the literature [15].

Data scraping was conducted for six months, incorporating suitable time delays between each scraped article to prevent any potential user of the website from experiencing Denial of Service (DoS). Our goal is to provide access to MuNfQuAD for non-commercial research endeavors. We have obtained authorization from the BBC to facilitate this initiative.

Table 2. Overview of MuNfQuAD statistics. The data spans from 2009 to 2024. We tokenized words using whitespace-based splitting and employed the ersatz library [37] for multilingual sentence segmentation.

Number of Languages	38	Number of QA pairs	578K	Number of Articles	294K
Number of Unique Questions	411K	Avg. Article Length (Word)	875	Avg. Paragraph Length (Word)	17
Avg. Answer Length (Word)	165	Avg. Question Length (Word)	6	Avg. Article Length (Sentence)	51
Avg. Paragraph Length (Sentence)	1.4	Avg. Answer Length (Sentence)	9.5	Avg. Question Length (Sentence)	1.0
Avg. Paragraphs per Article	35	Avg. Paragraphs per Answer	6.5		

Table 3. Distribution of QA pairs across languages (ISO 639-1 codes) in MuNfQuAD.

om	6.5k	am	2.6k	Fr$^{\alpha}$	6k	ha	3.1k	ig	1.8k	rw	3.4k	en$^{\beta}$	6k	so	6.7k	sw	7k	ti	3.8k	yo	2.7k	ky	2.3k	uz	3k	id	13k	ko	5.1k	vi	10k	bn	10k	gu	20k	hi	31k
mr	26k	ne	13k	ps	3.1k	pa	9k	si	4.2k	ta	20k	te	20k	ur	17k	az	4.6k	ru	23k	sr	24k	tr	19k	uk	19k	cy	12k	en	14k	pt	16k	es	21k	ar	13k	fa	13k

$^{\alpha}$African French $^{\beta}$Pidgin English

We used the multilingual checkpoint of ersatz [37] to segment the sentences in our dataset. A lexicon (list) of phrases was constructed for each language to exclude interrogative subheadings irrelevant to the article. For example, हे वाचलंत का? (*Did you read this?*) is a common interrogative subheading from news articles in Marathi language.

3.1 MuNfQuAD Statistics

The presented dataset encompasses over 411K unique question-answer pairs, establishing itself as the most extensive mQuAD. Table 2 provides an overview of diverse statistics related to this dataset. We conducted web crawling on the BBC news website for all supported languages, resulting in data collection from 38 languages out of the 43 supported. Language distribution is shown in Table 3.

We translated each question within MuNfQuAD to English. This translation was achieved using the *nllb-200-1.3B* model [35], which boasts the unique capability of translating across 200 languages through a single model. To categorize each English question into distinct classes, we utilized the non-factoid question classifier from Bolotova et al. [6], revealing that more than two-thirds of the questions in the proposed dataset were classified as non-factoid. The similarity in question category distribution between MuNfQuAD and NLQuAD [32] is illustrated in Fig. 2.

3.2 MuNfQuAD vs NLQuAD

An important distinction between the two datasets is that MuNfQuAD includes sentence-segmented multilingual data with question categories, which NLQuAD does not. Moreover, we do not assert that BBC articles as long-context because many LLMs capable of processing text longer than BBC articles have since been developed, which were not available when NLQuAD was proposed.

4 Answer Paragraph Selection

An Answer Paragraph Selection (APS) model assigns high confidence scores to paragraphs belonging to the silver answer. The APS model takes as input the concatenation of a question and the ith paragraph (p_i) from the context. The output is a probability value ranging from 0 to 1, indicating the likelihood of p_i being an answer to the provided question. The choice of employing an APS model, as opposed to a sliding window Reading-Comprehension model [32], stems from the APS model's alignment with the Answer Sentence Selection (AS2) approach, which is deemed to be more relevant than the RC approach [4]. Xu et al. [38] demonstrated the effectiveness of the APS component in automatically answering questions whose responses span across non-continuous text segments.

Our APS model is designed by fine-tuning multilingual pretrained encoders. Figure 1(c,d) shows an outline of the overall architecture and training methodology of our APS model. To ensure reproducibility, the dataset is released with predefined training, validation, and test splits. Given the inherent imbalance of the dataset, we adopted a weighted focal loss during the training process [19].

4.1 Implementation

The fine-tuning of our model was conducted across five GPU cards, employing a batch size of 12 on each GPU. We explored various pretrained encoders, including XLM-Roberta-base (XLM-R) [10], multilingual cased bert (mBERT) [12], multilingual LUKE (mLUKE) [30], and XLM-Vocabulary-base (XLM-V) [18], to serve as the backbone of our APS model. The fine-tuning layers of the APS model consisted of three linear layers with a dropout value of 0.2. Coupled with a linear scheduler, learning rates were set at 1e-5 and 3e-3 for the encoder and fine-tuning layers, respectively. All the models were fine-tuned for a single epoch, a process that lasted for 25-33 h.

4.2 Baselines

For establishing baselines, we employed the sentence-transformers library (sbert) [29] and calculated the cosine similarity between the vector embeddings of questions (E_q) and paragraphs (E_p). Another approach entailed obtaining E_q and E_p via training a TF-IDF vectorizer on the training set. Across all models, the threshold value was set to half the potential range of confidence scores. Specifically, a default threshold of 0.5 was adopted for the fine-tuned APS models and the TF-IDF baseline, as their output score spans 0 to 1. However, a default threshold of 0.0 was applied to the sbert baseline, which produces scores ranging from −1 to 1.

Table 4. Comparative performance of various models on the MuNfQuAD Test Set for APS task. Ones, Zeros, and Random denote an APS model that always predicts 1, 0, and random values of 0 or 1, regardless of the input. The APS model fine-tuned with the XLM-V encoder demonstrates the highest macro F1 and Label-1 F1 scores.

APS Model	Params.	Acc.	Macro	Label 0			Label 1			SR
			F1	P	R	F1	P	R	F1	
Ones	–	19	16	0	0	0	19	100	32	1.0
Zeros	–	81	45	81	100	90	0	0	0	0.0
Random	–	50	45	81	50	62	19	50	27	1.0
Ours (mLUKE)	585M	19	16	100	0	0	19	100	32	1.0
sbert (miniLM)	117M	22	20	87	4	7	19	97	32	0.99
TF-IDF	–	81	47	81	99	89	36	2	4	0.11
Ours (mBERT)	177M	74	67	93	73	82	40	77	53	0.93
Ours (XLM-R)	278M	79	71	93	80	86	46	73	**56**	0.91
Ours (XLM-V)	778M	80	**72**	92	83	87	48	68	**56**	0.90

4.3 Evaluation

For paragraphs not aligning with the silver answer, a ground truth label of 0 is assigned, while paragraphs that belong to the silver answer are given label 1. Our emphasis in this study is placed on the macro F1 and Label-1 F1, owing to the pronounced data imbalance where only 23% of samples fall under Label-1. Additionally, we incorporate the Success Rate (SR) metric, which calculates the ratio of successfully answered questions to the total question count [23]. A question is considered successfully answered if there is at least one paragraph common between the candidate and reference paragraphs.

Table 5. Performance of silver labels and best performing APS model (from Table 4) on the golden set. ISO 639-1 codes are used to represent a language. Cohens kappa is used as Inter Annotator Agreement (IAA) score. We do not compare the performance of our XLM-V based APS model with silver labels of the golden set because Table 4 already highlights the model performance on the entire MuNfQuAD testset having silver labels. It is observed that the silver labels achieve superior Label-0 F1 score (0), Label-1 F1 score (1), Macro F1 (M), and Success Ratio (SR) score on gold labels.

Lang	#Ann	#Qs	IAA	Silver Labels vs Gold Labels					APS Models vs Gold Labels									
									Ours (XLM-V)					C4AI Command-R				
				Acc	F1 score			SR	Acc	F1 score			SR	Acc	F1 score			SR
					0	1	M			0	1	M			0	1	M	
hi	2	100	0.26	75	83	49	66	1.0	69	77	51	64	0.95	72	81	47	64	0.97
bn	2	100	0.40	81	87	60	73	0.98	72	79	56	68	0.97	65	73	49	61	0.97
gu	2	100	0.42	83	90	55	72	1.0	72	81	49	65	0.95	69	79	43	61	0.98
te	2	40	0.58	84	91	50	70	0.95	76	85	48	66	0.97	54	66	30	48	1.0
tm	2	100	0.78	85	91	51	71	0.98	69	79	38	59	0.94	72	82	40	61	0.91
np	1	100	-	72	81	52	66	0.97	70	76	59	68	0.94	61	66	54	60	0.97
pa	1	100	-	87	93	52	72	0.98	79	87	45	66	0.93	73	83	31	57	0.92
ur	1	50	-	83	89	59	74	1.0	70	79	49	64	0.97	61	71	43	57	1.0
Average				**81**	**88**	**53**	**70**	**0.98**	72	81	50	66	0.95	68	77	43	60	0.96

5 Results

Our observations revealed that adopting a weighted focal loss with $\gamma = 2$ yielded superior results compared to other choices [19]. Furthermore, we noted an enhancement in performance when incorporating preceding paragraphs along with the provided paragraph. We observe from Table 4 that our APS model with XLM-V as a pretrained encoder yields the best results in terms of macro F1 and Label-1 F1. Other models achieved a better SR but a lower macro F1. During inference on the testset, excluding questions predicted as FACTOID minimally affects the performance of all fine-tuned APS models.

5.1 Golden Set

To evaluate the hypothesis that "*paragraphs succeeding an interrogative subheading contain its answer*" we employed human annotators to answer questions from a subset of MuNfQuAD, referred to as the golden set. Each annotator received a question along with its corresponding article (context) and was tasked with identifying paragraphs within the article that could answer the question. Importantly, annotators were not provided with silver labels. Native speakers of each language were chosen to serve as annotators. A compensation of 1 USD for every set of eight questions was given to each annotator. For languages with multiple annotators, the final gold annotations were derived by taking a union of the selected answers.

The data presented in Table 5 illustrates that silver labels exhibit a high Success Rate (~ 0.98) across various languages, indicating that the silver answer text in MuNfQuAD can address 98% of the questions. The fact that silver labels outperform the best APS model highlights the room for improvement in future APS model performance. Predictions generated by our XLM-V based APS model on Nepali golden set achieve a superior Macro-F1 and Label-1 F1 as compared to the silver labels, indicating the model's ability to generalize from the silver-labeled training data. We observed that the TF-IDF and sbert baselines exhibit lower F1 scores on Label-1.

Comparatively lower Inter-Annotator Agreement (IAA) score in Hindi (hi) can be attributed to the larger number of answer paragraphs per question in this language. On average, a Hindi question has 8.0 silver paragraphs in the golden set, whereas a Bengali (bn), Gujarati (gu), Telugu (te), and Tamil (tm) question has 7.8, 7.1, 5.9, and 5.8 silver paragraphs, respectively.

5.2 Large Language Models (LLMs)

Additionally, we investigated the potential of employing Large Language Models (LLMs) as Answer Paragraph Selection (APS) models. To ensure a fair comparison with our fine-tuned APS models, we prompted LLMs with a question and each paragraph from the context, directing it to output a binary value indicating whether the paragraph could answer the given question (1 for yes, 0 for no). We experimented with various LLMs but CohereForAI (C4AI) Command-R[3] performed the best. However, as shown in Table 5, C4AI's superior performance still falls short of that achieved by our fine-tuned APS model.

6 Discussion

We notice that the TF-IDF baseline yields a higher macro-F1 than the random baseline, indicating that silver answers frequently contain paragraphs with considerable word overlap with the given question. For instance, in the following Hindi question बैठना नुकसानदेह क्यों?

[3] https://huggingface.co/CohereForAI/c4ai-command-r-v01-4bit.

(*Why sitting is harmful?*), the silver answer begins with the sentence आखिर बैठना इतना नुकसानदेह क्यों है? चलिए इसे समझने की कोशिश करते हैं. (*Why is sitting so bad after all? Let's try to understand it.*). We observed that the XLM-R based APS model showcases the best performance at a threshold of 0.6, achieving a macro F1 of 73 and an accuracy of 84.

We assessed the validity of our APS model architecture by comparing its performance with the baseline APS model proposed by Clark et al. [9]. Upon fine-tuning both architectures on the TyDi training data, we observed that the average F1 score of our APS model was 0.73, in contrast to 0.66 for the baseline model.

In addition to its role as a resource, we hypothesize that the key strength of MuNfQuAD lies in the APS model that is fine-tuned upon it. Our experiments demonstrate that when the context associated with a question is reduced using the scores predicted by the MuNfQuAD-fine-tuned APS model, the performance of instructional LLMs, such as Gemma-2B and Gemma-7B [34], is enhanced. This improvement is observed across a sample from the evaluation splits of TyDi [9] and XQA [20]. Since TyDi and XQA are factoid-based question-answering datasets, we define the win ratio as the proportion of instances where the golden answer appears in the generated text, relative to the total number of generations. Our analysis shows that, on the TyDi evaluation dataset, the APS model fine-tuned on MuNfQuAD enhances the win ratio of Gemma-2B from 11% to 14%, and that of Gemma-7B from 19% to 25%. In contrast, on the XQA evaluation dataset, the win ratio for Gemma-2B increased from 1% to 6%, while Gemma-7B's win ratio improved from 3% to 4%. The relatively lower performance of the LLMs on XQA is expected, as XQA is a cloze-style question-answering dataset.

7 Conclusion

Question Answering (QA) in English has firmly established itself as a common task, backed by many tools and resources for answering factoid-based questions. Nonetheless, non-factoid QA have witnessed a significant expansion. Our study highlights the need for multilingual resources within this domain. In response, we introduce MuNfQuAD, a multilingual QA dataset addressing this gap. Comprised of non-factoid questions, MuNfQuAD spans across 38 languages, thus filling a critical gap in this area.

The compilation of the dataset involved scraping BBC news articles. The questions are identified through interrogative subheadings, while the subsequent paragraphs are taken as their corresponding silver answers. Notably, the news articles in MuNfQuAD predominantly revolve around the Asiatic subcontinent. A comparison with a manually curated golden set substantiates that nearly all of the silver answers can be used to answer the asked question. Additionally, our fine-tuned Answer Paragraph Selection (APS) model, trained using MuNfQuAD, yields a high Success Rate for both silver (0.91) and golden (0.96) labels. The results demonstrate that training the APS model with silver labels can effectively generalize some languages within the golden set.

The Question-Answer pairs of MuNfQuAD can be used for training generative techniques in question-answering across different low-resource languages. Our examination reveals that MuNfQuAD encompasses a substantial proportion of factoid-based questions. Therefore, a multilingual answer span extractor can be used to provide silver labels for the minimal answer spans within MuNfQuAD. In a multilingual RAG pipeline, the fine-tuned APS model can be used as a reranking module [22].

8 Limitations

The study conducted by Latham s [16] illustrated that BBC exhibits a left-of-center bias in its news coverage. Therefore, we recognize that MuNfQuAD will likely inherit a similar political bias. Lower F1 score for Label 1 suggests that the silver answers provide a response to the relevant question but include additional information as well. Additionally, nearly a third of questions in MuNfQuAD are classified as FACTOID, implying the potential presence of short-span answers within the silver paragraphs. It is imperative to approach the question categories with caution, primarily due to two reasons: (a) The classifier exhibits imperfections. We applied the same classifier to a non-factoid QuAD in English [32] and observed a comparable distribution of question categories, as depicted in Fig. 2. (b) The classifier processes automatically translated English questions, introducing the possibility of unnatural translations that may alter classifier predictions. For instance, the Hindi question was auto-translated as *What if I didn't?*, and the classifier predicted FACTOID. However, a more accurate translation of the question would be *What will happen if not done?*, for which the predicted category is EVIDENCE-BASED.

The findings derived from LLM experiments warrant cautious interpretation due to the potential influence of pre-training, where LLMs may have been exposed to BBC news articles. Therefore, there is a possibility of dataset contamination, as noted by [1].

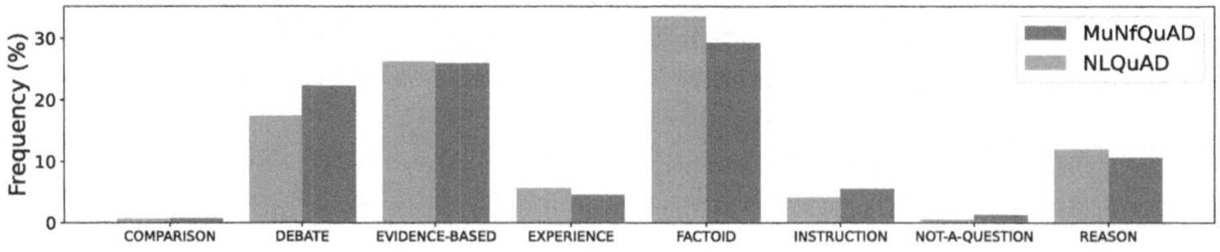

Fig. 2. An analysis of the distribution of question categories in the proposed MuNfQuAD and English NLQuAD [32] using the predictions from the question classifier model developed by [6].

Acknowledgments. Ritwik Mishra sincerely acknowledges the partial financial support provided by the University Grants Commission (UGC) of India through the UGC Senior Research Fellowship (SRF) program. Rajiv Ratn Shah expresses his gratitude for the partial funding received from the Infosys Center for AI, the Center for Design and New Media, and the Center of Excellence in Healthcare at IIIT-Delhi.

Disclosure of Interests. We declare no conflict of interest currently.

References

1. Ahuja, K., et al.: Mega: Multilingual evaluation of generative ai. In: Proceedings of EMNLP, pp. 4232–4267 (2023)
2. Artetxe, M., Ruder, S., Yogatama, D.: On the cross-lingual transferability of monolingual representations. In: Proceedings of the ACL, pp. 4623–4637 (2020)
3. Asai, A., Kasai, J., Clark, J., Lee, K., Choi, E., Hajishirzi, H.: XOR QA: Cross-lingual open-retrieval question answering. In: Proceedings of NAACL: Human Language Technologies, pp. 547–564. Online (Jun 2021). https://aclanthology.org/2021.naacl-main.46
4. Barlacchi, G., et al.: Focusqa: open-domain question answering with a context in focus. In: Findings of EMNLP 2022,pp. 5195–5208 (2022)
5. Baudiš, P., Šedivý, J.: Modeling of the question answering task in the YodaQA system. In: Mothe, J., Savoy, J., Kamps, J., Pinel-Sauvagnat, K., Jones, G., SanJuan, E., Cappellato, L., Ferro, N. (eds.) CLEF 2015. LNCS, vol. 9283, pp. 222–228. Springer, Cham (2015). https://doi.org/10.1007/978-3-319-24027-5_20
6. Bolotova, V., Blinov, V., Scholer, F., Croft, W.B., Sanderson, M.: A non-factoid question-answering taxonomy. In: Proceedings of the 45th International ACM SIGIR Conference on Research and Development in Information Retrieval, pp. 1196–1207 (2022)
7. Cambazoglu, B.B., Baranova, V., Scholer, F., Sanderson, M., Tavakoli, L., Croft, B.: Quantifying human-perceived answer utility in non-factoid question answering. In: Proceedings of the 2021 Conference on Human Information Interaction and Retrieval, pp. 75–84 (2021)
8. Chaturvedi, S., Castelli, V., Florian, R., Nallapati, R.M., Raghavan, H.: Joint question clustering and relevance prediction for open domain non-factoid question answering. In: Proceedings of the 23rd international conference on World Wide Web, pp. 503–514 (2014)

9. Clark, J.H., et al.: Tydi qa: a benchmark for information-seeking question answering in ty pologically di verse languages. Trans. Associat. Comput. Linguist. **8**, 454–470 (2020)
10. Conneau, A., et al.: Unsupervised cross-lingual representation learning at scale. CoRR arXiv: abs/1911.02116 (2019)
11. Cripwell, L., Legrand, J., Gardent, C.: Document-level planning for text simplification. In: Proceedings of EACL, pp. 993–1006 (2023)
12. Devlin, J., Chang, M.W., Lee, K., Toutanova, K.: BERT: Pre-training of deep bidirectional transformers for language understanding. In: Proceedings of NAACL. pp. 4171–4186. Association for Computational Linguistics, Minneapolis, Minnesota (Jun 2019). https://doi.org/10.18653/v1/N19-1423, https://aclanthology.org/N19-1423
13. Gupta, D., Ekbal, A., Bhattacharyya, P.: A deep neural network framework for English Hindi question answering. ACM Trans. Asian Low-Resource Lang. Inform. Process. (TALLIP) **19**(2), 1–22 (2019)
14. Gupta, D., Kumari, S., Ekbal, A., Bhattacharyya, P.: Mmqa: a multi-domain multilingual question-answering framework for English and Hindi. In: Proceedings of LREC (2018)
15. Hashemi, H., Aliannejadi, M., Zamani, H., Croft, W.B.: ANTIQUE: a non-factoid question answering benchmark. In: Jose, J.M., et al. (eds.) ECIR 2020. LNCS, vol. 12036, pp. 166–173. Springer, Cham (2020). https://doi.org/10.1007/978-3-030-45442-5_21
16. Latham, O.: Bias at the beeb? a quantitative study of slant in bbc online reporting (2012)
17. Lewis, P., Oguz, B., Rinott, R., Riedel, S., Schwenk, H.: Mlqa: evaluating cross-lingual extractive question answering. In: Proceedings of ACL, pp. 7315–7330 (2020)
18. Liang, D., et al.: Xlm-v: Overcoming the vocabulary bottleneck in multilingual masked language models (2023)
19. Lin, T.Y., Goyal, P., Girshick, R., He, K., Dollár, P.: Focal loss for dense object detection. In: Proceedings of the IEEE International Conference on Computer Vision, pp. 2980–2988 (2017)
20. Liu, J., Lin, Y., Liu, Z., Sun, M.: Xqa: a cross-lingual open-domain question answering dataset. In: Proceedings of ACL, pp. 2358–2368 (2019)
21. Longpre, S., Lu, Y., Daiber, J.: Mkqa: a linguistically diverse benchmark for multilingual open domain question answering. Trans. Associat. Comput. Linguist. **9**, 1389–1406 (2021)

22. Ma, X., Wang, L., Yang, N., Wei, F., Lin, J.: Fine-tuning llama for multi-stage text retrieval. In: Proceedings of the 47th International ACM SIGIR Conference on Research and Development in Information Retrieval, pp. 2421–2425 (2024)
23. Mishra, R., Singh, S., Kaur, J., Singh, P., Shah, R.: Hindi chatbot for supporting maternal and child health related queries in rural India. In: Proceedings of the 5th Clinical Natural Language Processing Workshop, pp. 69–77 (2023)
24. Molino, P.: Semantic models for answer re-ranking in question answering. In: Proceedings of the 36th international ACM SIGIR Conference on Research and Development in Information Retrieval, pp. 1146–1146 (2013)
25. Neerbek, J., Eskildsen, M., Dolog, P., Assent, I.: A real-world data resource of complex sensitive sentences based on documents from the monsanto trial. In: Proceedings of LREC, pp. 1258–1267 (2020)
26. Nowak, S., et al.: Transformer-based structuring of free-text radiology report databases. Eur. Radiol. **33**(6), 4228–4236 (2023)
27. Peñas, A., et al.: Overview of ResPubliQA 2009: question answering evaluation over european legislation. In: Peters, C., et al. (eds.) CLEF 2009. LNCS, vol. 6241, pp. 174–196. Springer, Heidelberg (2010). https://doi.org/10.1007/978-3-642-15754-7_21
28. Rajpurkar, P., Zhang, J., Lopyrev, K., Liang, P.: Squad: 100,000+ questions for machine comprehension of text. In: Proceedings of EMNLP, pp. 2383–2392 (2016)
29. Reimers, N., Gurevych, I.: Sentence-bert: sentence embeddings using siamese bert-networks. In: Proceedings of EMNLP. Association for Computational Linguistics (Nov 2019). http://arxiv.org/abs/1908.10084
30. Ri, R., Yamada, I., Tsuruoka, Y.: mluke: the power of entity representations in multilingual pretrained language models. In: Proceedings of ACL, pp. 7316–7330 (2022)
31. Rogers, A., Gardner, M., Augenstein, I.: Qa dataset explosion: a taxonomy of nlp resources for question answering and reading comprehension. ACM Comput. Surv. **55**(10), 1–45 (2023)
32. Soleimani, A., Monz, C., Worring, M.: Nlquad: a non-factoid long question answering data set. In: Proceedings of the 16th Conference of EACL, pp. 1245–1255 (2021)
33. Song, Z., et al.: Collecting natural sms and chat conversations in multiple languages: The bolt phase 2 corpus. In: LREC, pp. 1699–1704. Citeseer (2014)
34. Team, G., et al.: Gemma: Open models based on gemini research and technology. arXiv preprint arXiv:2403.08295 (2024)
35. Team, N., Costa-jussà, M.R., Cross, J., Çelebi, O., Elbayad, M., Heafield, K., Others: No language left behind: Scaling human-centered machine translation (2022)
36. Weston, J., et al.: Towards ai-complete question answering: a set of prerequisite toy tasks. In: ICLR (2016)

37. Wicks, R., Post, M.: A unified approach to sentence segmentation of punctuated text in many languages. In: Proceedings of the ACL-IJCNLP, pp. 3995–4007 (2021)
38. Xu, B., Xing, Z., Xia, X., Lo, D.: Answerbot: automated generation of answer summary to developers' technical questions. In: 2017 32nd IEEE/ACM International Conference on Automated Software Engineering (ASE), pp. 706–716. IEEE (2017)

Turning Uncertainty to Information by Intervals in Ensemble Classifiers

Mansoureh Maadi[1](✉), Uwe Aickelin[1], Hadi Akbarzadeh Khorshidi[2], and Michael Kirley[1]

[1] School of Computing and Information Systems, The University of Melbourne, Melbourne, Australia
{m.maadi,uwe.aickelin,mkirley}@unimelb.edu.au
[2] Melbourne School of Population and Global Health, The University of Melbourne, Melbourne, Australia
hadi.khorshidi@unimelb.edu.au

Abstract. Ensemble classifiers with the potential of improving classification by using multiple classifiers rather than one are constructed in two steps: generating individual classifiers and combining them. Homogeneous ensemble classifiers generate an ensemble of individual classifiers by training the same learning algorithm on different training datasets and aggregate them using combining methods. In this paper, we introduce an approach called Interval-Based Ensemble Learning (IBEL) for homogeneous ensemble classifiers using interval modelling. In IBEL, bagging and information granule concepts are used to generate an interval-based output named uncertainty interval for each individual classifier in the ensemble. Then, the Interval Agreement Approach (IAA), as an interval aggregation function, is used to combine uncertainty intervals to determine the prediction of the ensemble. Uncertainty intervals capture the uncertainty of individual classifiers on their predictions, and IAA minimises information loss by capturing uncertainty between individual classifiers. In an extensive experimental study, we show the superiority of the IBEL in improving ensemble classification performance for homogeneous ensemble classifiers for both synthetic and real benchmark datasets. Furthermore, we experimentally show that in the presence of more class uncertainty, IBEL is more successful in improving ensemble classification performance than typical point prediction homogeneous ensemble classifiers.

Keywords: Ensemble Classifiers · Uncertainty · Interval Modelling · IAA

1 Introduction

In classification, ensemble models (called ensemble classifiers) generate different classifiers (called individual classifiers) and combine their predictions to make an ensemble prediction. The superior performance of ensemble classifiers over individual classifiers for complex data mining problems has increased their popularity for solving real-world problems [1]. Ensemble classifiers are generally generated using two approaches: homogeneous and heterogeneous. Homogeneous ensemble classifiers train the same learning

X. Wu et al. (Eds.): PAKDD 2025, LNAI 15875, pp. 237–249, 2025.
https://doi.org/10.1007/978-981-96-8295-9_17

algorithm on different training datasets generated from the original dataset to create an ensemble of individual classifiers. Different learning algorithms are trained on the same training dataset in heterogeneous ensemble classifiers to generate an ensemble of individual classifiers. In ensemble classifiers, individual classifiers make predictions for a data sample in the form of posterior probabilities or class memberships [2, 3]. As the predictions of individual classifiers may be different, uncertainty is introduced in ensemble classifiers. This uncertainty is defined as the uncertainty between individual classifiers (inter-source uncertainty [10]). Combining methods use different strategies to aggregate the predictions of individual classifiers to capture this uncertainty and improve prediction. However, they usually overlook the uncertainty of individual classifiers in their predictions (intra-source uncertainty [10]), which affects the performance of ensemble classifiers. Majority Vote, Max, Min, Average, Oracle Rule [5], interval-valued fuzzy aggregation functions [6] and Meta-Learners [7, 8] are common combination strategies in ensemble classifiers.

This paper presents Interval-Based Ensemble Learning (IBEL), an interval modelling approach designed for homogeneous ensemble classifiers to capture both inter-source and intra-source uncertainties and improve ensemble prediction performance. While some previous studies have explored incorporating interval modelling to deal with uncertainty and improve the performance of ensemble classifiers, a significant gap exists: these studies focused on heterogeneous ensemble classifiers and suggested limited advantages for homogeneous ones [7, 8]. This motivated us to introduce IBEL specifically for homogeneous ensemble classifiers, aiming to enhance their classification performance by capturing uncertainty.

In the first step of IBEL, we generate an interval-based output called uncertainty interval for each individual classifier using bagging and information granularity concepts. These intervals capture intra-source uncertainty in homogeneous ensemble classifiers and serve as a form of uncertainty quantification. However, IBEL is not primarily focused on uncertainty quantification but rather on implementing these intervals to enhance ensemble prediction performance. Unlike confidence intervals, which rely on class memberships, mean, and variance, IBEL's uncertainty intervals are based on posterior probabilities and use the median as the key information granularity concept. In the second step of IBEL, we use the Interval Agreement Approach (IAA) introduced by Wagner et al. [9] to combine uncertainty intervals to determine the ensemble prediction. IAA minimises the information loss by capturing inter-source uncertainty through the aggregation of uncertainty intervals to a Type 1 Fuzzy Set to determine the ensemble prediction. In addition, although IBEL is designed to deal with intra-source and inter-source uncertainties, we show it can deal with some data-related uncertainty like noisy data as well.

To this end, we make three contributions. First, to the best of our knowledge, this is the first approach to model inter-source and intra-source uncertainties through interval modelling in homogeneous ensemble classifiers to improve ensemble performance. Second, we introduce an interval-based metric to measure the intra-source uncertainty in homogeneous ensemble classifiers using bagging and information granular concepts.

While ensemble classifiers work with a point prediction approach, we show how capturing intra-source uncertainty by uncertainty intervals can improve the ensemble prediction performance. The width of these intervals can indicate the extent of uncertainty for individual classifiers. Thus, the effect of the intra-source uncertainty on the ensemble prediction can be examined. Third, we use IAA to combine uncertainty intervals as an effective combining method to capture inter-source uncertainty and reduce information loss in the combining process. IAA has been applied as a technique to analyse uncertainty in human responses when these responses are represented as intervals. This is the first time that this aggregation method is used in ensemble classifiers.

Regarding the contributions of this paper, we define two research questions in this study to investigate the performance of IBEL. We address these research questions throughout the paper. They are defined as follows.

RQ1: Does the proposed IBEL improve prediction performance for homogeneous ensemble classifiers?
RQ2: Where does IBEL outperform typical point prediction modelling in homogeneous ensemble classifiers?

In the following section, we discuss some related work on using interval modelling to deal with uncertainty in ensemble classifiers. Then, the proposed IBEL for binary classification problems is described in detail in Sect. 3. Section 4 provides two experimental analyses on real benchmark datasets. The first experiment investigates the effect of capturing intra-source uncertainty through uncertainty intervals on the ensemble prediction performance. The second experiment investigates the performance of IBEL against its typical iterative point prediction ensemble classifiers to address *RQ1*. In Sect. 5, a comprehensive discussion on the IBEL performance to deal with class uncertainty is provided to address *RQ2*. Finally, Sect. 6 presents the concluding remarks.

2 Related Work

Interval modelling to capture uncertainty in ensemble classifiers is a new research topic. Studies on this subject generally focus on three research areas: developing ensemble classifiers to deal with uncertain data shown as intervals, dealing with missing values in ensemble classifiers using intervals, and interval modelling to improve ensemble prediction performance in ensemble classifiers. We review the papers presented in these research areas, which are as follows.

Liu et al. [10], presented the first ensemble classifier to deal with uncertain data shown as interval data using possible worlds called possible world-based AdaBoost. In this paper, a training set is sampled from the training dataset in each step of the training procedure, and a possible world is generated to train the individual classifier. Majority voting and weighted voting aggregated the outputs of individual classifiers to determine the ensemble prediction.

Wojtowicz et al. [11] captured the uncertainty related to missing values by intervals. In this paper, missing values were replaced with intervals using the minimum and maximum values of the related features for the ovarian tumour diagnosis problem. They used different interval aggregation functions to combine the predictions of classifiers for diagnosis. Better results were achieved for ovarian tumour diagnosis compared to classical models. In another study, Bentkowska [12] investigated the application of interval aggregation to optimise the performance of K-NN classifiers for missing values. In this paper, missing values on the test dataset were replaced by possible values from the training dataset using a selection strategy based on the Monte Carlo method. Generating intervals that considered different values for missing values by different numbers of K in K-NN and using interval aggregation methods showed that interval modelling of missing values and application of interval aggregation functions resulted in fewer performance drops for K-NN.

Interval modelling was introduced in some studies to capture uncertainty in heterogeneous ensemble classifiers and improve their prediction performance. Nguyen et al. [7, 8] presented two combining methods based on interval modelling in heterogeneous ensemble classifiers for the first time. These papers applied an optimisation approach using the information granularity concept to generate an interval for each class label using training dataset labels in [7] and test dataset labels in [8]. Then, distance metrics to calculate the distance between intervals and class labels in [7] and a transform function to transform intervals to numeric values in [8] were implemented to determine the ensemble prediction. Although these papers presented promising results for heterogeneous ensemble classifiers, they mentioned that their models are less likely to offer any significant benefits for homogeneous ensemble classifiers.

3 The Proposed Method

This section presents the development of IBEL for binary classification problems. IBEL consists of two steps: the first generates uncertainty intervals as outputs of individual classifiers, and the second uses IAA to aggregate uncertainty intervals to construct a Type 1 Fuzzy Set (T1 FS) and determine the ensemble prediction. These steps are described in detail in the following sections.

The prediction of classifiers on each sample is shown as posterior probabilities (soft labels) or class memberships (crisp labels). In IBEL, we use soft labels as the outputs of classifiers. For x as a sample and y_j as its label (j is 0 and 1 for binary classification), $p_C(y_j|x)$ is defined as the probability that x belongs to class with label y_j given by the classifier C.

3.1 Generating Uncertainty Intervals

We use the bagging strategy to generate uncertainty intervals as interval-based predictions for individual classifiers in a homogeneous ensemble classifier. Bagging is a sampling strategy in ensemble learning proposed by Breiman in 1996 [13]. The bagging strategy generates multiple training datasets (bags) from a training dataset by selecting a random subset of training samples with replacement. To continue, we describe how we generate an uncertainty interval for an individual classifier in the ensemble classifier. Consider H as the number of bags. By training an individual classifier on bags, H classifiers as CB_{h} ($h = 1..., H$) are generated. For a test data sample u, the predictions made by these classifiers generate a probability matrix as defined below.

$$P(u) = \begin{bmatrix} p_1(y_0|u) & p_1(y_1|u) \\ p_2(y_0|u) & p_2(y_1|u) \\ \cdots & \cdots \\ p_H(y_0|u) & p_H(y_1|u) \end{bmatrix} \tag{1}$$

Considering all test samples ($u_n, n = 1 \ldots, N$), the probability matrix is defined as Eq. (2).

$$\acute{p} = \begin{pmatrix} p_1(y_0|u_1) & p_1(y_1|u_1) & \cdots & p_1(y_1|u_N) \\ & \vdots & \ddots & \vdots \\ p_H(y_0|u_1) & p_H(y_1|u_1) & \cdots & p_H(y_1|u_N) \end{pmatrix} \tag{2}$$

Let define a certain class in the binary classification as the main class and the other class as the subordinate class. As we have $\sum_j p_h(y_j|x) = 1(j = 0, 1;\ h = 1 \ldots, H)$, we only consider the probability related to the main class ($y_1$here) in $\acute{p}$ and define a new probability matrix (P) related to the main class as follows.

$$P = \begin{pmatrix} p_1(y_1|u_1) & \cdots & p_1(y_1|u_N) \\ \vdots & \ddots & \vdots \\ p_H(y_1|u_1) & \cdots & p_H(y_1|u_N) \end{pmatrix} \tag{3}$$

The information granularity concept aims to find a simple and effective way to extract knowledge from data [14]. Zadeh [15] first introduced and discussed the issue of fuzzy information granulation. In this paper, we use the quartile-median granulation method and calculate the first quartile and the third quartile of the soft labels for each test sample in probability matrix P. We define an uncertainty interval ($UI_n, n = 1 \ldots, N$) for each test sample through calculating the first quartile (Q_1) and the third quartile (Q_3) of probabilities on each column of P. In Eq. (4), uncertainty intervals for all test data samples related to an individual classifier ($UI_{individual}$) are shown.

$$UI_{individual} = (UI_1 = [Lb_1, Ub_1], ..., UI_N = [Lb_N, Ub_N]) \tag{4}$$

In Eq. (4), Lb_n denotes the first quartile and Ub_n denotes the third quartile of the *nth* column of P ($n = 1 \ldots, N$). Let M be the number of individual classifiers in the ensemble. Uncertainty intervals for all individual classifiers and all test samples are defined as Eq. (5). $UI_{mn}(m = 1 \ldots, M, n = 1 \ldots, N)$ denotes uncertainty interval generated by *mth* individual classifier for the *nth* test sample.

$$UI_{all} = \begin{pmatrix} UI_{11} & \cdots & UI_{1N} \\ \vdots & \ddots & \vdots \\ UI_{M1} & \cdots & UI_{MN} \end{pmatrix} \tag{5}$$

We generated uncertainty intervals for individual classifiers regarding the main class (y_1). In the following proposition, we show that how we can generate uncertainty intervals for the subordinate class.

Proposition 1: If $UI_{y_1} = [Lb_{y_1}, Ub_{y_1}]$ is the uncertainty interval generated for a test sample by an individual classifier regarding the main class (y_1), the uncertainty interval related to the subordinate class is calculated as $UI_{y_0} = [1 - Ub_{y_1}, 1 - Lb_{y_1}]$.

Proof: Let C_{y_0} and C_{y_1} denote the main and subordinate class labels in a binary classification problem. As $C_{y_0} + C_{y_1} = 1$, then $UI_{y_0} + UI_{y_1} = 1$. So, we have

$$UI_{y_0} = [1 - Ub_{y_1}, 1 - Lb_{y_1}] \tag{6}$$

It is worth noting that we considered different *pth* percentiles (*p* = *0, 0.05, 0.1, 0.15, 0.20, 0.25*) to create uncertainty intervals. Empirically, the first quartile and the third quartile (p = 0.25) performed best as information granules in calculating uncertainty intervals.

3.2 Interval Agreement Approach (IAA) to Determine the Class Label

IAA is an aggregation function that generates fuzzy sets from interval data. This method is initially introduced to aggregate decision makers' opinions while opinions are presented as intervals. Here, we explain IAA [9], and then describe how we use IAA to aggregate uncertainty intervals in IBEL to determine the ensemble prediction for a test data sample. Suppose $I = \{I_1, \ldots, I_K\}$ is a set of intervals. A T1 FS with the membership function of μ_I can be generated from I to aggregate intervals based on where they are in agreement. μ_I is defined as follow [9].

$$\mu_I = \sum_{i=1}^{k} y_i \Big/ \left(\bigcup_{j_1=1}^{k-i+1} \bigcup_{j_2=j_1+1}^{k-i+2} \cdots \bigcup_{j_i=j_{i-1}+1}^{k} \left(I_{j_1} \cap \ldots \cap I_{j_i} \right) \right) \tag{7}$$

where $y_i = {}^{i}/_{k}$ is the degree of membership and '/' refers to the assignment of a given degree of membership. The degree of membership in IAA is related to the number of intervals overlapped in a point. So, y_i is equal to one when all intervals are in agreement. To simplify, for a real number like x, the degree of membership is the frequency of which x appears in all intervals divided by the number of intervals.

Let TFS show the generated T1 Fs by IAA. TFS is a list of tuples which tuples indicate different regions of change over the membership function as Eq. (8).

$$TFS = [R_1, R_2, \ldots, R_v] \quad R_i = ([R_{il}, R_{ir}], R_{ih}) \tag{8}$$

where R_{il} is the left point, R_{ir} is the right point and R_{ih} is the membership function value of region R_i. Centroid of the TFS is calculated using Eq. (9).

$$centroid(TFS) = \frac{\sum_{i=0}^{v} (R_{ih} \times R_{il}) + (R_{ih} \times R_{ir})}{\sum_{i=o}^{v} 2(R_{ih})} \tag{9}$$

In IBEL, IAA is used to aggregate uncertainty intervals generated for a test sample to a T1 FS. To determine the class label of the test sample, we calculate the centroid of the T1 FS and compare its value with 0.5. If the centroid is equal or more than 0.5, the test sample is assigned to the main class. If less than 0.5, the test sample belongs to the subordinate class. Each column in the UI_{all} matrix as in Eq. (5) is a set of uncertainty intervals generated by M individual classifiers in the ensemble for a test sample. We use IAA on each column of the matrix to determine the class label for the related test sample.

4 Experimental Studies

4.1 Datasets and Evaluation Metrics

We use 20 real benchmark datasets from [17–19] regarding different assumptions on data characteristics including various size, dimension, and class distribution, to have a fair analysis. Datasets characteristics are shown in Table 1. We use Accuracy, F1-score and G-mean as the evaluation metrics in our experiments, which are the widely used and appropriate criteria to evaluate the performance of classifiers. In all experiments, we report average of Accuracy, F1-score, and G-mean for 30 repetitions of randomly dividing datasets into training and testing subsets by ratios of 80% and 20% respectively.

Table 1. Real benchmark datasets characteristics.

Dataset	Name	Dimension	Size	Ratio	Dataset	Name	Dimension	Size	Ratio
D1	Pima	8	768	65-35	D11	Abalone	8	4177	68-32
D2	SA-Heart	9	462	65-35	D12	Spectf	44	267	79-21
D3	Statlog Heart	13	360	55-45	D13	Waveform	21	5000	67-33
D4	Liver Disorder	5	345	49-51	D14	Banknote	4	1372	55-45
D5	Parkinson	22	195	75-25	D15	Oil	49	937	96-04
D6	Biodeg	43	1055	66-34	D16	Poker	10	25010	50-50
D7	Vehicle Bus	18	846	76-24	D17	HCC Carcinoma	49	110	62-38
D8	Wisconsin	9	699	66-34	D18	Hepatitis	19	155	79-21
D9	Spam	57	4601	60-40	D19	Red Wine	11	1599	57-43
D10	Ring	20	7400	50-50	D20	White Wine	11	4898	55-45

4.2 Do Uncertainty Intervals Contain Information?

We present an approach for homogeneous ensemble classifiers where the uncertainty in individual classifiers can be captured using uncertainty intervals. The width of an uncertainty interval for an individual classifier shows how uncertain the classifier is on its prediction. The hypothesis is that this uncertainty negatively affects the classifiers' performance and, consequently, the performance of the ensemble. In this section, we design an experiment to examine this hypothesis using real benchmark datasets. In this experiment, we consider a decision tree as a learning algorithm to generate individual classifiers. We use the bagging strategy to generate uncertainty intervals (as described in Sect. 3) for individual classifiers and consider majority voting (the most practiced aggregation function) to predict labels. We initially select a subset of features from a training dataset to generate an ensemble of individual classifiers. Then, using the bagging method, we construct different training datasets and, subsequently, an ensemble of decision trees. Applying decision trees to each test data sample creates two outputs: the predicted class label (using majority voting) and an uncertainty interval. For each ensemble of decision trees, we calculate the average width of uncertainty intervals for all test samples, as well as the three evaluation metrics of accuracy, F1-score and G-mean. We calculate the Pearson correlation between the average width of uncertainty intervals and the metrics using all generated ensembles. We also consider naïve Bayes and logistic regression as the learning algorithm and calculate the correlations. In this experiment, the number of individual classifiers in an ensemble (the number of bags) is considered as 50. The number of bootstrapped samples in each bagging is equal to the number of training samples.

Table 2. The average of correlation values between the average of width of uncertainty intervals and performance measures for all datasets.

Base Classifier	Decision Tree			Naïve Bayes			Logistic Regression		
Performance metric	Accuracy	F1	G-mean	Accuracy	F1	G-mean	Accuracy	F1	G-mean
Average correlation	−0.49	−0.43	−0.45	−0.44	−0.39	−0.38	−0.48	−0.47	−0.44

The results show that the correlation between the average width of uncertainty intervals and performance metrics is negative for all datasets and all measures regarding all three learning algorithms. These negative correlations are statistically significant at a 95% confidence level. Table 2 demonstrates the average of correlation values by the direction (negatively or positively correlated) and the extent of the correlation calculated for all datasets. Thus, experimental results support the hypothesis that uncertainty negatively impacts ensemble prediction performance. In other words, a bigger width of uncertainty intervals results in weaker ensemble classifiers. However, ensemble classifiers ignore the information that can be provided by uncertainty intervals. IBEL is designed to capture this uncertainty in homogeneous ensemble classifiers and improve the ensemble prediction performance.

4.3 Experimental Analysis on the Performance of IBEL Using Real Benchmark Datasets

To evaluate the performance of IBEL in capturing uncertainty and improving the ensemble prediction performance (addressing *RQ1*), we design three experiments. In these experiments, we consider three learning algorithms, a decision tree, naïve Bayes and logistic regression, as the most important learning algorithms for generating homogeneous ensembles. In each experiment, we compare the performance of IBEL with its alternative Point Prediction Ensemble Learning approach called PPEL. In the PPEL approach, majority voting is used to aggregate the outputs of individual classifiers. Majority voting is the most commonly used aggregation function in ensemble classifiers. We compare IBEL and PPEL to investigate the effects of interval modelling in capturing uncertainty and improving ensemble prediction performance. To generate individual classifiers, we choose the random subspace method [20]. In the random subspace method, each individual classifier in the ensemble is made by selecting a random subset of features. We summarise the comparison results of IBEL and PPEL for a decision tree, naïve Bayes, and logistic regression as learning algorithms in Table 3. In this table, the better model regarding each dataset and performance metric is bolded. We set the number of bags as 30, the number of individual classifiers as 20, and the number of samples in a bag as the number of samples in the training dataset for these experiments. Parameters are determined using the trial-and-error approach.

According to Table 3, IBEL almost always performs better than PPEL across different datasets, learning algorithms, and performance metrics. For a decision tree as the learning algorithm, IBEL outperforms PPEL in 44 cases out of 60 (a combination of three measures and twenty datasets), and PPEL performs better than IBEL only in seven cases. The results show a significantly better performance of IBEL in comparison with PPEL in terms of F1-score and G-mean. When the learning algorithm is a decision tree, PPEL is a random forest. Random forest is a powerful and widely used homogeneous ensemble classifier. The superior performance of IBEL in comparison with random forest validates the effectiveness of the proposed interval modelling in homogeneous ensemble classifiers. So, IBEL paves the way for the introduction of new interval-based ensemble classifiers to improve the performance of typical ensemble classifiers. We also observe the outperformance of IBEL where naïve Bayes and logistic regression are learning algorithms. IBEL outperforms PPEL in 46 and 52 cases where naïve Bayes and logistic regression are the learning algorithms, respectively. There are no cases where PPEL performs better than IBEL.

To evaluate the statistical outperformance of IBEL over PPEL, we use a Bayesian signed-rank test as a new replacement for the frequent statistical signed and signed-rank tests [21]. This test compares two classification methods over multiple datasets and determines how high the probability is that one classifier is better than the other one or that two classifiers have the same performance. The output of this test can be shown as a posterior probability set. The test results for all three experiments are shown in Table 4.

Table 4 shows posterior probability sets calculated for each learning algorithm and performance metric. The first value in the posterior probability set is the outperformance probability of IBEL over PPEL. The second value shows the probability of the case that neither model is better, and the third value is the probability that PPEL outperforms IBEL. Bayesian signed-rank test results statistically confirm the outperformance of IBEL over PPEL in terms of F1-score and G-mean as the evaluation metrics. However, when considering accuracy, IBEL has a probability of at least 0.20 outperforming PPEL, while the probability of IBEL and PPEL having the same performance is at least 0.76. As our datasets are mostly imbalanced, the F1-score and G-mean are better measures. According to the results, the probability of PPEL outperformance is zero in all experiments. These results raise *RQ2* to investigate where IBEL outperforms PPEL. In the following section, we answer this research question.

Table 3. Comparison of IBEL and PPEL performance regarding three learning algorithms and three performance measures.

Dataset	Decision Tree						Naïve Bayes						Logistic Regression					
	Accuracy		F1-score		G-mean		Accuracy		F1-score		G-mean		Accuracy		F1-score		G-mean	
	PPEL	IBEL	PPEL	IBEL	PPEL	IBEL	PPEL	IBEL	PPEL	IBEL	PPEL	IBEL	PPEL	IBEL	PPEL	IBEL	PPEL	IBEL
D1	0.751	**0.760**	0.553	**0.573**	0.642	**0.661**	0.746	**0.756**	0.563	**0.573**	0.644	**0.654**	0.765	**0.770**	0.617	**0.618**	0.692	**0.695**
D2	**0.706**	**0.706**	0.450	**0.462**	0.558	**0.572**	0.697	**0.700**	0.565	**0.567**	**0.657**	**0.657**	0.723	**0.725**	0.524	**0.527**	**0.621**	**0.621**
D3	0.834	**0.839**	0.789	**0.808**	0.822	**0.831**	0.834	**0.836**	0.816	**0.819**	0.834	**0.836**	0.842	**0.844**	0.820	**0.824**	0.837	**0.840**
D4	0.771	**0.782**	0.865	**0.870**	0.443	**0.471**	0.765	**0.772**	0.857	**0.862**	0.462	**0.479**	0.753	**0.755**	0.851	**0.853**	0.382	**0.409**
D5	**0.874**	0.869	**0.918**	0.910	**0.777**	0.765	0.717	**0.718**	**0.775**	**0.775**	**0.767**	**0.769**	0.846	**0.848**	**0.902**	**0.902**	0.712	**0.716**
D6	0.842	**0.846**	0.743	**0.749**	0.802	**0.809**	0.690	**0.700**	0.666	**0.674**	0.731	**0.744**	0.862	**0.863**	0.787	**0.788**	0.836	**0.837**
D7	0.911	**0.926**	0.826	**0.835**	0.922	**0.929**	**0.661**	**0.661**	**0.535**	**0.535**	0.716	**0.717**	**0.968**	**0.968**	**0.934**	**0.934**	0.957	**0.959**
D8	**0.964**	**0.959**	**0.947**	0.937	**0.964**	0.959	0.952	**0.969**	0.934	**0.945**	0.957	**0.967**	0.967	**0.971**	0.952	**0.955**	0.963	**0.967**
D9	0.913	**0.914**	**0.886**	0.886	0.896	**0.900**	**0.827**	**0.827**	**0.811**	**0.811**	**0.842**	**0.842**	**0.920**	**0.920**	0.894	**0.895**	0.907	**0.909**
D10	0.763	**0.843**	0.802	**0.857**	0.736	**0.838**	0.971	**0.980**	0.973	**0.980**	0.970	**0.979**	0.756	**0.757**	0.772	**0.773**	0.752	**0.753**
D11	0.826	**0.828**	0.711	**0.712**	0.774	**0.778**	**0.860**	**0.860**	0.886	**0.887**	0.881	**0.882**	**0.880**	**0.886**	0.907	**0.909**	0.874	**0.8752**
D12	**0.805**	**0.805**	0.885	**0.885**	0.460	**0.465**	0.714	**0.721**	0.788	**0.791**	0.766	**0.769**	0.784	**0.785**	0.864	**0.866**	0.580	**0.584**
D13	0.869	**0.871**	**0.90**	**0.90**	0.873	**0.873**	**0.854**	**0.854**	0.881	**0.882**	0.875	**0.877**	**0.881**	**0.881**	0.909	**0.910**	0.878	**0.879**
D14	**0.949**	**0.949**	0.942	**0.943**	0.947	**0.947**	0.794	**0.830**	0.735	**0.798**	0.768	**0.819**	0.883	**0.932**	0.862	**0.922**	0.875	**0.929**
D15	0.954	**0.959**	0.334	**0.368**	0.289	**0.366**	0.927	**0.933**	0.388	**0.393**	**0.648**	**0.648**	0.958	**0.959**	0.367	**0.369**	**0.500**	**0.500**
D16	0.583	**0.604**	0.522	**0.644**	0.565	**0.590**	**0.500**	**0.500**	0.473	**0.476**	**0.493**	**0.493**	0.499	**0.501**	0.474	**0.476**	**0.491**	**0.495**
D17	0.639	**0.640**	0.746	**0.751**	0.457	**0.482**	0.621	**0.653**	0.733	**0.765**	0.373	**0.402**	0.652	**0.672**	0.748	**0.767**	0.505	**0.509**
D18	0.810	**0.813**	0.863	**0.889**	**0.592**	0.570	0.753	**0.756**	**0.824**	**0.824**	0.765	**0.772**	0.787	**0.789**	0.871	**0.873**	0.476	**0.479**
D19	0.735	**0.740**	0.665	**0.689**	**0.725**	**0.729**	0.721	**0.724**	0.674	**0.679**	0.715	**0.719**	0.714	**0.727**	0.630	**0.642**	0.685	**0.695**
D20	0.598	**0.599**	**0.538**	**0.538**	**0.584**	**0.584**	0.577	**0.578**	0.541	**0.543**	0.574	**0.575**	0.552	**0.555**	0.268	**0.290**	0.391	**0.418**

Table 4. Bayesian signed-rank test results for IBEL vs. PPEL.

Learning Algorithm	Accuracy	F1-score	G-mean
Decision Tree	(0.235, 0.765, 0)	(0.999, 0.001, 0)	(0.880, 0.120, 0)
Naïve Bayes	(0.205, 0.795, 0)	(0.682, 0.318, 0)	(0.563, 0.437, 0)
Logistic Regression	(0.236, 0.764, 0)	(0.662, 0.338, 0)	(0.574, 0.426, 0)

5 A Discussion on IBEL Performance: Where Does IBEL Outperform PPEL?

In this section, we use synthetic datasets to investigate where IBEL outperforms PPEL. We generate three balanced binary classification datasets using scikit-learn package functions in Python [22], simple classification datasets, circle datasets and two-moon datasets.

We generate them using the normal distribution. Simple classification datasets are linearly separable datasets. The degree of class separation specifies how easy two classes are to discriminate. The higher the degree, the easier the separation. Circle datasets generate binary classification data that falls into concentric circles. Two-moon datasets generate crescent-shaped data. Circle and two-moon datasets have non-linear class boundaries. In both two-moon and circle datasets, the noise parameter can control how much two classes overlap. The higher value for the noise parameter indicates the lower degree of class separation.

Class separation and noise also indicate the complexity of classification problems. So, these parameters are the key aspects in investigating the sensitivity of IBEL in comparison to PPEL. We consider values of 0.2, 0.4, 0.6, 0.8, and 2 for the degree of class separation and values of 0.1, 0.3, 0.5, 0.7, and 1 for the noise parameter. In all experiments, we set the number of individual classifiers as 10 and the number of bags as 20. We use the decision

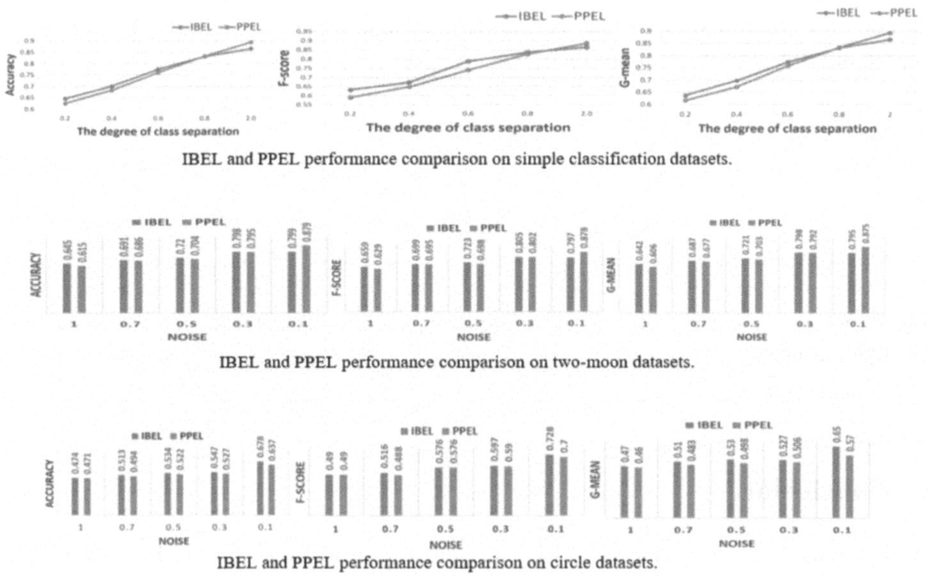

Fig. 1. IBEL and PPEL performance comparison on synthetic datasets regarding decision tree as the learning algorithm.

tree as the learning algorithm in the first experiment. Figure 1 shows the performance of IBEL compared to PPEL (random forest) on synthetic datasets over different values for degree of class separation and noise. As can be observed, IBEL generally outperforms PPEL, considering all performance metrics for all synthetic datasets where two classes are not easily separable. It shows that IBEL can handle complex classification problems and noisy data better than PPEL by capturing uncertainty.

In the second experiment, we consider naïve Bayes as the learning algorithm. We find that IBEL generally performs better than PPEL across all synthetic datasets, considering different values for the degree of class separation and noise. In this experiment, we observe a similar performance of IBEL to PPEL only in circle datasets when noise is equal to 0.1 (two classes are separated). Considering logistic regression as the learning algorithm in the third experiment, we observe the outperformance of IBEL over PPEL for simple classification datasets where the degree of class separation is low. For higher values of degree of class separation, IBEL and PPEL have similar performance. For two-moon datasets, IBEL outperforms PPEL for all noise values. When analysing circle datasets, IBEL demonstrates superior performance compared to PPEL across all levels of noise. However, the extent of IBEL's superiority diminishes as the noise level decreases. All the results are statistically analysed using the Bayesian signed-rank test.

We also calculate the average width of the uncertainty intervals for different values of the degree of class separation and noise in all experiments and show the results in Fig. 2. As can be observed, when the degree of class separation increases for simple classification datasets and noise values decrease for two-moon and circle datasets, the average width of uncertainty intervals decreases. It shows that noise and class inseparability increase uncertainty in ensemble learning. Therefore, IBEL performs better for complex classification problems (higher noise and lower separability) by capturing uncertainty through uncertainty intervals.

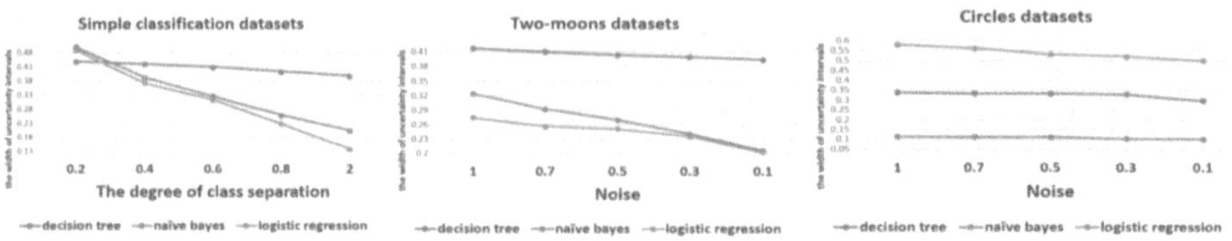

Fig. 2. Average of width of intervals vs. degree of class separation and noise in synthetic datasets for all experiments.

6 Conclusion

In this paper, we introduced IBEL, an interval modelling approach in homogeneous ensemble classifiers that captures intra-source and inter-source uncertainties and improves the ensemble prediction performance. In IBEL, we exploited the advantages of capturing intra-source uncertainty via uncertainty intervals for individual classifiers in the ensemble. We considered the width of uncertainty intervals as a measure of intra-source uncertainty in homogeneous ensemble classifiers. Using different experiments, we showed that this uncertainty negatively affects the performance of homogeneous ensemble classifiers, and IBEL can improve the classification task by capturing it. We also utilised the power of IAA as a new interval aggregation function for ensemble classifiers to capture inter-source uncertainty and improve classification.

References

1. Scholz, M., Wimmer, T.: A comparison of classification methods across different data complexity scenarios and datasets. Exp. Syst. Appl. **168**, 114217 (2021)
2. Forouzandeh, S., Berahmand, K., Rostami, M.: Presentation of a recommender system with ensemble learning and graph embedding: a case on MovieLens. Multimedia Tools Appl. **80**(5), 7805–7832 (2021)
3. Rostami, M., Berahmand, K., Forouzandeh, S.: A novel method of constrained feature selection by the measurement of pairwise constraints uncertainty. J. Big Data **7**(1), 1–21 (2020)
4. Kabir, S., Wagner, C., Ellerby, Z.: Towards handling uncertaintyat-source in AI – a review and next steps for interval regression. IEEE Trans. Artif. Intell. **5**, 3–22 (2023)
5. Kittler, J., Hatef, M., Duin, R.P.W., Matas, J.: On combining classifiers. IEEE Trans. Pattern Anal. Mach. Intell. **20**, 226–239 (1998)
6. Bentkowska, U.: New types of aggregation functions for interval-valued fuzzy setting and preservation of pos-B and nec-B-transitivity in decision making problems. Inf. Sci. (Ny) **424**, 385–399 (2018)
7. Nguyen, T.T., Nguyen, M.P., Pham, X.C., Liew, A.W.C., Pedrycz, W.: Combining heterogeneous classifiers via granular prototypes. Appl. Soft Comput. **73**, 795–815 (2018)

8. Nguyen, T.T., Pham, X.C., Liew, A.W.C., Pedrycz, W.: Aggregation of classifiers: a justifiable information granularity approach. IEEE Trans. Cybern. **49**, 2168–2177 (2019)
9. Wagner, C., Miller, S., Garibaldi, J.M., Anderson, D.T., Havens, T.C.: From interval-valued data to general type-2 fuzzy sets. IEEE Trans. Fuzzy Syst. **23**, 248–269 (2015)
10. Liu, H., Zhang, X., Zhang, X.: PwAdaBoost: possible world based AdaBoost algorithm for classifying uncertain data. Knowl. Based Syst. **186**, 1–15 (2019)
11. Wójtowicz, A., Zywica, P., Stachowiak, A., Dyczkowski, K.: Solving the problem of incomplete data in medical diagnosis via interval modeling. Appl. Soft Comput. J. **47**, 424–437 (2016)
12. Bentkowska, U., Bazan, J.G., Rząsa, W., Zaręba, L.: Application of interval-valued aggregation to optimization problem of k−NN classifiers for missing values case. Inf. Sci. (Ny) **486**, 434–449 (2019)
13. Breiman, L.: Bagging predictors. Mach. Learn. **24**, 123–140 (1996)
14. Pedrycz, W., Homenda, W.: Building the fundamentals of granular computing: a principle of justifiable granularity. Appl. Soft Comput. **13**, 4209–4218 (2013)
15. Zadeh, L.: Fuzzy sets and information granularity. Adv. Fuzzy Set Theor. Appl., 1–18 (1979)
16. Havens, T.C., Wagner, C., Anderson, D.T.: Efficient modeling and representation of agreement in interval-valued data. In: FUZZ-IEEE (2017)
17. Dua, D., Graff, C.: UCI Machine Learning Repository: Citation Policy
18. Kubat, M., Holte, R.C., Matwin, S.: Machine learning for the detection of oil spills in satellite radar images. Mach. Learn. **30**, 195–215 (1998)
19. Hastie, T., Tibshirani, R., Friedman, J.: The Elements of Statistical Learning, Data mining, Inference, and Prediction. Springer, New York (2009)
20. Ho, T.K.: The random subspace method for constructing decision forests. IEEE Trans. Pattern Anal. Mach. Intell. **20**, 832–844 (1998)
21. Benavoli, A., Corani, G., Demšar, J., Zaffalon, M.: Time for a change: a tutorial for comparing multiple classifiers through Bayesian analysis (2017)
22. Pedregosa, F., et al.: Scikit-learn: machine learning in Python (2011)

Determining the Need for Multi-label Classifiers by Measuring Unexplained Covariance

Laurence A. F. Park[1(✉)] and Jesse Read[2]

[1] Centre for Research in Mathematics and Data Science, Western Sydney University, Sydney, Australia
l.park@westernsydney.edu.au
[2] DaSciM Team, LIX Laboratory, École Polytechnique, 91120 Palaiseau, France
jesse.read@polytechnique.edu

Abstract. Multi-label classifiers make use of associations between labels in multi-label data to increase the accuracy of prediction. Before using a multi-label classifier, the data should be analysed to identify if there are associations between labels. If sets of independent labels are found, the data can be split in to multiple smaller data sets for analysis. Unfortunately, each label is dependent on the set of observations and so measuring label dependence is futile. What we actually seek is independence after taking the observations into account. In this article, we examine the concepts of explained and unexplained label covariance for measuring label dependence. We explore the use of a Normal copula model for modelling the label dependence/covariance and show that it is not able to measure conditional covariance directly. We then propose a new statistical model that allows direct measurement of label covariance (both constant and conditional). The model is validated using generated data and it is also used to examine the label covariance in real world data, allowing us to build simpler multi-label models.

1 Introduction

Multi-label classification is the process of predicting the binary state of a set of labels, conditioned on a state variable; a multi-label classifier is a model (function) that provides these classifications. Multi-label classifiers range from a set of binary classifiers, to one model for the powerset of labels, to a spectrum of models in between [3–5,11]. There are methods for measuring multi-label confidence [8,9] and many metrics for optimisation [7]. A set of binary classifiers is simple to fit to data, but does not model the label dependencies, while a model for the powerset models all dependencies, but requires a large amount of data and resources.

The first question that should be asked after obtaining multi-label data is "Which labels are independent of each other? and if they are dependent does their dependence change based on the state?" If all labels are independent, then a set of binary classifiers can be used. If no labels are independent then a full multi-label model should be used. But if some are independent, then multiple smaller

X. Wu et al. (Eds.): PAKDD 2025, LNAI 15875, pp. 250–261, 2025.
https://doi.org/10.1007/978-981-96-8295-9_18

multi-label models could be used on the dependent sets, depending on which sets of labels are independent. Gaining insight to the data helps in understanding the data and designing models for the data.

In this article, we investigate how to measure dependence between labels in a multi-label classifier. We then present the popular copula method for measuring label covariance and its shortcomings. A new model is proposed to directly model label covariance and we validate the model using simulated data. Our contributions are:

- a measure of binary independence (Sect. 2.1),
- a description of the limitations of the multivariate Normal copula model (Sect. 3.1),
- a model for measuring the unexplained dependence (Sect. 3.2).

The article proceeds as follows: Sect. 2 examines how to measure the independence of binary random variables and describes the concepts of explained and unexplained label dependence. Section 3 reviews the Probit model and demonstrates that it is not appropriate for modelling dependence, then presents a direct model of covariance. Section 4 presents experimental results on significance testing for the model coefficients.

2 Measuring Label Dependence

Multi-label data consists of a set of observations $\boldsymbol{x} \in \mathbb{R}^M$ with associated labels $\boldsymbol{y} \in \{0,1\}^L$. Given multi-label data, we can test if any pair of labels y_i and y_j are not independent by using a χ^2 test for independence. If the labels are independent, the data can be split in to smaller pieces to simplify the modelling [6]. A problem arises, however, due to the fundamental nature of multi-label data. Multi-label classifiers $\boldsymbol{y} = f(\boldsymbol{x})$ estimate the label set $\boldsymbol{y}$ conditioned on an observation $\boldsymbol{x}$. This implies that each label is dependent on the set of observations $\boldsymbol{x}$, and so any pair of labels should not be independent due to this confounding relationship with $\boldsymbol{x}$.

In this section, we show that the benefit of multi-label models come from modelling unexplained covariance, which is the covariance between labels conditioned on the observations.

2.1 Covariance and Independence

Covariance is a measure of the variance shared by two random variables. Covariance can be positive or negative, and is defined as

$$\mathrm{Cov}(Y_i, Y_j) = \mathbb{E}[(Y_i - \mu_i)(Y_j - \mu_j)] = \mathbb{E}[Y_i Y_j] - \mathbb{E}[Y_i]\mathbb{E}[Y_j]$$

where Y_i and Y_j are the random variables of interest and μ_i and μ_j are their means. A positive covariance shows that the two random variables are likely to increase and decrease together about their means, a negative covariance means that they are likely to vary in opposed directions and a covariance of zero implies that the variance about their mean is not related.

Independence is a statistical property that measures if two random variables have an association. Two random variables are considered independent if they satisfy the equation

$$P(Y_i, Y_j) = P(Y_i)P(Y_j) \tag{1}$$

where $P(Y_i, Y_j)$ is the joint probability of Y_i and Y_j and $P(Y_i)$ and $P(Y_j)$ are the marginal probabilities.

We now restrict the random variables to Bernoulli random variables. A Bernoulli random variable $Y \sim \text{Bern}(p)$ has the parameter $p \in [0, 1]$ and the domain $\{0, 1\}$. The probability of the variable being 1 is p and the probability of the variable being 0 is $1 - p$. The random variable Y has the properties $\mathbb{E}[Y] = p$ and $\text{Var}(Y) = p(1 - p)$. Given two Bernoulli random variables $Y_i \sim \text{Bern}(p_i)$ and $Y_j \sim \text{Bern}(p_j)$, we can derive their covariance.

$$\begin{aligned} \rho_{ij} = \text{Cov}(Y_i, Y_j) &= \mathbb{E}[Y_i Y_j] - \mathbb{E}[Y_i]\mathbb{E}[Y_j] \\ &= \mathbb{E}[Y_i Y_j] - p_i p_j \end{aligned}$$

To obtain $\mathbb{E}[Y_i Y_j]$, we use the definition of expected value to find $\mathbb{E}[Y_i Y_j] = P(Y_i = 1, Y_j = 1)$ (the probability that both Y_i and Y_j are both 1); we will call this probability p_{ij}, giving

$$\rho_{ij} = p_{ij} - p_i p_j \tag{2}$$

Note that for Bernoulli random variable $Y \sim \text{Bern}(p)$, $\mathbb{E}[Y] = P(Y = 1) = p$ giving the Bernoulli form of the independence equation from Eq. (1)

$$p_{ij} = p_i p_j \Rightarrow p_{ij} - p_i p_j = 0 \tag{3}$$

Comparing the Eqs. (2) and (3), we find that if one is zero, then both must be zero. So for Bernoulli random variables, if the random variables are independent, then the covariance is zero, and also if the covariance is zero, then the random variables are independent. Therefore covariance is a measure of independence for Bernoulli random variables. We also find that the covariance equation is equivalent to the independence equation, so *independence can be measured using covariance.*

2.2 Explained Dependence

Using the fact that $\mathbb{E}[Y] = p$, we can proceed with identifying the confounding effect of $\boldsymbol{x}$, which we call *explained dependence.* We can demonstrate this confounding with a simple example. Let $Y_1 \sim \text{Bern}(p = 2x_1 + x_2)$ and $Y_2 \sim \text{Bern}(p = x_1 + 2x_2)$, two labels that can be modelled independently (they are dependent on $\boldsymbol{x}$, but not on each other). Let's also assume that x_1 and x_2

are distributed Uniformly between 0 and 1/3. The expected value (and hence proportion) for Y_1 is

$$\mathbb{E}[Y_1] = \int_0^{1/3} \int_0^{1/3} 9(2x_1 + x_2)\, dx_1 dx_2 = \frac{1}{2}$$

The symmetry of the equations imply that the expected value of Y_2 is also 1/2. If Y_1 and Y_2 were independent, we expect the covariance to be zero ($\mathbb{E}[Y_1 Y_2] = \mathbb{E}[Y_1]\mathbb{E}[Y_2]$) and so $\mathbb{E}[Y_1 Y_2] = 1/4$, but we find

$$\mathbb{E}[Y_1 Y_2] = \int_0^{1/3} \int_0^{1/3} 9(2x_1 + x_2)(x_1 + 2x_2)\, dx_1 dx_2 = \frac{31}{108}$$

Showing that Y_1 and Y_2 are not independent, purely due to their dependence on x_1 and x_2 (dependence entirely explained by $\boldsymbol{x}$). Therefore, if we were to take this data and perform a χ^2 test on Y_1 and Y_2, we would conclude that they are not independent and so a multi-label classifier would be beneficial. But, in fact, knowing the underlying model shows us that estimating each Y_i conditioned on each x is sufficient to model this covariance explained by $\boldsymbol{x}$, implyng that two independent binary models can be used to model Y_1 and Y_2 and still exhibit dependence. A multi-label model is not needed in this case.

2.3 Unexplained Dependence

The previous section showed that independent models are sufficient to model each label, when the labels can be explained by the vector of observations $\boldsymbol{x}$. We know that multi-label models are able to provide better fitting models to multi-label data, so we need to ask, what are multi-label models capturing that independent models are not? The dependence between labels found in the previous section was entirely explained by the observations $\boldsymbol{x}$ (explained dependence); multi-label models take a further step and model the *unexplained dependence* between labels, which we will show is the label covariance conditioned on $\boldsymbol{x}$.

The notion of unexplained dependence can be demonstrated as correlated randomness between a pair of labels. Before examining the Bernoulli case, we first consider a Gaussian regression, having the form

$$y_j = \beta_0 + \sum_i \beta_i x_i + \epsilon \qquad \epsilon \sim \mathcal{N}(0, \sigma^2)$$

Two regressions would each have their own variance term ϵ, leading to two independent models, but if their error terms are correlated, then the two models are no longer independent. The same goes for Bernoulli models, where the randomness that resolves a label into a 1 or a 0 is dependent between models.

A simple method for resolving a Bernoulli random variable is to generate a Uniform random number between 0 and 1; if the number is less than p, the Bernoulli result is a 1, otherwise it is a 0. So the randomness in the Bernoulli variable can be attributed solely to this Uniform value. Given two Bernoulli

Table 1. A set of observations $\boldsymbol{x}$ with three labels y_i and the estimated proportion for each label for models f_i. Unexplained dependence can be seen between labels y_2 and y_3.

$\boldsymbol{x}$	[1,2,0]	[1,2,0]	[1,2,0]	[1,2,0]	[1,2,0]	[1,2,0]	[1,2,0]	[1,2,0]	[1,2,0]	[1,2,0]
$p_1 = f_1(\boldsymbol{x})$	0.6	0.6	0.6	0.6	0.6	0.6	0.6	0.6	0.6	0.6
y_1	1	0	1	0	1	1	1	1	0	0
$p_2 = f_2(\boldsymbol{x})$	0.6	0.6	0.6	0.6	0.6	0.6	0.6	0.6	0.6	0.6
y_2	1	0	0	1	1	1	0	1	0	1
$p_3 = f_3(\boldsymbol{x})$	0.6	0.6	0.6	0.6	0.6	0.6	0.6	0.6	0.6	0.6
y_3	1	1	0	0	1	1	0	1	0	1

random variables, two Uniform values must be generated. If the two Uniform random variables are independent, then the two Bernoulli random variables will be independent once conditioned on p. If the two Uniform variables are not independent, then the result from the two Bernoulli random variables will not be independent, regardless of p. Let's examine an example of this. Table 1 shows a set of observations $\boldsymbol{x}$ and labels y_i. The table presents ten observations with the same values $\boldsymbol{x} = [1, 2, 0]$, but the labels vary. The models for each label f_i provide the correct proportion of 0.6 for each label, since 0.6 of the observations are 1 for each label. If independent, we should expect 0.36 pairs of labels providing $(1, 1)$; the pair (y_1, y_2) provide 0.4 which is close to expected when independent, but the pair (y_2, y_3) provide 0.5. If we examine further, we find that only two of the observations have different labels when comparing (y_2, y_3) (they have high covariance). Therefore, the pair (y_2, y_3) is exhibiting *unexplained dependence since the dependence between the labels cannot be explained by* $\boldsymbol{x}$. It is the covariance conditioned on $\boldsymbol{x}$. It is this unexplained covariance that multi-label models capture, that provides benefit over a set of independent binary models.

3 Measuring Unexplained Covariance

We showed in the previous section that a pair of labels can exhibit explained and unexplained dependence and that dependence can be measured using covariance for Bernoulli random variables. In this section, we will examine how to account for the variation in each label due to $\boldsymbol{x}$, so that we can measure if there is additional covariance between labels that is unaccounted for (unexplained covariance). If this unaccounted covariance exists, then a multi-label model will be beneficial. First, the Probit model is examined and we find that it is not directly modelling the covariance. A joint probability model is then presented that models the covariance directly.

3.1 Probit Model for Unexplained Covariance

Copulas provide a means of sampling from a difficult distribution by mapping the problem to a simpler domain. Two dimensional Normal copulas (bivariate Normal distribution) are commonly used to model bivariate Bernoulli distributions. This allows us to fit bivariate Normal density functions to make inferences about the bivariate Bernoulli distributions.

Using a change of variables, we can write the bivariate Bernoulli joint label probability as

$$p_{ij} = P\left(\begin{bmatrix} N_i \\ N_j \end{bmatrix} > \begin{bmatrix} 0 \\ 0 \end{bmatrix}\right) \quad \begin{bmatrix} N_i \\ N_j \end{bmatrix} \sim \mathcal{N}\left(\mu = \begin{bmatrix} \Phi^{-1}(p_i) \\ \Phi^{-1}(p_j) \end{bmatrix}, \Sigma = \begin{bmatrix} 1 & \tau_{ij} \\ \tau_{ij} & 1 \end{bmatrix}\right) \tag{4}$$

where $p_{ij} = P(Y_i = 1, Y_j = 1)$, $\Phi^{-1}()$ is the inverse cumulative Normal distribution, $\mathcal{N}$ is Normally distributed with mean equal to the p_i and p_j quantiles and covariance τ_{ij}. If $\mathcal{N}$ is known, then the joint Bernoulli probabilities are simply each of the four quadrants of the Real sample space.

$$p_{\bar{i}\bar{j}} = P\left(\begin{bmatrix} N_i \\ N_j \end{bmatrix} < \begin{bmatrix} 0 \\ 0 \end{bmatrix}\right) \quad p_{i\bar{j}} = P\left(\begin{bmatrix} N_i \\ N_j \end{bmatrix} \begin{matrix} > \\ < \end{matrix} \begin{bmatrix} 0 \\ 0 \end{bmatrix}\right) \quad p_{\bar{i}j} = P\left(\begin{bmatrix} N_i \\ N_j \end{bmatrix} \begin{matrix} < \\ > \end{matrix} \begin{bmatrix} 0 \\ 0 \end{bmatrix}\right)$$

This leads us to the multi-variate Probit model [1,10].

$$\begin{aligned} \Phi^{-1}(p_i) &= \beta_{i0} + \boldsymbol{\beta}_i \boldsymbol{x} + \epsilon_i \\ \Phi^{-1}(p_j) &= \beta_{j0} + \boldsymbol{\beta}_j \boldsymbol{x} + \epsilon_j \end{aligned} \quad \begin{bmatrix} \epsilon_i \\ \epsilon_j \end{bmatrix} \sim \mathcal{N}\left(\mu = \begin{bmatrix} 0 \\ 0 \end{bmatrix}, \Sigma = \begin{bmatrix} 1 & \tau_{ij} \\ \tau_{ij} & 1 \end{bmatrix}\right)$$
$$\log\left(\frac{1+\tau_{ij}}{1-\tau_{ij}}\right) = \beta_{k0} + \boldsymbol{\beta}_k \boldsymbol{x} \tag{5}$$

Fitting the multivariate Probit model to data estimates the coefficients β_{i0} and $\boldsymbol{\beta}_i$ for p_i, the coefficients β_{j0} and $\boldsymbol{\beta}_j$ for p_j, and the coefficients β_{k0} and $\boldsymbol{\beta}_k$ for τ_{ij}.

Observing the fitted parameters β_{k0} and $\boldsymbol{\beta}_k$ tell us about the data covariance. β_{k0} is a measure of the constant covariance (not dependent on $\boldsymbol{x}$) and $\boldsymbol{\beta}_k$ is a measure of the covariance dependent on $\boldsymbol{x}$. Hypothesis tests on the coefficients reveal their significance.

Note that each of the coefficients that estimate p_i and p_j are estimated using only the marginal distribution for each label Y_i and Y_j, and so are not influenced by the estimation of the covariance τ_{ij}. So even though the model estimates all parameters at once, the estimate τ_{ij} is of the unexplained covariance. Note as well that the estimate of the covariance τ_{ij} is the covariance of the joint Normal distribution and not measuring the bivariate Bernoulli covariance ρ_{ij} directly.

The Probit model was constructed to identify if unexplained label covariance ρ_{ij} exists and if it is dependent on $\boldsymbol{x}$. Therefore, we must examine what we can deduce from τ_{ij} (the Normal copula covariance) about ρ_{ij}.

Figure 1 shows the relationship between τ_{ij} and p_i when ρ_{ij} is held constant. We can see that even though ρ_{ij} is constant, τ_{ij} changes when p_i changes. The dependence of τ_{ij} on p_i demonstrates that τ_{ij} will show dependence on $\boldsymbol{x}$ (since

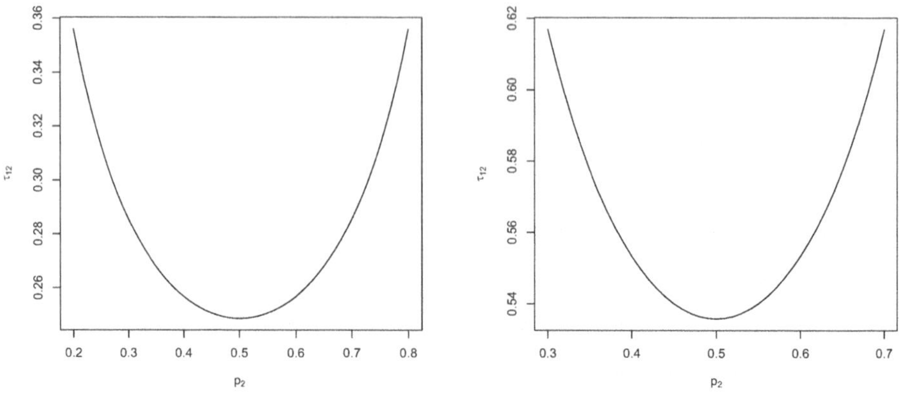

Fig. 1. The effect of changing p_2 on the Probit covariance τ_{ij} when $p_1 = 0.5$ and the binary covariance $\rho_{ij} = 0.04$ (left) and $\rho_{ij} = 0.09$ (right).

p_i is dependent on $\boldsymbol{x}$), even when ρ_{ij} is constant. This implies that *the Probit model can be used to determine if unexplained covariance between labels exist, but it cannot be used to identify if the unexplained covariance is dependent on the observations* $\boldsymbol{x}$.

3.2 Measuring Unexplained Covariance Using the Pairwise Joint Probability

The Probit model was not able to model label covariance dependence on $\boldsymbol{x}$ directly, so in this section we investigate a new model. We want to directly model the label correlation

$$\rho_{ij}|\boldsymbol{x} = f\left(\beta_0 + \sum_{k=1}^{n} \beta_k x_k\right) \tag{6}$$

doing so will provide estimates of β_0 and β_k where β_0 measures the constant dependence between labels i and j, and β_k measures the conditional dependence for each x_k. Conditional dependence is a dependence between a pair of labels that is conditional with respect to the observations $\boldsymbol{x}$. For some values of $\boldsymbol{x}$ the dependence might be strong, but for others it might be absent. The covariance equation from Eq. (2) can be written with a condition on $\boldsymbol{x}$

$$\rho_{ij}|\boldsymbol{x} = P(Y_i, Y_j|\boldsymbol{x}) - P(Y_i|\boldsymbol{x})P(Y_j|\boldsymbol{x}) \tag{7}$$

This allows us to model the covariance by modelling the joint probability and marginal probabilities, but this form contains multiple coefficients related to each of the $\boldsymbol{x}$ variables and so does not allow direct measure of the relationship between ρ_{ij} and $\boldsymbol{x}$.

The number of models can be reduced to one by observing that we want to model the difference between p_{ij} and $p_i p_j$. Using the estimate $p_i p_j$ as a baseline will reveal that Y_i and Y_j are not independent if p_{ij} departs from this baseline.

Therefore, we can examine the effect of each $\boldsymbol{x}$ on the dependence of Y_i and Y_j by modelling the joint probability, using the independent estimate as a baseline.

$$\hat{p}_{ij} = f\left(\beta_0 + \sum_{k=1}^{n} \beta_k x_k\right) + p_i p_j \tag{8}$$

where $f()$ is a function that ensures $\hat{p}_{ij} \in [0,1]$, x_k are the elements of $\boldsymbol{x}$, and β_k are the model coefficients. If the coefficients β_k are zero (or not statistically significant from zero), then there is no evidence that the Y_i and Y_j are not independent. If the coefficient β_0 is non-zero, then Y_i and Y_j are not independent regardless of each x_k. If β_k is non-zero, then the independence of Y_i and Y_j is effected by x_k.

After observing (8), it might seem that we can model the difference $p_{ij} - p_i p_j$, but remember that we do not know p_{ij}, we only have a sample from the Bernoulli distribution with parameter p_{ij}. A final adjustment to the model is needed to allow for this; adding $p_i p_j$ to the model is also difficult to perform, so instead, we model

$$\text{logit}(\hat{p}_{ij}) = \beta_0 + \sum_{k=1}^{n} \beta_k x_k + \text{logit}(\hat{p}_i \hat{p}_j) \tag{9}$$

Making $\text{logit}(\hat{p}_i \hat{p}_j)$ an offset to the linear portion of the model, where logit() is the log odds. This will provide different estimates for β_k, but they still retain the property that if they are nonzero, then the associated x_k has an effect on the dependence between Y_i and Y_j. *The model in Eq.* (9) *is a logistic regression over* $Y_i Y_j$ *using* $\text{logit}(\hat{p}_i \hat{p}_j)$ *for each* $\boldsymbol{x}$ *as an offset* and so simple to implement using any logistic regression library.

4 Experiments

We have described a model for detecting dependence between labels by testing the model coefficients for significance; in this section, we evaluate the model. There is data for evaluating the effectiveness of multi-label classification, but currently no data for evaluating the effectiveness of dependence detection. Therefore, in this section, a set of data models will be provided and data generated from them. The dependence model will then be used to detect dependencies between labels. Since the underlying models that generated the data are known, we are able to assess the effectiveness of the dependence detection models. The code for these experiments is available at `anonymised`.

4.1 Significance Investigation

Using logistic regression for estimating the unexplained covariance has the added benefit that there are known tests of significance for the regression coefficients. If a Wald test for each of the model coefficients (and intercept) provides a statistically significant result, we have evidence that the coefficient β_k is not zero, and hence the labels are not independent. Unfortunately, our model contains an estimate of $p_i p_j$ (estimated using another logistic regression) which itself contains variance that is not accounted for in the joint model. Therefore, the hypothesis test results for the coefficients are likely to under or overestimate the p-values.

To examine how effective the covariance model (Eq. (9)) is at finding effects on label covariance, we generate experimental data using three data models. The data models are

- Indep: There is no unexplained label covariance. The labels are generated purely from the observations $\boldsymbol{x}$.
- Cz: There is unexplained covariance, but it is constant (i.e. the unexplained covariance exists, but not dependent on $\boldsymbol{x}$). The constant covariance value is bound by $z/100$. We experiment using $z = 1, 4, 9$ and 16.
- Dz: There is unexplained covariance, and it is dependent on $\boldsymbol{x}$. The variable covariance value bound by $z/100$. We experiment using $z = 1, 4, 9$ and 16.

The experiment follows the process: 1) Generate the multi-label data containing N observations, where each observation consists of $n = 5$ standard Normal elements and two labels. Note only two labels are needed since the covariance model only models pairs of labels. 2) Generate the model coefficients β_0 to β_n for each label and the covariance. 3) Compute the proportions for each label using the observations $\boldsymbol{x}$ and coefficients β_k and randomly sample a set of labels from the proportions. Using this generated data, fit the covariance model (Eq. (9)) and report the results of the Wald test for each coefficient. This was repeated 1000 times for each configuration to obtain the probability of each covariance coefficient β_k being non-zero (i.e. having an effect on the covariance). The results are provided in Table 2.

Table 2 shows the probability of rejecting the Null Hypothesis for the Wald test on each model coefficient for the data models Indep, C1, C4, C9, C16, D1, D4, D9 and D16, where the Null Hypothesis is that the true coefficient value is zero. Therefore, the values in the table can be considered the probability that the covariance model is dependent on a constant (given by β_0) or it is dependent on a observation element x_k (given by β_k). For example, if we examine the values for C4 for $N = 500$, a constant relationship is detected 75% of the time and a dependent relationship 1% of the time.

Ideally, we want the probabilities for β_0 and β_k to be low for the Indep data (since the labels are independent), β_0 to be high and β_k to be low for the each of the Cz data (since the dependence is constant), and β_0 and β_k to be high for the Dz data (since the dependence depends on $\boldsymbol{x}$).

The Indep results (independent labels) show that the probabilities are accurate from the $N = 100$ data sample size onwards, implying that if there is no covariance, then the model will provide accurate results if the data contains at least 100 observations.

The Cz results (constant dependence) show that the accuracy is dependent on the size of the covariance (z) and the data sample size N. If the data label unexplained covariance is high (C16), then the model provides highly accurate results when the data sample size is at least 100; if the covariance is very small

Table 2. Probability of coefficient test showing that coefficient is nonzero ($\alpha = 0.05$). Indep, Cz and Dz contain no, constant and dependent covariance respectively. β_0 and β_k represent the offset and x value model coefficient respectively.

Model	Type	Sample Size (N)									
		10	20	50	100	200	500	1000	2000	5000	10000
Indep	β_0	0.812	0.917	0.315	0.025	0.010	0.011	0.007	0.015	0.013	0.006
	β_k	0.812	0.916	0.316	0.027	0.006	0.003	0.005	0.009	0.008	0.004
C1	β_0	0.801	0.584	0.015	0.007	0.010	0.026	0.101	0.212	0.643	0.908
	β_k	0.801	0.584	0.018	0.002	0.005	0.007	0.011	0.028	0.080	0.127
C4	β_0	0.771	0.318	0.010	0.055	0.197	0.750	0.986	1.000	1.000	1.000
	β_k	0.772	0.315	0.001	0.002	0.004	0.010	0.016	0.060	0.185	0.358
C9	β_0	0.731	0.196	0.135	0.559	0.963	1.000	1.000	1.000	1.000	1.000
	β_k	0.729	0.179	0.000	0.000	0.000	0.006	0.013	0.057	0.189	0.334
C16	β_0	0.700	0.147	0.792	0.998	1.000	1.000	1.000	1.000	1.000	1.000
	β_k	0.699	0.076	0.000	0.000	0.000	0.001	0.003	0.006	0.050	0.173
D1	β_0	0.778	0.443	0.004	0.003	0.004	0.009	0.010	0.016	0.071	0.148
	β_k	0.778	0.445	0.005	0.001	0.004	0.005	0.006	0.013	0.033	0.096
D4	β_0	0.773	0.241	0.003	0.007	0.009	0.053	0.161	0.309	0.464	0.584
	β_k	0.771	0.243	0.004	0.005	0.008	0.032	0.094	0.191	0.436	0.613
D9	β_0	0.746	0.152	0.009	0.040	0.111	0.307	0.490	0.586	0.752	0.827
	β_k	0.745	0.149	0.006	0.024	0.069	0.234	0.371	0.544	0.699	0.780
D16	β_0	0.761	0.079	0.043	0.180	0.356	0.579	0.695	0.774	0.842	0.890
	β_k	0.760	0.075	0.035	0.119	0.287	0.506	0.618	0.706	0.837	0.862

(C1), then a much larger sample size is needed (at least $N = 5000$ to get a 0.643 chance at correctly showing a constant covariance). For covariance of 0.04 and 0.09 (C4 and C9), high accuracy begins at $N = 1000$ and $N = 200$ respectively, but we also find that as the sample size increases to $N = 10000$ the chance of finding false dependence of the covariance on $\boldsymbol{x}$ increases. We are unsure why this *phantom* dependencies exists for large sample sizes when having constant covariance of 0.04 and 0.09, therefore further investigation is required.

Finally, the Dz results (label dependence depends on $\boldsymbol{x}$) are also dependent on the data sample size N and the magnitude of the covariance (z). Covariance of size 0.01 are difficult to detect, but larger covariances (such as 0.16) can be detected when the sample size is of the order of 1000. These results will be used to help interpret the results when analysing real data.

4.2 Detecting Label Relationships on Real Data

In this section, we examine the label dependence on real data. For this experiment, we obtained five data sets commonly used for multi-label classification evaluation from the MULAN and STARE[1] repositories. The covariance analysis will be used to determine the state of the dependence of each pair of labels in

[1] http://mulan.sourceforge.net/datasets-mlc.html, http://www.ces.clemson.edu/ aho over/stare/.

Table 3. Five multi-label data sets and the results of the covariance analysis on each. The numbers show the count of each pair of independent (Indep), constant (Const), and dependent (Dep) covariance labels.

Data Stats			Results		
Name	Size	Labels	Indep	Const	Dep
Stare	373	15	14	0	91
Emotions	593	6	8	6	1
Scene	2407	6	4	2	9
Yeast	2417	14	53	38	0
Slashdot	3782	20	4	0	186

each data set. Recall that there are three dependence states: Independent, Constant, Dependent. Note that there is no gold standard to evaluate the results obtained, therefore, the findings will be reported and compared to the Significance Investigation from Sect. 4.1. The results from the analysis are shown is Table 3.

For the data sample sizes, if a pair of labels shows to be independent, it is either because they are independent, or that the magnitude of the covariance was too small to detect. If constant covariance is detected, then it is likely to be true, and if dependent covariance is detected, then it is likely to be true. The sample size is not large enough for phantom dependent covariance to be of concern. Note that we also know which of the variables of $\boldsymbol{x}$ effect the dependent covariance, but the results are too many to include in this article. It is interesting to see that the majority of label pairs have dependent covariance for Stare and Slashdot, but the proportion is much lower for the other data. There are also a large number of label pairs that are independent in Yeast, suggesting that it might be possible to partition the label set for classification efficiency. Examining these results further will help gain a deeper insight into the generating process of the data.

There have been other attempts to model label dependency, such as work by [2,12] but they use surrogate functions and so do not directly measure the dependence between labels or measure the if there is a relationship between the label dependence and the observations. In this article, we have derived a model to directly measure the unexplained dependence between labels. Also note that regularisation can be used when fitting the models, but it leads to bias in the coefficients and so this bias must be taken into account when conducting the hypothesis tests.

5 Conclusion

Multi-label classifiers make use of associations between labels in multi-label data to increase the accuracy of prediction. Before using a multi-label classifier, the data should be analysed to make sure there is dependence between the labels (e.g. using a χ^2 test for independence). Unfortunately, all labels should show dependence due to their relationship to the explanatory variables.

We began this article by showing that dependence can be measured using covariance for binary labels. We then showed that labels have covariance due to their relationship to the explanatory variables (explained covariance). We presented an established model for unexplained covariance using a Normal copula and showed that the coefficients are not able to describe the source of the covariance. Finally, we derived a model to directly measure the unexplained covariance, that detects if the covariance is constant or if it depends on the explanatory variables. Our analysis shows that label dependence type can be accurately detected given a large sample size ($N > 100$) and large enough label covariance.

References

1. Ashford, J., Sowden, R.: Multi-variate probit analysis. Biometrics 535–546 (1970)
2. Gonçalves, A.R., Von Zuben, F.J., Banerjee, A.: Multi-label structure learning with ising model selection. In: Twenty-Fourth International Joint Conference on Artificial Intelligence (2015)
3. Katakis, I., Tsoumakas, G., Vlahavas, I.: Multilabel text classification for automated tag suggestion. ECML PKDD Discov. Chall. **75**, 2008 (2008)
4. Lanchantin, J., Sekhon, A., Qi, Y.: Neural message passing for multi-label classification. In: Machine Learning and Knowledge Discovery in Databases: European Conference, pp. 138–163. Springer (2020)
5. Liu, B., Tsoumakas, G.: Synthetic oversampling of multi-label data based on local label distribution. In: Machine Learning and Knowledge Discovery in Databases: European Conference, pp. 180–193. Springer (2020)
6. Park, L.A., Guo, Y., Read, J.: Assessing the multi-labelness of multi-label data. In: Joint European Conference on Machine Learning and Knowledge Discovery in Databases, pp. 164–179. Springer (2019)
7. Park, L.A., Read, J.: A blended metric for multi-label optimisation and evaluation. In: Machine Learning and Knowledge Discovery in Databases: European Conference, pp. 719–734. Springer (2019)
8. Park, L.A., Read, J.: Estimating multi-label expected accuracy using labelset distributions. IEEE Trans. Knowl. Data Eng. (2025). https://doi.org/10.1109/TKDE.2025.3545972
9. Park, L.A., Simoff, S.: Using entropy as a measure of acceptance for multi-label classification. In: Advances in Intelligent Data Analysis XIV: 14th International Symposium, pp. 217–228. Springer (2015)
10. Prentice, R.L.: Correlated binary regression with covariates specific to each binary observation. Biometrics 1033–1048 (1988)
11. Read, J., Pfahringer, B., Holmes, G., Frank, E.: Classifier chains for multi-label classification. In: Machine Learning and Knowledge Discovery in Databases: European Conference, pp. 254–269. Springer (2009)
12. Teisseyre, P.: Feature ranking for multi-label classification using markov networks. Neurocomputing **205**, 439–454 (2016)

Evaluating Generative Vehicle Trajectory Models for Traffic Intersection Dynamics

Yash Ranjan(✉), Rahul Sengupta, Anand Rangarajan, and Sanjay Ranka

University of Florida, Gainesville, FL, USA
{yashranjan,rahulseng,anandr,sranka}@ufl.edu

Abstract. Traffic Intersections are vital to urban road networks as they regulate the movement of people and goods. However, they are regions of conflicting trajectories and are prone to accidents. Deep Generative models of traffic dynamics at signalized intersections can greatly help traffic authorities better understand the efficiency and safety aspects. At present, models are evaluated on computational metrics that primarily look at trajectory reconstruction errors. They are not evaluated online in a 'live' microsimulation scenario. Further, these metrics do not adequately consider traffic engineering-specific concerns such as red-light violations, unallowed stoppage, etc. In this work, we provide a comprehensive analytics tool to train, run, and evaluate models with metrics that give better insights into model performance from a traffic engineering point of view. We train a state-of-the-art multi-vehicle trajectory forecasting model on a large dataset collected by running a calibrated scenario of a real-world urban intersection. We then evaluate the performance of the prediction models, online in a microsimulator, under unseen traffic conditions. We show that despite using ideally-behaved trajectories as input, and achieving low trajectory reconstruction errors, the generated trajectories show behaviors that break traffic rules. We introduce new metrics to evaluate such undesired behaviors and present our results.

Keywords: Trajectory Prediction · Generative Models · Deep Learning · Intelligent Transportation Systems · Traffic Safety · Model Reliability

1 Introduction

Traffic intersections play a critical role in the functioning of urban road networks. They ensure the smooth flow of vehicles and pedestrians by ensuring their trajectories do not conflict, avoiding congestion and accidents. Hence, dynamics seen at signalized urban intersections are complex, as various road users (vehicles and pedestrians) must not only interact with each other, but they must also pay attention to the traffic signal state as well as road markings indicating lanes and crosswalks. Computational-modeling of such scenarios was usually

X. Wu et al. (Eds.): PAKDD 2025, LNAI 15875, pp. 262–274, 2025.
https://doi.org/10.1007/978-981-96-8295-9_19

done on a case-by-case basis using agent-based micro-simulators, which can be computationally-heavy.

Intelligent Transportation Systems (ITS) has revolutionized the way sensor data from urban intersections is acquired, processed, and utilized. Various sensor modalities such as inductive loop detectors, signal state recorders, video cameras, LiDAR, radar etc. provide comprehensive high-resolution data.

Advances in Cloud Storage technologies have provided scalable data storage for vast amounts of multimodal sensor data. High-performance computing hardware (such as GPUs/TPUs) has accelerated the development of Deep Learning models trained on large datasets. Hence, there has been a great deal of published literature on the application of Deep Learning models to ITS.

One important area of application is the trajectory prediction of road users. Usually, trajectory data is generally obtained by performing object-tracking on video/LiDAR data. This often requires large data processing resources as well as advanced vision-processing algorithms [5].

1.1 Related Work

There are many works that deal with trajectory prediction for road users. Important among them include Trajectron (and its successor Trajectron++) [15], which represents a traffic scene as a directed spatio-temporal graph and produces dynamically-feasible trajectories using a graph-structured recurrent model. Another important model is TNT (target-driven trajectory) [18] which consists of three stages that are trained end-to-end: the first predicts an agent's potential target states, the second generates trajectory state sequences conditioned on targets and the final stage estimates trajectory likelihoods and selects a final compact set of trajectory predictions. Surveys such as [8,10], and [1] discuss various methods for vehicle trajectory prediction. However, they focus on autonomous vehicles and do not specifically deal with the modeling of intersection behaviors.

[6] surveys commonly used performance metrics and robustness strategies for vehicle trajectory prediction. [9] discusses commonly-used metrics and their drawbacks in the context of multimodal trajectory prediction. [3] introduces new metrics for trajectory prediction apart from commonly-used distance-based metrics. However, it does not focus on intersection-related behaviors, especially the effect of traffic signals and compliance thereof.

1.2 Our Contribution

While there is a good deal of relevant literature on deep generative models for trajectory prediction, there is a lack of metrics focusing on intersection dynamics in a multi-vehicle scenario. Present metrics are written from the point-of-view of autonomous vehicles, where the prediction of road users' trajectories revolves around the perception (via sensors) of the ego vehicle. This is not helpful to road traffic authorities who study the intersection dynamics as a whole, with trajectory data from intersection-mounted video cameras, LiDAR, DSRC/CV2X, drone-based cameras etc. These provide a birds-eye view of multiple vehicles at the intersection and their movements.

Further, the evaluation of these models is generally done using a pre-recorded test set. It is not possible to generate new arrival patterns or change other parameters, such as signal timing plan.

Given the critical importance of signalized intersections to road traffic authorities, our work makes the following contributions:

1. We develop an end-to-end framework for evaluating and benchmarking Deep Generative models for the task of multi-vehicle trajectory prediction at a signalized intersection. Given the trajectory prediction model architecture code, the system will train the model on the testbed intersection and vehicle flows, then evaluate the trained model and report its performance.
2. We introduce new evaluation metrics that address traffic-specific concerns such as traffic signal compliance, unnecessary stoppage, and braking behavior. These are novel metrics not usually seen in published trajectory prediction literature.
3. We include intersection-based vehicle position encoding and signal timing information in our trajectory prediction model, which is not usually seen in published literature.
4. We test the models using a simulation-in-loop real-world signalized urban intersection with calibrated vehicle flows.

2 Trajectory Prediction Model for Signalized Intersection Scenario

Our problem requires us to model the spatial-temporal relationship between an object's history with the state of the objects around it to predict the future position. The model used is largely based on [17] with key changes to incorporate intersection-specific details.

We discuss the different components of the model and how we incorporate the external factors which are the signal, intersection geometry, and the object position.

2.1 Trajectory Prediction

The objective of the trajectory prediction model is to generate plausible trajectory distributions for different agents (such as vehicles and pedestrians) that interact in a scene. This time-dependent number of agents is denoted by $n(t)$, representing the interacting agents $A_1, \ldots, A_{n(t)}$. In this work, we focus only on vehicles.

Each agent considers its own motion history, relevant traffic signals, potential interactions with neighboring agents, and its position relative to 'static points' (described later). The historical data for agent A_i at time t is expressed as

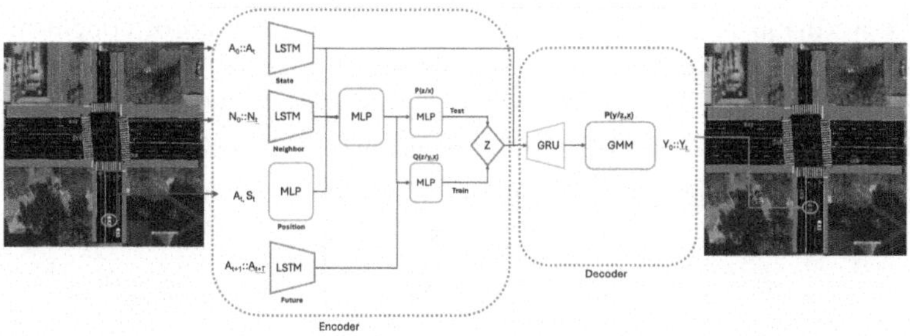

Fig. 1. The model architecture. The data comes from the simulator and is passed through the model to get the distribution over the future state. The mode of the distribution is sampled and sent back to the simulator, which visualizes it.

$X_i^{t-H+1,\dots,t}$, where H represents the number of history steps. The traffic signal for the route class c of agent A_i is denoted as s_c^t, its neighboring agents as N_i^t, and relative position R_i^t. The model outputs a probability distribution over the future trajectory $Y_i^{t+1,\dots,t+T}$ for the next T steps, formulated as $p(Y_i^{t+1,\dots,t+T}|X_i^{t-H+1,\dots,t}, N_i^t, R_i^t, s_c^t)$.

Here X consists of rectangular coordinates (x, y), velocity $(\dot{x}, \dot{y})$, acceleration $(\ddot{x}, \ddot{y})$ and Here Y consists of future rectangular coordinates (x, y).

2.2 Agent History Modeling

The agent history is modeled using an LSTM [7] that captures the sequential relationship of the object's past. We pass the object's position, velocity, acceleration, and one-hot-encoded signal for the H time stamp as input to the LSTM to generate the vector representation of the agent's state. The signal is the traffic light visible to the vehicle in its lane. Adding the signal in the object state helps us capture the transitional effect of the signal on the vehicle's trajectory. We denote the final encoded agent history as h_i^t.

$$h_i^t = LSTM\left(h_i^{t-1}, \mathbf{x}_i^t, W\right). \tag{1}$$

During the training phase, we incorporate a Node Future Encoder (NFE) to capture a node's true future trajectory [11]. This encoder is based on a bi-directional LSTM network with 32 hidden dimensions, producing outputs denoted as $h_{t+}^{i,\text{node}}$.

2.3 Agent Interaction Modeling

The agent-agent interactions in the scene are represented as a directed graph where the agents are the nodes and a directed edge is created between an agent and its neighbor if the neighbor is within the attention radius of the agent. This method is inspired from the Trajectron [11] model.

At any given timestep, the states of the connected neighbors of an object type (VEHICLE or PEDESTRIAN) will be added and concatenated with the vehicle's state. The addition of neighbors' states is preferred instead of concatenation or taking average as it can handle a variable number of neighbors without losing count information [12]. This is passed through an LSTM to capture the changing relation of the vehicle with the objects in its attention radius.

$$e_{i,k}^{t} = \left[\mathbf{x}_{i}^{t}, \sum_{j \in N_k(i)} \mathbf{x}_{j}^{t} \right], \tag{2}$$

$$h_{i,k}^{t} = LSTM\left(h_{i,k}^{t-1}, e_{i,k}^{t}; W\right). \tag{3}$$

We use additive attention mechanism to combine the embedding vectors and get a single neighbor influence embedding.

$$s_{ik}^{t} = \mathbf{v}_{C_i}^{T} \tanh\left(W_{1,C_i} h_{i,k}^{t} + W_{2,C_i} h_{i,\text{node}}^{t}\right), \tag{4}$$

$$a_{i}^{t} = \text{softmax}\left([s_{i1}^{t}, \ldots, s_{iK}^{t}]\right) \in \mathbb{R}^{K}, \tag{5}$$

$$h_{i,\text{edges}}^{t} = \sum_{k=1}^{K} a_{ik}^{t} \odot h_{i,k}^{t}. \tag{6}$$

2.4 Node Position Embedding

As the trajectory of the vehicle depends on the position of the vehicle at the intersection, we came up with a simple but effective way to embed the position of the vehicle. We divided the trajectories of vehicles into three regions as shown in Fig. 2. Each region has an endpoint that acts as a static point and the positional embedding of the vehicle is found relative to the static point based on the region that the vehicle belongs to in its trajectory. The idea behind this is that as a driver when we are approaching an intersection, we keep track of the end of the region in our mind. So we want to encode this distance in a vector. To get the positional embedding, we pass the current coordinate of the object, the coordinate of the static point based on the region that the vehicle belongs to, and the Euclidean distance between them. We also pass the one-hot encoding of the region as well as the one-hot encoding of the trajectory (or cluster) that the vehicle belongs to [17].

$$p_{i}^{t} = MLP(x_{i}^{t}, x_{st}, d_{i}^{t}, c_{i}, r_{i}^{t}). \tag{7}$$

Here x_{st} is the coordinate of the next static point, d_i^t is the distance between the object and the static point, c_i is the cluster that the vehicle belongs to and r_i^t is the region of the trajectory that the vehicle is in.

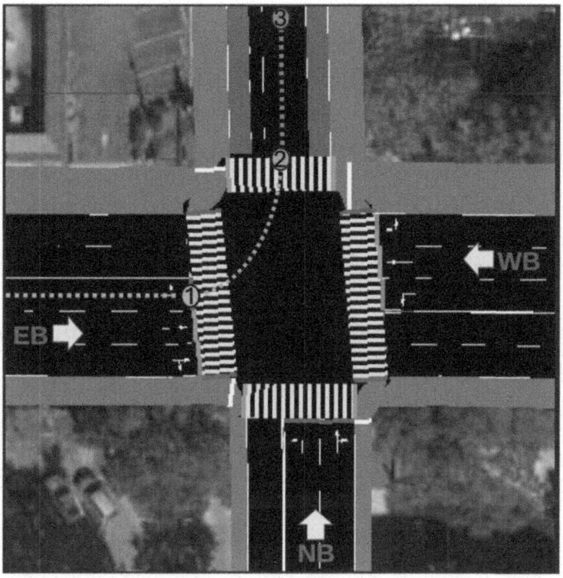

Fig. 2. The trajectory is divided into three regions. First is the incoming region, second is the in-between region, and third is the outgoing region. Each region has an end point which acts as the static point. The position embedding of the vehicle is found out relative to the static point based on the region the vehicle is

2.5 Conditional Variational AutoEncoder

We combine the three encoded vectors by concatenating them and passing them through a Multi-Layer Perceptron. This acts as the embedded information on which the model is conditioned to predict trajectories.

$$h_{i,enc}^{t} = MLP(h_i^t, h_{i,k}^t, p_i^t). \tag{8}$$

We followed Trajectron++ [15] to model the probability distribution over the trajectories $p(y \mid x)$ using the CVAE framework and introduced the discrete latent variable z to get the equation:

$$p(y|x) = \sum_{z \in Z} p_\theta(y|x,z) p_\phi(z|x)), \tag{9}$$

where discrete latent variable z models different modes for predicting future trajectories y given the input condition x. $p_\theta(y|x,z)$ and $p_\phi(z|x)$ are sub-networks with learnable parameters θ and ϕ. Instead of producing a deterministic trajectory prediction, the CVAE [13] model is able to predict a distribution of plausible future trajectories, accounting for the variability of vehicle and pedestrian behavior. For example, when drivers interact with a potentially conflicting pedestrian, some may opt to yield, while others may decide to accelerate and pass first. Under different situations, the probability of each behavior varies. We set $|Z| = 25$ in our experiments.

2.6 Decoder Network

During training the model, we get the latent variable probability from the output of the network $q(z|x, y)$. This is nothing but the mixture weights for the Gaussian Mixture Model that we use to model the future trajectories. During inference we sample from $p(z|x)$. We then send the sampled z and $h^t_{i,enc}$ to the decoder. We are using a GRU network with hidden dimensions of 128. The decoded vectors are then passed through mixture density network [2] to get the mean and covariance of the Gaussian for each z component for the next time-step. We do this for the next many time-steps.

For model training, we employ the CVAE's objective function, as described in [19] with the Adam optimizer:

$$\max_{\phi,\theta,\psi} \sum_{i=1}^{N} \mathbb{E}_{z \sim q_\phi(z|x_i,y_i)} \left[\log p_\psi(y_i \mid x_i, z)\right] - \beta D_{KL}(q_\phi(z \mid x_i, y_i) \parallel p_\theta(z \mid x_i)). \tag{10}$$

During inference, we sample the probability distribution $y \sim p_\theta(y|x, z)$ to obtain the future trajectories. The sampling can be performed by either finding the mixture weights of the GMM from the model $p(z \mid x)$ and then sampling the most likely output of the distribution $y \sim p_\theta(y|x, z)$ or sampling the individual Gaussians to obtain the trajectories for the different modes.

3 Intersection-Aware Evaluation Metrics

We summarize commonly-used metrics in Table 1, [1,8,10]. We then describe new intersection-aware evaluation metrics for intersection-focused scenarios.

3.1 Red Light Violation

This metric tracks the number of times a vehicle breaks the red light signal. The traffic phases for a 4-approach intersection are usually based on NEMA 8-phase numbering[1]. At an approach, a vehicle can take the following turn-movements:

1. Right-turn
2. Through (go straight)
3. Left-turn (and a U-turn)

It is important to note that taking a right-turn during a red light is often allowed at many intersections (after yielding). However, in our experiments, this sort of right turn on red light is not allowed.

1 https://ops.fhwa.dot.gov/publications/fhwahop08024/chapter4.htm.

Table 1. Summary of Evaluation Metrics for Trajectory Prediction Models.

Metric	Expression	Description
Average Displacement Error (ADE), minADE	$\frac{1}{T}\sum_{t=1}^{T}\sqrt{(x_t-\hat{x}_t)^2+(y_t-\hat{y}_t)^2}$	Mean Euclidean distance between the predicted trajectory locations and the corresponding ground truth locations over all time steps
Final Displacement Error (FDE), minFDE	$\sqrt{(x_T-\hat{x}_T)^2+(y_T-\hat{y}_T)^2}$	Euclidean distance between the predicted final position and the true final position of the trajectory
KDE Negative Log-likelihood (KDE-NLL)	$-\ \mathbb{E}_{i,t\in\tau}\log P(Y_t^i\vert\text{KDE}(\hat{Y}_t^{i,K}))$	Evaluates the mean log-likelihood of the ground truth trajectory for each future timestep using Kernel Density Estimation for handling multi-modal trajectory predictions

3.2 Mid-intersection Stoppage

This occurs when the vehicle unexpectedly stops in the middle of the intersection (i.e. in the area enclosed by the 4 crosswalks) and refuses to move. In some cases, this may be valid, as a left-turning vehicle may be waiting (yielding) for an on-coming vehicle to pass or a pedestrian to cross. Since in our experiments, there are no such conflicting movements, there should be no such stoppage. If a vehicle remains stopped at such a location for an extended period of time (5 mph/8 kmph or less, for 2 s at least), we flag that as an invalid stoppage.

3.3 Pre-stopbar Stoppage

This occurs when a vehicle does not respond properly to a signal transition from Red to Green light. The vehicle remains stopped or advances too slowly despite no vehicles ahead of it blocking its path. When the light is green, we take a queue discharge rate of 0.5 m/s, which would translate to the discharge of 1 vehicle (effective length including headway) of 7 m, in 14 s [16]. This is a low-end estimate; hence the vehicle should cover its distance to the stopbar faster than this during the green time it sees. If it does not, it could be a cause for concern and require further investigation. It was noticed that during the experiments, vehicles often encounter a queue, slow down/stop, and then are unable to start fast enough as the queue discharges. Thus, they block the lane despite having sufficient distance ahead of them to move.

3.4 Time-To-Collision (TTC) Encounters

TTC is the time it would take for two vehicles to collide, assuming their current speeds and paths remain unchanged. It is calculated based on the relative distance between the vehicles and their relative speed. TTC is a well-known metric in traffic safety analysis but is rarely used in trajectory models[2].

[2] https://www.fhwa.dot.gov/publications/research/safety/03050/02.cfm.

Let:

- d be the distance between two vehicles.
- v_{rel} be the relative speed between the two vehicles.

Then, the Time to Collision (TTC) is given by:

$$\mathrm{TTC} = \frac{d}{v_{rel}}, \tag{11}$$

where:

- d is the distance between the two vehicles.
- $v_{rel} = |v_1 - v_2|$ is the absolute value of the relative speed between the vehicles, where v_1 and v_2 are the speeds of the two vehicles.

A high TTC indicates a safer situation, as there is more time before a potential collision. We use SUMO simulator's in-built TTC calculation module[3] to extract the TTC values. SUMO reports 'conflict' events where the TTC is low enough to be of concern. We use 4 s [14] as the threshold, below which TTC values are of concern. Each 'conflict' event can be considered a concerning encounter. Different encounters may include the same vehicles but at different points in their trajectories.

4 Results and Discussion

In order to evaluate our metrics under 'live' simulation settings, we first train our model on a dataset obtained by simulating a traffic intersection in SUMO[4] microscopic traffic simulator. The dataset was based on 36 h of simulated data, based on a real-world intersection, calibrated with real-world traffic flows seen during the day. The average vehicle flow is between 900–1100 vehicles per hour. Signal cycle length was set at 90 s with actuated phases. The intersection simulated consists of 4 approaches as shown in Fig. 2. Note that the approach on the upper portion is a 1-way street, where vehicles can only leave the intersection. Hence, only at the other 3 approaches i.e. North-bound (NB), East-bound (EB) and West-bound (WB) approaches can the vehicles approach the intersection, stop (if required) and then cross the intersection. There are multiple lanes on these approaches:

- NB: 2 lanes, that allow Through/Right (TR) and only Left (L) turns.
- WB: 3 lanes, that allow Through/Right (TR), only Through (T) and only Left (L) turns.
- EB: 3 lanes, that allow Through/Right (TR), only Through (T) and only Left (L) turns.

[3] https://sumo.dlr.de/docs/Simulation/Output/SSM_Device.html.
[4] https://sumo.dlr.de/docs/SUMO_at_a_Glance.html.

Table 2. Table of Results for Improved Model. Note that different TTC Encounters can include the same vehicles but at different points in their trajectories.

Trajectory Cluster	Total Count	Red Light Violation	Mid-Intersection Stoppage	Pre-Stopbar Stoppage	TTC Encounters
L on EBL	102	0 (0.0%)	0 (0.0%)	69 (67.6%)	139
L on NBL	133	0 (0.0%)	1 (0.8%)	47 (35.3%)	207
L on WBL	127	25 (19.7%)	0 (0.0%)	1 (0.8%)	109
R on EBTR	99	0 (0.0%)	0 (0.0%)	0 (0.0%)	139
R on NBTR	115	0 (0.0%)	0 (0.0%)	20 (17.4%)	118
R on WBTR	106	69 (65.1%)	0 (0.0%)	3 (2.8%)	57
T on EBT	57	0 (0.0%)	0 (0.0%)	35 (61.4%)	36
T on EBTR	61	0 (0.0%)	0 (0.0%)	7 (11.5%)	85
T on NBTR	111	0 (0.0%)	0 (0.0%)	15 (13.5%)	123
T on WBT	36	13 (36.1%)	0 (0.0%)	0 (0.0%)	12
T on WBTR	79	45 (57.0%)	0 (0.0%)	0 (0.0%)	22
Total	1026	152 (14.8%)	1 (0.0%)	197 (19.2%)	1047

While a vehicle on 'only Through' lane can go only straight, one on the 'Through/Right' may go straight or take a right turn. This leads to 11 possible trajectory clusters, based on which direction and lane the vehicle approaches the intersection, and then which turn the vehicle makes. For example, a trajectory cluster could be of a West-bound vehicle on the 'Through/Right' lane, deciding to take a right turn: R on WBTR. A vehicle can only belong to one cluster.

This intersection is a real-world intersection in an urban area of North American city, bordering a large university campus. The simulation model has been calibrated based on flow rates, speeds etc.

Once the generative model has been trained to convergence (with reasonable accuracy based on pre-existing metrics in Table 1), we freeze the weights. The PyTorch[5] model is coupled with SUMO simulator via TRACI interface. The first 2 s of ground-truth trajectory is provided for each vehicle as it enters the simulation, as the initial condition to start the model prediction. Then the model predicts the rest of the trajectory till the vehicle exits the simulation [17]. We use 4000 s (1 h 6 min) of simulation data, for 1026 vehicles in total. Multiple vehicles may enter the scene at any given time, and the model must unroll all the trajectories simultaneously. Once the predicted trajectories are unrolled, simulation logs from SUMO are collected. MovingPandas library [4] was used for calculating the metrics.

We train our novel 'Improved Model' in which we have incorporated an intersection-based vehicle position encoder (generalizable across any intersection geometry) as well as the traffic signal state, while predicting future trajectories.

[5] pytorch.org.

Table 3. Table of Results for Baseline Model. Note that different TTC Encounters can include the same vehicles but at different points in their trajectories.

Trajectory Cluster	Total Count	Red Light Violation	Mid-Intersection Stoppage	Pre-Stopbar Stoppage	TTC Encounters
L on EBL	102	37 (36.3%)	0 (0.0%)	13 (12.7%)	55
L on NBL	133	3 (2.3%)	1 (0.8%)	66 (49.6%)	202
L on WBL	127	13 (10.2%)	35 (27.6%)	3 (2.4%)	118
R on EBTR	99	49 (49.5%)	0 (0.0%)	4 (4.0%)	37
R on NBTR	115	18 (15.7%)	1 (0.9%)	27 (23.5%)	175
R on WBTR	106	0 (0.0%)	0 (0.0%)	29 (27.4%)	89
T on EBT	27	15 (55.6%)	0 (0.0%)	0 (0.0%)	9
T on EBTR	91	8 (8.8%)	0 (0.0%)	4 (4.4%)	65
T on NBTR	111	4 (3.6%)	2 (1.8%)	43 (38.7%)	208
T on WBT	61	9 (14.8%)	8 (13.1%)	13 (21.3%)	49
T on WBTR	54	0 (0.0%)	0 (0.0%)	12 (22.2%)	34
Total	1026	156 (15.2%)	47 (4.5%)	214 (20.8%)	1041

The 'Baseline' model (based on [17]) only incorporates signal timing information. The original Trajectron++ paper [15] incorporates neither, and is not included in the analysis.

We tabulate the results for the Improved Model in Table 2 and for the Baseline Model in Table 3, stratified by trajectory clusters. We see that both these models show significant Red Light Violation (around 15 %) and Pre-Stopbar Stoppage (around 20 %). However, the 'Improved Model' virtually eliminates all Mid-Intersection Stoppage violations. These are incredibly dangerous, as stopping in the middle of the intersection can lead to extreme accidents, due to the high speeds involved. The Time-To-Collision Encounters of concern remain significantly high. These phenomena would not have been captured by the existing metrics (Table 1). With these new metrics, we are able to detect and characterize previously unstudied issues.

5 Conclusion

In this work, we trained two models on vehicle trajectories near an intersection and then evaluated them in an unseen 'live' simulation scenario. We calculated the intersection-aware metrics: Red Light Violation, Mid-Intersection Stoppage, Pre-Stopbar Stoppage, and Time-To-Collision Encounters. The 'Improved Model' included a novel intersection-based vehicle position encoding, which virtually eliminated Mid-Intersection Stoppage events. However, significant issues remain; they pose a serious public hazard and need to be fixed. We hope that the findings in this work, along with the metrics presented, will help researchers and policy-makers deploy safer and more robust multi-vehicle trajectory models.

References

1. Bharilya, V., Kumar, N.: Machine learning for autonomous vehicle's trajectory prediction: a comprehensive survey, challenges, and future research directions. arXiv preprint arXiv:2307.07527 (2023)
2. Bishop, C.M.: Mixture Density Networks, Technical Report, Aston University, Birmingham (1994)
3. Chen, C., Pourkeshavarz, M., Rasouli, A.: Criteria: a new benchmarking paradigm for evaluating trajectory prediction models for autonomous driving. In: 2024 IEEE International Conference on Robotics and Automation (ICRA), pp. 8265–8271 (2024). https://doi.org/10.1109/ICRA57147.2024.10610911
4. Graser, A.: MovingPandas: efficient structures for movement data in Python. GIForum J. Geogr. Inf. Sci. **7**(1), 54–68 (2019). https://doi.org/10.1553/giscience2019_01_s54
5. Gupta, A., Anpalagan, A., Guan, L., Khwaja, A.S.: Deep learning for object detection and scene perception in self-driving cars: survey, challenges, and open issues. Array **10**, 100057 (2021). https://api.semanticscholar.org/CorpusID:233562996
6. Hagenus, J., Mathiesen, F.B., Schumann, J.F., Zgonnikov, A.: A survey on robustness in trajectory prediction for autonomous vehicles. arXiv preprint arXiv:2402.01397 (2024)
7. Hochreiter, S., Schmidhuber, J.: Long short-term memory. Neural Comput. **9**(8), 1735–1780 (1997). https://doi.org/10.1162/neco.1997.9.8.1735
8. Huang, R., Xue, H., Pagnucco, M., Salim, F., Song, Y.: Multimodal trajectory prediction: a survey. arXiv preprint arXiv:2302.10463 (2023)
9. Huang, R., Xue, H., Pagnucco, M., Salim, F., Song, Y.: Multimodal trajectory prediction: a survey. arXiv preprint arXiv:2302.10463 (2023)
10. Huang, Y., Du, J., Yang, Z., Zhou, Z., Zhang, L., Chen, H.: A survey on trajectory-prediction methods for autonomous driving. IEEE Trans. Intell. Veh. **7**(3), 652–674 (2022). https://doi.org/10.1109/TIV.2022.3167103
11. Ivanovic, B., Pavone, M.: The Trajectron: probabilistic multi-agent trajectory modeling with dynamic spatiotemporal graphs. arXiv preprint arXiv:1810.05993 (2019)
12. Jain, A., Zamir, A.R., Savarese, S., Saxena, A.: Structural-RNN: deep learning on spatio-temporal graphs. arXiv preprint arXiv:1511.05298 (2016)
13. Pagnoni, A., Liu, K., Li, S.: Conditional variational autoencoder for neural machine translation. arXiv preprint arXiv:1812.04405 (2018)
14. Ramezani-Khansari, E., Moogehi, S., Moghadas Nejad, F.: Comparing time to collision and time headway as safety criteria. Pamukkale Univ. J. Eng. Sci. **27**(6), 669–675 (2020). https://doi.org/10.5505/pajes.2020.79837
15. Salzmann, T., Ivanovic, B., Chakravarty, P., Pavone, M.: Trajectron++: dynamically-feasible trajectory forecasting with heterogeneous data. arXiv preprint arXiv:2001.03093 (2021)
16. Sengupta, R., Karnati, Y., Rangarajan, A., Ranka, S.: TQAM: temporal attention for cycle-wise queue length estimation using high-resolution loop detector data. In: 2021 IEEE International Intelligent Transportation Systems Conference (ITSC), pp. 3313–3320 (2021). https://doi.org/10.1109/ITSC48978.2021.9564900
17. Wu, A., Ranjan, Y., Sengupta, R., Rangarajan, A., Ranka, S.: A data-driven approach for probabilistic traffic prediction and simulation at signalized intersections. In: 2024 IEEE Intelligent Vehicles Symposium (IV), pp. 3092–3099 (2024). https://doi.org/10.1109/IV55156.2024.10588424

18. Zhao, H., et al.: TNT: target-driven trajectory prediction. arXiv preprint arXiv:2008.08294 (2020)
19. Zhao, S., Song, J., Ermon, S.: InfoVAE: information maximizing variational autoencoders. arXiv preprint arXiv:1706.02262 (2018)

Trustworthiness

Inversion Triplet - A Contrastive Backdoor Mitigation Method for Self-Supervised Vision Encoders

Hiep Vo[1](✉), Zhiyi Tian[1], Chenhan Zhang[2], Xi James Zheng[2], and Shui Yu[1]

[1] Faculty of Engineering and Information Technology, University of Technology Sydney, Sydney, Australia
{hiep.k.vo,zhiyi.tian}@student.uts.edu.au, shui.yu@uts.edu.au
[2] School of Computing, Macquarie University, Sydney, Australia
{chenhan.zhang,james.zheng}@mq.edu.au

Abstract. Traditional mitigation approaches for backdoored neural networks, such as unlearning and finetuning, face significant challenges when applied to self-supervised learning (SSL) models. Unlearning through neuron pruning often causes catastrophic loss of legitimate features, while finetuning fails to effectively update backdoor neurons due to their dormant nature during clean sample processing. Both approaches distort the learned feature distributions due to the lack of label supervision in SSL settings. We propose Inversion Triplet, a novel mitigation method that leverages contrastive learning principles to overcome these limitations. Our key insight is that by treating clean samples as positive pairs and backdoored samples as negative pairs in a triplet loss formulation, we can isolate and eliminate poison feature clusters while preserving the semantic structure of clean representations. This approach requires neither white-box model access nor label supervision, making it practical for real-world deployment. Extensive experiments demonstrate that Inversion Triplet reduces attack success rates compared to existing methods while maintaining best model utility on clean data.

Keywords: Self-Supervised Learning · Backdoor Mitigation · Trojan Detection · Contrastive Learning

1 Introduction

Self-supervised learning (SSL) models have achieved remarkable success in computer vision through contrastive learning approaches [3,5,9]. However, these models are vulnerable to backdoor attacks that can be triggered to produce malicious behaviors while maintaining normal performance on clean inputs. Recent works [4,17] have demonstrated that attackers can implant backdoors during the SSL training phase by inserting trigger patterns that link to target distributions, compromising the model's security.

X. Wu et al. (Eds.): PAKDD 2025, LNAI 15875, pp. 277–289, 2025.
https://doi.org/10.1007/978-981-96-8295-9_20

Effective backdoor defense typically requires three key stages: detecting the presence of backdoors, identifying their trigger patterns, and mitigating their effects [18]. While DECREE [7] has established state-of-the-art methods for the first two stages in SSL vision models through trigger inversion techniques, the critical final stage of mitigation remains a significant challenge. The key difficulty lies in removing backdoor effects while preserving the model's learned representations after backdoor detection and identification:

Finetuning approaches [14,16] attempt to adapt models to poisoned samples through additional training. However, as demonstrated by Fowl et al. [8], these methods significantly degrade model performance on clean data in SSL settings. This degradation occurs because backdoor neurons remain dormant during clean sample processing, providing no gradient signals for effective updates.

Machine unlearning techniques [13,15] aim to selectively remove the influence of poisoned training data. However, Du et al. [6] show that without label supervision, these methods often result in "catastrophic unlearning," disrupting the model's learned patterns and significantly distorting feature distributions.

To address these limitations, we propose Inversion Triplet, a novel mitigation method that complements existing backdoor detection methods for SSL models. Building upon DECREE's successful trigger inversion capabilities, our approach focuses specifically on the critical mitigation stage. Our key insight is that contrastive learning principles, which have proven effective for SSL training, can be repurposed for backdoor mitigation. By leveraging triplet loss formulation, we treat clean samples as positive pairs and DECREE-identified backdoored samples as negative pairs, enabling our method to selectively eliminate poison feature clusters while preserving the semantic structure of clean representations. This approach provides a natural solution for SSL models, as it maintains the same contrastive learning principles used during their original training. While our method can serve as a plug-in mitigation module for any backdoor detection method, we demonstrate its effectiveness in conjunction with DECREE, the current state-of-the-art detection approach for pre-trained SSL vision encoders, showing significant improvements over traditional mitigation strategies.

The main contributions of our work are:

- A novel contrastive mitigation method that leverages triplet loss to preserve clean feature distributions during backdoor removal, addressing the fundamental limitations of existing approaches in SSL settings.
- A theoretically motivated approach that explicitly maintains clean feature consistency by treating poison samples as negative pairs, rather than modifying clean distributions to match poison samples.
- Comprehensive empirical validation demonstrating remarkable reduction in attack success rates while maintaining high clean data accuracy, significantly outperforming traditional mitigation methods.

We organize the remainder of the paper as follows: Sect. 2 reviews backdoor attacks and defenses in SSL. Section 3 further clarifies our research motivation. Section 4 details our Inversion Triplet method. Section 5 provides experimental validation. Section 6 discusses limitations and future work.

2 Related Work

2.1 Backdoor Attacks in Self-Supervised Learning

The success of self-supervised learning (SSL) models has been accompanied by increased vulnerability to backdoor attacks that can compromise their learned representations. Recent works have demonstrated that attackers can inject malicious triggers that significantly impact model behavior while remaining stealthy [12,17].

The BadEncoder attack [10] represents a particularly sophisticated threat, establishing connections between trigger patterns and target class representations that persist through downstream tasks. This attack creates strong feature-space associations that traditional mitigation approaches struggle to break. While other attacks like SSLBackdoor [17] and PoisonedEncoder [12] have also been proposed, BadEncoder's robust feature-space manipulation makes it an important benchmark for evaluating mitigation strategies.

2.2 Limitations of Existing Mitigation Approaches

Current approaches to mitigating backdoor attacks in SSL models primarily fall into two categories: finetuning and unlearning. However, both face fundamental limitations when applied to self-supervised learning contexts.

Limitations of Finetuning. Finetuning approaches attempt to adapt models through adversarial training on mixed clean and poisoned samples. Recent work by Bai et al. [2] and Addepalli et al. [1] has proposed various enhancements to this approach. However, as identified in comprehensive evaluations [8,11], finetuning suffers from two critical limitations in SSL settings:

1) **Feature Distribution Distortion**: Without label supervision, finetuning significantly distorts the learned feature distributions of clean samples, degrading model utility.
2) **Ineffective Neuron Updates**: Backdoor neurons remain dormant during clean sample processing, providing no gradient signals for effective weight updates. This leaves the backdoor largely intact while damaging clean feature representations.

Limitations of Unlearning. Machine unlearning approaches aim to selectively remove the influence of poisoned training data. While methods like gradient ascent [13] and crafted poison samples [15] have shown promise in supervised settings, they face severe challenges in SSL contexts:

1) **Catastrophic Forgetting**: Without label supervision to guide the unlearning process, these methods often remove both malicious and beneficial features, severely degrading model performance [11].

2) **Feature Entanglement**: In SSL models, poison features often share significant overlap with clean features in the representation space. Traditional unlearning approaches lack mechanisms to disentangle these features, leading to complete feature destruction rather than selective removal [11].

These limitations highlight the need for mitigation approaches specifically designed for SSL settings that can: - Preserve clean feature distributions while removing backdoor effects - Selectively target backdoor features without damaging clean representations - Operate effectively without requiring label supervision

Our proposed Inversion Triplet method addresses these challenges by leveraging contrastive learning principles to maintain clean feature consistency while explicitly pushing away poison features through triplet loss. This approach provides a more principled solution for backdoor mitigation in SSL models compared to existing methods.

3 Research Motivation

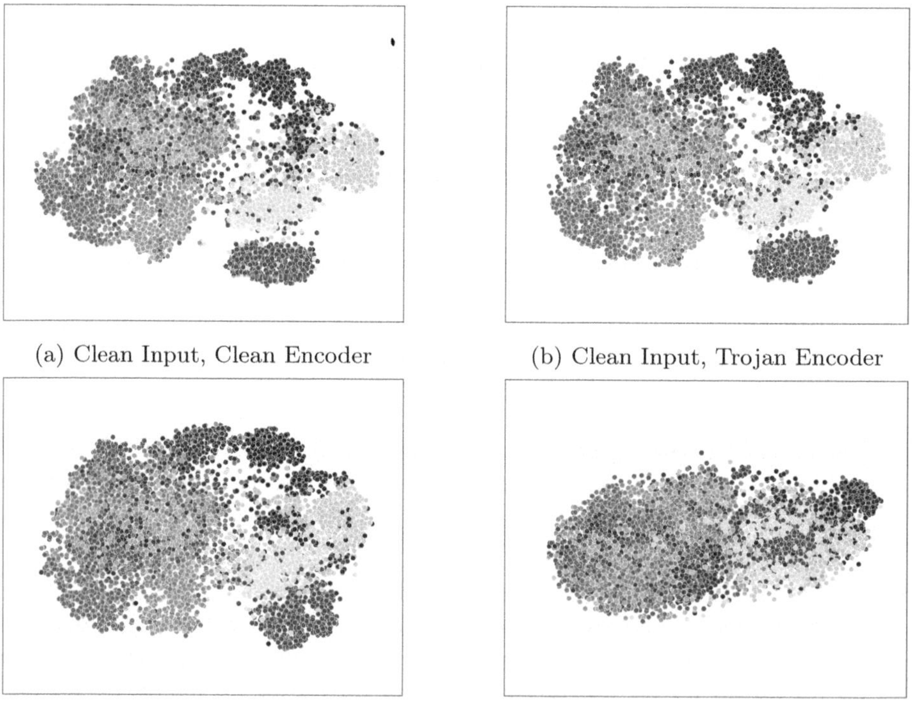

(a) Clean Input, Clean Encoder

(b) Clean Input, Trojan Encoder

(c) Poison Input, Clean Encoder

(d) Poison Input, Trojan Encoder

Fig. 1. Embedding Space distribution comparison. When comparing the pairs, poison embeddings show denser clustering in the center with trojaned encoder.

3.1 Threat Model

The attacker aims to inject backdoors into pre-trained image encoders that cause targeted misclassification in downstream tasks while maintaining normal performance on clean inputs. We assume the attacker has access to a clean pre-trained encoder and can inject a small portion ($\leq 5\%$) of poisoned samples into the training set during the backdoor injection phase. However, the attacker's capabilities are constrained by having no access to downstream datasets or training processes, and being limited to using static, patch-based triggers while maintaining model utility (accuracy drop $< 2\%$ on clean inputs).

From the defender's perspective, our model assumes black-box access to the potentially compromised encoder and limited access to either a small subset of the pre-training dataset or data from a similar distribution. The defender must detect and mitigate backdoors without any knowledge of attack targets or trigger patterns, and without access to downstream task labels or datasets.

3.2 Intuition Behind Contrastive Mitigation

The fundamental challenge in backdoor mitigation for SSL models lies in selectively removing backdoor effects while preserving clean feature distributions. To understand why contrastive learning provides an effective solution, we analyze the feature embedding spaces of clean and trojaned encoders (Fig. 1dcba).

In clean encoders, both clean and poisoned inputs form distinct, well-separated clusters in the feature space, maintaining proper semantic relationships. However, in trojaned encoders, while clean inputs retain their natural distribution, inputs stamped with the backdoor trigger coalesce into a dense, centralized cluster. This centralization occurs because the backdoor attack forces all triggered inputs to map to a specific target representation, regardless of their original semantic content.

Traditional mitigation approaches like finetuning and unlearning struggle with this scenario because:

- Finetuning attempts to modify all feature distributions to accommodate poison samples, disrupting clean feature relationships
- Unlearning often removes both poison and clean features due to their proximity in the embedding space

Our key insight is that contrastive learning principles provide a natural solution to this challenge. By leveraging triplet loss with clean samples as positive pairs and poison samples as negative pairs, we can:

- Explicitly push poison features away from clean feature clusters
- Maintain semantic relationships between clean samples
- Preserve the original distribution of clean features

This approach aligns with how SSL models originally learn representations through contrastive learning, making it particularly suitable for backdoor mitigation in SSL contexts while avoiding the feature distortion issues.

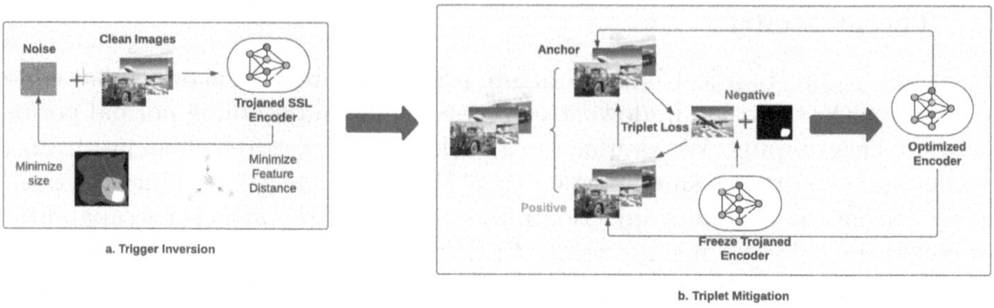

Fig. 2. Inversion Triplet workflow: (a) Trigger Inversion using DECREE [7] for backdoor detection and trigger reconstruction, and (b) Triplet Mitigation where pairs of clean images serve as positive examples while clean images with inverted triggers serve as negative examples, enabling selective backdoor removal while preserving clean feature distributions.

4 Inversion Triplet

4.1 Overview

Our Inversion Triplet method provides a principled approach to backdoor mitigation in SSL models by leveraging contrastive learning principles. As illustrated in Fig. 2, the method consists of two main stages: (a) trigger reconstruction using DECREE [7], and (b) contrastive mitigation using triplet loss. The key innovation lies in our triplet mitigation stage, which maintains clean feature distributions while selectively removing backdoor effects.

4.2 Trigger Reconstruction Stage

We leverage DECREE's trigger inversion capability to identify potential backdoor patterns. Given a potentially compromised encoder E and a shadow dataset $\mathcal{X}$ with similar distribution to the training data, DECREE reconstructs a trigger pattern that approximates the behavior of the attacker's trigger. While this reconstructed trigger may not exactly match the original, it captures the essential feature-space manipulation that characterizes the backdoor attack. This reconstructed trigger serves as the foundation for our mitigation strategy.

4.3 Contrastive Mitigation

The core of our method lies in the triplet mitigation stage, which uses contrastive learning to selectively remove backdoor effects while preserving clean features. We designate E_{opt} as the encoder being mitigated and maintain E_{poi} as a frozen copy to ensure consistent poison features. For each clean image x, we create positive pairs using two augmented versions (x_u, x_v) of the clean image, while negative examples are formed by combining clean images with the inverted trigger $(x_c + t)$.

4.4 Loss Function Design

Our mitigation approach leverages two complementary loss functions that work together to remove backdoor effects while preserving clean feature representations. The first component is a triplet loss that explicitly pushes backdoored samples away from clean feature distributions:

$$\mathcal{L}_{\text{triplet}} = \sum_{x \in \mathcal{X}} \max\Bigg(0, \|E_{opt}(x_u) - E_{opt}(x_v)\|_2^2 - \|E_{opt}(x_u) - E_{poi}(x_c + t)\|_2^2 + \epsilon\Bigg) \tag{1}$$

This triplet loss formulation minimizes the distance between clean sample pairs while ensuring backdoored samples are pushed away by at least margin ϵ. Using a frozen copy E_{poi} ensures stable negative examples during training, and the entire process works without requiring class labels, making it suitable for SSL settings.

The second component is the InfoNCE loss, which maintains the model's original representation learning capability:

$$\mathcal{L}_{\text{InfoNCE}} = -\mathbb{E}\left[\log \frac{\exp(sim(x_u, x_v)/\tau)}{\sum_{k=1}^{N} \exp(sim(x_u, x_k)/\tau)}\right] \tag{2}$$

This loss preserves the core contrastive learning objective used in SSL training, maintains discriminative power for downstream tasks, and prevents feature collapse during mitigation. The temperature parameter τ provides control over the concentration of feature distributions.

The total loss combines these objectives:

$$\mathcal{L}_{\text{total}} = \mathcal{L}_{\text{InfoNCE}} + \lambda \mathcal{L}_{\text{triplet}} \tag{3}$$

This dual objective method provides several key advantages over traditional mitigation approaches. Unlike finetuning, which modifies all feature distributions indiscriminately, our approach selectively pushes away only backdoor features while preserving clean features. In contrast to unlearning methods that often cause catastrophic forgetting, our InfoNCE component explicitly maintains useful feature representations. The approach operates without requiring class labels, making it particularly suitable for SSL settings. Additionally, the frozen E_{poi} and margin-based triplet loss provide stable optimization targets, while requiring only forward passes through the frozen encoder for negative examples reduces computational overhead.

5 Performance Evaluation

5.1 Experimental Setup

We evaluate our mitigation method on three datasets following the BadEncoder [10] benchmark setup: STL-10 (10 classes, 96×96 RGB images), GTSRB (43 classes of traffic signs), and SVHN (digit recognition). The encoders are pre-trained on CIFAR-10, which contains 50,000 training and 10,000 test images of 32×32 color images across 10 classes.

For fair comparison with existing mitigation methods, we use a shadow dataset containing 10% of the pre-training dataset size, representing a realistic scenario where defenders have limited access to training data. This shadow dataset is used both for trigger reconstruction using DECREE and for the subsequent mitigation process.

We compare our approach against two baseline mitigation methods:

- **Finetuning** [18]: Train on mixed clean (50%) and poisoned (50%) samples
- **Unlearning** [13]: Gradient ascent on poisoned samples

Performance is evaluated using two metrics: - Backdoored Accuracy (BA): Model utility on clean samples - Attack Success Rate (ASR): Percentage of successful backdoor attacks.

5.2 Experimental Results

Feature Space Analysis

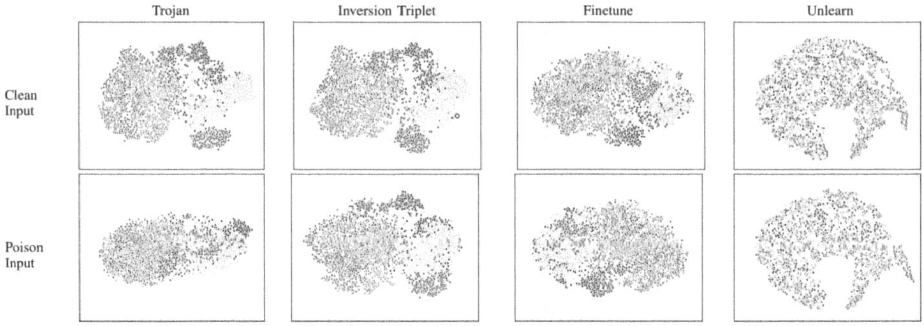

Fig. 3. Embedding Distribution of Mitigation Strategies.

To understand how different mitigation strategies affect learned representations, we analyze the embedding space distributions before and after mitigation. Figure 3 visualizes the feature embeddings of clean and poisoned inputs across four scenarios: the original trojaned encoder and its optimized versions using Inversion Triplet, Finetuning, and Unlearning.

In the original trojaned encoder, clean samples maintain distinct, well-separated clusters while poisoned samples form a dense, centralized cluster - the hallmark of a successful backdoor attack. This centralization occurs because the backdoor forces all triggered inputs to map to a specific target representation, regardless of their original semantic content.

The three mitigation strategies exhibit markedly different effects on these feature distributions:

Finetuning attempts to accommodate poisoned samples by adapting clean feature distributions, resulting in clean features being pulled toward the poison cluster center. While this reduces ASR, it significantly compromises the model's ability to maintain semantic relationships between clean samples, leading to decreased classification accuracy.

Unlearning produces a chaotic feature distribution where both clean and poisoned features are severely distorted. This "catastrophic forgetting" occurs because poison samples share crucial semantic features with clean samples, and aggressive removal of poison features inevitably damages clean representations.

In contrast, **Inversion Triplet** achieves two key objectives:

1) Preserves the original clustering structure of clean features, maintaining semantic relationships learned during pre-training
2) Successfully breaks apart the centralized poison cluster while guiding poisoned features back toward their natural distributions

This analysis quantitatively demonstrates why our approach outperforms existing methods: it selectively removes backdoor effects while preserving the fundamental feature relationships that make SSL models effective. The results highlight the importance of maintaining clean feature distributions during mitigation, especially in SSL contexts where these learned representations directly impact downstream task performance.

Quantitative Performance Analysis

We conduct comprehensive experiments to evaluate our method's effectiveness across multiple downstream tasks. Figure 4 presents the Attack Success Rate (ASR) and Backdoored Accuracy (BA) during training

Attack Success Rate Reduction. Starting from an initial ASR of 99.32% in the compromised models, all mitigation strategies achieve significant reductions, but with varying degrees of effectiveness and stability. Inversion Triplet demonstrates consistent performance across all datasets, reducing ASR to 6.40% on GTSRB, 13.46% on STL-10, and 46.17% on SVHN. In contrast, Finetuning shows inconsistent performance, with particularly poor results on SVHN (64.67% ASR) and doubled ASR compared to our method on STL-10. Unlearning, while achieving low ASRs on SVHN and STL-10, exhibits unstable behavior with a high ASR of 32.69% on GTSRB.

Model Utility Preservation. While reducing ASR is crucial, maintaining model utility on clean data is equally important. Our method uniquely maintains clean data performance, reaching the baseline model accuracy (74.18%) on both GTSRB and STL-10 datasets. This stands in stark contrast to Finetuning, which shows moderate degradation in clean accuracy, and Unlearning, which suffers severe performance drops below 40% accuracy.

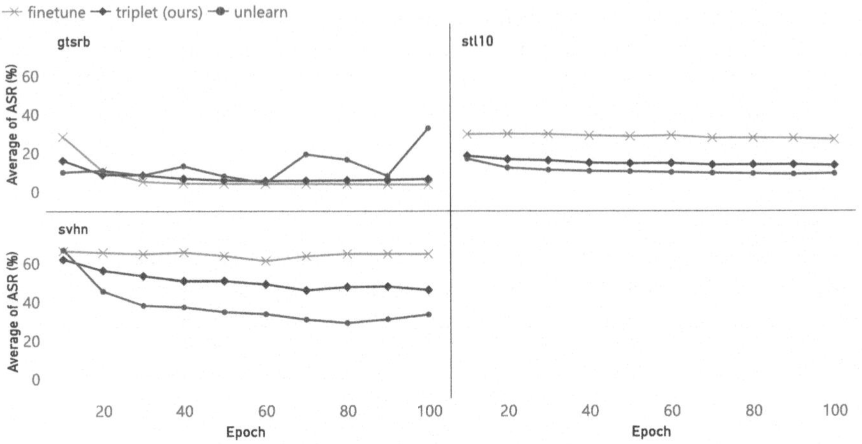

(a) ASR on downstream datasets

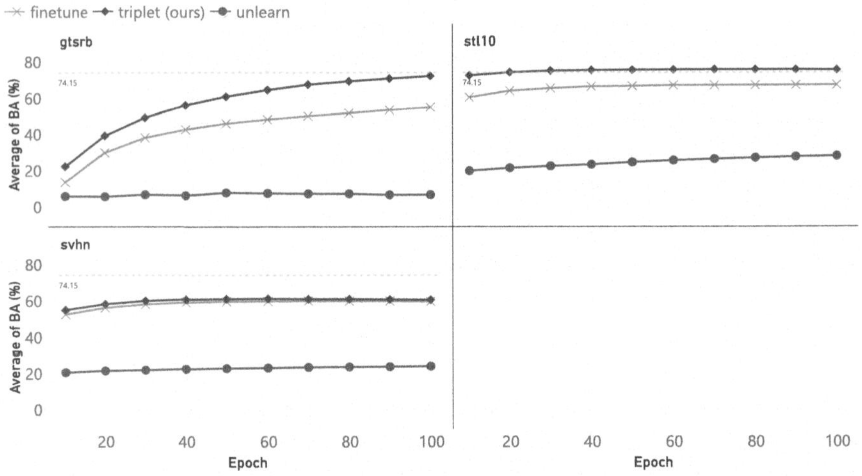

(b) BA on downstream datasets

Fig. 4. Performance comparison of mitigation strategies on downstream datasets.

The superior performance of our method on STL-10 compared to GTSRB and SVHN can be attributed to domain similarity - both CIFAR-10 (pre-training dataset) and STL-10 focus on object detection, while GTSRB and SVHN represent different visual domains (traffic signs and digits respectively). This observation highlights an important consideration for deployment: mitigation effectiveness is influenced by the relationship between pre-training and downstream task domains.

Training Stability Analysis. Our method demonstrates stable convergence behavior across all datasets, maintaining consistent performance throughout training. This stability is particularly notable compared to Unlearning's fluctuating performance on GTSRB and Finetuning's variable effectiveness across datasets. The consistent performance can be attributed to our contrastive learning approach, which provides stable optimization objectives through the combination of triplet and InfoNCE losses.

These results quantitatively demonstrate the key advantages of our approach:

- Consistent ASR reduction across different downstream tasks
- Superior preservation of clean data performance
- Stable training dynamics and reliable convergence
- Robustness to domain differences between pre-training and downstream tasks

6 Conclusion and Future Work

We have presented Inversion Triplet, a novel mitigation method for backdoored SSL encoders that addresses the limitations of existing approaches. Our method uniquely preserves clean feature distributions during mitigation by leveraging contrastive learning principles. Through comprehensive experiments, we demonstrate that our approach provides superior stability and effectiveness compared to traditional methods like finetuning and unlearning.

While our results are promising, several important directions remain for future work:

- **Advanced Attack Scenarios**: Our current approach focuses on patch-based triggers. Extending the method to handle pervasive triggers and stealthy poisoning attacks would increase its practical applicability.
- **Domain Adaptation**: Performance varies with domain similarity between pre-training and downstream tasks. Investigating domain adaptation techniques could improve robustness across different application domains.
- **Broader SSL Applications**: The principles demonstrated here for contrastive learning could be extended to other SSL paradigms such as autoencoders and generative models.

These limitations point to exciting research directions that could further advance the field of secure self-supervised learning. The success of our contrastive learning-based approach suggests that leveraging SSL principles themselves may be key to developing more robust defenses against emerging threats.

Acknowledgments. This work is partially supported by Australia ARC LP190100676.

References

1. Addepalli, S., Jain, S., Venkatesh Babu, R.: Efficient and effective augmentation strategy for adversarial training. In: Koyejo, S., Mohamed, S., Agarwal, A., Belgrave, D., Cho, K., Oh, A. (eds.) Advances in Neural Information Processing Systems, vol. 35, pp. 1488–1501. Curran Associates, Inc. (2022). https://proceedings.neurips.cc/paper_files/paper/2022/file/09d22e4155aa4fdadf3dac8c6bd940fe-Paper-Conference.pdf
2. Bai, T., Luo, J., Zhao, J., Wen, B., Wang, Q.: Recent advances in adversarial training for adversarial robustness. arXiv preprint arXiv:2102.01356 (2021)
3. Caron, M., et al.: Unsupervised learning of visual features by contrasting cluster assignments. In: Advances in Neural Information Processing Systems, vol. 33, pp. 9912–9924 (2020)
4. Chen, H., Fu, C., Zhao, J., Koushanfar, F.: ProFlip: targeted trojan attack with progressive bit flips. In: Proceedings of the IEEE/CVF International Conference on Computer Vision (ICCV), pp. 7718–7727 (2021)
5. Chen, T., Kornblith, S., Norouzi, M., Hinton, G.: A simple framework for contrastive learning of visual representations. In: Daumé III, H., Singh, A. (eds.) Proceedings of the 37th International Conference on Machine Learning. Proceedings of Machine Learning Research, vol. 119, pp. 1597–1607. PMLR (2020)
6. Du, M., Chen, Z., Liu, C., Oak, R., Song, D.: Lifelong anomaly detection through unlearning. In: Proceedings of the 2019 ACM SIGSAC Conference on Computer and Communications Security, pp. 1283–1297. CCS '19, Association for Computing Machinery, New York, NY, USA (2019). https://doi.org/10.1145/3319535.3363226
7. Feng, S., et al.: Detecting backdoors in pre-trained encoders. In: Proceedings of the IEEE/CVF Conference on Computer Vision and Pattern Recognition (CVPR), pp. 16352–16362 (2023)
8. Fowl, L., Goldblum, M., Chiang, P.Y., Geiping, J., Czaja, W., Goldstein, T.: Adversarial examples make strong poisons. In: Ranzato, M., Beygelzimer, A., Dauphin, Y., Liang, P., Vaughan, J.W. (eds.) Advances in Neural Information Processing Systems, vol. 34, pp. 30339–30351. Curran Associates, Inc. (2021). https://proceedings.neurips.cc/paper_files/paper/2021/file/fe87435d12ef7642af67d9bc82a8b3cd-Paper.pdf
9. He, K., Fan, H., Wu, Y., Xie, S., Girshick, R.: Momentum contrast for unsupervised visual representation learning (2020)
10. Jia, J., Liu, Y., Gong, N.Z.: BadEncoder: backdoor attacks to pre-trained encoders in self-supervised learning. In: IEEE Symposium on Security and Privacy (2022)
11. Kaviani, S., Sohn, I.: Defense against neural trojan attacks: a survey. Neurocomputing **423**, 651–667 (2021). https://doi.org/10.1016/j.neucom.2020.07.133
12. Liu, H., Jia, J., Gong, N.Z.: PoisonedEncoder: poisoning the unlabeled pre-training data in contrastive learning. In: 31st USENIX Security Symposium (USENIX Security 22), pp. 3629–3645. USENIX Association, Boston, MA (2022). https://www.usenix.org/conference/usenixsecurity22/presentation/liu-hongbin
13. Liu, Y., et al.: Backdoor defense with machine unlearning. In: IEEE INFOCOM 2022–IEEE Conference on Computer Communications, pp. 280–289. IEEE Press (2022). https://doi.org/10.1109/INFOCOM48880.2022.9796974
14. Madry, A., Makelov, A., Schmidt, L., Tsipras, D., Vladu, A.: Towards deep learning models resistant to adversarial attacks. In: International Conference on Learning Representations (2018). https://openreview.net/forum?id=rJzIBfZAb

15. Marchant, N.G., Rubinstein, B.I.P., Alfeld, S.: Hard to forget: poisoning attacks on certified machine unlearning. In: AAAI Conference on Artificial Intelligence (2021). https://api.semanticscholar.org/CorpusID:237562940
16. Pang, T., Yang, X., Dong, Y., Su, H., Zhu, J.: Bag of tricks for adversarial training. In: International Conference on Learning Representations (2021). https://openreview.net/forum?id=Xb8xvrtB8Ce
17. Saha, A., Tejankar, A., Koohpayegani, S.A., Pirsiavash, H.: Backdoor attacks on self-supervised learning. In: 2022 IEEE/CVF Conference on Computer Vision and Pattern Recognition (CVPR), pp. 13327–13336 (2021). https://api.semanticscholar.org/CorpusID:233207068
18. Wang, B., et al.: Neural cleanse: identifying and mitigating backdoor attacks in neural networks. In: 2019 IEEE Symposium on Security and Privacy (SP), pp. 707–723 (2019). https://doi.org/10.1109/SP.2019.00031

Beyond Uniformity: Robust Backdoor Attacks on Deep Neural Networks with Trigger Selection

Shixiong Li[1], Xingyu Lyu[1], Ning Wang[2], Tao Li[3], Danjue Chen[4], and Yimin Chen[1](✉)

[1] Miner School of Computer and Information Sciences, UMass Lowell, Lowell, USA
{shixiong_li,xingyu_lyu,ian_chen}@uml.edu
[2] Department of Computer Science and Engineering, University of South Florida, Tampa, USA
ningw@usf.edu
[3] Department of Computer and Information Technology, Purdue University, West Lafayette, USA
li4270@purdue.edu
[4] Department of Civil, Construction, and Environmental Engineering, NC State University, Raleigh, USA
dchen33@ncsu.edu

Abstract. Backdoor attacks have been extensively explored in recent years which attack deep neural networks (DNNs) by poisoning their training set and causing targeted mis-classification. Research on such attacks is critical for today's widespread applications based on DNNs due to their low-cost and high efficacy. While many backdoor attacks have been proposed, they usually rely on using a static and fixed trigger for attacks, which not only lacks adaptability but also renders them easier to detect. To address such a limitation, we introduce `OpenTrigger` in this paper, a novel backdoor attack framework employing dynamic triggers for enhancing attack flexibility and robustness. Unlike traditional approaches that rely on a single fixed trigger, our proposed attack learns a generalized consistent feature across a built trigger pool, hence enabling even the use of unseen triggers during testing that differ from those used during training. Extensive experiments across multiple datasets and model architecture confirm the high effectiveness and robustness of `OpenTrigger` against state-of-the-art and even adaptive backdoor defenses, establishing it as a versatile and practical backdoor attack strategy.

Keywords: Backdoor Attacks · Dynamic Triggers · Adaptive Defense

1 Introduction

Deep Neural Networks (DNNs) are being widely deployed for various machine learning tasks such as image classification, owing to their exceptional learning

X. Wu et al. (Eds.): PAKDD 2025, LNAI 15875, pp. 290–302, 2025.
https://doi.org/10.1007/978-981-96-8295-9_21

and generalization capacity. However, many studies show that they are highly vulnerable towards low-cost backdoor attacks [1], whereas an attacker embeds a hidden "trigger"–such as a specific pattern or characteristic–within the victim model (hence creating a "backdoor") by introducing poisoned samples into the training dataset. After the poisoned model is deployed, such a trigger in any input will cause it to behave as how the attacker wants, often misclassifying the input. Given that many models could be trained from unverified third-party or public data, backdoor attacks thus pose significant security risk, particularly in security-critical applications like malware detection and autonomous driving.

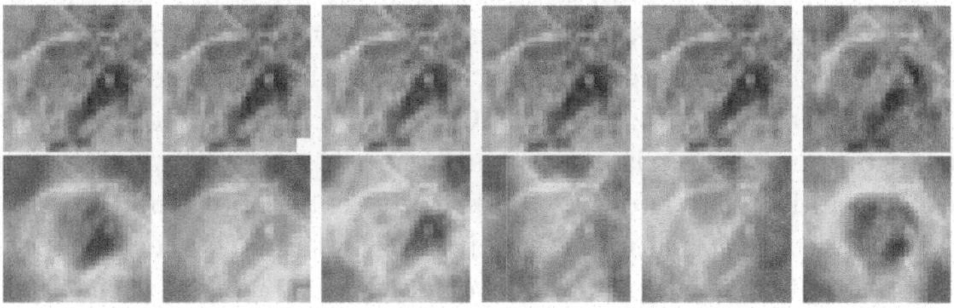

Clean BadNets[2] LF[3] FTrojan[4] CTRL[5] **Ours**

Fig. 1. Visualizing poisoned samples of different backdoor attacks including ours using Grad-CAM [6].

As expected, the design of trigger is the most crucial for any backdoor attack and correspondingly defense against them. The ideal design is two-folded: the trigger needs to be sufficiently easy to memorize by the victim model in order to result in a high attack success rate (ASR) while the trigger also needs to be dynamic enough so that the poisoned samples would not be easily detected. Though a lot has been done exploring backdoor attacks particularly in image domain [2], the triggers used are found to be mostly **static**, resulting in high ASR but low resilience towards detection. For instance, Fig. 1 shows the localization map of the poisoned samples of the same clean one under different backdoor attacks including ours using Grad-CAM [6]. The figure suggests that existing attacks tend to result in poisoned samples that are significantly different from the one without attack, rendering them easier to detect.

To address such a limitation, in this paper, we propose `OpenTrigger`, a novel backdoor attack on DNNs that utilizes dynamic triggers from a trigger pool for attacks rather than sticking to the same trigger as in earlier work. Our goals for `OpenTrigger` is that it is *low-cost*, *effective*, and *robust* (i.e., resilient towards existing defenses). To that end, `OpenTrigger` first adopts the popular image blend strategy from existing attacks [7] for trigger planting while we expect other strategies will work on `OpenTrigger` as well. The adopted strategy is straightforward; an attacker blends the trigger t, which is nothing but a selected image, into a clean image x to generate a poisoned sample $\tilde{x}$. Next, the key difference in `OpenTrigger` is that the attacker samples a small set of images from the same category (e.g., images of 'panda') as a trigger pool Ω by optimization and selects a t for a given x to generate the corresponding $\tilde{x}$. In doing so, we aim to exploit the consistent feature across Ω as the trigger rather than the single fixed one, hence improving attack versatility and stealthiness. As illustrated in Fig. 1, the localization map under our attack is much more similar to that without any attack, suggesting that `OpenTrigger` is expected to be more robust.

Intuitively, using different ts in Ω rather than a fixed t as a trigger introduces significant challenges. Meanwhile, our attack even allows to use unseen triggers to activate the planted backdoor during test phase. We summarize our contributions when solving such challenges and validating the effectiveness of `OpenTrigger` as follows. **1.** We propose `OpenTrigger`, a novel backdoor attack that learns dynamic and generalized backdoor triggers rather than a fixed trigger. **2.** We improve the effectiveness of `OpenTrigger` by designing a trigger selection method based Particle Swarm Optimization (PSO). **3.** We perform comprehensive evaluations, performance comparison, and ablation study for `OpenTrigger` across various datasets and victim models. The results show that `OpenTrigger` consistently achieves high ASR without sacrificing prediction accuracy for clean inputs. Most importantly, our attack performs well under state-of-the-art (SOTA) backdoor defenses. We evaluate our attack even under a strong adaptive defense as well, confirming its robustness towards strong defenses. Our code is available at https://github.com/Shixiong-Li/openTrigger.

2 Related Work

Backdoor Attacks. In the literature, backdoor attacks are often categorized into two types: *sample-agnostic* and *sample-specific*, depending on whether the embedded trigger differs from one sample to another. The majority of backdoor attacks are sample-agnostic ones. For example, BadNets [2] inserted a white square placed in the bottom-right corner of any poisoned image as the trigger. Other examples include [4,8–10], where different types of static triggers were used for backdoor attacks. Meanwhile, sample-specific attacks are to generate an individual trigger for a given victim sample and are usually based on generative models [11]. As a result, these attacks tend to be expensive and less effective while being stealthier. There are other ways of categorizing backdoor attacks as well. For instance, depending on whether the attacker changes the label of a poisoned sample or not, backdoor attacks can be divided into *dirty-label* [2] and *clean-label* attacks [8], among which dirty-label attacks are far more common.

Our proposed attack falls into sample-specific and dirty-label attack category. The main advantage in `OpenTrigger` is that it selects highly effective triggers via lightweight optimization and even unseen triggers can activate the planted backdoor during test phase, thus rendering the attack versatile and stealthy.

Backdoor Defenses. In general, backdoor defenses consist of two lines: backdoor detection and robust training. The underlying principle behind backdoor detection is that clean samples without a trigger and poisoned samples with the trigger are separable in certain embedding space. Following this idea, researchers had exploited different methods to find such an embedding space for backdoor detection, including training data sanitization [12] and Gram matrices [13]. Meanwhile, researchers had explored other characteristics of backdoor samples for detection such as [3,14,15]. Another line of defense focuses on enhancing model robustness during training in order to mitigate the impact of backdoor attacks. For instance, NAD [16] relies on aligning the attention maps for robust training. BNP [17] isolates backdoor neurons by leveraging batch normalization statistics while EP [17] detects abnormal entropy patterns in pre-activation distributions. ASD [18] adaptively separates clean and backdoor samples in different periods of training process. Other methods [19] looked into alternative defenses. In this paper, we use recent popular defenses to evaluate the robustness of `OpenTrigger`.

3 Methodology

3.1 Threat Model and Notions

Threat Model. In this paper, we adopt the standard threat model from the literature. First, our proposed attack is a **black-box** one, meaning that the attacker has no control over the training process or knowledge of the victim model. Second, the attacker can insert a small number of poisoned samples labeled with a targeted class into the training set of the victim model. This can be achieved by the victim model obtaining training samples either from unverified crowd-sourcing or simply from compromised public dataset. Meanwhile, the three goals of our attack are: *(1) Functionality preservation*: A backdoor attack should have little impact on the prediction accuracy of on clean samples to remain as stealthy as possible. *(2) Effectiveness*: A backdoor sample should be misclassified as the targeted class with a high probability, that is, a high ASR. *(3) Robustness*: A backdoor attack should remain effective even when existing popular defenses are deployed.

Notions. For simplicity, we introduce the main notions used in this paper as below. We use x, t, and $\tilde{x}$ to denote a clean sample, a backdoor trigger, and the corresponding poisoned (backdoored) sample, respectively. Assume that we adopt the image blend strategy $\mathcal{P}(\cdot,\cdot)$ for trigger planting, we then have

$$\tilde{x} = \mathcal{P}(x,t) = \alpha \cdot t + (1-\alpha) \cdot x, \tag{1}$$

where $\alpha \in [0,1]$ is the blend ratio. $\alpha = 0$ corresponds to no attack on x.

Regarding datasets, at the attacker side, we use $D_{a,\mathrm{clean}}$ and D_{poison} to denote the small clean dataset and the corresponding poisoned dataset, respectively. That is, the attacker obtains D_{poison} by poisoning $D_{a,\mathrm{clean}}$ through `OpenTrigger`. Assume the set of triggers, i.e., the trigger pool, is Ω. We further divide Ω into two non-overlapping trigger sets by a ratio of $\rho_t : (1-\rho_t)$, i.e., Ω_{train} and Ω_{test}, dedicated for training and test phases accordingly. Similarly, at the victim model side, we use D_{train} to denote the large poisoned training dataset.

As a result, $\rho_p = \frac{|D_{\text{poison}}|}{|D_{\text{train}}|}$ is the poisoning ratio. Regarding models, we denote the parameters of the substitute model and the victim model by θ_s and θ_v, respectively. Correspondingly, we have $y = f_s(x; \theta_s)$ and $y = f(x; \theta_v)$, where y is the output prediction. We use *target* to denote the target label of our attack.

3.2 Attack Overview

Figure 2 illustrates the workflow of `OpenTrigger`, which consists of the following five steps. **Step 1:** Building a trigger pool. The attacker builds a trigger pool Ω with the help of a substitute model θ_s and a custom PSO algorithm. The attacker further divides Ω into Ω_{train} and Ω_{test} dedicated to training and test phases, respectively. **Step 2:** Trigger planting for training. To obtain $\tilde{x}$ for a given $x \in D_{a,\text{clean}}$, the attacker randomly selects a t from Ω_{train} and computes $\tilde{x}$ following Equation (1). In the end, the attacker obtains D_{poison}. **Step 3:** Victim model training. The victim model θ is trained from D_{train} which the attacker inserts D_{poison} into. **Step 4:** Trigger planting for test. Similarly, during test phase, the attacker obtains $\tilde{x}$ for a given x by randomly selecting a t from Ω_{test} and computing $\tilde{x}$ with Equation (1). **Step 5:** Testing backdoor attack. The attacker uses clean and poisoned samples to evaluate the attack performance. Note that Ω_{train} and Ω_{test} are *two non-overlapping sets*, meaning that the triggers used for training and test phase are completely different. On the contrary, most prior backdoor attacks would instead use the same trigger for both phases, which many existing defenses rely on for detecting backdoor samples.

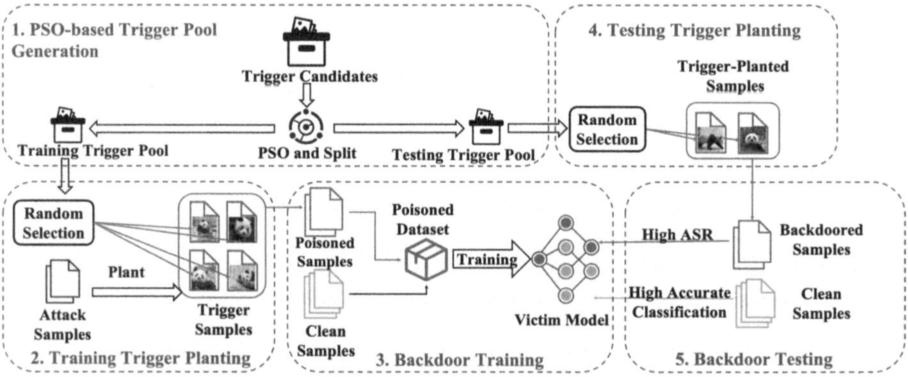

Fig. 2. The workflow of our proposed `OpenTrigger` attack.

3.3 Building a Trigger Pool

In general, our attack relies on the trigger pool Ω rather than a single trigger for planting the backdoor so as to improve its versatility and stealthiness. The main challenges of building Ω are that: **1.** the victim model θ_v needs to learn the mapping relationship between all ts in Ω_{train} and the target label $target$, and **2.** any t in Ω_{test} can trigger the planted backdoor in θ_v and lead to targeted misclassification with a high probability. To that end, we incorporate two methods to address such challenges: training a substitute model θ_s for θ and PSO to obtain the trigger pool Ω. Specifically, we start the process by first crawling images from the same category from the Internet and then refining them down to Ω through PSO. In our experiments, we use images of 'panda' as our candidates.

PSO Iterations. Particle Swarm Optimization (PSO) [20] has been known as an effective gradient-free optimization algorithm. As shown below, we implement a custom PSO to refine crawled images and build Ω. The search process of a general PSO is to select the best K particles from a swarm of M particles according to a penalty function while the M particles follow the standard Ho-McKayrate kinetics model. When applying PSO to our proposed attack, we treat each candidate trigger, i.e., a crawled image of 'panda', as a particle while the end result from PSO, i.e., the best K particles, is the refined Ω we use as our trigger pool. In `OpenTrigger`, we use the following loss as the penalty function, which is to evaluate the effectiveness of a given t as a backdoor trigger:

$$\text{loss}_t = \sum_{x \in D_{poison}} \text{CrossEntropy}(f_s(x;\theta_s), target), \tag{2}$$

where CrossEntropy$(\cdot,\cdot)$ computes cross-entropy loss and $target$ is the target label. A smaller loss_t indicates that $\tilde{x}$ embedded with trigger t has a higher probability to result in mis-classification as $target$, hence the more effective t is.

Our PSO works as follows. Denote the size of Ω and the set of crawled panda images by K and M, respectively. We also use Ω_K to denote the best trigger set of the current round. First, we randomly initialize a set of particles including their positions t_i and velocity v_i. t_i of a particle, i.e., a selected image, is initialized by its index among all M particles. After initialization, we proceed to iteratively update the particles selected for T rounds, adjusting their velocity and positions based on the adopted kinetics model and penalty function. As pointed out in [8], small noise can help to have the backdoor model rely more on the trigger for prediction without damaging the original features of the poisoned sample. Therefore, after we narrow down to Ω_K, we continue to use PSO to search for the optimal Gaussian noise level for each t in Ω_K. Finally, we obtain Ω, i.e., the output Ω_K from PSO, which is used to generate D_{poison}.

Substitute Model. Equation (2) suggests that we need a substitute model θ_s to evaluate how effective a trigger is. Note that the attacker is assumed to only have access to D_{poison} rather than D_{train}. Here we use the training outcome from the initial few epochs from a training subset for such a purpose. That is, we simply train θ_s from D_{poison} and use it for Equation (2).

Table 1. ASR comparison between `OpenTrigger` and baselines.

Attack	Dataset									
	CIFAR-10		CIFAR-100		GTSRB		TinyImageNet		Average	
	ASR	ACC	ASR	ACC	ASR	ACC	ASR	ACC	ASR	ACC
BadNets	89.52	92.35	83.75	56.73	90.96	92.55	**99.99**	39.30	91.06	70.23
Blend	99.31	93.76	95.58	59.78	99.19	93.41	98.13	39.85	98.05	71.70
Input-Aware	93.67	91.10	89.69	57.54	82.63	90.48	97.23	38.55	90.81	69.42
LF	97.29	93.51	89.68	58.96	93.67	90.87	93.97	39.25	93.65	70.65
WaNet	93.39	91.45	1.60	57.54	92.84	79.85	94.34	41.78	70.54	67.66
Bpp	99.60	91.32	0.67	58.71	91.61	91.10	99.41	38.67	72.82	69.95
FTrojan	**100.00**	93.47	12.22	55.38	**100.00**	93.14	99.88	41.67	78.03	70.92
CTRL	**100.00**	93.57	**99.99**	58.43	**100.00**	92.87	99.90	40.45	**99.97**	71.33
(**Ours**)	**100.00**	93.70	81.66	58.75	98.14	90.81	**99.99**	34.49	94.95	69.44

4 Performance Evaluation

4.1 Experimental Setup

Datasets, Models, and Hyper-Parameters. We conduct evaluations on CIFAR-10, CIFAR-100, GTSRB, and TinyImageNet datasets, using PreactResNet18, VGG19, DenseNet161, and MobileNetV3-Large as the corresponding victim model, respectively. By default, $\alpha = 0.3$, $\rho_t = 0.8$, $K = 50$, and $\rho_p = 5\%$.

Evaluation Metrics. In the experiments, we use the following four metrics for evaluating `OpenTrigger`: *(1) Model accuracy (ACC)*, i.e., the prediction accuracy of the victim model, i.e., θ_v, on clean test samples; *(2) Attack success rate (ASR)*, i.e., the proportion of backdoor samples misclassified into the target class; *(3) True Positive Rate (TPR)*, i.e., the proportion of backdoor samples correctly detected by a backdoor detection method while lower TPR indicates the higher robustness of a backdoor attack; and *(4) False Positive Rate (FPR)*, i.e., the proportion of clean samples misclassified as backdoor samples while lower FPR indicates the better performance of a detection method.

Baselines. We include the following backdoor attacks as baselines: BadNets [2], Blend [7], Input-Aware [11], LF [3], WaNet [21], Bpp [22], FTrojan [4] and CTRL [5]. Note that in our experiments, all configurations of detection methods and baseline attacks are the same as their default settings in [1].

Defenses. Here we evaluate `OpenTrigger` against three detection methods, i.e., Spectre [12], Beatrix [13], and Asset [14], three robust training schemes, i.e., NAD [16], BNP [17], and EP [17], and finally a recent adaptive defense, ASD [18], which aims to adaptively separate clean and backdoor samples.

Table 2. Robustness under backdoor detection (CIFAR-10 and CIFAR-100)

Datasets →	CIFAR-10								CIFAR-100							
Detection →	Spectre		Beatrix		Asset		Average		Spectre		Beatrix		Asset		Average	
Metrics →	TPR	FPR	TPR	FPR	TPR	FPR	TPR	FPR	TPR	FPR	TPR	FPR	TPR	FPR	TPR	FPR
BadNets	7.76	7.49	96.32	0.65	35.84	64.58	46.64	24.24	1.40	7.93	39.84	0.78	100.00	30.19	47.08	12.97
Blend	2.76	7.75	0.20	1.43	2.44	57.09	1.80	22.09	**1.24**	7.93	0.16	0.32	100.00	56.54	33.80	21.60
Input-Aware	5.70	7.69	1.09	2.66	23.83	47.93	10.21	19.43	27.99	5.49	9.85	1.41	13.59	28.02	17.14	11.64
LF	0.52	7.87	0.04	1.46	63.96	31.05	21.51	13.46	1.44	7.92	**0.08**	0.57	96.96	10.24	32.83	6.24
WaNet	7.76	7.46	4.01	2.44	21.96	23.18	11.24	11.03	10.11	7.19	2.06	0.76	36.06	5.46	16.08	4.47
Bpp	6.07	7.73	3.03	2.47	40.80	60.18	16.63	23.46	10.48	7.13	2.49	0.78	0.50	0.10	4.49	2.67
FTrojan	**0.04**	7.89	**0.00**	1.23	**1.72**	40.41	**0.59**	16.51	1.52	7.92	8.40	0.89	**0.00**	**0.00**	3.31	2.94
CTRL	0.08	7.89	**0.00**	1.39	89.64	53.59	29.91	20.96	1.52	7.92	100.00	0.41	100.00	61.29	67.17	23.21
Ours	**0.04**	7.87	0.08	1.49	3.28	38.21	1.13	15.86	1.52	7.92	0.40	0.45	**0.00**	**0.00**	**0.64**	2.79

Table 3. Robustness under backdoor detection (GTSRB and TinyImageNet)

Datasets →	GTSRB						TinyImageNet							
Detection →	Beatrix		Asset		Average		Spectre		Beatrix		Asset		Average	
Metrics →	TPR	FPR	TPR	FPR	TPR	FPR	TPR	FPR	TPR	FPR	TPR	FPR	TPR	FPR
BadNets	1.68	2.39	88.93	10.87	45.31	6.63	**0.00**	0.08	10.18	7.58	100.00	45.30	36.73	17.65
Blend	**0.36**	2.38	4.23	8.11	2.30	5.25	0.64	7.97	39.42	6.86	100.00	48.10	46.69	20.98
Input-Aware	3.51	4.99	22.39	14.90	12.95	9.95	3.17	8.06	37.55	6.77	63.37	45.25	34.70	20.03
LF	6.28	3.56	20.46	10.56	13.37	7.06	0.66	7.97	1.40	7.68	100.00	41.90	34.02	19.18
WaNet	4.62	4.34	10.03	12.28	7.33	8.31	4.74	8.07	6.84	9.32	72.24	44.90	27.64	20.76
Bpp	3.55	3.78	19.60	11.26	11.58	7.52	4.21	8.15	5.34	7.59	67.34	51.00	25.63	22.25
FTrojan	20.77	2.16	31.43	4.08	26.10	3.12	**0.00**	0.08	1.02	7.45	100.00	48.40	33.67	18.64
CTRL	45.72	2.22	45.60	10.04	45.66	6.13	**0.00**	0.08	0.24	7.70	100.00	45.38	33.41	17.72
Ours	3.21	3.84	**0.46**	9.17	**1.84**	6.51	0.16	7.99	**0.00**	8.39	**0.00**	5.08	**0.05**	7.15

4.2 Attack Effectiveness

We summarize ASR and ACC of our attack and baselines across various datasets in Table 1. Note that the higher ASR and ACC, the better backdoor attack. As seen from the results, `OpenTrigger` is able to achieve high ASR across all datasets investigated, demonstrated its wide applicability. In comparison to the baselines, `OpenTrigger` achieves slightly lower yet comparable attack performance with CTRL, which is the best among all backdoor attacks in our experiments. As expected, the adverage ASR and ACC of `OpenTrigger` are lower than those of Blend simply because Blend used the same fixed trigger for all victim samples during training and test phase. As a result, the victim model learns the feature of the fixed trigger in Blend better. However, as will be shown later, our proposed `OpenTrigger` demonstrates much better resilience towards backdoor defenses than Blend does, showcasing the advantage of our trigger planting strategy.

Overall, the experimental results indicate the effectiveness of our attack in most settings, rendering it a lightweight and effective backdoor attack.

4.3 Attack Robustness Under Defenses

Attack Under Detection Methods. Here we present the TPR and FPR results of three popular detection methods when deployed to detect backdoor attacks. The results on CIFAR-10 and CIFAR-100 are summarized in Table 2 while those on GTSRB and TinyImageNet in Table 3. Note that we mainly focus on TPR performance while the lower TPR, the better attack resilience against the detection methods. Meanwhile, since Spectre is currently inapplicable to non-iid datasets, we did not evaluate it on GTSRB. As can be seen from both tables, the results indicate that `OpenTrigger` achieves consistently high robustness against the three detection methods investigated. For instance, our proposed attack achieves the lowest average TPR on CIFAR-100, GTSRB, and TinyImageNet, outperforming all baselines. Even on CIFAR-10, `OpenTrigger` achieves the second lowest TPR. Particularly, our attack achieves lower TPR than Blend consistently, which clearly demonstrates that using dynamic triggers instead of the same one enhances attack robustness against backdoor defenses.

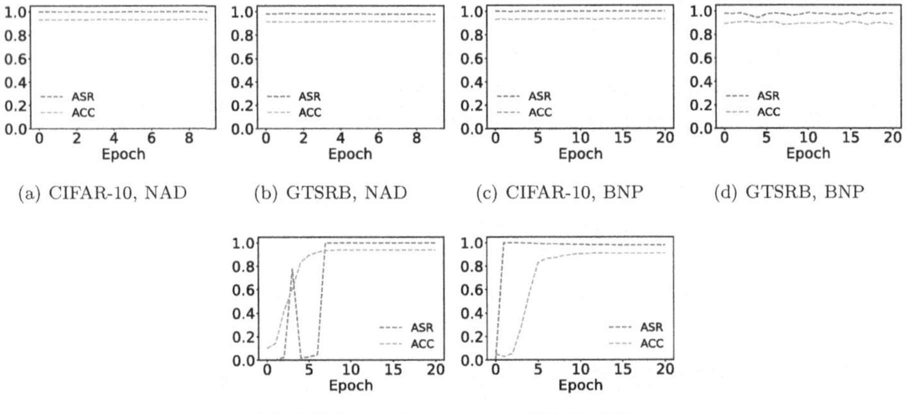

(a) CIFAR-10, NAD (b) GTSRB, NAD (c) CIFAR-10, BNP (d) GTSRB, BNP

(e) CIFAR-10, EP (f) GTSRB, EP

Fig. 3. Robustness of `OpenTrigger` under NAD [16], BNP [17], and EP [17].

Attack Under Robust Training Methods. Here we show the ASR and ACC of `OpenTrigger` when three popular robust training methods are deployed to protect models from backdoor attacks. Note that the higher ASR and ACC, the better backdoor attack. Figure 3a and Fig. 3b show the attack performance of `OpenTrigger` on CIFAR-10 and GTSRB, respectively, when NAD is deployed for backdoor defense. As we can see from the figures, our proposed attack is

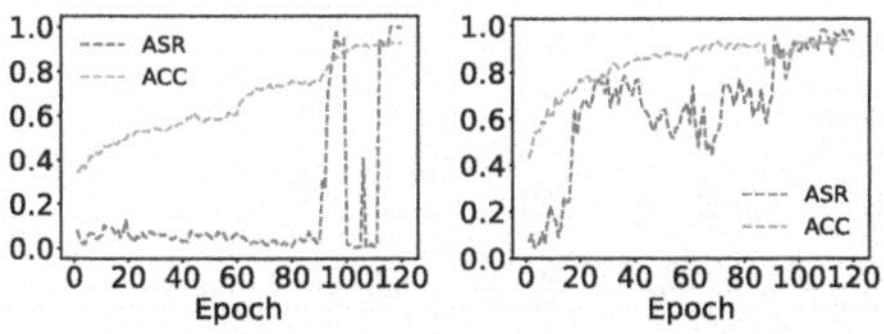

CIFAR-10 GTSRB

Fig. 4. Robustness under ASD [18].

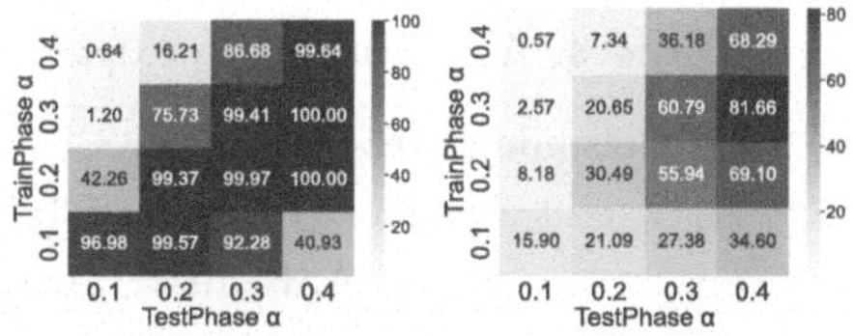

CIFAR-10 CIFAR-100

Fig. 5. Impact of α.

able to achieve high ASR and ACC simultaneously, indicating that NAD cannot defend against the proposed attack. The results of BNP and EP as defenses are illustrated in Fig. 3c, Fig. 3d, Fig. 3e, and Fig. 3f, correspondingly, which are consistent to those of NAD. Our results here confirm that the proposed backdoor attack remains effective even when robust training methods are deployed.

Attack Under an Adaptive Defense. Here we show the ASR and ACC of `OpenTrigger` when the SOTA adaptive backdoor defense, ASD [18], is deployed to protect models from backdoor attacks. Note that we are the only one among all baseline attacks to evaluate attack performance under adaptive defenses. The results on CIFAR-10 and GTSRB are illustrated in Fig. 4, highlighting that our proposed `OpenTrigger` still achieves very high ASR even when ASD is deployed. The strong resilience of `OpenTrigger` towards the adopted adaptive defense further confirms the benefits of using dynamic triggers for planting backdoor, rendering the proposed attack an effective and robust backdoor method.

5 Ablation Study

Impact of Poisoning Rate ρ_p. Intuitively, the higher ρ_p, the more poisoned samples in D_{train}, the higher ASR a backdoor attack can achieve. We summarize the results of ASR and ACC of our proposed attack under different ρ_p on various datasets in Table 4. ASR of our attack increases when ρ_p increases, which is as expected. The results suggest that even with a small ρ_p of 3%, `OpenTrigger` can already achieve a high ASR. When $\rho_p \geq 5\%$, ASR starts to saturate, indicating that a small poisoning rate is sufficient for the proposed attack.

Impact of K. We summarize the results of ASR and ACC of our proposed attack under different K in Table 5. We observe that ASR first increases when K increases. When K exceeds a certain threshold, e.g., 50, ASR decreases slightly when K increases. We anticipate that when K is small, a larger K can result in a higher probability of selecting a better trigger, therefore the increase of ASR. However, when K is large enough, a larger K cannot further increase the probability of getting a better trigger. Instead, a larger K will lead to a more diverse Ω, hence the decrease of ASR. The results suggest that a small Ω, e.g., $K = 50$, is sufficient for maintaining `OpenTrigger` low-cost and effective.

Table 4. ACC and ASR of `OpenTrigger` with different poisoning rates

Poisoning rate	Dataset							
	CIFAR-10		CIFAR-100		GTSRB		TinyImageNet	
	ASR	ACC	ASR	ACC	ASR	ACC	ASR	ACC
No attack	–	93.20	–	59.33	–	89.02	–	39.36
3%	100.00	92.48	71.62	59.66	97.16	90.78	93.99	39.59
5%	100.00	93.70	81.66	58.75	98.14	90.81	99.99	34.49
8%	99.99	93.46	89.81	58.01	99.42	89.24	100.00	40.37
10%	99.99	93.36	89.53	57.63	99.36	90.41	99.99	37.86

Table 5. ACC and ASR of `OpenTrigger` with different K

K	Dataset							
	CIFAR-10		CIFAR-100		GTSRB		TinyImageNet	
	ASR	ACC	ASR	ACC	ASR	ACC	ASR	ACC
20	99.96	93.88	67.40	59.33	91.26	90.43	97.70	39.31
50	100.00	93.70	81.66	58.75	98.14	90.81	99.99	34.49
100	99.99	93.62	81.16	59.08	99.08	90.03	100.00	41.44
200	100.00	93.95	78.41	59.00	99.49	89.41	97.62	38.05

Impact of Blend Rate α. Intuitively, the higher α, the more trigger portion in a poisoned sample, therefore the higher ASR a backdoor attack can achieve. We summarize the results of ASR and ACC of our proposed attack under different α on CIFAR-10 and CIFAR-100 in Fig. 5, where we vary α of training and test phase between 0.1 and 0.4. The figure shows that ASR of our attack tends to increase when α increases, which is as expected. The results also suggest that the optimal α could depend on the specific dataset under attack. Nonetheless, α between 0.2 and 0.3 serves as proper choices for practical backdoor attacks.

6 Conclusion

We developed `OpenTrigger`, a flexible and robust backdoor attack on DNNs that utilizes dynamic triggers instead of a fixed trigger. Our attack is shown to be robust against existing backdoor defenses including recent adaptive ones. As a result, our attack allows that the attacker uses unseen triggers to activate the planted backdoor, which most existing attacks cannot. Extensive experiments and performance comparison confirm the better performance of `OpenTrigger`.

Acknowledgments. This paper is supported in part by the US National Science Foundation under grants CNS-2422863 and CMMI-2401555 and the Office of Naval Research subawarded through Virginia Tech under Grant N00014-24-1-2730. We would like to thank anonymous reviewers for their constructive comments and helpful advice.

References

1. Wu, B., et al.: Backdoorbench: a comprehensive benchmark of backdoor learning. In: Proceedings of the 36th International Conference on Neural Information Processing Systems, NIPS 2022, Curran Associates Inc., Red Hook, NY, USA (2022)
2. Gu, T., Liu, K., Dolan-Gavitt, B., Garg, S.: Badnets: evaluating backdooring attacks on deep neural networks. IEEE Access **7**, 47230–47244 (2019). https://doi.org/10.1109/access.2019.2909068
3. Zeng, Y., Park, W., Mao, Z.M., Jia, R.: Rethinking the backdoor attacks' triggers: a frequency perspective. In: 2021 IEEE/CVF International Conference on Computer Vision (ICCV). IEEE (Oct 2021). https://doi.org/10.1109/iccv48922.2021.01616
4. Wang, T., Yao, Y., Xu, F., An, S., Tong, H., Wang, T.: An Invisible Black-Box Backdoor Attack Through Frequency Domain, pp. 396-413. Springer Nature Switzerland (2022). https://doi.org/10.1007/978-3-031-19778-9_23
5. Li, C., Pang, R., Xi, Z., Du, T., Ji, S., Yao, Y., Wang, T.: An embarrassingly simple backdoor attack on self-supervised learning. In: 2023 IEEE/CVF International Conference on Computer Vision (ICCV), pp. 4344-4355. IEEE (Oct 2023).https://doi.org/10.1109/iccv51070.2023.00403
6. Selvaraju, R.R., Cogswell, M., Das, A., Vedantam, R., Parikh, D., Batra, D.: Grad-cam: visual explanations from deep networks via gradient-based localization. In: 2017 IEEE International Conference on Computer Vision (ICCV), pp. 618-626. IEEE (Oct 2017). https://doi.org/10.1109/iccv.2017.74
7. Chen, X., Liu, C., Li, B., Lu, K., Song, D.: Targeted backdoor attacks on deep learning systems using data poisoning (2017). https://arxiv.org/abs/1712.05526
8. Turner, A., Tsipras, D., Madry, A.: Label-consistent backdoor attacks (2019). https://arxiv.org/abs/1912.02771
9. Singha, A., Bi, Z., Li, T., Chen, Y., Zhang, Y.: Securing contrastive mmwave-based human activity recognition against adversarial label flipping. In: Proceedings of the 17th ACM Conference on Security and Privacy in Wireless and Mobile Networks, WiSec 2024, pp. 31-41. , ACM (May 2024). https://doi.org/10.1145/3643833.3656123
10. Guo, H., Chen, X., Guo, J., Xiao, L., Yan, Q.: Masterkey: practical backdoor attack against speaker verification systems. In: Proceedings of the 29th Annual International Conference on Mobile Computing and Networking, ACM MobiCom 2023. pp. 1-15. ACM (Oct 2023). https://doi.org/10.1145/3570361.3613261
11. Nguyen, T.A., Tran, T.A.: Input-aware dynamic backdoor attack. In: Proceedings of the 34th International Conference on Neural Information Processing Systems, NIPS 2020, Curran Associates Inc., Red Hook, NY, USA (2020)
12. Hayase, J., Kong, W., Somani, R., Oh, S.: Spectre: defending against backdoor attacks using robust statistics. In: Meila, M., Zhang, T. (eds.) Proceedings of the 38th International Conference on Machine Learning. Proceedings of Machine Learning Research, 18–24 Jul, vol. 139, pp. 4129–4139. PMLR (2021). https://proceedings.mlr.press/v139/hayase21a.html
13. Ma, W., Wang, D., Sun, R., Xue, M., Wen, S., Xiang, Y.: The "beatrix" resurrections: robust backdoor detection via gram matrices. In: 30th Annual Network and Distributed System Security Symposium, NDSS (2023)
14. Pan, M., Zeng, Y., Lyu, L., Lin, X., Jia, R.: ASSET: robust backdoor data detection across a multiplicity of deep learning paradigms. In: 32nd USENIX Security Symposium (USENIX Security 23), pp. 2725–2742. USENIX Association, Anaheim, CA (Aug 2023). https://www.usenix.org/conference/usenixsecurity23/presentation/pan

15. Wang, N., Xiao, Y., Chen, Y., Hu, Y., Lou, W., Hou, Y.T.: Flare: defending federated learning against model poisoning attacks via latent space representations. In: Proceedings of the 2022 ACM on Asia Conference on Computer and Communications Security, ASIA CCS 2022, pp. 946-958. ACM (May 2022). https://doi.org/10.1145/3488932.3517395
16. Li, Y., Lyu, X., Koren, N., Lyu, L., Li, B., Ma, X.: Neural attention distillation: erasing backdoor triggers from deep neural networks. In: International Conference on Learning Representations (2021). https://openreview.net/forum?id=9l0K4OM-oXE
17. Zheng, R., Tang, R., Li, J., Liu, L.: Pre-activation distributions expose backdoor neurons. In: Koyejo, S., Mohamed, S., Agarwal, A., Belgrave, D., Cho, K., Oh, A. (eds.) Advances in Neural Information Processing Systems, vol. 35, pp. 18667–18680. Curran Associates, Inc. (2022)
18. Gao, K., Bai, Y., Gu, J., Yang, Y., Xia, S.T.: Backdoor defense via adaptively splitting poisoned dataset. In: 2023 IEEE/CVF Conference on Computer Vision and Pattern Recognition (CVPR). IEEE (Jun 2023). https://doi.org/10.1109/cvpr52729.2023.00390
19. Guo, J., Li, Y., Chen, X., Guo, H., Sun, L., Liu, C.: SCALE-UP: an efficient black-box input-level backdoor detection via analyzing scaled prediction consistency. In: The Eleventh International Conference on Learning Representations (2023). https://openreview.net/forum?id=o0LFPcoFKnr
20. Eberhart, R., Kennedy, J.: A new optimizer using particle swarm theory. In: MHS 1995. Proceedings of the Sixth International Symposium on Micro Machine and Human Science, MHS 1995, pp. 39-43. IEEE. https://doi.org/10.1109/mhs.1995.494215
21. Nguyen, T.A., Tran, A.T.: Wanet - imperceptible warping-based backdoor attack. In: International Conference on Learning Representations (2021). https://openreview.net/forum?id=eEn8KTtJOx
22. Wang, Z., Zhai, J., Ma, S.: Bppattack: stealthy and efficient trojan attacks against deep neural networks via image quantization and contrastive adversarial learning. In: 2022 IEEE/CVF Conference on Computer Vision and Pattern Recognition (CVPR), pp. 15054-15063. IEEE (Jun 2022). https://doi.org/10.1109/cvpr52688.2022.01465

Defense Against Multi-target Multi-trigger Backdoor Attacks

Haripriya Harikumar[1(✉)], Santu Rana[1], Kien Do[1], Sunil Gupta[1], Wei Zong[2], Willy Susilo[2], and Svetha Venkatesh[1]

[1] Applied Artificial Intelligence Institute, Deakin University, Geelong, Australia
haripriyaaharikumar@gmail.com,
{santu.rana,k.do,sunil.gupta,svetha.venkatesh}@deakin.edu.au
[2] University of Wollongong, Wollongong, Australia
{wei_zong,wsusilo}@uow.edu.au

Abstract. Neural backdoor attacks present a critical vulnerability in deep-learning systems. In this work, we introduce a more general form of trigger-based attacks that bypass existing defense methods. To counter this, we propose a novel defense mechanism grounded in an information-theoretic analysis of how backdoors are most efficiently encoded in the feature space. Specifically, we prove that the most information-efficient way to represent multiple trigger-based backdoors is to construct layered manifolds, where each layer corresponds to a unique backdoor trigger and the separation between layers is encoded radially (assuming the usual high-dimensional hyper-spherical data distribution). This structure enables the main classifier to maintain its original classification boundary while accessing backdoors through simple radial separations. Our defense exploits this insight by searching for perturbations that exhibit global behaviour, indicative of the presence of such layered manifolds and, therefore, backdoors. Extensive experiments using a variety of image datasets demonstrate that our method successfully identifies backdoors that are missed by state-of-the-art detection methods.

Keywords: multi-trigger · multi-target · intrinsic backdoor · backdoor attack · backdoor defense

1 Introduction

Deep learning models are susceptible to backdoor attacks [5,7,10,12], where adversaries poison the training data to manipulate the model's predictions when presented with inputs containing a hidden trigger. While such models perform as expected on clean data, they misclassify test inputs embedded with triggers. These triggers are often small, inconspicuous, hidden and cause no issues for human users or other models, making detection highly challenging. Identifying backdoors therefore requires specialised testing methods.

Existing detection methods, however, are constrained by their threat model assumptions. Early approaches such as Neural Cleanse (NC) [15] and Gangsweep

X. Wu et al. (Eds.): PAKDD 2025, LNAI 15875, pp. 303–314, 2025.
https://doi.org/10.1007/978-981-96-8295-9_22

(GS) [16] operate under the assumption that only a small number of classes are poisoned, focusing on identifying anomalous behaviour in those classes. Subsequent methods, including STS [6] (improves scalability) and FreeEagle [3] (reduces data requirements), enhance detection capabilities but still fail when backdoors are widespread, i.e., when many target classes are present.

Some other methods aim to identify triggered samples at test time (e.g., STRIP [2,4]) or purify models to remove backdoors (e.g., NAD [8] and Fine-Pruning [9,11]). However, STRIP is ineffective for small triggers and is generally unattractive due to its high inference cost. Similarly, NAD and Fine-Pruning fail when backdoors are plentiful because the triggers lose their global behaviour, resulting in a lack of unique attention maps (NAD) and a failure to localise activations to a small subset of neurons (Fine-Pruning).

In this paper, we introduce the Multi-Target Backdoor (MTB) attack, a generalised and more practical variant of the classic BadNet [5] attack. Unlike single-trigger, single-target attacks, MTB uses unique triggers to target multiple classes simultaneously. In our attack setup, each target class is assigned a distinct trigger. These triggers can be placed at arbitrary locations within the input image, enhancing the attack's stealth and flexibility while rendering existing detection methods ineffective through these multiple strategies.

To counter the MTB attack, we present a theoretical analysis demonstrating that, when backdoors are learned by the classifier, the most information-efficient encoding involves creating simple radial offsets in the feature space for each trigger. This approach minimises the complexity of the decision boundary, resulting in only a small entropy increase. Empirically, this efficiency is reflected in the low sample complexity required to train a backdoor—often achievable with just a handful of triggered examples.

Building on this insight, we propose a novel detection mechanism called Multi Target backdoor Detection, MTD that performs trigger reverse engineering using only a small number of example images. The method evaluates the global transferability of a reverse-engineered trigger: a trigger is considered valid if it consistently maps other clean images onto the same trigger-specific manifold, following a radial pattern in the feature space.

We evaluate the efficacy of our proposed MTB attack and detection mechanism across five diverse image datasets—MNIST, CIFAR-10, GTSRB, YouTube Face Recognition (YTF), and TinyImagenet—with class sizes ranging from 10 to 1,283. Our detection method is benchmarked against eight state-of-the-art detection and mitigation methods, showcasing its superior performance. Additionally, we demonstrate the adaptability of our detection mechanism against single-target and adaptive backdoor attacks. Our results highlight the robustness and effectiveness of the proposed detection method in addressing both conventional and advanced backdoor threats, particularly in scenarios involving multiple triggers and widespread poisoning.

2 Method

A Deep Neural Network (DNN) can be defined as a parameterized function $f_\theta : \mathcal{X} \rightarrow \mathbb{R}^C$ that recognizes the class label probability of an image $x \in \mathcal{X}$ where $\mathcal{X} \sim P_\mathcal{X}$ and θ represents the learned function's parameter. Let us consider the probability vector for the image x by the function f_θ as $[p_1....p_C]$, thus the class corresponding to x will be $\text{argmax}_{i\in[1..C]} p_i$.

2.1 Multi Target Backdoor Attack

The MTB attack setting has three key features: 1) multiple unique triggers, $[\Delta x_1....\Delta x_N]$; 2) multiple target classes, $[t_1....t_N]$; where $N \leq C$ and 3) trigger can be placed anywhere in the image. We have a unique trigger associated with each target class, i.e. trigger Δx_1 for target class t_1. The following shows the composition method for the trigger and the images.

Definition 1: A trigger is formally defined as a small square image patch Δx of size s that is either physically or digitally overlaid onto the input image x at a location (i^*, j^*) to create a modified image $x^{'}$. Concretely, an image of index k of the dataset x_k is altered into an image $x_k^{'}$ by,

$$x_k'(i,j) = \begin{cases} a & \text{if } i \in [i^*, i^*+s], j \in [j^*, j^*+s] \\ b & \text{elsewhere.} \end{cases} \tag{1}$$

where $a = (1 - \alpha(i^{'}, j^{'}))x_k(i,j) + \alpha(i^{'}, j^{'})\Delta x(i^{'}, j^{'})$, $b = x_k(i,j)$, $(i^{'}, j^{'})$ denote the local location on the patch $(i^{'}, j^{'}) = (i - i^*, j - j^*)$ as defined in [6]. The transparency of the trigger is controlled by the weight, α. This parameter can be considered as a part of the trigger, and we will be inclusively mentioning it as Δx. Meanwhile, the rest of the image is kept the same. In our setting, (i^*, j^*) can be at any place as long as the trigger stays fully inside the image. The backdoor model can be trained with the following loss function:

$$\min_\theta \sum_{i=0}^{N_{\text{train}}} \mathcal{L}\left(f_\theta\left(x_i\right), y_i\right) + \sum_{l=0}^{N} \sum_{j=0}^{N_{\text{backdoor}}} \mathcal{L}\left(f_\theta\left(x_{l,j}'\right), t_{l,j}\right) \tag{2}$$

The first component of the loss function uses the pure training data, the second component uses the trigger implanted subset of training dataset along with their classes modified to their corresponding target classes.

2.2 Analysis of the Backdoor Manifold

We here show the case for a binary classifier but the general idea should transfer to the multi-class classification scenario. We also assume the data being distributed on the high-dimensional feature space that naturally leads to hyperspherical data manifold.

Consider a binary classification problem in a high-dimensional feature space. The data are mapped by a feature extraction function

$$g : \mathcal{X} \to \mathbb{R}^{n+1}$$

onto the surface of a hypersphere S^n of radius r. Assume that Class 1 corresponds to the upper hemisphere S^n_+ (where $x_{n+1} > 0$) and Class 2 to the lower hemisphere S^n_-. Let

$$f : \mathbb{R}^{n+1} \to \{1, 2\}$$

be the original classifier defined on S^n such that

$$f(g(x)) = 1 \quad \text{if} \quad g(x) \in S^n_+ \quad \text{and} \quad f(g(x)) = 2 \quad \text{if} \quad g(x) \in S^n_-.$$

We introduce two global perturbation vectors δ_1 and δ_2 in $\mathbb{R}^{n+1}$ with the property that, 1) For any x from Class 1, $f'(g(x) + \delta_1) = 2$, and 2) For any x from Class 2, $f'(g(x) + \delta_2) = 1$.

Here, f' is the new classifier that must maintain the original classification on unperturbed data ($f'(g(x)) = f(g(x))$) but also implement the class-flipping behaviour induced by δ_1 and δ_2.

Lemma 1 (Radial Perturbation Encoding): If the perturbations δ_1 and δ_2 are encoded as uniform radial expansions of the hypersphere, then representing each perturbed distribution on a separate concentric hypersphere $S^n_{\delta_i}$ of radius $r + \epsilon_i$ preserves the original angular separability while requiring only a minimal increase in entropy.

Proof Sketch: By extending the hypersphere's radius from r to $r + \epsilon_i$, the perturbed data remain uniformly distributed on $S^n_{\delta_i}$. The entropy of a uniform distribution on S^n scales as $\log(r^n)$, so changing the radius from r to $r + \epsilon_i$ increases entropy by approximately

$$n \log\left(1 + \frac{\epsilon_i}{r}\right) \approx n\frac{\epsilon_i}{r}$$

for small ϵ_i.

The classification rule can be extended by adding a simple radial condition without altering the underlying angular decision boundary, maintaining low complexity and minimal additional parameters. Thus, the perturbation is encoded with a negligible increase in information compared to the original classifier f.

Theorem 1 (Minimal Information Encoding): Let f' be a classifier that, in addition to performing the original binary classification on S^n, also encodes the two perturbations δ_1, δ_2 as described above. If f' is required to misclassify Class 1 data under δ_1 and misclassify Class 2 data under δ_2, then the information-theoretically minimal increase in complexity from f to f' is achieved by representing the perturbed data distributions on two separate concentric hyperspheres $S^n_{\delta_1}, S^n_{\delta_2}$ rather than relearning a single decision boundary on S^n.

Proof Sketch:

1. **Concentric Hyperspheres Approach:** By Lemma 1, each perturbation δ_i is represented as a small radial displacement, resulting in perturbed data lying on $S^n_{\delta_i}$. The classifier f' can then be described by a small number of additional parameters (the ϵ_i's) and a piecewise definition of the radial classification rule. The resulting entropy increase is linear in ϵ_i/r and does not require complex angular transformations.
2. **Single Hypersphere Approach:** Attempting to encode both perturbations on the original hypersphere S^n necessitates non-linear adjustments to the decision boundary. This creates overlapping and irregular class partitions that must be resolved by f'. The complexity of describing such a boundary grows significantly, increasing entropy and mutual information well beyond the linear scaling of the concentric hypersphere approach.

Since the additional information content needed in the concentric hyperspheres method is minimal (proportional to ϵ_i/r) and no non-linear parameters need to be introduced, this approach dominates the single-boundary solution in terms of information efficiency. Hence, the concentric hyperspheres configuration is information-theoretically optimal.

2.3 Multi Target Backdoor Detection

We propose a backdoor attack detection for this Multi-Target Backdoor attack called **M**ulti **T**arget backdoor **D**etection as **MTD**. This mechanism is designed to reverse engineer trigger for a small sets of images and later verify its universality using other sets of images. Specifically, as the backdoor manifold is global and resides at an orthogonal shift of the original manifold, we will expect that reverse engineering triggers to be easy and that checking the triggers to be universally effective. The absence of universality should also prove otherwise that is, such a backdoor is not present.

We use the validation dataset of pure images for trigger reverse engineering and transferable trigger detection by splitting it into two separate datasets: a) *Data_ Trigger* - for trigger reverse engineering, and b) *Data_ Transfer* - for checking transferability of the triggers. For each image in the *Data_ Trigger* we find a set of perturbations for each class. We do not need a large number of data in the *Data_ Trigger* as our trigger optimisation is over a single image. Each reverse engineered trigger is then used on the images of *Data_ Transfer* to compute the class-distribution entropies. If a perturbation is a backdoor trigger, it will cause all images in the *Data_ Trigger* set to be classified into its target class, creating a sharply skewed class distribution and resulting in a low entropy value. We provide a mechanism to compute the entropy threshold (Eqn. 7, right side) which a perturbation is termed target trigger. Algorithm 1 gives a step by step process of our proposed backdoor defense, MTD.

Algorithm 1. Multi Target backdoor Detection (MTD).

Inputs : $x \in$ *Data_Trigger*, *Data_Transfer*, C, $f_\theta(.)$, threshold (Eqn. 7, right side).
Outputs: target_class_list, **Boolean** $f_\theta(.)$ *is backdoored*
for each *class* in C **do**
  Compute optimized image for x ie x' for *class* using Eq 4.
  Compute reverse engineered trigger, Δx_{rev} with Eq. 5.
  Compute entropy, $\mathrm{H}(\Delta x_{rev})$ on *Data_Transfer* with Eq 6.
    if $(\mathrm{H}(\Delta x_{rev})) \leq$ threshold **then** #target class detected
      target_class_list.append(*class*)
    end if
end for
if length (target_class_list) ≥ 1 **then**
    $f_\theta(.)$ *is backdoored* = **True**
else
    $f_\theta(.)$ *is backdoored* = **False**
end if

Trigger Reverse Engineering. Given an image, $x \in \mathbb{R}^{Ch\times H\times W}$, where Ch, H, W are the number of channels, height, and width with a target label y, we define $B(x)$ as the mask that only keeps inside pixels active for the optimization i.e.,

$$B(x) = x \odot B, \tag{3}$$

where $\odot$ is the element-wise product and B is a binary matrix. B has a value of 1 across a region $H/4 \times W/4$ across all Ch channels, and can be positioned anywhere, as long as it is fully within the image. Alternatively, one can use a mask of size $H \times W$ with L1-regularization on $B(x)$ (initial $x^{'}$) to control the sparsity level of perturbations. We have shown some analysis based on both types in the experiments. We then minimize the cross-entropy loss between the predicted class for $B(x)$ and the given class y:

$$x^{'} = \mathrm{argmin}_{B(x)} \mathcal{L}\left(f_\theta(B(x)), y\right). \tag{4}$$

We denote the new optimized value of $B(x)$ as $x^{'}$. The reverse engineered trigger which we denote as Δx_{rev} is the difference between $x^{'}$ and $B(x)$:

$$\Delta x_{rev} = x^{'} - B(x). \tag{5}$$

Transferability Detection. To check for transferability, we compute the entropy [14] of the class distribution for each reverse engineered trigger when used on all the images of the *Data_transfer* as follows,

$$\mathrm{H}(\Delta x_{rev}) = -\sum_{i=1}^{C} p_i \log_2(p_i), \tag{6}$$

where $\{p_i\}$ is the class probability for the i-th class for using that perturbation. The entropy of a target trigger will be zero if the attack success accuracy is

100%. However, in real-world situation, we assume a slightly less success rate that leads to a non-zero entropy value. The following lemma shows how to compute an upper bound on the value of this score for the backdoor models in specific settings, which then can be used as a threshold for detecting backdoors.

Lemma 2: Let the accuracy of backdoored model on data with embedded target triggers to be $(1-\delta)$, where $\delta << 1$, and let there be C different classes. If Δx_{rev} is a target trigger then the entropy computed by Eq. 6 will be bounded by

$$\mathrm{H}(\Delta x_{rev}) \leq -(1-\delta) * \log_2(1-\delta) - \delta * \log_2(\frac{\delta}{C-1}). \tag{7}$$

The above lemma can easily be proved by observing that the highest entropy of class distribution in this setting happens when $(1-\delta)$ fraction of the images go to the target class t' and the rest δ fraction of the images gets equally distributed in the remaining $(C-1)$ classes. This entropy score is independent of the type and size of triggers used and is universally applicable. This threshold (right side) computation has been adopted from STS [6].

3 Experiments

Datasets: We evaluate our proposed defense method on five datasets namely, MNIST, CIFAR-10, German Traffic Sign Recognition Benchmark (GTSRB), YouTube Face Recognition (YTF) and TinyImagenet dataset (a modified subset of the original ImageNet dataset). We use PreactResnet-18 [13] and VGG11 as the classifier for CIFAR-10, PreactResnet-18 for GTSRB and Resnet-18 for YTF and TinyImagenet (details in Table 1).

Table 1. Datasets, model architecture, Pure and MTB model accuracies.

Dataset	Input Size	Model	#target class	Pure Model	Pure acc.		Attack success acc.	
				Pure acc.	Pure acc.	4×4		8×8
					4×4	8×8	4×4	8×8
MNIST	1×28×28	2 conv, 2 fc	7	99.53	98.83	99.24	99.76	99.98
CIFAR-10	3×32×32	vgg11	7	91.41	90.81	91.15	100.0	100.0
		PreActResnet18		94.55	93.93	94.39	100.0	100.0
GTSRB	3×32×32	PreActResnet18	30	98.85	98.84	98.76	100.0	100.0
YTF	3×55×47	Resnet18	384	99.70	99.55	99.34	96.73	99.79
TinyImagenet	3×64×64	Resnet18	60	64.69	–	63.01	–	99.99

MTB (Attack) Model Training Setup: We trained the **MTB attack** model using multiple random target classes, each with unique triggers. For most datasets, we used a 70:30 target-to-non-target class ratio (Table 1), but for YTF and TinyImagenet, a 30:70 ratio was adopted to maintain clean accuracy comparable to the pure model. The Trojan data ratio per batch was set to 10% (MNIST,

CIFAR-10), 2% (GTSRB), 0.2% (YTF), and 1% (TinyImagenet). Square triggers of size 4×4 and 8×8 with transparency 1.0 were used, along with a 16×1 trigger to test detection effectiveness against long, irregular triggers. Triggers used random pixel values to demonstrate both attack potency and the robustness of our detection method to unstructured triggers.

MTD (Defense) Setup: We use Adam optimiser with a constant learning rate of 0.01 for trigger reverse engineering with MTD. The number of data we used for the trigger reverse engineering varies between 20 to 128.

Baseline Defense Mechanisms: We use eight state-of-the-art defense and mitigation strategies, i.e., STRIP [4], FinePruning [11], RNP [9], STS [6], NAD [8], Neural Cleanse [15], and FreeEagle [3] to compare the proposed attack. However, STRIP is a testing time defenses, and Finepruing, RNP and mitigation techniques and thus do not admit the same metric of attack model detection techniques as Neural Cleanse, FreeEagle, STS, and MTD. We do not compare with DeepInspect [1] or GangSweep [16] as we believe that it would be unreasonable to train a GAN (used in both) with the small number of trigger reverse-engineering data (20–128) for different datasets.

3.1 Effectiveness of Our Proposed MTB Attack

Accuracy: We define **pure accuracy** as the accuracy on clean data (without triggers) using ground-truth labels and **attack success accuracy** as the accuracy on backdoored data (trigger-added and labelled as target classes). Table 1 reports the mean attack success accuracy across all target classes and the pure accuracy for MTB attack models on different datasets. Results demonstrate that the MTB attack achieves **high attack success accuracy** (~100%) while maintaining **pure accuracy** close to that of a model trained solely on clean data.

Robustness to Misplacement: To further highlight the strengths of the **MTB attack**, we analyse its robustness properties on the CIFAR-10 MTB model (8×8 trigger). Specifically, we evaluate *how the attack success accuracy is impacted by* image translations, which mimic the misplacement of an object detection bounding box. We test the robustness of MTB under slight image window misplacements and compare it against state-of-the-art backdoor attacks: the fixed-location single-target BadNet and the input-specific Input-aware attack.

From Fig. 1(a,b), it is evident that image translations disproportionately affect the baseline attacks. Translating the image upward causes a significant drop in BadNet's [5] attack success accuracy, while translating downward heavily impacts the Input-aware attack [13]. In contrast, the MTB attack demonstrates greater robustness, as its attack success accuracy is only slightly affected under these conditions.

Resistance to Test-Time and Remedial Defense: We investigate the robustness of MTB attack against state-of-the-art *test time defenses* such as STRIP, and *Trojan mitigation mechanisms* such as Fine-pruning. Figure 2 demonstrates the

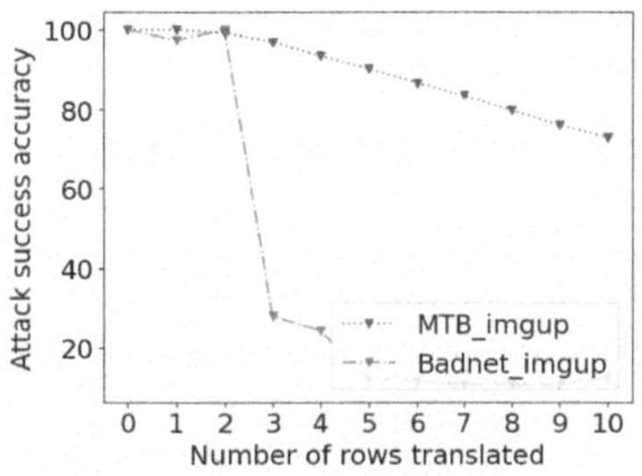

(a) MTB vs BadNet

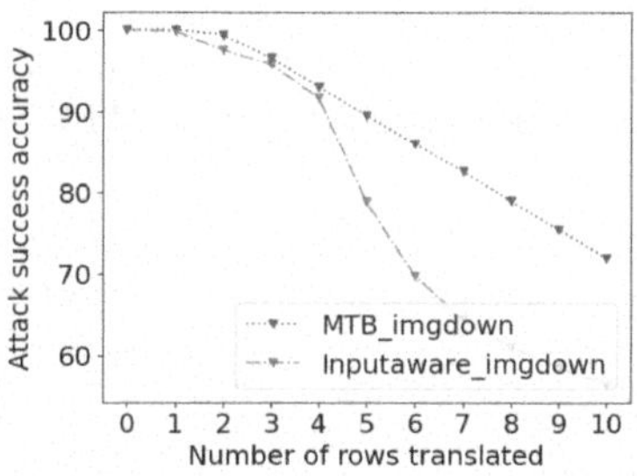

(b) MTB vs Input-aware Attack

Fig. 1. Robustness of MTB attack under trigger misplacement: (a) MTB vs BadNet, and (b) MTB vs Input-aware attack.

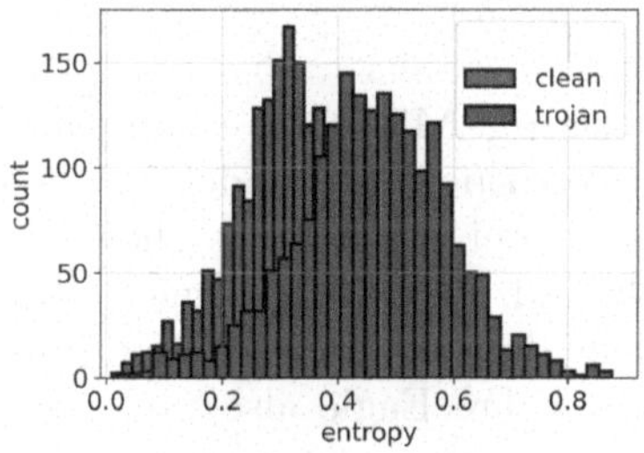

(a) STRIP Histogram

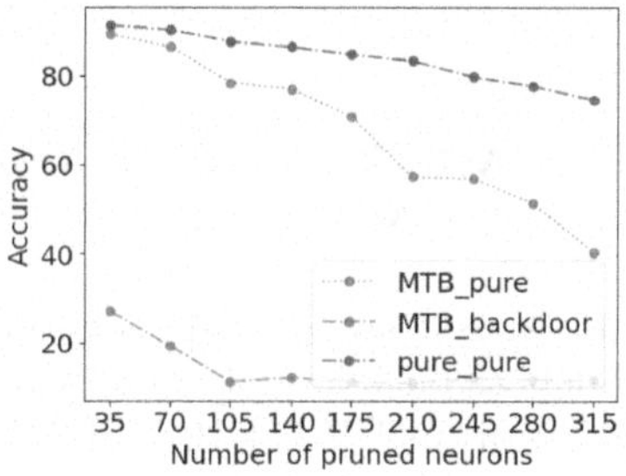

(b) Fine-pruning (RNP) Results

Fig. 2. MTB attack against test-time defenses: (c) STRIP (inseparability of the histograms), and (d) Fine-pruning (inability to make trigger accuracy 0 (MTB_backdoor) and incurring higher pure accuracy drop (MTB_pure) compared to a pure model (pure_pure)).

failure of existing defenses against the MTB attack. STRIP struggles to separate pure and backdoored images, as shown by the overlapping entropy distributions and the ineffective threshold (red dotted line). RNP-based Fine-Pruning [9] is also ineffective, with both attack success and pure accuracy remaining intact during pruning, only degrading together when nearly all neurons are pruned, an undesirable outcome. On the same model NAD reduced the pure accuracy by 5% and the attack success accuracy by 45% - achieving only partial success.

3.2 Detection of MTB Attack by Our Proposed MTD

We evaluate the performance of our proposed backdoor detection method **MTD** against our proposed backdoor attack **MTB**. MTD first identifies whether a model is backdoored and then detects the target classes and their corresponding triggers. A class is flagged as a target class if any recovered trigger produces class-distribution entropy below a set threshold ($\delta = 0.1$). A model is classified as backdoored if at least one target class is detected.

Table 2. Detection performance of MTD (90% Trojan effectiveness for entropy threshold computation), NC (anomaly detection threshold 2.0), STS and FreeEagle (anomaly detection threshold 2.0) on MTB attack models.

Dataset	Model	MTD model detection				MTD target class detection					
						4×4	8×8	4×4	8×8	4×4	8×8
		NC	STS	FreeEagle	MTD	NC	FreeEagle	MTD	NC	FreeEagle	MTD
MNIST	2 conv, 2 fc	0.0	0.8	0.0	**1.0**	0.0	0.0	**0.92**	0.0	0.0	**1.0**
CIFAR-10	vgg11	0.0	0.8	0.0	**1.0**	0.0	0.0	**0.83**	0.0	0.0	**1.0**
CIFAR-10	PreActResnet18	0.0	0.8	0.67	**1.0**	0.0	0.22	**1.0**	0.0	0.44	**1.0**
GTSRB	PreActResnet18	0.0	**1.0**	0.5	**1.0**	0.0	0.0	**0.81**	0.0	0.06	**0.77**
YTF	Resnet18	0.0	**1.0**	0.8	**1.0**	0.0	0.05	**0.43**	0.0	0.01	**0.46**
TinyImagenet	Resnet18	0.0	**1.0**	0.67	**1.0**	-	-	-	0.0	0.03	**0.7**

MTB Attack Backdoor Model Detection Using MTD. We compared MTD with **Neural Cleanse (NC)**, **STS**, and **FreeEagle** on models trained with 4×4 and 8×8 triggers. The F1-score for model detection is shown in Table 2. NC failed to detect MTB attack models and incorrectly flagged pure models as backdoored. STS, while accurate for some datasets, misclassified pure models from CIFAR-10 and MNIST as backdoored. FreeEagle also underperformed overall. In contrast, our proposed MTD achieved an **F1-score of 1.0 across all datasets**, successfully distinguishing backdoored from pure models.

MTB Attack Target Class Detection Using MTD. We further evaluated MTD's ability to detect target classes, as shown in Table 2. MTD performed well on **MNIST** and **CIFAR-10**, accurately identifying the target classes. However, for **GTSRB, YTF, and TinyImagenet**, there was a drop in class-wise detection performance. This drop occurs due to **intrinsic backdoors**-where similar classes share shortcuts introduced by target classes, causing false positives. For example, when a target class exists among a group of similar classes, the shortcut also serves nearby classes. As expected, NC failed completely due to its reliance on outlier detection, which fails for multiple target classes. STS was excluded from this evaluation because it can detect at most one target class. Overall, MTD effectively detects backdoored models and identifies target classes, outperforming existing methods.

Single Target Class Detection Using MTD. For a single-target attack (*class 0*), MTD successfully identified the model as backdoored, correctly flagging the target class as compromised while detecting all non-target classes as clean. The entropy plots in Fig. 3aand Fig. 3bhighlight the difference in entropy distributions between the target class (*class 0*) and a randomly sampled non-target class (*class 2*), demonstrating the effectiveness of MTD in distinguishing target classes.

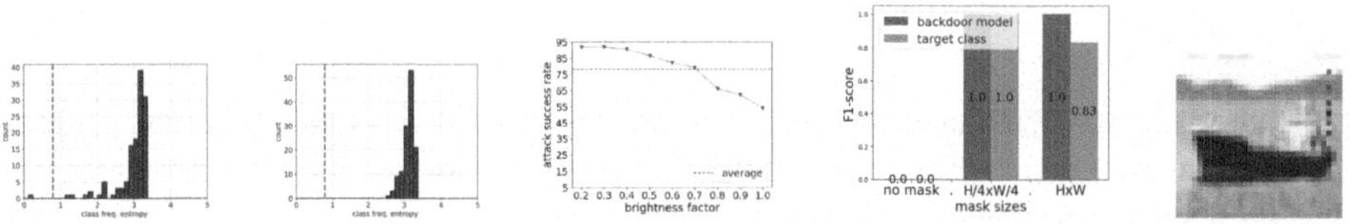

(a) (b) (c) (d) (e)

Fig. 3. a) Target class, b) non-target class, c) Adaptive Attack, d) F1-score of backdoor model and target class detection and e) a sample of 16×1 trigger.

MTD Against Adaptive Attacks. To evaluate MTD's robustness, we considered two adaptive attack strategies: a) using a large trigger, or b) targeting only a subset of the data distribution to invalidate trigger transferability. While a larger trigger reduces stealth, it may not be practical. Targeting a subset requires dividing a class into two non-overlapping sub-distributions, which is unlikely for real data. To test this, we divided the GTSRB STOP sign class into two subsets: darker images (cloudy conditions) and brighter images (sunny conditions). A backdoored model trained on the darker subset achieved an **attack success accuracy of 78%** on the brighter subset, as shown in Fig. 3e. This demonstrates that the trigger still transfers to a large portion of the data.

Performance Without, with (Quarter), and Full-Size Mask. We evaluated the performance of MTD on a CIFAR-10 MTB model trained with a 16×1 long trigger (Fig. 3e), while varying the mask size during the trigger reverse engineering process. L1-regularisation was used to control perturbation sparsity. The F1-scores for detecting the MTB model, target classes, and the corresponding entropy scores are reported in Fig. 3d. The results confirm the utility of our proposed defense method against MTB models trained with triggers of varying shapes and lengths, without assuming prior knowledge of the trigger size.

4 Conclusion

We introduced the Multi-Target Backdoor (MTB) attack, a more general variant that employs multiple unique triggers to target multiple classes, bypassing existing defenses. To counter this, we proposed a trigger reverse engineering-based detection called Multi Target backdoor Detection (MTD), grounded in our theoretical analysis that backdoors are most efficiently encoded as radial offsets in feature space. Extensive evaluations across diverse datasets demonstrate our method's effectiveness in detecting MTB and outperforming state-of-the-art defenses, even in adaptive attack scenarios.

Acknowledgement. This research was partially supported by the Australian Government through the Australian Research Council's Discovery Projects funding scheme (project DP210102798). The views expressed herein are those of the authors and are not necessarily those of the Australian Government or Australian Research Council.

References

1. Chen, H., Fu, C., Zhao, J., Koushanfar, F.: DeepInspect: a black-box trojan detection and mitigation framework for deep neural networks. In: International Joint Conference on Artificial Intelligence (2019)
2. Do, K., et al.: Towards effective and robust neural trojan defenses via input filtering. In: European Conference on Computer Vision (2022)
3. Fu, C., et al.: FreeEagle: detecting complex neural trojans in data-free cases. In: USENIX Security Symposium (2023)
4. Gao, Y., Xu, C., Wang, D., Chen, S., Ranasinghe, D.C., Nepal, S.: Strip: a defence against trojan attacks on deep neural networks. In: Annual Computer Security Applications Conference (2019)
5. Gu, T., Dolan-Gavitt, B., Garg, S.: Badnets: Identifying Vulnerabilities in the Machine Learning Model Supply Chain. arXiv preprint (2017)
6. Harikumar, H., Le, V., Rana, S., Bhattacharya, S., Gupta, S., Venkatesh, S.: Scalable backdoor detection in neural networks. In: Hutter, F., Kersting, K., Lijffijt, J., Valera, I. (eds.) ECML PKDD 2020. LNCS (LNAI), vol. 12458, pp. 289–304. Springer, Cham (2021). https://doi.org/10.1007/978-3-030-67661-2_18
7. Harikumar, H., Do, K., Rana, S., Gupta, S., Venkatesh, S.: Semantic host-free trojan attack. arXiv preprint (2021)
8. Li, Y., Lyu, X., Koren, N., Lyu, L., Li, B., Ma, X.: Neural attention distillation: erasing backdoor triggers from deep neural networks. In: International Conference on Learning Representations (2021)
9. Li, Y., et al.: Reconstructive neuron pruning for backdoor defense. In: International Conference on Machine Learning (2023)
10. Li, Y., Jiang, Y., Li, Z., Xia, S.T.: Backdoor learning: a survey. IEEE Trans. Neural Netw. Learn. Syst. (2022)
11. Liu, K., Dolan-Gavitt, B., Garg, S.: Fine-pruning: defending against backdooring attacks on deep neural networks. In: International Symposium on Research in Attacks, Intrusions, and Defenses (2018)
12. Liu, Y., Xie, Y., Srivastava, A.: Neural Tojans. In: International Conference on Computer Design (2017)
13. Nguyen, T.A., Tran, A.: Input-aware dynamic backdoor attack. Adv. Neural Inform. Process. Syst. (2020)
14. Shannon, C.E.: A mathematical theory of communication. Bell Syst. Tech. J. (1948)
15. Wang, B., et al.: Neural cleanse: identifying and mitigating backdoor attacks in neural networks. In: IEEE Symposium on Security and Privacy (2019)
16. Zhu, L., Ning, R., Wang, C., Xin, C., Wu, H.: Gangsweep: sweep out neural backdoors by GAN. In: International Conference on Multimedia (2020)

How to Backdoor Consistency Models?

Chengen Wang[1(✉)] and Murat Kantarcioglu[2]

[1] University of Texas at Dallas, Dallas, TX 75080, USA
chengen.wang@utdallas.edu
[2] Virginia Tech, Blacksburg, VA 24061, USA
muratk@vt.edu

Abstract. Consistency models are a new class of models that generate images by directly mapping noise to data, allowing for one-step generation and significantly accelerating the sampling process. However, their robustness against adversarial attacks has not yet been thoroughly investigated. In this work, we conduct the first study on the vulnerability of consistency models to backdoor attacks. While previous research has explored backdoor attacks on diffusion models, those studies have primarily focused on conventional diffusion models, employing a customized backdoor training process and objective, whereas consistency models have distinct training processes and objectives. Our proposed framework demonstrates the vulnerability of consistency models to backdoor attacks. During image generation, poisoned consistency models produce images with a Fréchet Inception Distance (FID) comparable to that of a clean model when sampling from Gaussian noise. However, once the trigger is activated, they generate backdoor target images. We explore various trigger and target configurations to evaluate the vulnerability of consistency models, including the use of random noise as a trigger. This novel trigger is visually inconspicuous, more challenging to detect, and aligns well with the sampling process of consistency models. Across all configurations, our framework successfully compromises the consistency models while maintaining high utility and specificity. We also examine the stealthiness of our proposed attack, which is attributed to the unique properties of consistency models and the elusive nature of the Gaussian noise trigger. Our code is available at https://github.com/chengenw/backdoorCM

Keywords: Backdoor Attacks · Consistency Models · Diffusion Models

1 Introduction

In recent years, diffusion models [9,25,28,29] have significantly advanced deep learning-based generation techniques across multiple domains, including image, audio [15] and video [10]. One disadvantage of these models is that they generate samples through hundreds or thousands of iterative steps, which require much longer processing time compared to previous generative models such as GAN [6] or VAE [14]. Although researchers have made efforts to address this issue,

X. Wu et al. (Eds.): PAKDD 2025, LNAI 15875, pp. 315–327, 2025.
https://doi.org/10.1007/978-981-96-8295-9_23

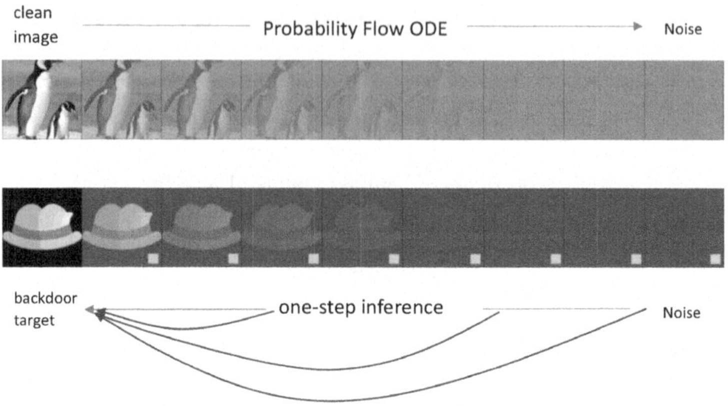

Fig. 1. Training backdoor consistency model using both clean images and backdoor targets. In this example, the target is a hat and the trigger is a square box at the bottom right corner.

diffusion models still require numerous steps to generate high-quality images using improved sampling techniques [18,19], or require retraining through the distillation of pre-trained diffusion models [24].

To address the above challenge, [27] proposes consistency models (CM), which can generate high-quality samples in one step. They also support multi-step sampling, allowing for a trade-off between computation and quality. Although consistency models can be trained by the distillation of diffusion models, they can also be trained independently. This make consistency models a new family of generative models.

Given the advantages of consistency models and their potential for a wide range of applications, we seek to gain a deeper understanding of their security implications, as these models can serve as foundational building blocks for developing downstream applications [22]. In this work, we focus on the vulnerability of consistency models to backdoor attack within the image domain. If these base models are compromised with a backdoor during training, they can generate unintended or malicious images when the backdoor trigger is activated, potentially leading to serious consequences.

Our objective is to explore how consistency models can be backdoored and to evaluate their performance across various backdoor settings. While several studies have examined the vulnerabilities of diffusion models to backdoor attacks [2–4], they primarily concentrate on basic diffusion models and do not cover consistency models, which have distinct training processes and objectives. As a result, the potential risks associated with consistency models remain under-explored.

Backdoor attacks on clean models typically have two primary goals. First, the tampered model should function normally and produce clean images when the backdoor is not triggered, ensuring high utility. Second, the tampered model should generate target images, as defined by the attacker, once the hidden trigger is activated, ensuring high specificity.

Figure 1 illustrates how the backdoor consistency model is trained, while Fig. 2 shows how sampling on the compromised model behaves before and after the trigger is activated.

In previous work on diffusion model backdoor attacks, common triggers include special shapes or objects, such as a box or glasses. In this work, we propose also using noise as a trigger, leveraging the fact that consistency models generate images by sampling from Gaussian noise, making the noise trigger less noticeable. The noise trigger is a sample from a Gaussian distribution, which remains fixed during both the model backdoor training and inference stages.

Our main contributions are as follows:

1. To the best of our knowledge, we are the first to propose a backdoor attack tailored to consistency models,
2. We conducted extensive experiments to reveal the vulnerabilities of consistency models under various trigger, target, and poison rate settings.
3. We propose using Gaussian noise as a trigger, which is less noticeable and well-suited to the sampling process of consistency models. This trigger is easy to generate and difficult to detect.

Our paper is organized as follows: we begin by introducing the background and motivation for investigating the vulnerability of consistency models in Sect. 1, followed by a discussion of related work in Sect. 2. Next, we present the preliminary knowledge necessary for the understanding of this work in Sect. 3, and describe how to backdoor consistency models in Sect. 4. We detail the experiments in Sect. 5, and discuss the stealthiness of the proposed attack in Sect. 6. Finally, we address the limitations of our work in Sect. 7 and conclude the paper in Sect. 8.

2 Related Work

2.1 Consistency Models

Consistency models [26,27] significantly accelerate the sampling process of diffusion models, enabling one-step sample generation. These models have been extended in various ways. For instance, [7,13] extend the models to multi-step sampling, while [31] introduces conditional consistency models. Additionally [20] extends the approach to latent space. Notably, [5] substantially improves the training efficiency of consistency models by starting from a pre-trained diffusion model and progressively refining the full consistency conditions throughout the training process.

2.2 Backdoor Attacks on Diffusion Models

[2,3] are the first work to investigate backdoor attack on diffusion models. They introduced a customized training schedule specifically designed to backdoor

denoising diffusion models [4,9] extends backdoor attack to a more generalized version, although it is still not a universal one. Their work focuses on standard diffusion models, analyzing the reversed process primarily in the form of $q(x_{t-1}|x_t, x_0)$, similar to the approach in [9], although with a more generalized training schedule. Additionally, it does not provide an explicit correction term for a universal backdoor training schedule.

To make the trigger less noticeable, [17] proposes using random noise as a trigger through a bi-level optimization approach. While this trigger is similar to the noise trigger introduced in our work, our approach involves directly sampling from a Gaussian distribution and fixing this noise trigger during backdoor training. This choice is based on our understanding that any Gaussian noise sample is generally inconspicuous in consistency model inference settings.

3 Preliminaries

3.1 Diffusion Models

Since consistency models are inspired by the theory of diffusion models, we first introduce diffusion models. Let us denote the data distribution by $p_{data}(\mathbf{x})$. The diffusion process perturbs data via Gaussian perturbations and is described by a stochastic differential equation (SDE) [29]

$$\mathrm{d}\mathbf{x}_t = \boldsymbol{f}(\mathbf{x}, t)\mathrm{d}t + g(t)\mathrm{d}\mathbf{w}, \tag{1}$$

where $t \in [0, T]$, $T > 0$ is a predefined constant, set to 80 in [11], $\boldsymbol{f}(\cdot, t)$ is the drift coefficient, $g(\cdot)$ the diffusion coefficient of $\mathbf{x}_t$, and $\mathbf{w}$ represents the standard Brownian motion. This SDE has a corresponding ordinary differential equation (ODE) known as Probability Flow (PF) ODE [29], which shares the same marginal probability $p_t(\mathbf{x})$:

$$\mathrm{d}\mathbf{x}_t = \left[\boldsymbol{f}(\mathbf{x}, t) - \frac{1}{2}g(t)^2\nabla_{\mathbf{x}} \log p_t(\mathbf{x})\right] \mathrm{d}t, \tag{2}$$

where the term $\nabla_{\mathbf{x}} \log p_t(\mathbf{x})$ is known as the score function [28,29].

When adopting a Gaussian perturbation schedule defined as $\mathbf{x}_t \sim \mathcal{N}(\mathbf{x}; t^2\boldsymbol{I})$ [11], where $\mathbf{x} \sim p_{data}$, the PF ODE is simplified as

$$\mathrm{d}\mathbf{x}_t = -t\nabla_{\mathbf{x}} \log p_t(\mathbf{x})\mathrm{d}t. \tag{3}$$

3.2 Consistency Models

Consistency models are a new type of models designed for single-step generation. These models are trained to map any point on a PF ODE trajectory back to the origin of the same trajectory. For a given solution trajectory of the PF ODE, the consistency function is defined as $\boldsymbol{f} : (\mathbf{x}_t, t) \mapsto \mathbf{x}$, where $t \in [\epsilon, T]$. This means the outputs are consistent for any pairs of $(\mathbf{x}_t, t)$ on the same PF ODE trajectory [27].

As shown in [27], an unbiased estimator for the score function in the PF ODE is

$$\nabla \log p_t(\mathbf{x}_t) = -\mathbb{E}\left[\frac{\mathbf{x}_t - \mathbf{x}}{t^2} \middle| \mathbf{x}_t\right], \tag{4}$$

thereby eliminating the need to learn a score function during consistency model training. This allows consistency models to be trained independently without relying on the distillation of a pre-trained diffusion model.

The consistency training loss in [27] is defined as follows:

$$\mathcal{L}_{CT}^{N}(\boldsymbol{\theta}, \boldsymbol{\theta}^{-}) = \\ \mathbb{E}[\lambda(t_n) d(\boldsymbol{f}_{\boldsymbol{\theta}}(\mathbf{x}_{n+1}, t_{n+1}), \boldsymbol{f}_{\boldsymbol{\theta}^{-}}(\mathbf{x}_n, t_n))], \tag{5}$$

where $\lambda(\cdot)$ is a positive weighting function, $d(\cdot)$ is a distance function, $\boldsymbol{\theta}$ represents the model parameters, and $\boldsymbol{\theta}^{-}$ is the running average of the past values of $\boldsymbol{\theta}$ during training. The function $\boldsymbol{f}_{\boldsymbol{\theta}}(\cdot, \cdot)$ is a learned consistency function, while $\mathbf{x}_{n+1}$ and $\mathbf{x}_n$ are two adjacent samples at times t_{n+1} and t_n, respectively. These samples are defined as $\mathbf{x}_{n+1} = \mathbf{x} + t_{n+1}\boldsymbol{\epsilon}$ and $\mathbf{x}_n = \mathbf{x} + t_n\boldsymbol{\epsilon}$, with $\boldsymbol{\epsilon} \sim \mathcal{N}(0, \boldsymbol{I})$.

4 Methods and Algorithms

4.1 Threat Model

Due to the increasing training cost, it is a common practice to use third-party models. Following [3,4], there are two parties involved: (1) the user, who utilizes the off-the-self third-party model to perform a specific task, and (2) the attacker, who releases a compromised third-part model for malicious purposes. In a backdoor attack on consistency models, these models behave in a predetermined way when a specific trigger is activated and behave normally otherwise. Specifically, they generate clean image when sampling from Gaussian noise, but produce backdoor targets once a trigger is added to the noise. A trigger is a predefined image.

There are two types of metrics to measure the quality of the models: utility and specificity. Utility metrics, such as Fréchet Inception Distance (FID) [8] and Inception Score (IS) [23], assess the quality of the generated images. Specificity metrics, like Mean Square Error (MSE), used solely by the attacker, evaluate how similar the generated backdoor images are to the backdoor target. The attacker will release the backdoor-compromised model if both the utility and specificity meet the requirements, while the user will accept the model if its utility meets their requirements.

In the backdoor attack scenario, we consider a white-box attack, where the attacker has full access to the training data, training schedule, hyperparameters and loss function. In contrast, the user has access only to the released model parameters and a subset of the *clean* training data, allowing them to evaluate the model's performance.

Algorithm 1. Backdoor Consistency Model Training

Input: clean dataset $\mathcal{D}_c$, backdoor target dataset $\mathcal{D}_p$, pretrained diffusion/consistency model $\boldsymbol{\theta}_0$, log-normal distribution $p(t)$, mapping function $p(r|t,k)$, weighting function $\lambda(t)$, $\mathbf{R}$ as defined in Eq. (8), and poison rate ρ.

$\boldsymbol{\theta} \leftarrow \boldsymbol{\theta}_0, k = 0$
repeat
  Sample $\boldsymbol{\epsilon} \sim \mathcal{N}(0, \boldsymbol{I}), t \sim p(t), r \sim p(r|t,k)$
  Sample u from a uniform distribution $\mathcal{U}[0,1]$,
  if $u < \rho$ **then**
    Sample $\mathbf{x}_0' \sim \mathcal{D}_p$
    Compute $\mathbf{x}_t = \mathbf{x}_0' + t\mathbf{R} + t\epsilon$
    Compute $\mathbf{x}_r = \mathbf{x}_0' + r\mathbf{R} + r\epsilon$
  else
    sample $\mathbf{x}_0 \sim \mathcal{D}_c$
    Compute $\mathbf{x}_t = \mathbf{x}_0 + t\epsilon$
    Compute $\mathbf{x}_r = \mathbf{x}_0 + r\epsilon$
  end if
  $\mathcal{L}(\boldsymbol{\theta}, \boldsymbol{\theta}^-) = \lambda(t) \cdot d(\boldsymbol{f}_{\boldsymbol{\theta}}(\mathbf{x}_t), \boldsymbol{f}_{\boldsymbol{\theta}^-}(\mathbf{x}_r))$
  $\boldsymbol{\theta} \leftarrow \boldsymbol{\theta} - \eta \nabla_{\boldsymbol{\theta}} \mathcal{L}(\boldsymbol{\theta}, \boldsymbol{\theta}^-)$
  $k = k + 1$
until convergence

4.2 Backdoor Consistency Models

For training clean image consistency models, as discussed in Sect. 3, the training schedule is defined as

$$\mathbf{x}_t = \mathbf{x} + t\epsilon, \epsilon \sim \mathcal{N}(0, \boldsymbol{I}), \tag{6}$$

where $\mathbf{x}$ is the training image sampled from $p_{data}(\mathbf{x})$ and $\mathbf{x}_t$ is the sample at time t.

Correspondingly, we propose that the backdoor consistency model training schedule be defined as

$$\mathbf{x}_t' = \mathbf{x}' + t\mathbf{R} + t\epsilon, \epsilon \sim \mathcal{N}(0, \boldsymbol{I}), \tag{7}$$

where $\mathbf{x}'$ is the backdoor target sampled from the target distribution $p_{target}(\mathbf{x}')$, $\mathbf{x}_t'$ is the sample at time t, and

$$\mathbf{R} = \mathbf{M} \odot \mathbf{g} + (1 - \mathbf{M}) \odot \mathbf{x}, \tag{8}$$

where $\mathbf{g}$ is the trigger, and $\mathbf{M} \in \{0, 1\}$ is a binary mask.

We propose that the backdoor training loss function be defined as

$$\mathcal{L}'(\boldsymbol{\theta}, \boldsymbol{\theta}^-) = \mathbb{E}[\lambda(t_n) d(\boldsymbol{f}_{\boldsymbol{\theta}}(\mathbf{x}_{n+1}', t_{n+1}), \boldsymbol{f}_{\boldsymbol{\theta}^-}(\mathbf{x}_n', t_n))], \tag{9}$$

where $\mathbf{x}_{n+1}' = \mathbf{x}' + t_{n+1}\mathbf{R} + t_{n+1}\boldsymbol{\epsilon}$, $\mathbf{x}_n' = \mathbf{x}' + t_n\mathbf{R} + t_n\boldsymbol{\epsilon}$, and other symbols are defined in Eq. (5).

During the consistency model training, the two adjacent times t_{n+1} and t_n are sampled as suggested in [5]. Specifically, t_{n+1} is sampled from a distribution $p(t) \sim \text{LogNormal}(\mu, \sigma)$, while t_n is sampled from a mapping function $p(r|t)$. Initially, $r = 0$; As training progresses, $\frac{r}{t} \to 1$. Readers are referred to [5] for detailed parameter settings and the mapping function designs. The backdoor training algorithm is summarized in Algorithm 1.

5 Experiments

5.1 Experimental Setup

We use the benchmark dataset CIFAR-10 (32×32) [16] and the FFHQ [12] dataset, which is converted to a resolution of 64×64. In the following text, CIFAR-10 is used as the default dataset, unless specified otherwise.

We use Box and Glasses as triggers, and Hat and Cat as targets, following [3]. Additionally, we introduce noise as a trigger, which is less noticeable since consistency models sample from Gaussian noise when generating images. The noise trigger is obtained by sampling from a Gaussian distribution, and is fixed throughout the backdoor training process. The triggers and targets used in this paper are summarized in Table 1. By default, noise is used as the trigger and Hat as the target, unless specified otherwise.

Table 1. Triggers and targets

Triggers			Targets	
Noise	Box	Glasses	Hat	Cat

We train the backdoor consistency model by *fine-tuning a pre-trained base model*, which can be either a diffusion model or a consistency model [5,11]. We use a pre-trained diffusion model by default. The poison rate, defined as the fraction of training time allocated for backdoor training, is set to 0.1 by default.

5.2 Evaluation Metrics

We use Fréchet Inception Distance (FID) [8] to measure the utility of the generated clean images. The FID is calculated between 50,000 generated clean images and all available training images, which also totals 50,000 for the CIFAR-10 dataset. A lower FID indicates better quality of the generated images.

To assess the specificity of the generated backdoor images, we use Mean Square Error (MSE). MSE is computed as the mean pixel value difference between 64 generated backdoor images and the ground truth targets. A lower MSE indicates better backdoor image generation.

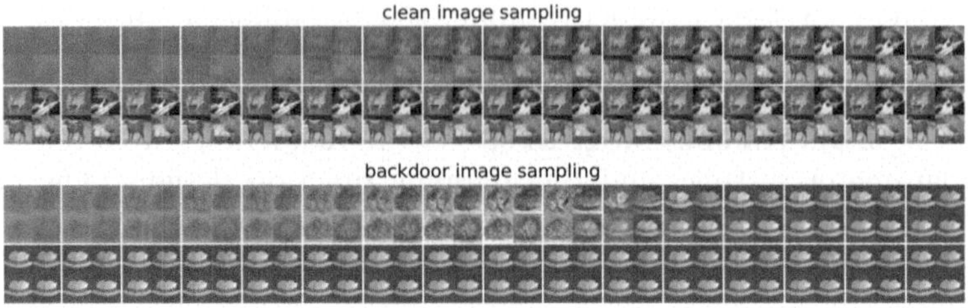

Fig. 2. Sampling of clean and backdoor images as training progresses (fine-tuning a pre-trained *diffusion* model on the CIFAR-10 dataset).

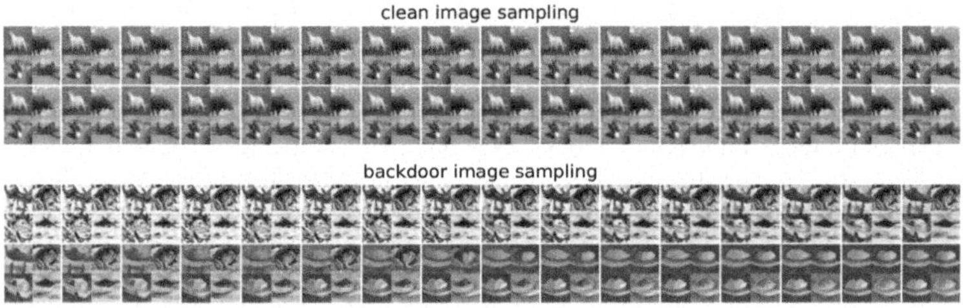

Fig. 3. Sampling of clean and backdoor images as training progresses (fine-tuning a pre-trained *consistency* model on the CIFAR-10 dataset). Note that the clean image sampling exhibits no visual difference since the training starts from a well-trained clean *consistency* model.

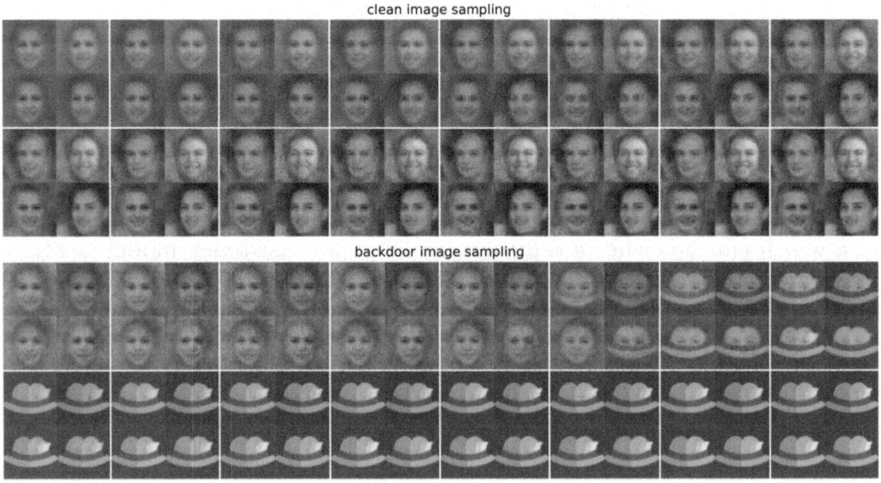

Fig. 4. Sampling of clean and backdoor images as training progresses (fine-tuning a pre-trained *diffusion* model on the FFHQ dataset). Note that the clean images samples are initially blurred, but their quality progressively improves with further training, eventually reaching a level comparable to samples generated from a clean CM model (not shown here).

5.3 Generating Samples as Training Progresses

We visualize the generated clean images and backdoor targets as training progresses on the CIFAR-10 dataset, as shown in Fig. 2. The default noise trigger is used and the base model is a pre-trained *diffusion* model. It is evident as training continues, the consistency model improves, gradually learning the data distribution.

We also visualize the generated clean images and backdoor targets as training progresses during the fine-tuning of a pre-trained *consistency* model on the CIFAR-10 dataset, as illustrated in Fig. 3. The generated clean image exhibit no visual difference since *the base model is a pre-trained consistency model.* It is observed that the backdoor training in this setting converges more slowly.

For the FFHQ dataset, we visualize the generated clean images and backdoor targets as training progresses during the fine-tuning of a pre-trained *diffusion* model, as shown in Fig. 4. Due to space constraints, the visualization of the training process for fine-tuning a pre-trained *consistency* model is included in the Supplement.

5.4 Varying Poison Rate

We investigate how the utility and specificity vary with different poison rates under various trigger and target settings. We combine FID and MSE into a single figure, as shown in Fig. 5. The poison rate for clean images is set to 0. As the poison rate increases, the FID rises, indicating a gradual decline in the quality of the generated images. When the poison rate reaches 0.4, the FID increases by approximately 20%, signaling a significant drop in quality. Therefore, we did not record the FID for higher poison rates. Conversely, as the poison rate increases, the MSE decreases, indicating a gradual improvement in the quality of the generated backdoor images. The data reveal that the FID and MSE values are well-balanced at poison rates ranging from 0.05 to 0.2 for this setting. Additionally it is evident that a 5% poison rate can successfully backdoor a clean consistency model.

Figure 6 illustrates the FID and MSE values obtained by fine-tuning a pre-trained *consistency* model, revealing a similar pattern. It is observed the balanced poison rate in this setting ranges from 0.05 to 0.1, where we achieve both low MSE and low FID, resulting in high utility and high specificity. Note that it can achieve a lower FID score under the same poison rate.

Additional experimental results with various trigger and target combinations can be found in the Supplement, which demonstrate a *similar pattern.*

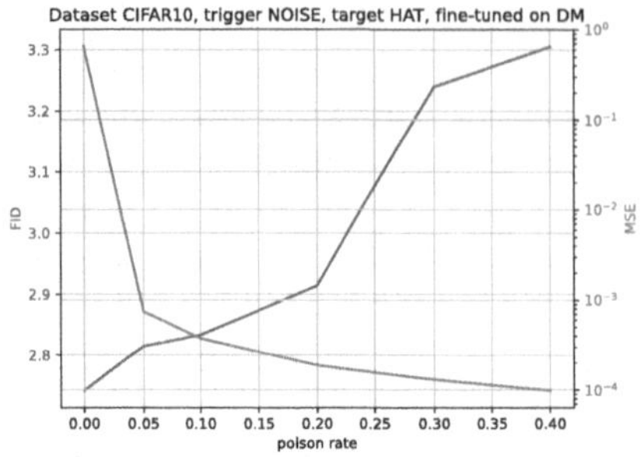

Fig. 5. FID and MSE values with respect to poison rate, fine-tuning on a pre-trained *diffusion* model.

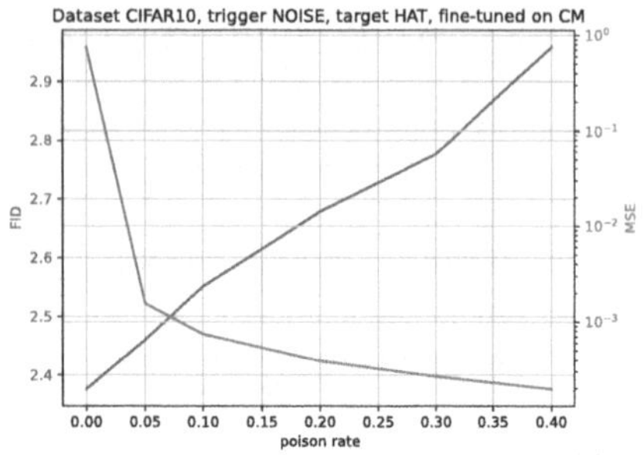

Fig. 6. FID and MSE values with respect to poison rate, fine-tuning on a pre-trained *consistency* model.

6 Defense

Several studies have proposed countermeasures against backdoor attacks on diffusion models [1,21,30]. Due to the unique properties of consistency models and the stealthiness of the Gaussian noise trigger, most of them do not work in our setting. For instance, [30] observes that the distribution of poisoned noise containing a trigger differs from that of clean noise. However, this assumption does not hold for our novel Gaussian noise trigger, which shares the same distribution as clean noise. The insight of [1] is that a backdoored diffusion model exhibits a distribution shift, which is dependent on the backdoor trigger. Specifically, if the input at time t contains a trigger, the output of the compromised diffusion model will include an additional backdoor term at time $t-1$, where $t-1>0$ represents an intermediate step. In contrast, the output of consistency models is a clean images at time $t=0$, making the Eq. (4) in their paper inapplicable to backdoored consistency models.

The defense proposed in [21] introduces an optimization method to approximate the hidden trigger from pure noise, based on the insight that a backdoored model make better predictions from Gaussian noise combined with the hidden trigger than from Gaussian noise with random nonzero noise. While this method is likely to succeed in approximating the hidden trigger in diffusion models by manipulating their noise predictions, it is not directly applicable to consistency models, as consistency models predict clean images. Additionally, our novel Gaussian trigger can bypass the element-wise distribution-based detection method described in Eq. (22) of their work. This is because the distribution of two Gaussian variables remains Gaussian. Although the variance of the combined Gaussian distributions doubles, this can be mitigated by scaling down the value of R and ϵ in Eq. (7).

7 Limitations

The proposed backdoor attack is a white-box attack, requiring the attacker to have access to the training code. The attack could be an insider within an organization or may release a compromised model directly, making the attack feasible in these scenarios.

8 Conclusion

In this work, we present the first backdoor attack tailored specifically for consistency models. Our extensive experimental results clearly demonstrate the vulnerability of these models. Given the inherent strengths of consistency models, their potential for wide applications, and the stealthiness introduced by combining consistency models with a Gaussian noise trigger, our findings underscore the potential risks associated with applications built upon these frameworks.

Acknowledgment. The research reported herein were supported in part by NSF awards DMS-2204795, OAC-2115094, CNS-2331424, ARL/Army Research Office awards W911NF-24-1-0202 and W911NF-24-2-0114.

References

1. An, S., et al.: Elijah: Eliminating backdoors injected in diffusion models via distribution shift. In: Proceedings of the AAAI Conference on Artificial Intelligence, vol. 38, pp. 10847–10855 (2024)
2. Chen, W., Song, D., Li, B.: Trojan attacks on diffusion models with diverse targets, Trojdiff (2023)
3. Chou, S.-Y., Chen, P.-Y., Ho, T.-Y.: How to backdoor diffusion models? (2023)
4. Chou, S.-Y., Chen, P.-Y., Ho, T.-Y.: Villandiffusion: a unified backdoor attack framework for diffusion models. In: NeurIPS 2023 Workshop on Backdoors in Deep Learning - The Good, the Bad, and the Ugly (2024)
5. Geng, Z., Pokle, A., Luo, W., Lin, J., Zico Kolter, J.: Consistency models made easy (2024)
6. Goodfellow, I., et al.:Generative adversarial nets. Adv. Neural Inform. Process. Syst. **27** (2014)
7. Heek, J., Hoogeboom, E., Salimans, T.: Multistep consistency models. arXiv preprint, arXiv: 2403.06807 (2024)
8. Heusel, M., Ramsauer, H., Unterthiner, T., Nessler, B., Hochreiter, S.: Gans trained by a two time-scale update rule converge to a local nash equilibrium. In: Advances in Neural Information Processing Systems, volume 30. Curran Associates, Inc. (2017)
9. Ho, J., Jain, A., Abbeel, P.: Denoising diffusion probabilistic models. In: Larochelle, H., Ranzato, M., Hadsell, R., Balcan, M.F., Lin, H. (eds.) Advances in Neural Information Processing Systems, vol. 33, pp. 6840–6851. Curran Associates, Inc. (2020)

10. Ho, J., Salimans, T., Gritsenko, A., Chan, W., Norouzi, M., Fleet, D.J.: Video diffusion models. In: Advances in Neural Information Processing Systems, vol. 35, pp. 8633–8646 (2022)
11. Karras, T., Aittala, M., Aila, T., Laine, S.: Elucidating the design space of diffusion-based generative models. In: Oh, A.H., Agarwal, A., Belgrave, D., Cho, K. (eds.) Advances in Neural Information Processing Systems (2022)
12. Karras, T., Laine, S., Aila, T.: A style-based generator architecture for generative adversarial networks. In: Proceedings of the IEEE/CVF Conference on Computer Vision and Pattern Recognition, pp. 4401–4410 (2019)
13. Kim, D., et al.: Consistency trajectory models: Learning probability flow ode trajectory of diffusion. arXiv preprint, arXiv: 2310.02279 (2023)
14. Kingma, D.P., Welling, M.: Auto-encoding variational bayes. arXiv preprint, arXiv: 1312.6114 (2013)
15. Kong, Z., Ping, W., Huang, J., Zhao, K., Catanzaro, B.: Diffwave: A versatile diffusion model for audio synthesis. arXiv preprint, arXiv:2009.09761 (2020)
16. Krizhevsky, A., Hinton, G. Learning multiple layers of features from tiny images. Technical Report 0, University of Toronto, Toronto, Ontario (2009)
17. Li, S., Ma, J., Cheng, M.: Invisible backdoor attacks on diffusion models. arXiv preprint, arXiv:2406.00816 (2024)
18. Liu, L., Ren, Y., Lin, Z., Zhao, Z.: Pseudo numerical methods for diffusion models on manifolds. arXiv preprint, arXiv:2202.09778 (2022)
19. Cheng, L., Zhou, Y., Bao, F., Chen, J., Li, C., Zhu, J.: Dpm-solver: a fast ode solver for diffusion probabilistic model sampling in around 10 steps. Adv. Neural. Inf. Process. Syst. **35**, 5775–5787 (2022)
20. Luo, S., Tan, Y., Huang, L., Li, J., Zhao, H.: Latent consistency models: Synthesizing high-resolution images with few-step inference. arXiv preprint, arXiv:2310.04378 (2023)
21. Mo, Y., Huang, H., Li, M., Li, A., Wang, Y.: Terd: a unified framework for safeguarding diffusion models against backdoors. arXiv preprint, arXiv:2409.05294 (2024)
22. Rombach, R., Blattmann, A., Lorenz, D., Esser, P., Ommer, B.: High-resolution image synthesis with latent diffusion models (2022)
23. Salimans, T., Goodfellow, I., Zaremba, W., Cheung, V., Radford, A., Chen, X.: Improved techniques for training gans. In: Advances in Neural Information Processing Systems, vol. 29 (2016)
24. Salimans, T., Ho, J.: Progressive distillation for fast sampling of diffusion models. arXiv preprint, arXiv:2202.00512 (2022)
25. Sohl-Dickstein, J., Weiss, E., Maheswaranathan, N., Ganguli, S.: Deep unsupervised learning using nonequilibrium thermodynamics. In: International Conference on Machine Learning, pp. 2256–2265. PMLR (2015)
26. Song, Y., Dhariwal, P.: Improved techniques for training consistency models. arXiv preprint, arXiv:2310.14189 (2023)
27. Song, Y., Dhariwal, P., Chen, M., Sutskever, I.: Consistency models. arXiv preprint, arXiv:2303.01469 (2023)
28. Song, Y., Ermon, S¿: Generative modeling by estimating gradients of the data distribution. In: Wallach, H., Larochelle, H., Beygelzimer, A., d'Alché-Buc, F., Fox, E., Garnett, R. (eds.) Advances in Neural Information Processing Systems, vol. 32. Curran Associates, Inc. (2019)
29. Song, Y., Sohl-Dickstein, J., Kingma, D.P., Kumar, A., Ermon, S., Poole, B.: Score-based generative modeling through stochastic differential equations. In: International Conference on Learning Representations (2021)

30. Sui, Y., et al.: Disdet: exploring detectability of backdoor attack on diffusion models. arXiv preprint, arXiv:2402.02739 (2024)
31. Xiao, J., et al.: Ccm: adding conditional controls to text-to-image consistency models. arXiv preprint, arXiv:2312.06971 (2023)

Multi-granularity Policy Explanation of Deep Reinforcement Learning Based on Saliency Map Clustering

Yujiao Wang, Hai Huang(✉), and Xingquan Zuo

Beijing University of Posts and Telecommunications, Beijing 100876, China
{dorothea,hhuang,zuoxq}@bupt.edu.cn

Abstract. Explainable Deep Reinforcement Learning (XDRL) holds significant potential for clarifying the decision-making logic of agents in complex tasks. Current XDRL research increasingly focuses on multi-granularity policy explanation methods that integrate both local and global decision-making insights. However, most of them treat local and global explanations as separate processes, overlooking their interdependencies and compromising logical consistency. This paper introduces a Multi-granularity Policy eXplanation method based on Saliency map Clustering (MPXSC), which computes both local and global policy explanations for a DRL agent in a unified, end-to-end process. MPXSC begins by employing superpixel perturbation to generate saliency maps for all the agent's states, representing its local explanation. These maps are then categorized by agent's actions and clustered based on local saliency features. Subsequently, the key states within each cluster are identified and serve as decision rules for their respective actions. Collectively, these results constitute the global explanation for the agent's decision-making. The superiority of MPXSC is validated through objective experiments on both local and global explanations, with a case study visually demonstrating the method's compelling explainable evidence regarding DRL model prediction outcomes.

Keywords: Explainable AI · Deep Reinforcement Learning

1 Introduction

Deep Reinforcement Learning (DRL) holds significant promise in tackling complex tasks, yet its decision-making process remains obscure due to the nonlinearity of deep neural networks and the distinctiveness of reinforcement learning. This opacity is intolerable in critical applications such as autonomous driving, financial risk assessment and intelligent healthcare, which drives research in Explainable Deep Reinforcement Learning (XDRL) to improve model security and reliability.

The predominant approach in XDRL is the post hoc explanation method, which treats the model as a black box and uses feature attribution to elucidate its decision-making basis. This method is divided into local decision and global policy explanation based on the granularity of their goals [19]. Local decision explanations clarify the

X. Wu et al. (Eds.): PAKDD 2025, LNAI 15875, pp. 328–339, 2025.
https://doi.org/10.1007/978-981-96-8295-9_24

model's reasoning in specific states. For instance, [7] use Gaussian noise to perturb states, identifying feature importance through changes in Q values. [8] propose Specificity and Relevance for scene-focused explanations, while [10] integrate object detection models, using object-level perturbations for precise saliency estimation. However, local explanations cannot capture the model's overall strategy, limiting their ability to convey global behavior across scenarios.

In contrast, global explanation methods analyze the model's overall strategy, helping users understand the model's general logic. HIGHLIGHTS [1] selects key states to summarize agent behavior, aiding in strategy understanding. However, this approach may lack full coverage of decision rules and could suffer from redundancy. Topin [18] created an abstract state transition diagram based on Markov decision processes, determining the probability of agent actions in state groups to predict behavior. This method offers a more comprehensive global policy explanation with higher comprehensibility but it also introduces increased complexity, requiring extensive state space coverage in decision-making. Additionally, global explanations fail to capture detailed behavior in specific instances, possibly missing local anomalies or subtle differences.

To address the limitations of local and global explanations, multi-granularity explanation methods combine their strengths, offering explanations at various levels to help users understand the model's behavior from coarse to fine-grained [19]. Among them, [9] apply LRP-based local decision explanation to the key states identified by HIGHLIGHTS-DIV [1], calculating their saliency maps. However, this method treats the calculations for two types of explanations as independent processes and merely integrates them, without considering the connections between them.

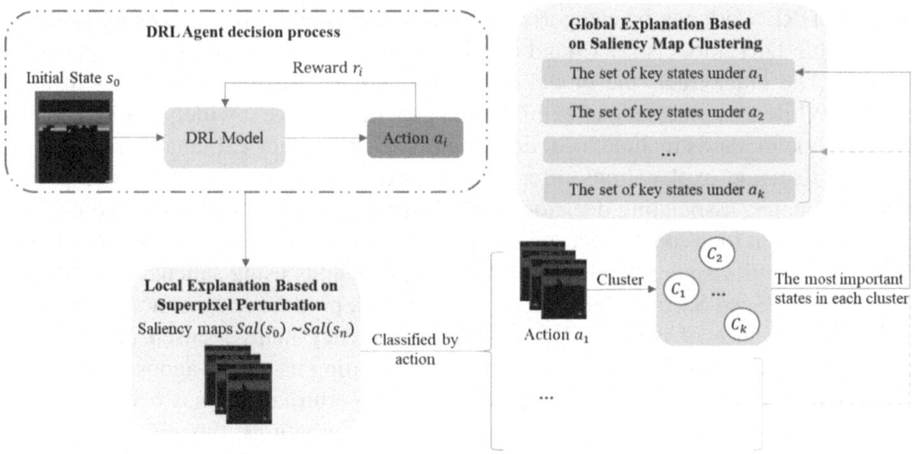

Fig. 1. The overall architecture of the MPXSC.

This paper proposes the Multi-granularity Policy eXplanation method based on Saliency map Clustering (MPXSC), which simultaneously calculates the local decision explanation and the global policy explanation for the DRL agent in a single end-to-end process. The global explanations fully utilize the results of the local explanations, ensuring logical coherence in the multi-granularity policy explanations. Figure 1 illustrates the overall architecture of MPXSC. Specifically, MPXSC starts by applying super-pixel perturbation to generate saliency maps for all the agent's states, representing its local explanation. These maps are then categorized according to the agent's actions and clustered based on local saliency features. Finally, the key states within each cluster are identified and regarded as decision rules for their respective actions, collectively forming the global explanation of the agent's decision-making. The main contributions of this paper are outlined below:

- We propose a multi-granularity policy explanation method for DRL models that offers both local and global explanations in a single end-to-end process. At the local level, super-pixel perturbation generates precise saliency maps with low complexity. At the global level, clustering is applied to the state feature saliency maps from local explanations, enhancing the comprehensiveness of global rules and reducing redundancy while maintaining logical coherence.
- We conduct extensive experiments in two DRL gaming scenarios and evaluate the performance of our method's local and global explanations using multiple objective metrics. The experimental results demonstrate that our method outperforms others in most aspects, demonstrating superior performance.
- Through case studies, we visually demonstrate that our approach outperforms in identifying effective explainable evidence for the DRL model's decision.

2 Related Work

Current XDRL work can be categorized into two approaches [15]: designing naturally interpretable DRL models [2, 13]and developing model-agnostic post-hoc explanation methods. The latter treats the model as a black box and attributes decisions to input features. While less precise, its flexibility and generality make it widely applicable [10, 17]. Post-hoc methods include local decision and global policy explanations [19].

Local decision explanations offer fine-grained insights into an agent's responses in specific states, associating decisions with environmental features. As noted by [16], explanations that focus on specific evidence are preferred. Motivated by this, [9] adapted LRP [3] to identify state regions that influencing decisions using saliency maps. Similarly, [8] introduced Sarfa by building on [7] and [10], proposing the concepts of specificity and relevance of perturbation to generate saliency maps focused on the target action. The LRP method relies on model weights, limiting its model-agnosticism, while perturbation-based methods are more post-hoc. Pixel perturbations may overlook object contours, leading to inaccuracies, whereas object perturbations, though more precise, require complex detection models.

Global policy explanations clarify an agent's overall decision-making logic, offering essential coarse-grained insights into its behavior and performance. HIGHLIGHTS [1] summarize behavior by selecting key states where incorrect actions significantly affect rewards, while HIGHLIGHTS-DIV avoids selecting similar states. [18] proposed Abstract Policy Graphs (APG), which utilize Markov chains of abstract states to predict actions. Global policy explanations fall into two types: key state summaries and strategy diagrams. The former is simpler and transferable but requires task expertise and may lack comprehensive rules, causing redundancy. The latter, using policy or state transition diagrams, though more detailed and intuitive, is complex and requires conditions such as binary features or broad state space coverage.

Single-granularity explanations often fail to adequately capture an agent's decision logic. To address this, [9] proposed the first multi-granularity explanation method combining global and local insights. By integrating HIGHLIGHTS-DIV [1] with LRP-based local explanation, they incorporated saliency maps into key states in the summary, highlighting regions that influence decisions. However, this approach limits the unity of the agent's decision-making logic, as the two granular explanations are computed independently.

3 Problem Statement

In a DRL framework with a state set $\mathcal{S}$ and an action set $\mathcal{A}$, the agent selects an action $a \in \mathcal{A}$ based on the current state $s \in \mathcal{S}$ and the Q-value function $Q(s, a)$, which is approximated by a deep neural network with parameters W, representing the expected cumulative reward for each action in the state.

This paper focuses on DRL tasks with visual input and a discrete action space such as games like Breakout and MsPacman (see Fig. 2), A state is represented by a screenshot of a specific game frame, denoted as $s \in \mathbb{R}^{c \times h \times w}$, where c, h and w represent the number of channels, height, and width of s respectively. At state s, $\hat{a}$ denotes the action selected by the agent according to Eq. (1).

$$\hat{a} = \underset{a}{argmax} Q(s, a) \tag{1}$$

The multi-granularity explainability problem in this DRL task can be described as follows:

Given: A DRL model with state set $\mathcal{S}$, action set $\mathcal{A}$, and Q-value function $Q(s, a)$.

Find: A multi-granularity explanation scheme including:

(1) **Local Explanation**: Generate a saliency map $Sal(s) : (s, a, Q) \rightarrow \mathbb{R}^{h \times w}$ that highlights the key features influencing the decision for a specific action a in state s.
(2) **Global Explanation**: For each action a, identify a set of key states $s_a^1, s_a^2, \cdots, s_a^k$ that characterize the action. Each key state is represented by its corresponding saliency map $Sal\left(s_a^i\right)$. These saliency maps collectively form the rules that elucidate the global behavior of the agent.

4 Method

4.1 Local Explanation Based on Superpixel Perturbation

We design a superpixel perturbation-based local explanation method for MPXSC by adopting the multi-level superpixel perturbation approach from MFPP [20] and integrating the specificity and relevance concepts from Sarfa [8].

Specifically, for the agent's state s at time t, we first perform L ($L > 0$) levels of superpixel segmentation on s, producing a superpixel fragment list of length L, denoted as Seg, where $g(Seg_l)$ is the number of superpixel fragments in the $l-th$ group. For each Seg_l, we randomly generate $g(Seg_l)$ binary masks m, with the values of the superpixel blocks set to 0 or 1. The number of masks m is G. Then, each mask is multiplied by the original image, yielding G perturbed inputs s', which are then batch-input into the Q-values function to obtain the $Q(s', a)$ for each s'.

Next, we need to score each mask m. In existing work [10], the mask score is typically represented by the difference in Q-values before and after perturbation. However, Sarfa suggests that for DRL tasks, perturbation should also consider specificity and relevance to the target action. Specificity refers to a feature's high saliency when its removal significantly impacts the Q-value of the target action $\hat{a}$, more than other actions $a'(a' \neq \hat{a})$. This is expressed as a larger expected relative Q-value difference before and after perturbation, as shown in Eq. (2):

$$\Delta p = \frac{exp(Q(s,\hat{a}))}{\sum_a exp(Q(s,a))} - \frac{exp(Q(s',\hat{a}))}{\sum_a exp(Q(s',a))} \tag{2}$$

Relevance refers to ensuring minimal impact on the relative expected Q-values of a'. This is represented by the high similarity in the distribution of the relative expected Q-values for a before and after the perturbation, as shown in Eq. (3) and Eq. (4).

$$P_{\text{rem}}(s, a) = \frac{\exp(Q(s,a))}{\sum_{a' \neq \hat{a}} \exp(Q(s,a'))} \forall a \neq \hat{a} \tag{3}$$

$$K = \frac{1}{1+D_{KL}} \tag{4}$$

where D_{KL} denotes the KL divergence between $P_{rem}(s', a)$ and $P_{rem}(s, a)$. The harmonic mean of Δp and K represents the specificity-relevance score of m, denoted as w. Further, we assign weights w to each mask m and compute their weighted average. Using Monte Carlo sampling, we can obtain the saliency map $Sal(s)$ for the state image s, as shown in Eq. (5).

$$Sal(s) = \frac{1}{E[m] \cdot \sum_{l=1}^{L} g(Seg_l)} \sum_{l=1}^{L} \sum_{i=1}^{g(Seg_l)} w_{l,i} \cdot m_{l,i} \tag{5}$$

where $E[m]$ represents the expected value of the random variable m, which in this context is the average value of the masks obtained through multiple samples. The saliency map explanation based on superpixel perturbation is presented in Algorithm 1.

Algorithm 1: Saliency Map Explanation Based on Superpixel Perturbation

```
Input: The state s at time t, the action set 𝒜, and the Q-value function Q(s, a).
Parameter: The levels of superpixel segmentation L
Output: The saliency map Sal(s) of s
Perform L levels of superpixel segmentation on s to obtain the fragment list Seg.
For Seg_l in Seg:
    Random generate g(Seg_l) binary mask m_l.
    For every mask m_{l,i}:
        s' = s * m_{l,i}
        Calculate w_{l,i}
    end for
end for
Calculate Sal(s) according to eq. (5)
return Sal(s)
```

4.2 Global Explanation Based on Saliency Map Clustering

At the global level, we cluster saliency maps from local explanations to identify key states for each action, which represent the agent's overall behavior. This enhances the comprehensiveness of global rules, reduces redundancy, and ensures consistency with local explanations. From the local method, we obtain the saliency maps $Sal(s)$ for each decision during a task and classify them by action, resulting in a set of saliency maps $D(a)$ for each a. Next, we apply a clustering algorithm to $D(a)$. Since the number of decision rules for each a cannot be predetermined, we need a clustering method that does not require specifying the number of clusters, such as DBSCAN, a density-based clustering algorithm [5]. It automatically determines the number of clusters by examining point neighborhoods, making it suitable for abundant, noisy, and irregularly shaped data.

Each resulting cluster represents a scenario type for action a, where the agent exhibits similar decision logic, which can be interpreted as following the same decision rule. We then calculate the importance $I(s)$ of each state in the clusters, using the saliency map of the state with the highest importance (key state) to represent the decision rule. The calculation of state's importance follows [9], using the Q-value difference between the selected action and the suboptimal action, as shown in Eq. (6). This approach mitigates issues related to the meaningless low Q-values often observed in undertrained models.

$$I(s) = max_a Q(s, a) - secondhighest_a Q(s, a) \tag{6}$$

Ultimately, we can identify a set of saliency maps of key states $R_a = \{Sal(s_a^1), Sal(s_a^2), \cdots, Sal(s_a^k)\}$ where k is the number of clusters for each a. This set illustrates the decision rationale for each a taken by the agent. The algorithm for this process is presented as Algorithm 2.

Algorithm 2: Global Policy Explanation Based on Saliency Map Clustering

Input: The states set S along with saliency map set Sal, and the action set $\mathcal{A}$.
Parameter: The minimum number of data points within a cluster, $minPts$, and the maximum neighborhood radius between points, ε.
Output: Decision rules set $R = \{R_{a1}, R_{a2}, \cdots, R_{a|\mathcal{A}|}\}$ for all actions.

Classify Sal by action, resulting in the set D.
For $D(a)$ ***in*** D
  Given $minPts$ and ε , obtain the list of clusters $dbscan$ of $D(a)$ by DBSCAN.
  For every cluster db_i ***in*** $dbscan$
    $I_{max} = 0$, $i_{max} = 0$
    For the state of the saliency map **in** db_i
      calculate $I(s_k)$ according to eq. (6)
      if $I(s_k) > I_{max}$:
        $I_{max} = I(s_k)$, $i_{max} = k$
      end if
    end for
    Add $Sal(s_{ai}^{i_{max}})$ into R_{ai}
  end for
end for
return R

5 Experiments

5.1 Experiment Settings

We conducted experiments on two Atari games, Breakout and MsPacman, using agents trained with the OpenAI Gym library [4], where both environments use image data as input and output rewards for each action. The Breakout agent, which selects from four actions, was trained for 10,000 episodes and achieved an average score of 409. The MsPacman agent, with nine actions, was trained for 15,000 episodes and reached an average score of 2,680. Both agents employed the Deep Q-Network (DQN) model, with three convolutional layers and two fully connected layers. The code is publicly available at https://github.com/Thea-wyj/MPXSC.

Baselines and Evaluation Metrics: For local explanation, we compared the five metrics of MPXSC with two pixel-perturbation-based saliency methods: Goh [7], which uses Gaussian noise for saliency, and Sarfa [8], which incorporates specificity and relevance. The objective evaluation metrics for local explanations are as follows:

- **Infidelity** [21] measures the alignment between saliency scores and the correlation between state and decision by introducing perturbations to the state and analyzing changes in Q-values.
- **Action Sensitivity (AS)** [21] measures the specificity of the explanation to the target action.

- **Stability** [12] assesses whether small perturbations to the state cause significant changes in the explanation when the model's decision remains unchanged.
- **Sufficiency** [11] measures whether the explanation method can completely reflect the areas related to the decision.
- **Validity** [6] measures whether the explanation method can correctly reflect the areas unrelated to the decision.

For global explanation, we evaluated MPXSC against HIGHLIGHTS-DIV [1] and HIGHLIGHTS-LRP [9] using three metrics. HIGHLIGHTS-DIV identifies key states, whereas HIGHLIGHTS-LRP provides saliency maps. In contrast, MPXSC integrates decision rules for action-key scenarios. The objective evaluation metrics for global explanations are based on prior work [14], as outlined below:

- **Accuracy** evaluates whether the explanation correctly identifies key states by analyzing overall reward changes ($\text{Accuracy}_{\text{total}}$) and the average reward changes per step ($\text{Accuracy}_{\text{step}}$) when suboptimal actions are applied at key states.
- **Completeness** measures how well the explanation covers the action space by determining whether the decisions of key states comprehensively represent all actions.
- **Redundancy** evaluates whether key states under the same decision rule exhibit excessive similarity, with the aim of minimizing redundancy in the explanation.

5.2 Experiment Results

We conducted 10 game rounds with the agent in both scenarios and generated explanations using three local and three global methods. The evaluation metrics described above were applied to assess the performance of each method.

Table 1. Metric values for three local explanation methods across two tasks. All metric values range from 0 to 1, with the best-performing values highlighted in bold.

	Breakout			MsPacman		
	Goh	Sarfa	MPXSC	Goh	Sarfa	MPXSC
Infidelity	0.512	0.507	**0.504**	0.535	0.516	**0.512**
AS	----	0.582	**0.393**	----	0.477	**0.405**
Stability	**0.981**	0.952	0.934	**0.996**	0.993	0.971
Sufficiency	0.899	0.902	**0.912**	0.803	0.830	**0.863**
Validity	0.501	0.502	**0.530**	0.568	0.570	**0.596**

The results for local methods are summarized in Table 1. For these metrics, lower values of infidelity and AS indicate better performance, whereas higher values of stability, sufficiency, and validity are desirable. Notably, Goh does not support action-specific explanations, and thus AS cannot be calculated. The results demonstrate that MPXSC outperforms the baselines in both scenarios, with the exception of stability. It excels particularly in AS, sufficiency, and validity, emphasizing its focus on the target action

Table 2. Metric values for three global explanation methods across two tasks.

	Breakout			MsPacman		
	HIGHLIGHTS-DIV	HIGHLIGHTS-LRP	MPXSC	HIGHLIGHTS-DIV	HIGHLIGHTS-LRP	MPXSC
$\text{Accuracy}_{\text{total}}$	0.854	0.866	**0.885**	0.855	0.858	**0.899**
$\text{Accuracy}_{\text{step}}$	0.808	0.846	**0.857**	0.845	0.899	**0.919**
Completness	0.750	0.813	**0.956**	0.667	0.750	**0.950**
Redundacy	0.518	0.364	**0.203**	0.447	0.408	**0.299**

and its ability to accurately identify regions of interest during decision-making. This improvement results from the use of superpixel perturbation, which captures target contours more effectively than pixel perturbation, leading to more precise saliency regions. The slightly lower stability score of MPXSC, which still exceeds 93%, can be attributed to the inherent randomness of superpixel perturbation. Nevertheless, MPXSC maintains a reasonable level of stability, further supporting its overall effectiveness.

Table 2 presents the metric results for the three global methods. Higher accuracy and completeness values indicate better performance, whereas lower redundancy values are preferred. MPXSC outperforms the baselines across all metrics in both scenarios, with particular excellence in completeness and redundancy. By leveraging saliency map clustering in action classification, MPXSC generates decision rules that comprehensively cover the action space while minimizing similarity among key states for the same action. In contrast, baselines relying solely on state importance for policy summarization fall short of achieving this. Additionally, MPXSC demonstrates superior accuracy in metrics based on overall reward changes and single-step reward changes, further validating the effectiveness of its saliency map clustering in identifying key states.

5.3 Case Study

This section provides single-sample explanations to visually validate the effectiveness of MPXSC. Figure 2(a) presents local explanations for the Breakout scenario. Goh highlights broad areas including irrelevant regions, such as the top-right corner. Sarfa improves action specificity, though it remains relatively broad. In contrast, MPXSC aligns saliency precisely with the ball's position, demonstrating enhanced accuracy through superpixel perturbation, which effectively captures the ball's shape. A similar conclusion can be drawn for the MsPacman scene. As shown in Fig. 2(b), MPXSC provides a precise annotation of the energy pellet's location, whereas the baselines do not, further supporting the effectiveness of MPXSC's local explanations.

Figure 3 presents a global explanation for Breakout, with saliency maps of key states for each action. It illustrates the agent's decision logic: NOOP is chosen when the ball's position or trajectory overlaps with the paddle or hits inside the bricks. LEFT or RIGHT is chosen based on the ball's direction relative to the paddle. FIRE targets a gap in the upper bricks when the ball is on the paddle, aiming for a higher score.

Original image predict: NOOP   Original image predict: RIGHT

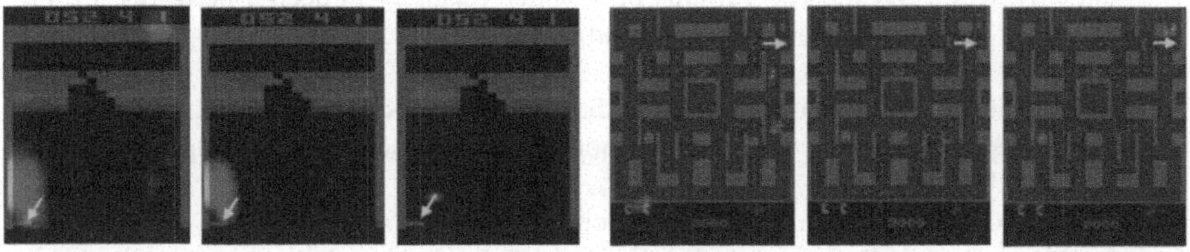

(a) Goh (a) Sarfa (a) MPXSC (b) Goh (b) Sarfa (b) MPXSC

Fig. 2. (a) When the ball bounces to the bottom-left and the agent selects NOOP, saliency maps from Goh, Sarfa, and MPXSC are compared. Yellow-green regions highlight salient areas, with a yellow arrow indicating the ball's direction. (b) When MsPacman approaches the energy pellet in the top-right corner and the agent chooses the RIGHT action, MPXSC accurately highlights the energy pellet's location. In contrast, baseline explanations focus on irrelevant areas. (Color figure online)

Action of agent	Key State 1	Key State 2	Action of agent	Key State 1	Key State 2
NOOP	The ball's trajectory overlaps with the paddle's position. 	The ball is inside the bricks, no movement is required for now. 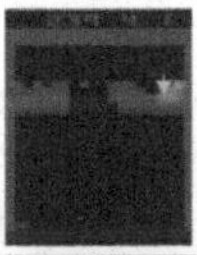	RIGHT	The ball's trajectory is on the right side of the paddle. 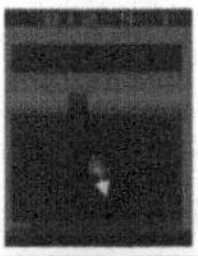	The ball's trajectory is on the right side of the paddle. 
LEFT	The ball's trajectory is on the left side of the paddle. 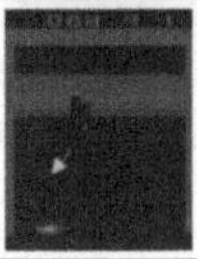	The ball has been launched, and after bouncing off the block, its trajectory remains on the left side of the paddle. 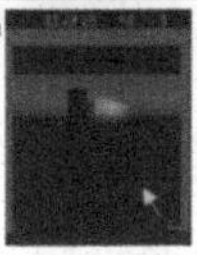	FIRE	The ball is on the paddle and needs to be launched towards the block. 	

Fig. 3. MPXSC's global policy explanation for Breakout presents a decision rule table showing key states for each action. For instance, the *NOOP* action is frequently chosen when the ball's trajectory overlaps with the paddle or hits inside the bricks, indicating that interception is unnecessary.

6 Conclusion

This paper proposes MPXSC, a multi-granularity explanation method for DRL that utilizes saliency map clustering. By leveraging superpixel perturbations, MPXSC generates saliency maps for local explanations and clusters them to derive global decision rules. Objective evaluations demonstrate its superior performance compared to existing methods. However, clustering increases computational complexity in high-dimensional scenarios, highlighting the need for research into more efficient techniques. Refining evaluation standards for reliable and comprehensive assessment of DRL explanations remains a key direction for future research.

References

1. Amir, D., Amir, O.: Highlights: Summarizing agent behavior to people. In: Proceedings of the 17th International Conference on Autonomous Agents and Multiagent Systems, pp. 1168–1176 (2018)

2. Annasamy, R.M., Sycara, K.: Towards better interpretability in deep q-networks. In: Proceedings of the AAAI Conference on Artificial Intelligence, vol. 33, pp. 4561– 4569 (2019)
3. Bach, S., Binder, A., Montavon, G., Klauschen, F., Müller, K.R., Samek, W.: On pixel-wise explanations for non-linear classifier decisions by layer-wise relevance propagation. PLoS ONE **10**(7), e0130140 (2015)
4. Brockman, G., et al.: Openai gym. arXiv preprint arXiv:1606.01540 (2016)
5. Ester, M., Kriegel, H.P., Sander, J., Xu, X., et al.: A density-based algorithm for discovering clusters in large spatial databases with noise. In: kdd, vol. 96, pp. 226–231 (1996)
6. Fan, M., Wei, W., Xie, X., Liu, Y., Guan, X., Liu, T.: Can we trust your explanations? sanity checks for interpreters in android malware analysis. IEEE Trans. Inf. Forensics Secur. **16**, 838–853 (2020)
7. Greydanus, S., Koul, A., Dodge, J., Fern, A.: Visualizing and understanding atari agents. In: International Conference on Machine Learning, pp. 1792–1801. PMLR (2018)
8. Gupta, P., Puri, N., Verma, S., Kayastha, D., Deshmukh, S., Krishnamurthy, B., Singh, S.: Explain your move: Understanding agent actions using specific and relevant feature attribution. In: International Conference on Learning Representations (ICLR) (2020)
9. Huber, T., Weitz, K., André, E., Amir, O.: Local and global explanations of agent behavior: Integrating strategy summaries with saliency maps. Artif. Intell. **301**, 103571 (2021)
10. Iyer, R., Li, Y., Li, H., Lewis, M., Sundar, R., Sycara, K.: Transparency and explanation in deep reinforcement learning neural networks. In: Proceedings of the 2018 AAAI/ACM Conference on AI, Ethics, and Society, pp. 144–150 (2018)
11. Kumarakulasinghe, N.B., Blomberg, T., Liu, J., Leao, A.S., Papapetrou, P.: Evaluating local interpretable model-agnostic explanations on clinical machine learning classification models. In: 2020 IEEE 33rd international symposium on computer- based medical systems (CBMS), pp. 7–12. IEEE (2020)
12. Li, X.H., Shi, Y., Li, H., Bai, W., Cao, C.C., Chen, L.: An experimental study of quantitative evaluations on saliency methods. In: Proceedings of the 27th ACM SIGKDD Conference on Knowledge Discovery & Data Mining, pp. 3200–3208 (2021)
13. Liu, G., Schulte, O., Zhu, W., Li, Q.: Toward interpretable deep reinforcement learning with linear model u-trees. In: Machine Learning and Knowledge Discovery in Databases: European Conference. ECML PKDD 2018, Dublin, Ireland, September 10–14, 2018, Proceedings, Part II, pp. 414–429. Springer-Verlag, Berlin, Heidelberg (2019).https://doi.org/10.1007/978-3-030-10928-8_25
14. Luss, R., Dhurandhar, A., Liu, M.: Local explanations for reinforcement learning. In: Proceedings of the AAAI Conference on Artificial Intelligence, vol. 37, pp. 9002–9010 (2023)
15. Milani, S., Topin, N., Veloso, M., Fang, F.: Explainable reinforcement learning: a survey and comparative review. ACM Comput. Surv. **56**(7), 1–36 (2024)
16. Miller, T.: Explanation in artificial intelligence: insights from the social sciences. Artif. Intell. **267**, 1–38 (2019)
17. Rizzo, S.G., Vantini, G., Chawla, S.: Reinforcement learning with explainability for traffic signal control. In: 2019 IEEE Intelligent Transportation Systems Conference (ITSC), pp. 3567–3572. IEEE (2019)
18. Topin, N., Veloso, M.: Generation of policy-level explanations for reinforcement learning. In: Proceedings of the AAAI Conference on Artificial Intelligence, vol. 33, pp. 2514–2521 (2019)

19. Vouros, G.A.: Explainable deep reinforcement learning: state of the art and challenges. ACM Comput. Surv. **55**(5), 1–39 (2022)
20. Yang, Q., Zhu, X., Fwu, J.K., Ye, Y., You, G., Zhu, Y.: Mfpp: morphological fragmental perturbation pyramid for black-box model explanations. In: 2020 25th International Conference on Pattern Recognition (ICPR), pp. 1376–1383. IEEE (2021)
21. Yeh, C.K., Hsieh, C.Y., Suggala, A., Inouye, D.I., Ravikumar, P.K.: On the (in) fidelity and sensitivity of explanations. Adv. Neural Inform. Process. Syst. 32 (2019)

FACROC: A Fairness Measure for Fair Clustering Through ROC Curves

Tai Le Quy[1(✉)], Long Le Thanh[2], Lan Luong Thi Hong[3], and Frank Hopfgartner[1,4]

[1] University of Koblenz, Koblenz, Germany
{tailequy,hopfgartner}@uni-koblenz.de
[2] Hanoi University of Science and Technology, Hanoi, Vietnam
[3] Hanoi University of Industry, Hanoi, Vietnam
lanlhbk@haui.edu.vn
[4] University of Sheffield, Sheffield, UK

Abstract. Fair clustering has attracted remarkable attention from the research community. Many fairness measures for clustering have been proposed; however, they do not take into account the clustering quality w.r.t. the values of the protected attribute. In this paper, we introduce a new visual-based fairness measure for fair clustering through ROC curves, namely FACROC. This fairness measure employs AUCC as a measure of clustering quality and then computes the difference in the corresponding ROC curves for each value of the protected attribute. Experimental results on several popular datasets for fairness-aware machine learning and well-known (fair) clustering models show that FACROC is a beneficial method for visually evaluating the fairness of clustering models.

Keywords: clustering · fair clustering · fairness measure · ROC curve · fairness-aware datasets

1 Introduction

Clustering is a fundamental problem in unsupervised learning, and fairness in clustering has garnered significant attention within the machine learning (ML) community, starting with the foundational work of Chierichetti et al. [7]. Fair clustering techniques aim to ensure equitable representation or treatment of groups or individuals within clusters. Researchers have focused on two main challenging problems in fair clustering: defining and enforcing fairness constraints [6]. Hence, a number of fairness notions and techniques were introduced to ensure fairness constraints in clustering and can be applied in many domains, such as healthcare [4], education [13], etc.

There are more than 20 fairness notions in fair clustering [6]. Since fairness notions can be turned into measures [17], we will use the terms "fairness notion" and "fairness measure" interchangeably. Fairness notions are defined based on the

X. Wu et al. (Eds.): PAKDD 2025, LNAI 15875, pp. 340–352, 2025.
https://doi.org/10.1007/978-981-96-8295-9_25

group-level, individual-level, algorithm agnostic, and algorithm specific. At the group-level, the algorithms should not discriminate against or unfairly favor any group of individuals in the predictions. For instance, *balance* notion [7], the most popular fairness notion used for fair clustering, requires a ratio balance between the protected group, e.g., female, and the non-protected group, e.g., male. Unlike group-level fairness, individual-level fairness ensures that similar individuals are treated similarly by the clustering model. The fairness notion of *proportionality* for centroid clustering [5] is an example of individual-level fairness. However, to our knowledge, all the defined fairness notions do not consider the clustering quality w.r.t. values of the protected attribute.

Regarding the clustering quality, it is determined by several measures, such as the silhouette coefficient, sum of squared error (SSE), Dunn index (DI), and others [13,15]. However, most of these metrics are difficult to visualize for comparison between clustering models. To address this issue, AUCC (Area Under the Curve for Clustering) [10] has been introduced as a measure of clustering quality. This is a visual-based measure that utilizes the Receiver Operating Characteristics (ROC) analysis of clustering results. In addition, ABROCA, a visual-based fairness measure, has been proposed for the classification problem [8]. This method evaluates the fairness of classification models through slicing analysis based on the ROC curves. ABROCA measures the absolute value of the area between the ROC curves of the protected and non-protected groups.

To this end, we propose a new fairness measure for FAir Clustering through ROC curves (shortly: FACROC) which takes into account the clustering quality on each value of the protected attribute. In particular, we use AUCC to measure the clustering quality and then compute the FACROC by the deviation between the ROC curves corresponding to each value of the protected attribute. Afterward, we perform the experiments on five popular datasets used in fairness-aware ML with three prevalent fair clustering models to evaluate the performance of FACROC versus other popular fairness measures.

The rest of our paper is structured as follows: Sect. 2 overviews the related work. The computation of the FACROC measure is described in Sect. 3. Section 4 presents the details of our experiments on various datasets and clustering models. Finally, the conclusion and outlook are summarized in Sect. 5.

2 Related Work

In this work, we focus on two main types of fairness measures for clustering including group-level fairness and individual-level fairness. Regarding group-level, *balance* is the first group fairness measure introduced by [7]. Next, the *bounded representation* measure was proposed with the aim of reducing imbalances in cluster representations of protected attributes [1]. This measure was generalized with two parameters α and β in the study of [3]. Subsequently, *social fairness* notion was introduced by [9], which aims to provide equitable costs for different clusters. Additionally, *diversity-aware* fairness was initiated by [16] which ensures a minimum number of cluster centers in the clustering are selected from each group. Based on the summarization task, the *fair summaries* measure was used to ensure that the data summary for each group is represented equally [12].

In contrast to group-level fairness, individual-level fairness notions aim to ensure that similar individuals are treated similarly. For example, Chen et al. [5] proposed the individual-level fairness notion of proportionality for centroid clustering to guarantee that points are treated equally. An individual-level concept that establishes a fair radius for clusters in center-based clustering objectives was presented by Jung et al. [11]. Chakrabarti et al. [4] provided algorithms for the k-center objective and proposed the idea of individual fairness, which guarantees that points receive comparable quality of service.

However, unlike other fairness measures for clustering based on the representation of protected attribute values in clusters, we propose a new visual-based fairness measure that takes into account the difference in the clustering performance for each group w.r.t. the protected attribute.

3 FACROC: A Fairness Measure for Fair Clustering

Given a dataset $\mathcal{X} = \{x_1, \ldots, x_n\}$ with n data points and a clustering $\mathcal{C} = \{C_1, C_2, ..., C_k\}$ with k clusters, the corresponding AUCC value is computed with the following steps [10]:

1. Compute a similarity matrix of the objects in the original dataset.
2. Acquire two arrays that present the pairwise relationship for each pair of objects:
   (a) Similarity: get from the similarity matrix.
   (b) Clustering: 1 if the pair belongs to the same cluster; otherwise, 0.
3. AUCC is obtained from the ROC analysis procedure with two above arrays, in which pairwise clustering memberships correspond to the "true classes" concept in classification.

Given a binary protected attribute $P = \{p, \overline{p}\}$; e.g., gender={female, male}, we inherit the concept of fairness from the ABROCA fairness measure for classification [8]: "*equal model performance across subgroups*". To this end, FACROC is defined as the absolute value of the area between the protected (ROC_p) and non-protected group ($ROC_{\overline{p}}$) curves across all possible thresholds $t \in [0, 1]$ of False Positive Rate (FPR) and True Positive Rate (TPR). The absolute difference between the two curves is measured to capture the case that the curves may cross each other.

$$\int_0^1 \mid ROC_p(t) - ROC_{\overline{p}}(t) \mid dt. \tag{1}$$

The value range: $FACROC \in [0, 1]$. The lower value indicates a lower difference in the clustering quality between the two groups and, therefore, a fairer model. Figure 1 visualizes the FACROC slice plot on the COMPAS dataset.

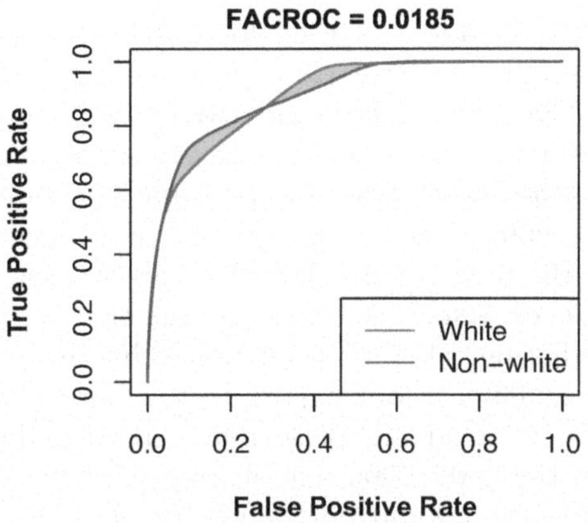

Fig. 1. An example of the FACROC slice plot.

4 Evaluation

In this section, we evaluate the performance of (fair) clustering models on well-known fairness measures for clustering and our proposed FACROC measure on prevalent fairness-aware datasets.

4.1 Datasets

We perform the experiments on five popular datasets for fairness-aware ML [14], as summarized in Table 1[1].

Table 1. An overview of the datasets

Datasets	#Instances	#Instances (cleaned)	#Attributes	Protected attribute	k*
Adult	48,842	45,222	15	Gender (F: 14,695; M: 30,527)	2
COMPAS	4,743	4,020	51	Race (NW: 2,561; W: 1,459)	7
Credit card	30,000	30,000	24	Gender (F: 18,112; M: 11,888)	2
German credit	1000	1000	21	Gender (F: 310; M:690)	2
Student-Mat.	395	395	33	Gender (F: 208, M: 187)	9
Student-Por.	649	649	33	Gender (F: 383; M: 266)	9

In particular, the ***Adult dataset***[2] is one of the most prevalent datasets for fairness-aware ML research. The class attribute (whether the income is greater than 50,000$) is removed because this study uses the dataset for clustering

[1] Abbreviations: F (Female), M (Male), W (White), NW (Non-White).
[2] https://archive.ics.uci.edu/dataset/2/adult.

tasks. The ***COMPAS dataset***[3] is used for crime recidivism risk prediction. We convert the protected attribute *Race* into a binary attribute with values *{White, Non-White}*. We remove datetime attributes (*compas_screening_date*, *dob*, etc.), defendants' name/ID, and the class label *two years recidivism*. The ***Credit card clients dataset***[4] contains information about 30,000 customers in Taiwan in October 2005. The prediction task is to forecast whether a customer will default in the next month. We remove the class label for clustering. The ***German credit dataset***[5] contains information about 1000 customers, with the goal of predicting whether a customer has good or bad credit. We also eliminate this class label in our experiments. The ***Student performance dataset***[6] details students' academic performance in secondary education at two Portuguese schools in the 2005–2006 school year, covering two subsets: Mathematics (shortly: Student-Mat.) and Portuguese (shortly: Student-Por.).

4.2 Experimental Setups

Clustering Models. We evaluate the performance of traditional k-means and hierarchical clustering, as well as well-known fair clustering models.

- **Fair clustering through fairlets** [7] (shortly: Fairlet): This is the first work on fair clustering at the group-level to ensure an equal representation of each value of the protected attribute in each cluster. A two-phase approach was proposed: 1) Fairlet decomposition: grouping all instances into "fairlets" which are small clusters that satisfy the fairness constraint; 2) Clustering on fairlets: applying standard clustering methods, such as k-center, k-median, to these fairlets to produce the final fair clusters.
- **Scalable fair clustering** [2] (shortly: Scalable): This is an extended investigation of the fair k-median clustering problem [7], with a new practical approximate fairlet decomposition algorithm that runs in nearly linear time. Therefore, this proposed approach can be applied to large datasets.
- **Proportionally fair clustering** [5] (shortly: Proportionally): The authors define proportional fairness: any group of n/k points should have the right to form their cluster if there exists a center closer for all n/k points. The goal is to find clustering where no subset of points has a justified complaint about their assigned cluster without assuming predefined protected groups.

Fairness Measures. We compare the proposed FACROC measure with the following well-known fairness measures:

- **Balance** [7]: Given a clustering $\mathcal{C} = \{C_1, C_2, ..., C_k\}$ with k clusters, $\mathcal{X}$ be a set of data points, $P = \{p, \bar{p}\}$ be the protected attribute, $\psi : \mathcal{X} \rightarrow P$ denotes

[3] https://github.com/propublica/compas-analysis.
[4] https://archive.ics.uci.edu/ml/datasets/default+of+credit+card+clients.
[5] https://archive.ics.uci.edu/dataset/144/statlog+german+credit+data.
[6] https://archive.ics.uci.edu/dataset/320/student+performance.

the demographic group to which the data point belongs, i.e., male or female, the balance of clustering $\mathcal{C}$ is computed by:

$$balance(\mathcal{C}) = min_{i=1}^{k} balance(C_i). \quad (2)$$

where:

$$balance(C_i) = \min\left(\frac{|\{x \in C_i \mid \psi(x) = p\}|}{|\{x \in C_i \mid \psi(x) = \bar{p}\}|}, \frac{|\{x \in C_i \mid \psi(x) = \bar{p}\}|}{|\{x \in C_i \mid \psi(x) = p\}|}\right). \quad (3)$$

- **Proportionality** [5]: Given $\mathcal{X}$ be a set of n data points, $\mathcal{Y}$ be a set of feasible cluster centers m and a number $\rho > 1$, we call a clustering $\mathcal{C} \subseteq \mathcal{Y}$ ($|\mathcal{C}| = k$) is *ρ-proportional* if $\forall S \subseteq \mathcal{X}$ with $|S| \geq \lceil \frac{n}{k} \rceil$ and for all $y \in \mathcal{Y}$, there exists $i \in S$ with $\rho \cdot d(i, y) \geq D_i(\mathcal{C})$, where $d(i, y)$ is the distance between points i and y, and $D_i(\mathcal{C}) = min_{x \in \mathcal{C}} d(i, x)$. When $\rho = 1$, we call this proportional fairness.

Parameter Selection. We denote the optimal number of clusters as k^*, which is determined based on the AUCC values using the implementation provided by [10]. Figure 2 illustrates the value of k^* for the COMPAS dataset; other values of k^* are presented in Table 1. The minimum balance threshold is set to 0.4 because the minimum balance score of all datasets is 0.4493 (corresponding to the German credit dataset).

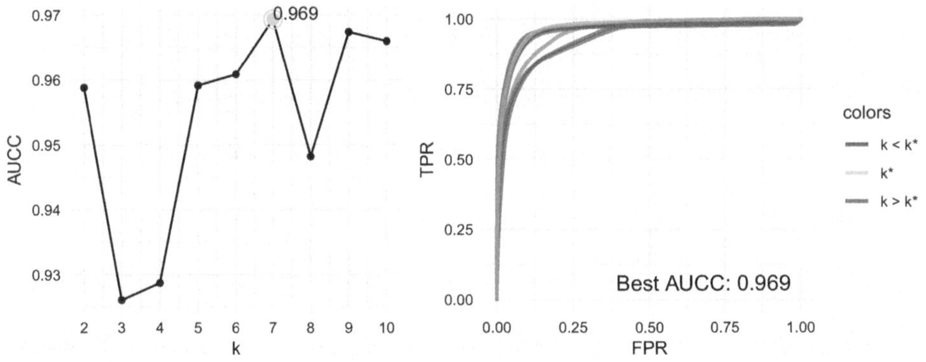

Fig. 2. COMPAS: selecting the optimal number of cluster (k^*) by AUCC.

4.3 Experimental Results

We present the results of clustering models regarding performance (Silhouette coefficient and AUCC) and fairness (Balance, Proportionality, and FACROC). The source code is available at https://github.com/tailequy/FACROC.

Adult Dataset. The performance of the clustering models is shown in Table 2 and Fig. 3, with the best results highlighted in **bold**. In terms of clustering

Table 2. Adult: performance of (fair) clustering models

Measures	k-means	Hierarchical	Fairlet	Scalable	Proportionally
Silhouette coefficient	**0.9861**	**0.9861**	0.4062	0.4377	0.3711
AUCC	**1.0000**	0.9998	0.6607	0.6569	0.8503
Balance	0.1684	0.1926	**0.5001**	0.4396	0.2966
Proportionally	1.0000	1.3274	1.4701	1.5994	**1.6321**
FACROC	**0.0000**	0.0054	0.0509	0.0602	0.0760

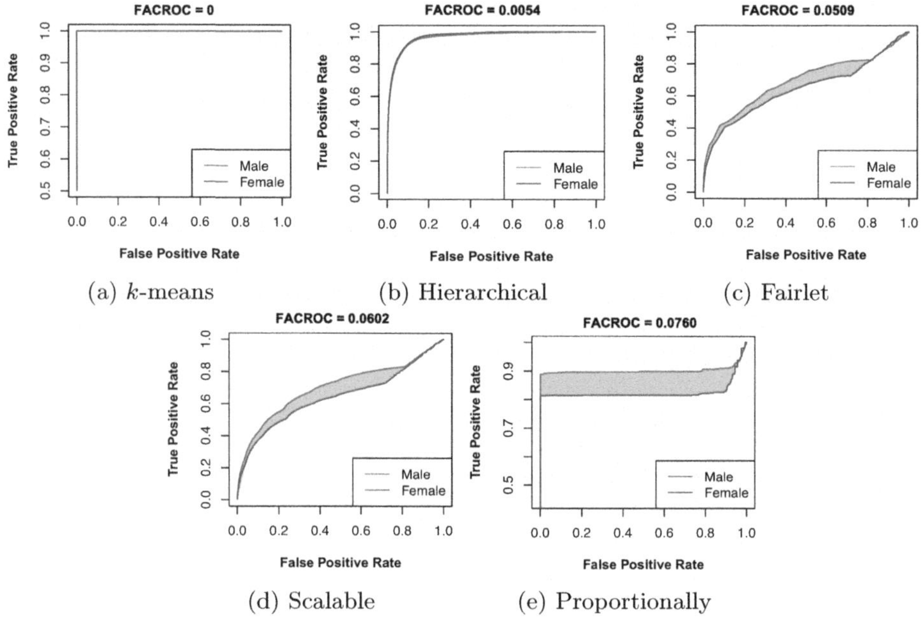

(a) k-means (b) Hierarchical (c) Fairlet (d) Scalable (e) Proportionally

Fig. 3. Adult: FACROC slice plots.

performance, obviously, k-means outperforms fair clustering methods. However, regarding fairness constraint, fair clustering models outperform k-means w.r.t. the fairness measure that they optimize. In detail, *Fairlet* and *Scalable* have better balance scores, while the *Proportionally* model is better in terms of the proportionality measure. This is explained by the fact that the definition of fairness is different in all models. Interestingly, the FACROC value of k-means is perfect, while the *Proportionally* model shows the worst result, i.e., the performances of observed fair clustering models are biased toward groups of people.

Table 3. COMPAS: performance of (fair) clustering models

Measures	k-means	Hierarchical	Fairlet	Scalable	Proportionally
Silhouette coefficient	**0.6110**	0.6082	0.3827	0.3583	0.3903
AUCC	**0.9690**	0.9689	0.9233	0.9236	0.8933
Balance	0.0000	0.1017	0.4186	**0.4291**	0.3226
Proportionality	0.8554	1.1728	0.9443	1.0937	**1.2861**
FACROC	**0.0029**	0.0044	0.0097	0.0064	0.0185

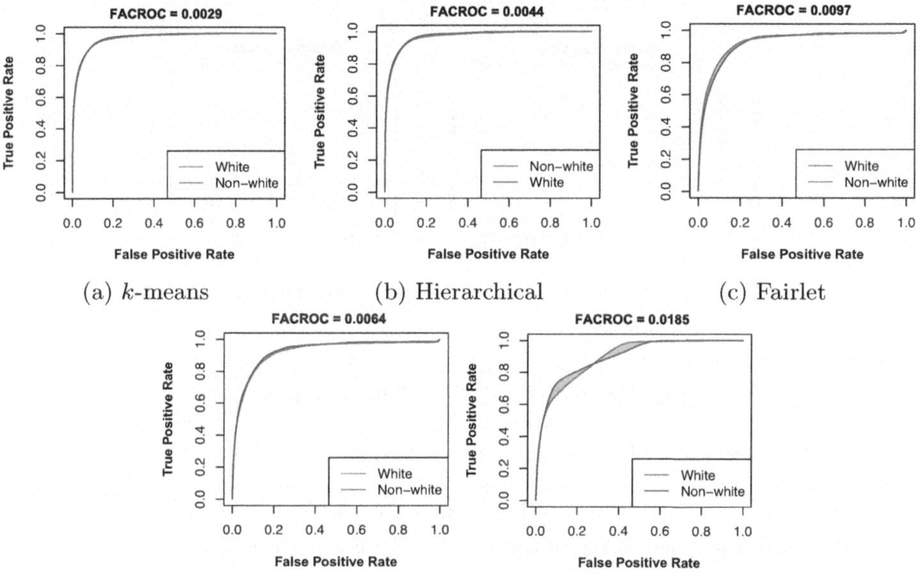

(a) k-means (b) Hierarchical (c) Fairlet

(d) Scalable (e) Proportionally

Fig. 4. COMPAS: FACROC slice plots.

Table 4. Credit card clients: performance of (fair) clustering models

Measures	k-means	Hierarchical	Fairlet	Scalable	Proportionally
Silhouette coefficient	0.5778	**0.6014**	0.3390	0.3476	0.5528
AUCC	0.9247	**0.9422**	0.7456	0.7607	0.9127
Balance	0.6145	0.5927	**0.6957**	0.6440	0.6477
Proportionality	1.0003	0.9939	1.0909	1.0828	**1.4935**
FACROC	**0.0060**	0.0083	0.0374	0.0405	**0.0060**

COMPAS Dataset. An interesting trend is also observed in the results of the clustering models in the COMPAS dataset (Table 3 and Fig. 4) when the FACROC of the k-means algorithm is significantly better than that of other models. This can be attributed to fair clustering models prioritizing fairness constraints, which can result in variations in clustering quality among groups.

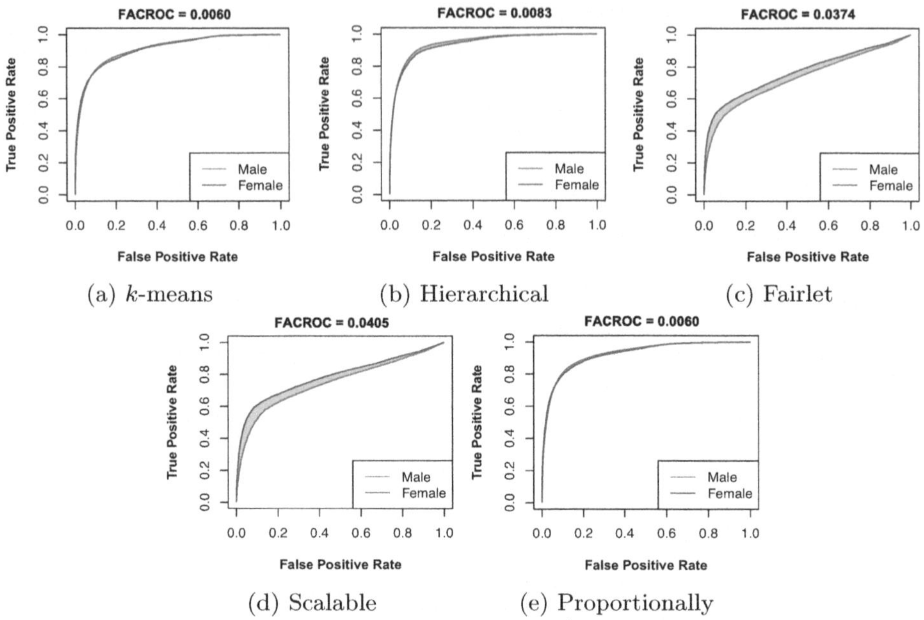

(a) k-means (b) Hierarchical (c) Fairlet (d) Scalable (e) Proportionally

Fig. 5. Credit card: FACROC slice plots.

Table 5. German credit: performance of (fair) clustering models

Measures	k-means	Hierarchical	Fairlet	Scalable	Proportionally
Silhouette coefficient	**0.7222**	0.6963	0.1387	0.3053	0.5302
AUCC	**0.9672**	0.9523	0.8046	0.8192	0.8440
Balance	0.3008	0.3314	**0.4545**	0.4045	0.3669
Proportionality	0.9469	0.9874	1.2964	1.2975	**1.3829**
FACROC	**0.0115**	0.0121	0.0875	0.0753	0.0625

Table 6. Student-Mat: performance of (fair) clustering models

Measures	k-means	Hierarchical	Fairlet	Scalable	Proportionally
Silhouette coefficient	**0.1814**	0.1526	0.0456	0.0552	0.1111
AUCC	**0.9117**	0.8931	0.8201	0.7705	0.8823
Balance	0.0000	0.1294	0.4231	**0.6176**	0.3182
Proportionality	0.9970	1.1295	1.2782	1.1510	**1.4167**
FACROC	**0.0098**	0.0153	0.0260	0.0484	0.0222

Credit Card Clients Dataset. Compared with fair clustering models, k-means is still the model with the highest clustering quality (Table 4 and Fig. 5). How-

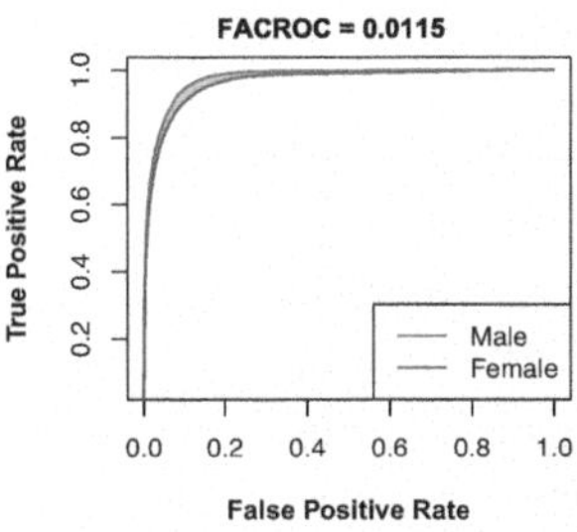

(a) k-means

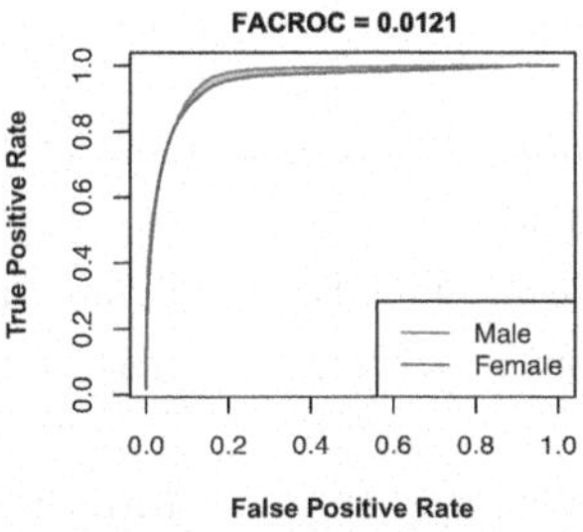

(b) Hierarchical

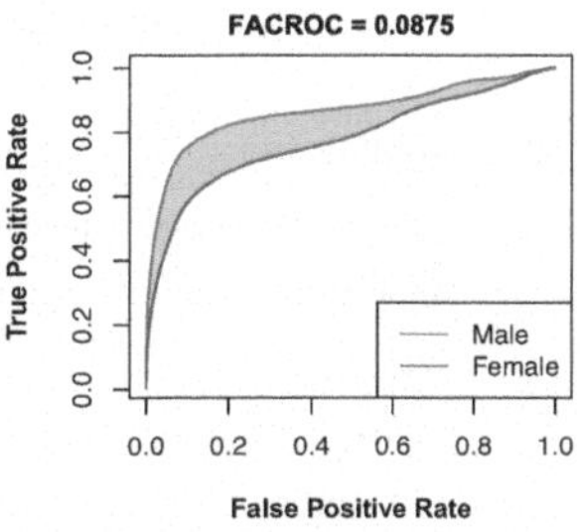

(c) Fairlet

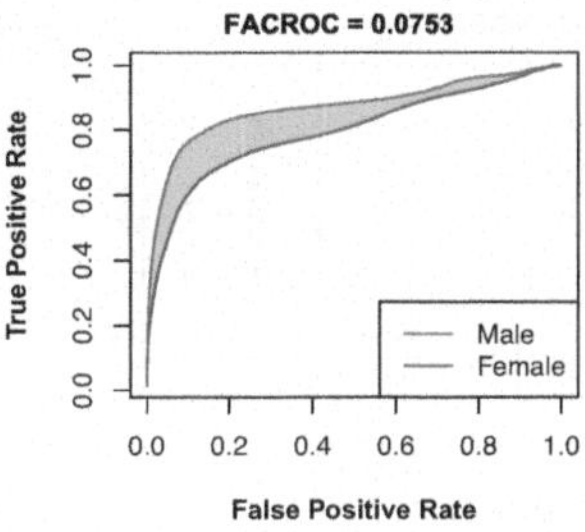

(d) Scalable

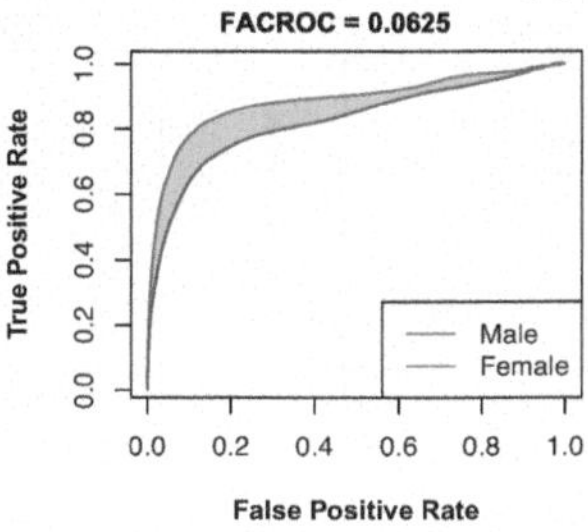

(e) Proportionally

Fig. 6. German credit: FACROC slice plots.

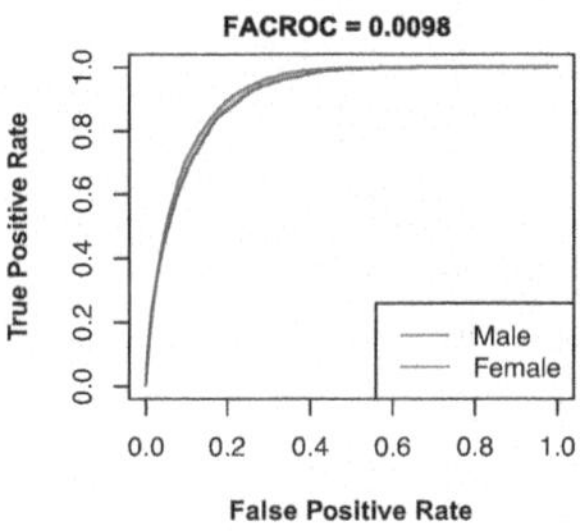

(a) k-means

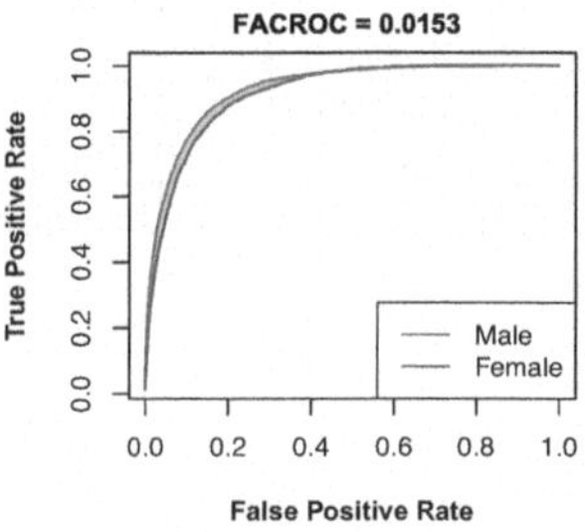

(b) Hierarchical

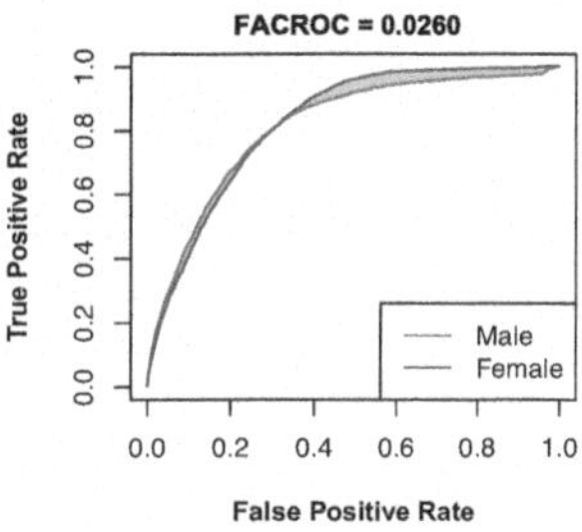

(c) Fairlet

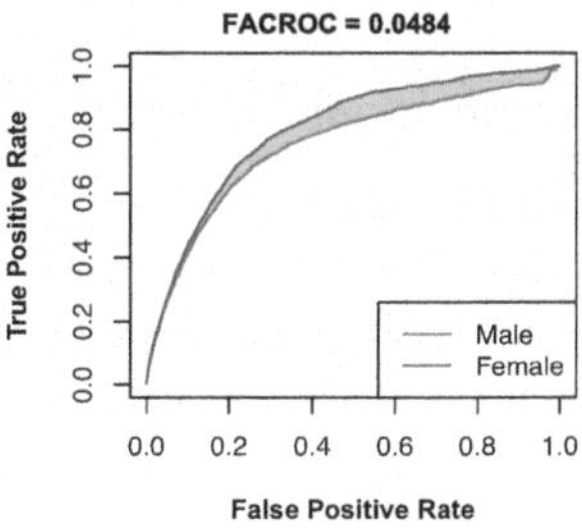

(d) Scalable

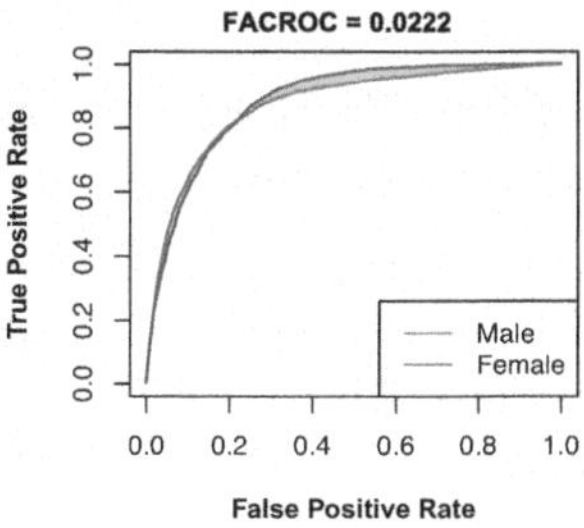

(e) Proportionally

Fig. 7. Student-Mat: FACROC slice plots.

Table 7. Student-Por: performance of (fair) clustering models

Measures	k-means	Hierarchical	Fairlet	Scalable	Proportionally
Silhouette coefficient	**0.1345**	0.1136	0.0242	0.0591	0.0618
AUCC	**0.8935**	0.8641	0.7389	0.8270	0.8459
Balance	0.3881	0.3972	0.4184	0.4231	**0.4483**
Proportionality	1.0341	1.3728	1.2142	1.1623	**1.4320**
FACROC	0.0083	0.0108	**0.0068**	0.0484	0.0244

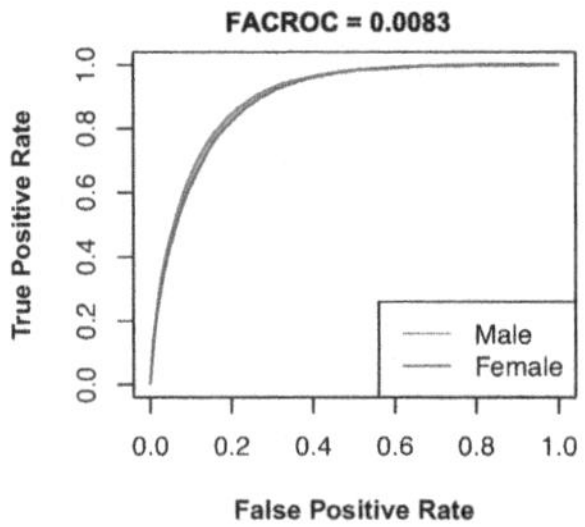

(a) k-means

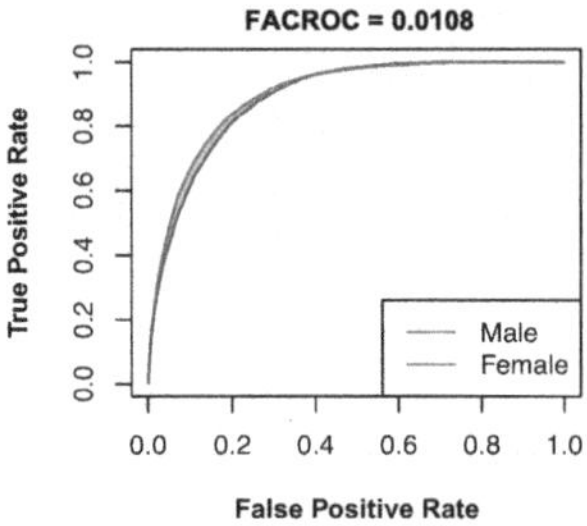

(b) Hierarchical

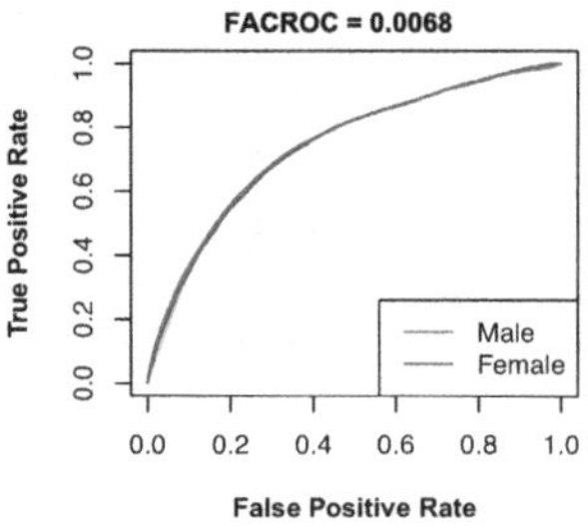

(c) Fairlet

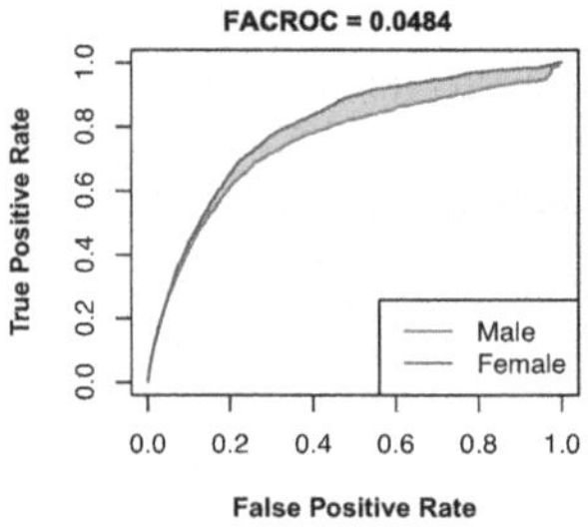

(d) Scalable

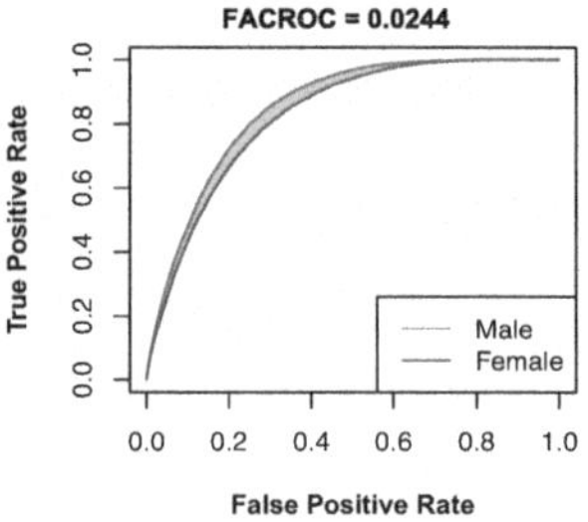

(e) Proportionally

Fig. 8. Student-Por: FACROC slice plots.

ever, in terms of the FACROC measure, *Propotionally* and k-means share the top rank with a very low value.

German Credit. In this dataset, a similar trend is observed in Table 5 and Fig. 6. k-means achieves the best results according to the FACROC measure, while fair clustering models have better results on the fairness measure they are designed to optimize.

Student Performance Dataset. In the Student-Mat subset (Table 6 and Fig. 7), in terms of clustering performance, k-means outperforms other clustering models, although its silhouette coefficient is low. Interestingly, *Scalable* fair clustering achieves the highest balance score due to having the lowest AUCC score.

Moreover, the k-means model once again achieves the highest FACROC value, followed by *Hierarchical clustering* and *Proportionally* fair clustering. However, a different trend emerges in the Student-Por subset (Table 7 and Fig. 8), where *Fairlet* outperforms other models in terms of FACROC, while *Proportionally* achieves the best balance score and proportionality.

5 Conclusions and Outlook

In this work, we introduced FACROC, a new fairness notion for fair clustering that leverages the ROC curves of clustering analysis w.r.t. a protected attribute. We evaluated our proposed fairness measure on several datasets and (fair) clustering models. The results demonstrate that our measure is effective in visualizing and analyzing the fairness of clustering models using ROC curves. Furthermore, the evaluation highlights significant variations among fairness measures due to differences in their definitions and objective functions. In the future, we plan to extend this work by developing a method to optimize the FACROC value and adapt it for multiple protected attributes.

References

1. Ahmadian, S., Epasto, A., Kumar, R., Mahdian, M.: Clustering without over-representation. In: KDD 2019, pp. 267–275 (2019)
2. Backurs, A., Indyk, P., Onak, K., Schieber, B., Vakilian, A., Wagner, T.: Scalable fair clustering. In: ICML 2019, pp. 405–413. PMLR (2019)
3. Bera, S., Chakrabarty, D., Flores, N., Negahbani, M.: Fair algorithms for clustering. Adv. Neural Inform. Process. Syst. **32** (2019)
4. Chakrabarti, D., Dickerson, J.P., Esmaeili, S.A., Srinivasan, A., Tsepenekas, L.: A new notion of individually fair clustering: α-equitable k-center. In: AISTATS 2022, pp. 6387–6408. PMLR (2022)
5. Chen, X., Fain, B., Lyu, L., Munagala, K.: Proportionally fair clustering. In: ICML 2019, pp. 1032–1041. PMLR (2019)
6. Chhabra, A., Masalkovaitė, K., Mohapatra, P.: An overview of fairness in clustering. IEEE Access **9**, 130698–130720 (2021)
7. Chierichetti, F., Kumar, R., Lattanzi, S., Vassilvitskii, S.: Fair clustering through fairlets. Adv. Neural Inform. Process. Syst. **30** (2017)
8. Gardner, J., Brooks, C., Baker, R.: Evaluating the fairness of predictive student models through slicing analysis. In: LAK 2019, pp. 225–234 (2019)
9. Ghadiri, M., Samadi, S., Vempala, S.: Socially fair k-means clustering. In: ACM FAccT 2021, pp. 438–448 (2021)
10. Jaskowiak, P.A., Costa, I.G., Campello, R.J.: The area under the roc curve as a measure of clustering quality. Data Mining Knowl. Dis. **36**(3), 1219–1245 (2022). https://github.com/pajaskowiak/clusterConfusion
11. Jung, C., Kannan, S., Lutz, N.: Service in your neighborhood: fairness in center location. Foundations Responsible Comput. (FORC) (2020)
12. Kleindessner, M., Awasthi, P., Morgenstern, J.: Fair k-center clustering for data summarization. In: ICML 2019, pp. 3448–3457. PMLR (2019)

13. Le Quy, T., Friege, G., Ntoutsi, E.: A review of clustering models in educational data science toward fairness-aware learning. Educational data science: Essentials, approaches, and tendencies: Proactive education based on empirical big data evidence, pp. 43–94 (2023)
14. Le Quy, T., Roy, A., Iosifidis, V., Zhang, W., Ntoutsi, E.: A survey on datasets for fairness-aware machine learning. Wiley Interdisciplinary Rev. Data Mining Knowl. Dis. **12**(3), e1452 (2022)
15. Saxena, A., et al.: A review of clustering techniques and developments. Neurocomputing **267**, 664–681 (2017)
16. Thejaswi, S., Ordozgoiti, B., Gionis, A.: Diversity-aware k-median: clustering with fair center representation. In: Oliver, N., Pérez-Cruz, F., Kramer, S., Read, J., Lozano, J.A. (eds.) ECML PKDD 2021. LNCS (LNAI), vol. 12976, pp. 765–780. Springer, Cham (2021). https://doi.org/10.1007/978-3-030-86520-7_47
17. Žliobaitė, I.: Measuring discrimination in algorithmic decision making. Data Min. Knowl. Disc. **31**(4), 1060–1089 (2017). https://doi.org/10.1007/s10618-017-0506-1

Learning on Complex Data

Action Sequence Analysis Using Temporal Commonsense Knowledge

Steven Lynden[1(✉)], Kyoung-Sook Kim[1], Akiyoshi Matono[1], Hai-Tao Yu[2], and Xin Liu[1]

[1] Artificial Intelligence Research Center, National Institute of Advanced Industrial Science and Technology (AIST), Tsukuba, Japan
{steven.lynden,ks.kim,a.matono,xin.liu}@aist.go.jp

[2] Institute of Library, Information and Media Science, Tsukuba University, Tsukuba, Japan
yuhaitao@slis.tsukuba.ac.jp

Abstract. Reasoning about temporal commonsense knowledge in action sequences is important for applications involving task planning, automation, anomaly detection, etc. This paper explores the fine-tuning of various language models including RoBERTa, SmolLM2, DistilBERT, and the edge-optimized smaller TinyBERT for effectively capturing and predicting temporal commonsense knowledge and action relationships while maintaining energy efficiency and suitability for deployment on low-resource devices. Leveraging an existing commonsense temporal action knowledge dataset, we modified and extended the data to evaluate the ability of several language models to classify action-describing natural language sentences into various temporal categories in terms of the time they take to perform and make accurate predictions about the relationships between action sequence goal/steps. Our results demonstrate that even relatively compact models such as TinyBERT can achieve competitive performance in predicting commonsense knowledge about action-describing sentences and also perform anomaly detection tasks. The findings highlight the trade-offs between model size and prediction accuracy, providing insights into the viability of using lightweight models for real-world applications.

Keywords: Temporal Commonsense Knowledge · Edge Computing · Action Sequence Planning · Low-Energy AI

1 Introduction

Planning action sequences often consists of a series of tasks where each action has an inherent duration and may affect subsequent steps. Understanding these temporal properties allows systems to anticipate outcomes, and detect deviations from expected behavior. In this work, we present a study on using edge-optimized language models such as TinyBERT [3] – lightweight models designed to achieve effective performance while maintaining the computational efficiency required

X. Wu et al. (Eds.): PAKDD 2025, LNAI 15875, pp. 355–367, 2025.
https://doi.org/10.1007/978-981-96-8295-9_26

for deployment on edge devices, such as personal assistant robotic devices, wearable devices etc. We demonstrate how temporal commonsense knowledge allows us to predict how long actions take and the longevity of their effects. Furthermore, categorizing actions into two basic temporal categories – being singular or periodic, allows automated reasoning about action sequences. We also experiment with introducing and injecting artificial anomalies to test the robustness of models in detecting various action sequence irregularities. The contributions of this paper are as follows:

- **Commonsense Temporal Knowledge Prediction**: We extend an existing commonsense temporal knowledge dataset and assess the ability of predicting temporal commonsense action knowledge using small to medium sized language models. We perform experiments on several language models in a variety of commonsense temporal action prediction tasks, showing results that highlight the ability of even small language models to make accurate predictions.
- **Action Sequence Prediction**: In addition to classifying action-describing sentences into temporal categories, we show how relationships such as whether an action relates to a specific goal or an action's position in a sequence of action can be predicted by the models, we refer to this task as action sequence prediction. For this task, we also investigate whether the temporal models can aid the action sequence models by adding additional parameters to the data used for fine-tuning.
- **Temporal/Sequence Anomaly Prediction**: We demonstrate how the trained models can facilitate anomaly detection by using models trained during the previous two tasks.

Our experiments highlight the importance of efficient models capable of maintaining reliable performance while operating under constrained computational resources. The results contribute to advancing capabilities towards processing temporal aspects of action sequences, ensuring enhanced decision-making and planning in edge-computing scenarios.

2 Temporal Commonsense Knowledge for Action Sequences

Understanding the temporal characteristics of actions is an integral part of human cognition, enabling seamless interaction with the world through well-coordinated plans and responses. Temporal commonsense reasoning encompasses the ability to assess how long an action takes, how persistent its effects are, and how often it occurs, allowing for accurate and adaptive task planning. In this paper, we focus on analyzing plans where each plan consists of a sequence of short action-describing natural language sentences, with an associated goal or objective that is also described as a short natural language sentence. Such plans may be formulated by humans or automated systems. We consider temporal commonsense knowledge as represented through three primary aspects:

- **Perform Duration**: the length of time an action takes to be performed, from initiation to completion. Capturing the perform duration of actions is vital for understanding the immediate temporal requirements of tasks. Actions may range from taking mere *seconds* (e.g., pressing a button) to lasting *days* or even *longer* (e.g., growing a plant).
- **Effect Duration**: The effect duration refers to the persistence of the consequences or outcomes of an action – how long the impact of an action remains after its completion. For instance, turning off a light switch has an effect duration that might be described as *minutes* or *hours*, while planting a tree has an effect that could be classified as *days* or *longer*.
- **Occurrence Type**: This aspect denotes whether an action occurs as a singular event or is part of a repeated or continuous sequence. For example, taking daily medication represents a **periodic** occurrence, whereas submitting a final project might be a **singular** event.

Each temporal property is categorized into distinct classes: **seconds**, **minutes**, **hours**, **days**, and **longer** for durations, while the **occurrence type** is classified as either **singular** or **periodic**. Understanding and predicting these properties in action sequences is crucial for systems that need to plan, reason, and adapt in dynamic environments. The challenge of incorporating temporal commonsense reasoning into computational models lies in accurately predicting these properties for a given sequence of actions and using this information to identify anomalies.

3 Datasets and Models

This section describes the datasets used, the enhancements made for our study and the models used.

3.1 WikiHow and the Temporal Commonsense Action Knowledge (CoTAK) Dataset

The base dataset utilized in our paper is the Commonsense Temporal Action Knowledge (CoTAK) dataset [5], which in turn builds on data extracted from WikiHow, an online how-to repository comprising over hundreds of thousands of articles over diverse domains ranging from lifestyle improvements to specific tasks e.g. the usage of specific software etc. These articles are curated by both professionals and community contributors, and the volume of data available allows generalization of the nature of action sequences. Action sequences involve goals followed by a sequence of steps. The CoTAK dataset extracts action sequences from WikiHow articles, in the form of goal titles and step titles, each consisting of short natural language sentences, and provides annotations for each step in these sequences with two key temporal attributes: perform duration and effect duration. The perform duration refers to the approximate amount of time required to carry out an action. For example, actions like "chop an onion" may take a few minutes, while "boil water" could take several minutes to half an hour. The effect duration, on the other hand, indicates the expected duration of the results or impact of an action. For instance, the effect of "cleaning a room" might last for several days or weeks, depending on usage, while "sending an email" could have an effect lasting indefinitely as the information remains available to the recipient. Labels are provided by human annotators. Furthermore, the dataset provides a high confidence subset of the dataset, subsequently referred to as *high-confidence*, on which all annotators agree, representing probable annotations which are more objective rather than subjective. The v1.0-beta version of the dataset was downloaded and split (80/20) randomly into test/train datasets for the purpose of study for both the complete dataset (all) and the high confidence part of the dataset. Individual items in this dataset consist of a **goal** corresponding to the WikiHow article title (or method where articles are divided into different methods for achieving the same goal), and a **step**, representing an individual action in the sequence. Additionally each item is labeled with the perform/effect temporal durations of the individual action. In this sense, items represent individual actions and the goal represents the context of the action. Some actions are identical (although attached to different goals, there are no identical goal + step pairs). Resulting dataset statistics are summarized in Table 1.

3.2 Enhancements Made for Our Study

We are motivated in this study to delve deeper into the temporal commonsense understanding of actions by introducing an additional temporal attribute: *occurrence type*. This attribute categorizes actions as either *singular* or *periodic*, depending on whether the action typically occurs once or is repeated over time. For example, an action like "bake a cake" would be annotated as *singular*, while "water the plants" would be annotated as *periodic*. We also introduce an *unknown* class for sentences describing actions that are not classifiable due to inherent subjectivity etc. To incorporate this enhancement, we employed GPT-4o[1] to annotate the dataset with occurrence types. To ensure the reliability of these annotations, we conducted a manual verification process on a random sample of 100 annotations. The results of this validation indicated a high degree of accuracy. The only disagreements between the validator and the GPT-4o classifications occurred in highly subjective action-describing sentences, confirming the utility of the automated approach for generating occurrence type annotations. This enhancement adds another dimension to the CoTAK dataset, enabling a richer understanding of temporal patterns in action sequences and supporting more nuanced applications in task planning and temporal reasoning. More details about the enhancements made, GPT-4o prompts utilized and a description of the datasets used in our subsequent experimental sections can be found online[2].

---

[1] https://openai.com/gpt-4.
[2] https://github.com/nsjl/aakg-data.

Table 1. Summary of dataset sizes and unique values for selected columns in All Train, All Test, high-confidence Train, and high-confidence Test datasets.

Dataset	Size	Goal (Unique)	Step (Unique)
All Train	271,286	38,405	245,944
All Test	33,818	22,006	32,728
High-Confidence Train	98,640	31,667	92,869
High-Confidence Test	12,360	10,077	12,125

3.3 Models Used in Our Experiments

This section introduces the models used and their characteristics and suitability for deployment in different environments. Below we explain in detail each model used and the rationale for their adoption.

Transformer-based models, notably BERT [2] and its numerous derivatives, have demonstrated high performance in NLP tasks involving context-dependent reasoning. RoBERTa [4], an optimized variant of BERT, has been shown to improve pre-training efficiency and performance on downstream tasks. To adapt these models for edge deployment, smaller and more efficient variants have been developed, such as DistilBERT [7], which retains 97% of BERT's performance while being 60% faster. Similarly, TinyBERT [3] and SmolLM2 [1] have been designed for low-latency, energy-efficient use cases, making them suitable for real-time anomaly detection in constrained environments. Models used in our study are compared in Table 2 and described below.

Table 2. Comparison of language models used for temporal commonsense reasoning, showing parameter count, inference time, power consumption, and deployment suitability.

Model	Params (M)	Type	Suitability
SmolLM2	135	Lightweight	Edge/mobile
RoBERTa-Base	125	Full Model	Server/cloud
TinyBERT	14	Distilled	Edge/mobile/embedded
DistilBERT	66	Distilled	Edge/server

SmolLM2: SmolLM2 is a lightweight language model with approximately 135 million parameters. It is designed for efficient inference and has low computational overhead. The inference time and power consumption for SmolLM2 are very low, making it highly suitable for deployment on edge and mobile devices. This model strikes a balance between performance and energy efficiency, which is essential for real-time applications in constrained environments.

RoBERTa-Base: RoBERTa-Base is a robust, full-sized transformer model with around 125 million parameters. It employs an optimized BERT-like architecture that enhances performance across a range of NLP tasks. However, the increased complexity leads to higher inference times and power consumption. Due to these characteristics, RoBERTa-Base is more suitable for deployment on powerful servers or cloud infrastructures where computational resources are available.

TinyBERT: With only about 14 million parameters, TinyBERT is a distilled version of BERT, specifically optimized to retain as much of BERT's performance as possible while significantly reducing size and computational cost. It has the lowest power consumption of the models in this study, making it ideal for edge, mobile, and embedded applications.

DistilBERT: DistilBERT, with approximately 66 million parameters, is another distilled model aimed at providing a balance between performance and efficiency, with power consumption and inference times somewhere in between TinyBERT and the other models.

4 Commonsense Temporal Knowledge Prediction

In this section we describe experiments for predicting commonsense temporal action knowledge using the four models, followed by a presentation and discussion of their results. Prediction of perform and effect period corresponds to a multiclass classification problem with five categories and the occurrence type experiments correspond to a multiclass classification problem with three categories due to the introduction of the unknown class.

4.1 Experiments

We performed experiments aimed at predicting perform/effect duration and occurrence type. Two types of experiments were performed:

- Step Only: the models are fine-tuned and evaluated by an input that consists only of the short sentence describing the action being performed.
- Goal + Step: the models are fine-tuned and evaluated by an input that consists of the goal of the overall action sequence and the short sentence describing the action.

The full and high-confidence training sets were used to fine tune each model with the initial split shown in Table 1. The implementation of each training/test process utilized the models and Transformers library from the HuggingFace Model Hub[3] with default configurations and input sequences tokenized using

[3] https://huggingface.co/HuggingFaceTB/SmolLM2-135M-Instruct, https://huggingface.co/FacebookAI/roberta-base, https://huggingface.co/huawei-noah/TinyBERT_General_4L_312D, https://huggingface.co/distilbert-base-uncased.

the model-specific tokenizers from HuggingFace, ensuring compatibility with each model. A fixed length of 128 tokens was chosen to balance computational efficiency and context preservation, preventing excessive truncation while maintaining batch consistency. The temporal attributes (e.g., perform duration categories) were encoded as categorical labels. Hyperparameters (2.6×10^{-5} learning rate, batch size 32, weight decay 0.05, 6 epochs) were selected via exploratory tuning to balance convergence speed and generalization. Performance differences across models were minimal during tuning, so the aforementioned hyperparameters were adopted as a unified set across all models. To assess model performance, we measured accuracy and F1 scores for predicting the temporal attributes and action perform/effect/occurrence type prediction.

4.2 Results

Results, shown in Table 3, show the performance of each model on the three prediction tasks, where the close alignment between accuracy and F1 scores observed in the results reflects the robustness and consistency of the model's performance across tasks. This indicates that the dataset is well-balanced, with no significant bias toward any particular class. Furthermore, the close values suggest that the model effectively handles both precision and recall, maintaining a good trade-off between the two. Overall, it can be seen that perform period prediction is more accurate, owing to the objective nature of the labeling of the duration of perform tasks, whereas effect labeling is more subjective. Occurrence type prediction was shown to be accurately captured and predicted by the models.

5 Action Sequence Prediction

In this section we analyze how well predictions can be made of the relative ordering of actions and the relations between goal/step. We also investigate whether utilizing the output of the trained temporal prediction models can enhance the prediction of action sequence information.

5.1 Experiments

We perform two experiments:

- Goal/Step Alignment: this consists of a binary classification task, where the input to the models is in the form of a goal and step with the aim being to predict whether a described goal matches a step. Data was created by selecting goal/step matches from the high confidence portion of the CoTAK dataset and randomly inserting goal/step mismatches. The number of matches/mismatches was made equal to ensure a balanced dataset for training and evaluation.

Table 3. Accuracy and F1 scores for predicting *perform*, *effect*, and *occurrence* using different models, datasets, and configurations.

Perform Period Prediction			
Model	**Dataset**	**Acc (F1)**	
		Step only	**Goal + Step**
SmolLM2	high-confidence	0.64 (0.63)	0.64 (0.63)
	All	**0.76** (0.74)	0.72 (0.70)
TinyBERT	high-confidence	0.63 (0.61)	0.7 (0.68)
	All	0.7 (0.69)	0.7 (0.69)
DistilBERT	high-confidence	0.65 (0.64)	0.65 (0.63)
	All	0.71 (0.69)	0.72 (0.71)
RoBERTa-Base	high-confidence	0.66 (0.64)	**0.74** (0.73)
	All	0.74 (0.72)	**0.74** (0.71)

Effect Period Prediction			
Model	**Dataset**	**Acc (F1)**	
		Step only	**Goal + Step**
SmolLM2	high-confidence	0.55 (0.53)	0.57 (0.54)
	All	**0.64** (0.62)	**0.61** (0.6)
TinyBERT	high-confidence	0.53 (0.52)	0.57 (0.56)
	All	0.59 (0.57)	0.58 (0.56)
DistilBERT	high-confidence	0.55 (0.53)	0.56 (0.55)
	All	0.59 (0.58)	**0.61** (0.6)
RoBERTa-Base	high-confidence	0.57 (0.55)	0.56 (0.55)
	All	0.61 (0.6)	**0.61** (0.6)

Occurrence Type Prediction		
Model	**Dataset**	**Acc (F1)**
		Goal + Step
SmolLM2	high-confidence	0.78 (0.8)
	All	0.79 (0.81)
TinyBERT	high-confidence	0.72 (0.75)
	All	0.75 (0.78)
DistilBERT	high-confidence	0.8 (0.78)
	All	**0.82** (0.81)
RoBERTa-Base	high-confidence	0.8 (0.82)
	All	**0.82** (0.84)

- Step position prediction: The second part of the study focuses on prediction the position that an action occurs in a sequence, where the possible values are first step, mid step, and end step encompassing a multiclass classification problem with three categories. For the positional experiments, goal/step pairs were extracted from the first, last, and middle positions within sequences. Since middle positions were more frequent, the dataset was balanced by randomly removing excess middle-position samples.

For each of the above, we perform the experiments using both basic and enhanced methods. For the basic method, only the goal/step input is used to fine-tune and evaluate the model. For the enhanced methods, the input incorporates temporal predictions generated by the models fine-tuned in the previous section from the input goal and step (for goal/step alignment) and step (for step position prediction), along with perform duration, occurrence type, and effect duration, all formatted as part of the input sequence, encoded into a numerical representation for the class labels and separated using the special `[SEP]` token. The enhanced approach aims to provide richer contextual information, leveraging both textual and temporal features to improve predictions for alignment and positional accuracy. Train/test splits, text preprocessing, including tokenization, padding, and truncation, was performed in the same manner as described earlier in the commonsense temporal knowledge prediction experiments.

5.2 Results

The results, shown in Table 4, highlight that goal/step prediction, essentially showing an alignment between the goal objective of an action sequence and the step performed can be predicted to be true/false very reliably with accuracy up to 90%. Furthermore positional location (first/last/middle) can also be captured by the models to a lesser extent showing the added difficulty of the task.

Table 4. Results for prediction of goal/step alignment and step position

Goal/Step Alignment		
Metric	**Acc (F1)**	
	Basic	**Enhanced**
SmolLM2	0.89 (0.89)	**0.9** (0.9)
TinyBERT	0.82 (0.81)	0.82 (0.81)
DistilBERT	0.89 (0.87)	0.88 (0.85)
RoBERTa	**0.9** (0.9)	**0.9** (0.9)

Step Position Prediction		
Metric	**Acc (F1)**	
	Basic	**Enhanced**
SmolLM2	0.61 (0.58)	0.62 (0.62)
TinyBERT	0.69 (0.61)	0.69 (0.61)
DistilBERT	**0.7** (0.62)	0.69 (0.63)
RoBERTa	**0.7** (0.62)	**0.7** (0.65)

6 Temporal/Sequence Anomaly Prediction

In this section we present a study that focuses on the categorization of anomalies artificially introduced into a set of action sequences. We use only the high confidence data from the CoTAK dataset, and use only the TinyBERT trained models described in the previous section, as we intend to demonstrate how a small model optimized for low-resource settings performs. We focus on temporal analysis of the perform period and the prediction of the goal/step alignment of actions only, utilizing the best performing models for temporal/sequence prediction achieved in the previous two sections.

6.1 Experiments

In this section, we present the results of our action sequence anomaly prediction experiment by artificially inserting anomalies into the dataset.

Anomaly Types and Rationale. We define four main types of anomalies for action sequences:

- **No Anomaly:** The action sequence is executed as expected, and both the predicted label and perform period align with their observed counterparts.
- **Goalstep Anomaly:** The step taken does not align with the expected label but matches the expected perform period.
- **Temporal Anomaly:** The step matches the expected label but is performed over an unexpected duration (i.e., the predicted perform period differs from the observed one).
- **Goalstep Temporal Anomaly**: Both the step label and the perform period differ from the expected values.

No anomaly is predicted if the perform classification model and the goal/step prediction model both agree. Goalstep anomaly is predicted if the perform period model agrees, but the goal/step model differs from the ground truth. Temporal Anomaly is predicted if the goal/step prediction matches the ground truth, but the predicted perform period deviates from the actual duration. Goalstep Temporal Anomaly is predicted when both models disagree with the ground truth. To evaluate the effectiveness of our approach, we randomly injected anomalies into an existing set of action sequences. This dataset contained four distinct types of anomalies that mirror potential real-world deviations in planned actions. The distribution of the anomaly types in the dataset is as follows:

- Goalstep Temporal Anomaly: 1,310 instances
- No Anomaly: 1,280 instances
- Temporal Anomaly: 1,254 instances
- Goalstep Anomaly: 1,218 instances

6.2 Results

Table 5 summarizes the performance of the TinyBERT-trained models on anomaly type prediction. The results demonstrate that reasonable prediction accuracy is achieved, in particular the overall prediction of anomaly type is accurately predicted, and the results show that action sequence prediction models using the prediction of goal/step alignment are more effective than the temporal models.

Table 5. Performance Metrics for Anomaly Type Prediction, Goalstep Prediction, and Perform Period Prediction

Anomaly Detection		
Metric	F1 Score	Accuracy
Overall Anomaly Type Prediction	0.6691	0.6106
Goalstep Anomaly Prediction	0.8202	0.8202
Temporal Anomaly Prediction	0.47	0.38

7 Discussion and Results Significance

The results indicate that while TinyBERT achieves relatively high performance on the goalstep prediction task, the accuracy and F1 score for perform period predictions are lower due to the more difficult nature of the prediction task. Anomaly type prediction, which combines the results of both goal/step and perform period predictions, achieves a moderate F1 score and accuracy. Overall, these results demonstrate that TinyBERT can be effectively utilized for lightweight anomaly detection tasks on edge devices, paving the way for practical applications such as real-time monitoring in robotics and automated systems.

8 Related Work

Various works have explored the automation of reasoning about the temporal properties of individual actions. Notably, the authors of [10] crowdsource the McTACO dataset, with a wide-range of annotation categories including event frequency and duration. In contrast to our focus they consider individual atomic actions in contrast to action sequences, however their work could complement our study in the practical application of automated action sequence planning etc. Various question answering approaches, e.g. [6] could also complement our work. Research in the wider field of general commonsense reasoning in automated systems such as COMET [8] are related and complementary to our work.

Most closely related to our work, Zhang and Lyu [9] propose a framework for reasoning about goals, steps, and temporal ordering using data extracted from

WikiHow. Their approach focuses on identifying and organizing action steps required to achieve a specific goal, with particular attention to the sequential dependencies and logical ordering of steps. While their work primarily emphasizes the structural organization and temporal alignment of action sequences, our study extends beyond sequence modeling by incorporating detailed temporal attributes such as the perform and effect duration of actions. Both studies leverage WikiHow as a resource, but our focus on temporal annotations expands its applicability to tasks requiring nuanced temporal reasoning such as anomaly detection as demonstrated in our paper.

9 Conclusions and Future Work

This study explored the capabilities of edge-optimized language models, including SmolLM2, TinyBERT, and DistilBERT, for temporal commonsense reasoning and anomaly detection in action sequences. Using the Commonsense Temporal Action Knowledge (CoTAK) dataset, we extended the data with artificially injected anomalies to evaluate model robustness. The results demonstrate that lightweight models can effectively reason about temporal properties such as perform duration, effect duration, and occurrence type. Additionally, these models show promise in detecting and classifying anomalies in action sequences. Our findings highlight the potential of compact models for low-energy applications, balancing performance and computational efficiency.

Acknowledgement. This paper is based on results obtained from a project, JPNP20006, commissioned by the New Energy and Industrial Technology Development Organization (NEDO) and the policy-based project "R&D on Generative AI Foundation Models for the Physical Domain".

References

1. Allal, L.B., et al.: SmolLM2 - with great data, comes great performance (2024). https://huggingface.co/HuggingFaceTB/SmolLM2-1.7B
2. Devlin, J., Chang, M.W., Lee, K., Toutanova, K.: BERT: pre-training of deep bidirectional transformers for language understanding. In: Proceedings of the 2019 Conference of the North American Chapter of the Association for Computational Linguistics: Human Language Technologies, vol. 1, pp. 4171–4186. Association for Computational Linguistics (2019). https://doi.org/10.18653/v1/N19-1423. https://aclanthology.org/N19-1423
3. Jiao, X., et al.: TinyBERT: distilling BERT for natural language understanding. In: Findings of the Association for Computational Linguistics: EMNLP 2020, pp. 4163–4174. Association for Computational Linguistics (2020). https://doi.org/10.18653/v1/2020.findings-emnlp.372
4. Liu, Y., et al.: RoBERTa: a robustly optimized BERT pretraining approach. arXiv preprint arXiv:1907.11692 (2019)

5. Lynden, S., et al.: Commonsense temporal action knowledge (coTAK) dataset. In: Proceedings of the 32nd ACM International Conference on Information and Knowledge Management (CIKM 2023), pp. 5361–5365 (2023)
6. Ning, Q., Wu, H., Roth, D.: TORQUE: a temporal ordering and reasoning corpus explored. In: Proceedings of the 2018 Conference on Empirical Methods in Natural Language Processing (EMNLP), pp. 2543–2553. Association for Computational Linguistics (2018)
7. Sanh, V., Debut, L., Chaumond, J., Wolf, T.: DistilBERT, a distilled version of BERT: smaller, faster, cheaper and lighter. In: 5th Workshop on Energy Efficient Machine Learning and Cognitive Computing - NeurIPS 2019 (2019)
8. Zellers, R., Holtzman, A., Bisk, Y., Farhadi, A., Choi, Y.: COMET: commonsense transformers for automatic knowledge graph construction. In: Proceedings of the 58th Annual Meeting of the Association for Computational Linguistics (ACL), pp. 4762–4779. Association for Computational Linguistics (2020)
9. Zhang, L., Lyu, Q.: Reasoning about goals, steps, and temporal ordering with WikiHow. In: Proceedings of the Association for Computational Linguistics (2023)
10. Zhou, B., Khashabi, D., Ning, Q., Roth, D.: Going on a vacation takes longer than going for a walk: a study of temporal commonsense understanding. In: Proceedings of the 2019 Conference on Empirical Methods in Natural Language Processing and the 9th International Joint Conference on Natural Language Processing (EMNLP-IJCNLP), pp. 3363–3369 (2019)

Foundation Model for Lossy Compression of Spatiotemporal Scientific Data

Xiao Li(✉), Jaemoon Lee, Anand Rangarajan, and Sanjay Ranka

University of Florida, Gainesville, FL 32611, USA
{xiao.li,j.lee1}@ufl.edu, {anand,ranka}@cise.ufl.edu

Abstract. We present a foundation model (FM) for lossy scientific data compression, combining a variational autoencoder (VAE) with a hyper-prior structure and a super-resolution (SR) module. The VAE framework uses hyper-priors to model latent space dependencies, enhancing compression efficiency. The SR module refines low-resolution representations into high-resolution outputs, improving reconstruction quality. By alternating between 2D and 3D convolutions, the model efficiently captures spatiotemporal correlations in scientific data while maintaining low computational cost. Experimental results demonstrate that the FM generalizes well to unseen domains and varying data shapes, achieving up to 4× higher compression ratios than state-of-the-art methods after domain-specific fine-tuning. The SR module improves compression ratio by 30% compared to simple upsampling techniques. This approach significantly reduces storage and transmission costs for large-scale scientific simulations while preserving data integrity and fidelity.

Keywords: Foundation Models · Data Compression · Spatiotemporal Scientific Data

1 Introduction

Foundation models (FMs) have attracted significant attention across various domains due to their remarkable versatility and adaptability. In recent years, models like GPT [2] and SAM [12] have been proposed to tackle a wide range of tasks, from natural language processing to image task. At their core, these models leverage transfer learning, which allows them to be pre-trained on large, diverse datasets and then fine-tuned for specific tasks. This paper develops a FM for scientific data compression, an area where these models have largely been unexplored. Scientific simulations often generate vast amounts of data per run, especially on high-performance computing systems. Current popular lossy compressors for scientific data largely rely on traditional methods such as wavelet transforms, singular value decomposition, or domain-specific algorithms [7,15].

This work was partially supported by DOE RAPIDS2 DE-SC0021320 and DOE DE-SC0022265.

X. Wu et al. (Eds.): PAKDD 2025, LNAI 15875, pp. 368–380, 2025.
https://doi.org/10.1007/978-981-96-8295-9_27

While these approaches have proven effective, they often struggle to fully capture the intricacies and high-dimensional relationships inherent in complex simulation data. In contrast, machine learning-based methods have demonstrated remarkable success across multiple domains, suggesting that similar approaches could yield significant improvements in scientific data compression. Given the increasing demand for efficient data storage and processing in scientific simulations, there is a pressing need for research into the application of FMs in the scientific domain and is the main focus of this paper. Foundational models for lossy scientific data compression have to address the following key challenges:

- Varying Physics and Spatiotemporal Correlations: Scientific simulations involve complex physics with different spatiotemporal correlations that vary across domains. These variations make it difficult for a single FM to effectively capture all patterns.
- Generalization Across Domains: Scientific data varies significantly across fields, making it hard for FMs to generalize without extensive domain-specific fine-tuning.
- Wide Data Range and Distribution: Unlike images, scientific datasets have a broader range of values and more complex distributions, including outliers, which pose difficulties for model training and compression.
- Guaranteed Error Bounds: Lossy compression in scientific data must ensure that errors do not compromise downstream analyses or simulations. Achieving high compression while maintaining guaranteed error bounds on the primary data.
- Multi-resolution output: Decompression should allow for variable output sizes with different resolution depending on the requirement of the application and available hardware resources.

The proposed foundation model (FM) consists of two main components: a variational autoencoder (VAE) with a hyper-prior structure and a super-resolution (SR) module. The VAE framework incorporates a hyper-prior structure to capture dependencies in the latent space, enhancing compression efficiency. The SR module improves reconstruction quality by refining low-resolution representations. By integrating 3D convolutional layers, the model effectively captures spatiotemporal correlations, enabling efficient representation and compression of complex scientific data with temporal dependencies. The contributions of this paper are as follows:

- We propose a foundation model that combines VAE and super-resolution architectures for lossy scientific data compression, addressing all the described challenges.
- We demonstrate that adapting the SR module in the decoder improves compression performance by up to 30% compared to a standard upsampling decoder.
- Experimental results show that the proposed model generalizes well to unseen domains and varying data shapes, achieving up to 4× higher compression ratios compared to state-of-the-art compressors.

2 Related Work

Error-bounded lossy compression provides reliable error control, which is crucial for scientific applications. These methods can be broadly classified into traditional approaches and machine learning-based techniques.

Traditional Methods: Transform-based methods include ZFP [7], which partitions data into 4D blocks and applies nearly orthogonal transformations. MGARD [3] employs a multigrid approach to transform data into multilevel coefficients, enabling efficient error-bounded compression. Prediction-based methods focus on predicting data points based on adjacent values. SZ [15] predicts each point and encodes the residuals, while DPCM [22] encodes differences between sequential samples. FAZ [18] integrates prediction models and wavelets, providing a flexible framework for compression.

Machine Learning-Based Methods: Machine learning-based methods offer dynamic and adaptive compression strategies, improving performance and efficiency. AE-SZ [17] combines SZ with autoencoder-based compression, while SRN-SZ [16] integrates machine learning-driven super-resolution with SZ for hierarchical, error-controlled compression. [13] introduces a region-adaptive compression approach by combining a UNet ROI detector with an autoencoder, and [14] proposes an attention-based network to effectively capture key information from hyper-blocks. Despite their promise, machine learning-based methods for scientific data compression remain underexplored.

Recent advances in natural image compression, such as Variational Autoencoders (VAEs) [5,21], have demonstrated significant potential for lossy data compression. VAEs are particularly effective for modeling the distortion-rate trade-off during training, making them popular for lossy compression. SR models [4,6], which reconstruct high-resolution data from low-resolution inputs, also show promise for compression by recovering fine details and improving reconstruction quality.

However, existing machine learning-based methods predominantly focus on 2D image compression using simplistic decoders, which fail to fully capture the spatiotemporal correlations present in scientific datasets. To overcome this limitation, we extend VAE [21] from 2D image compression to spatiotemporal data and incorporate a SR module as the decoder. This approach not only preserves the temporal dynamics inherent in the data but also enhances the reconstruction fidelity, making it more suitable for complex scientific datasets with intricate spatiotemporal patterns.

3 Methodology

3.1 VAE with Hyperprior

When using transform-based compression, the encoder $\mathcal{E}_x$ computes the latent space $\boldsymbol{y} = \mathcal{E}_x(\boldsymbol{x})$ from the input data $\boldsymbol{x}$. To enhance compression efficiency, the latent space $\boldsymbol{y}$ is quantized by rounding, yielding $\tilde{\boldsymbol{y}} = \text{Round}(\boldsymbol{y})$, and then further compressed using lossless entropy coding. The decoder $\mathcal{D}_x$ reconstructs the data as $\hat{\boldsymbol{x}} = \mathcal{D}_x(\tilde{\boldsymbol{y}})$. Compression performance is evaluated by distortion D, which measures the difference between $\boldsymbol{x}$ and $\hat{\boldsymbol{x}}$, and bit-rate R, which represents the number of bits required to encode $\tilde{\boldsymbol{y}}$.

To optimize latent space compression, [21] models each element $\boldsymbol{y}_i$ as a Gaussian distribution $\mathcal{N}(\boldsymbol{\mu}_i, \boldsymbol{\sigma}_i^2)$, with mean $\boldsymbol{\mu}_i$ and variance $\boldsymbol{\sigma}_i^2$. A hyperprior autoencoder (AE) captures the latent space distribution using a hyperencoder $\mathcal{E}_h$ and hyperdecoder $\mathcal{D}_h$. The hyperlatent space $\boldsymbol{z} = \mathcal{E}_h(\boldsymbol{y})$ is quantized to $\tilde{\boldsymbol{z}}$, and $\mathcal{D}_h(\tilde{\boldsymbol{z}})$ estimates $(\boldsymbol{\mu}, \boldsymbol{\sigma})$ for $\boldsymbol{y}$. Each quantized element $\tilde{\boldsymbol{y}}_i$ is then modeled as:

$$\tilde{\boldsymbol{y}}_i \sim \mathcal{N}(\boldsymbol{\mu}_i, \boldsymbol{\sigma}_i^2) * \mathcal{U}\,(-0.5, 0.5)\,, \tag{1}$$

where $\mathcal{N}(\boldsymbol{\mu}_i, \boldsymbol{\sigma}_i^2)$ denotes the normal distribution with mean $\boldsymbol{\mu}_i$ and variance $\boldsymbol{\sigma}_i^2$, and $\mathcal{U}\,(-0.5, 0.5)$ represents the uniform distribution introduced by the quantization process, with $*$ denoting convolution. The conditional probability $p(\tilde{\boldsymbol{y}} \mid \boldsymbol{\mu}, \boldsymbol{\sigma})$ is then given by:

$$p(\tilde{\boldsymbol{y}} \mid \boldsymbol{\mu}, \boldsymbol{\sigma}) = \prod_i \left(\mathcal{N}(\boldsymbol{\mu}_i, \boldsymbol{\sigma}_i^2) * \mathcal{U}\,(-0.5, 0.5)\right)(\tilde{\boldsymbol{y}}_i). \tag{2}$$

Since there is no prior belief about the hyperlatent space, $\tilde{\boldsymbol{z}}$ is modeled with the non-parametric, fully factorized density model $p(\tilde{\boldsymbol{z}})$ [5]. Entropy models, such as arithmetic coding [23], can losslessly compress $\tilde{\boldsymbol{y}}$ given the probability model. Therefore, the bit-rate for $\tilde{\boldsymbol{y}}$ can be expressed as $R_{\tilde{y}} = \mathbb{E}_{\tilde{y}}\left(-\log_2(p(\tilde{\boldsymbol{y}} \mid \boldsymbol{\mu}, \boldsymbol{\sigma}))\right)$. The bit-rate for $\tilde{\boldsymbol{z}}$ can be expressed as $R_{\tilde{z}} = \mathbb{E}_{\tilde{z}}\left(-\log_2(p(\tilde{\boldsymbol{z}}))\right)$. The total bit-rate can be calculated as $R = R_{\tilde{y}} + R_{\tilde{z}}$.

3.2 FM Architecture

The FM consists of two main components: VAE with a hyperprior structure and a SR module. First, we extend the hyperprior-based VAE from 2D to 3D to effectively capture spatio-temporal correlations in the data. Next, a 2D SR module is applied to enhance the fidelity of each time slice. These components are described in detail below.

Variational 3D Autoencoder. We construct a variational 3D autoencoder combining 2D and 3D convolutions to balance efficiency and capacity. While 3D convolutions capture spatio-temporal correlations effectively, they are computationally expensive for large inputs. Therefore, we apply downsampling to the spatial dimensions first and capture temporal correlations at lower-dimensional features. For input dimensions $[T, H, W]$, where T is temporal frames and H, W are spatial, we start with 2D convolutions and downsampling on each temporal slice, producing features $\left[C, \frac{H}{4}, \frac{W}{4}\right]$ per frame, where C is the channel count. Then, We merge temporal dimensions (MTD) into $\left[C, T, \frac{H}{4}, \frac{W}{4}\right]$. Next, 3D convolutions with downsampling reduce dimensions to $\left[2C, \frac{T}{4}, \frac{H}{16}, \frac{W}{16}\right]$, denoted $\boldsymbol{y}$. This latent representation is quantized and passed to the decoder. Decoding involves

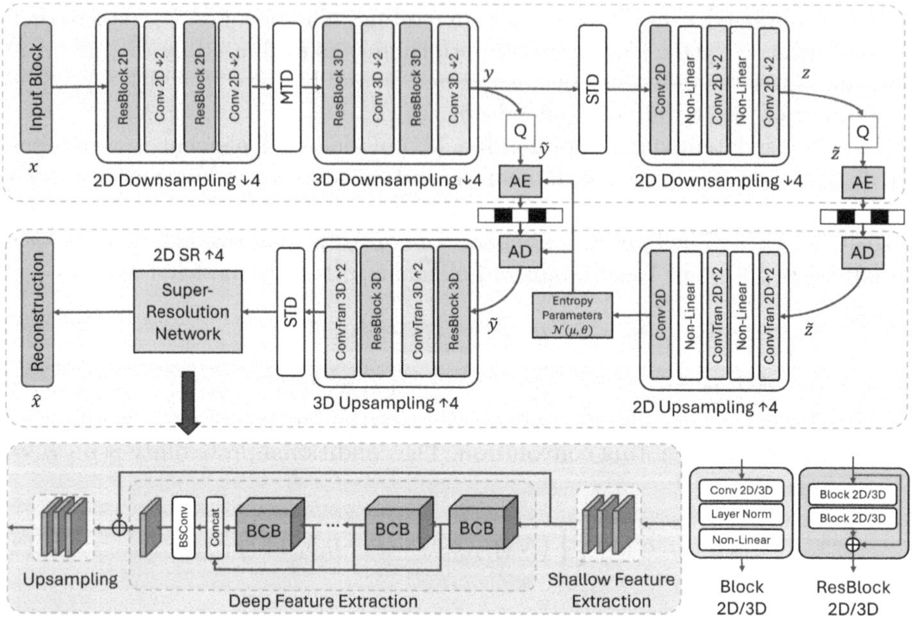

Fig. 1. Overview of the architecture of the foundation model (FM). 'Conv 2D/3D ↓' denotes a convolution operation with stride 2, while 'ConvTran 2D/3D ↑' denotes a transposed convolutional layer. 'MTD' refers to merging the temporal dimension, and 'STD' refers to splitting the temporal dimension. 'Q' represents rounding quantization. 'AE' and 'AD' denote arithmetic encoding and decoding, respectively. Leaky ReLU is used for nonlinearity.

transposed 3D convolutions for upsampling back to $\left[2C, T, \frac{H}{4}, \frac{W}{4}\right]$, followed by an SR module for independent temporal slice upsampling, reconstructing the data to $[T, H, W]$.

To model the quantized latent space $\tilde{\boldsymbol{y}}$, we split the temporal dimension (STD) of $\boldsymbol{y}$ into $\frac{T}{4}$ slices. Each slice is independently compressed by a hyper-prior autoencoder (AE), resulting in a hyper-latent representation $\boldsymbol{z}$ with dimensions $\left[4C, \frac{H}{64}, \frac{W}{64}\right]$. The hyper-prior decoder upsamples the features to match the original shape of $\boldsymbol{y}$ while doubling the number of channels. Half of these channels model the mean, and the remaining channels model the standard deviation of the latent space.

SR Module. In lossy data compression, the decoding phase is essentially an upsampling and interpolation process that closely aligns with tasks in SR. Recent advancements in SR [4,6] emphasize the effectiveness of more sophisticated architectures compared to conventional upsampling layers, as complex network structures capture finer details and enhance output fidelity.

For our decoder, we adopted a SR architecture inspired by recent work in [4]. As shown in Fig. 1, the design begins with a shallow feature extractor followed by a sequence of BCB blocks, each enhancing the representation by focusing on hierarchical features. Each BCB block consists of stacked components: an efficient BSConv layer [9], a ConvNeXt block [19], and two channel attention modules, ESA and CCA. Both BSConv and the ConvNeXt block are widely used in lightweight networks due to their computational efficiency (Fig. 2).

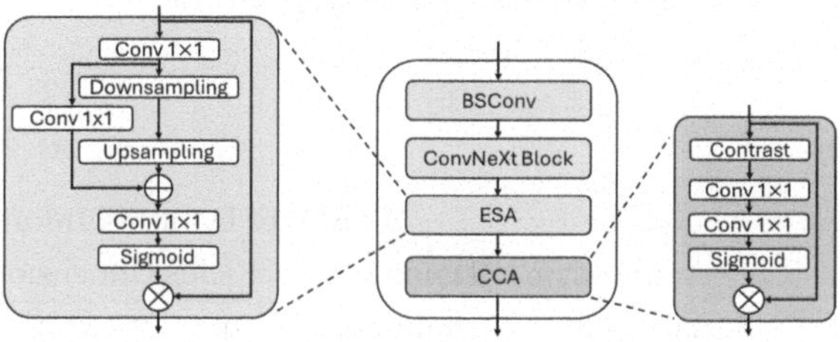

Fig. 2. The architecture of the BCB module.

The ESA module begins by applying a 1×1 convolution to reduce the input features' channel dimensions. It then employs stepwise convolutions and max pooling to downsample the feature map. Interpolation-based upsampling restores the feature map to its original spatial dimensions, followed by another 1×1 convolution to recover the channel dimensions. Finally, a sigmoid layer generates the attention mask. The CCA module, originally proposed in [11], starts by combining the mean and standard deviation to generate contrast information. It applies a 1×1 convolution to reduce the channel dimensions, followed by another 1×1 convolution to restore the original channel size. A sigmoid activation function is then used to generate the attention mask.

As shown in Fig. 1, the output of each BCB block is concatenated, forming a progressive feature extraction pipeline that captures both shallow and deep representations. This multi-layer aggregation enables the model to retain broad structural information while preserving intricate details. A skip connection merges shallow and deep feature maps, reinforcing the final output by combining low-level texture details with high-level semantic information. The architecture concludes with a channel shuffle layer, used as the primary upsampling mechanism. By rearranging and expanding feature channels, this layer increases the image resolution by a factor of 4.

3.3 Error Bound Guarantee

To ensure the reconstruction error meets a user-defined error bound, we apply a block-based method. We begin by splitting both the original and reconstructed data into non-overlapping blocks, aiming to minimize the ℓ_2-norm error for each block, $\left\|\boldsymbol{x} - \boldsymbol{x}^G\right\|_2$, where $\boldsymbol{x}$ represents the original data, and $\boldsymbol{x}^G$ represents the error-corrected reconstruction. To achieve this, we use Principal Component Analysis (PCA) on the residuals of the data to extract the principal components, or basis matrix, $\boldsymbol{U}$, with basis vectors ordered by their eigenvalues. We then project each block's residual onto this basis and select the leading coefficients such that the ℓ_2 norm of the corrected residual is within a specified threshold, τ. The residual coefficients are calculated as: $\boldsymbol{c} = \boldsymbol{U}^T(\boldsymbol{x} - \boldsymbol{x}^R)$. The most significant coefficients are chosen to satisfy the error bound τ. To reduce storage costs, we then compress these selected coefficients, $\boldsymbol{c}_s$, using quantization and entropy coding. The final corrected reconstruction is given by: $\boldsymbol{x}^G = \boldsymbol{x}^R + \boldsymbol{U}_s\boldsymbol{c}_q$, where $\boldsymbol{c}_q$ is the quantized set of selected coefficients, and $\boldsymbol{U}_s$ is the corresponding set of basis vectors. We add coefficients until $\left\|\boldsymbol{x} - \boldsymbol{x}^G\right\|_2 \leq \tau$.

4 Experimental Results

Table 1. Datasets Information

Application	Domain	Fields	Dimensions	Total Size
S3D [25]	Combustion	40	$40 \times 50 \times 640 \times 640$	6.5 GB
E3SM [8]	Climate	5	$5 \times 1440 \times 240 \times 1440$	10.0 GB
Hurricane [1]	Weather	3	$3 \times 48 \times 100 \times 500 \times 500$	14.4 GB
JHTDB [24]	Turbulence	3	$3 \times 512 \times 512 \times 512$	1.6 GB
ERA5 [10]	Meteorology	1	$1 \times 4096 \times 300 \times 402$	2.0 GB
Sunquake [20]	Solar Physics	1	$1 \times 12888 \times 256 \times 256$	3.4 GB

Datasets. We utilize several datasets to train and evaluate our FM. The E3SM dataset, known for its diverse climate variables and patterns, is specifically used for evaluation, while the other datasets are employed for training. **S3D Dataset** [25] is a combustion simulation dataset that models the ignition of hydrocarbon fuels under homogeneous charge compression ignition (HCCI) conditions. It captures detailed chemical and thermal dynamics. **E3SM Dataset** [8] is a climate simulation dataset that models various atmospheric and environmental variables. The dataset provides detailed climate dynamics and is used to simulate and predict climate patterns across a wide range of scales. **Hurricane Dataset** [1] simulates the dynamics of tropical cyclones, providing time-varying scalar and vector fields. **JHTDB Dataset** [24] is a turbulence simulation dataset that provides high-resolution data on fluid dynamics. **ERA5 Dataset** [10] is a global reanalysis dataset that provides detailed information about the Earth's atmosphere, land, and oceans. **Sunquake Dataset** [20] is an astrophysical dataset that captures seismic-like waves generated by solar flares on the Sun's surface. We summarize the details of each dataset in Table 1. The dimensions include the number of fields, the number of temporal slices, and the spatial resolution. For 3D spatial resolution datasets, such as the Hurricane dataset, we use the last two spatial dimensions along with the temporal dimension to construct spatio-temporal blocks.

Training Setup. The training loss consists of two components: distortion D and bit-rate R. We measure the distortion between the original data $\boldsymbol{x}$ and the reconstructed data $\hat{\boldsymbol{x}}$ using the Mean Squared Error (MSE). The bit-rate includes the bit-rate for both the latent space $\tilde{\boldsymbol{y}}$ and the hyper-latent space $\tilde{\boldsymbol{z}}$. A factor λ is applied to the bit-rate term to balance the trade-off between distortion and bit-rate. The overall training loss L can be formulated as:

$$L = \mathrm{MSE}(\boldsymbol{x}, \hat{\boldsymbol{x}}) + \lambda \left(\mathbb{E}_{\tilde{\boldsymbol{y}}} \left[-\log_2(p(\tilde{\boldsymbol{y}} \mid \boldsymbol{\mu}, \boldsymbol{\sigma})) \right] + \mathbb{E}_{\tilde{\boldsymbol{z}}} \left[-\log_2(p(\tilde{\boldsymbol{z}})) \right] \right) \quad (3)$$

Evaluation Metrics. We employ the Normalized Root Mean Square Error (NRMSE) as a relative error criterion to evaluate the quality of reconstruction, taking into account that different datasets may span various data ranges. The NRMSE is defined as: $\mathrm{NRMSE}\left(\Omega, \Omega^G\right) = \frac{\sqrt{\|\Omega - \Omega^G\|_2^2 / N_d}}{\max(\Omega) - \min(\Omega)}$, where N_d is the number of data points in the dataset, and Ω and Ω^G represent the entire original dataset and the reconstructed datasets after error bound guarantee post-processing, respectively.

Configuration. The FM is trained using PyTorch on two A100-80GB GPUs with the Adam optimizer. Training is divided into two stages: FM training and domain-specific fine-tuning. In the foundation training stage, we run 500,000 iterations, initially setting the weight parameter λ to 1×10^{-5}, which is adjusted to 1×10^{-4} after 250,000 iterations. The learning rate starts at 1×10^{-3} and is halved every 100,000 iterations. For the domain-specific fine-tuning stage, we conduct an additional 100,000 iterations, setting λ at 1×10^{-4}. Here, the learning rate starts at 1×10^{-4} and is reduced by half every 20,000 iterations. The batch size is set to 8 for both stages. This model is then evaluated on the E3SM application with five climate variables: one for sea level pressure (PSL), two for atmospheric temperature (T200, T500), and two for wind speed (VBOT, UBOT). After obtaining the reconstructed data, we apply an error bound guarantee post-processing step to ensure data quality. Different error bounds are set during post-processing, resulting in varying compression ratios and reconstruction errors.

Data Pre-processing. Given that each dataset has a unique data range, typically spanning from 1×10^{-10} to 1×10^{10}, we normalize each field individually within each domain. This normalization is performed by subtracting the mean and dividing by the range of each field, resulting in a normalized data range of $[0, 1]$ with a mean of 0. Our model, based on convolutional layers, is designed to accommodate flexible input shapes and can handle spatial dimensions of $64N$, where $N \in \mathbb{Z}^+$. During training, we prepare data blocks with dimensions $8 \times 256 \times 256$, where 8 represents consecutive time slices, and 256×256 denotes the spatial resolution. These blocks are generated through random cropping from the full data. For datasets with spatial dimensions smaller than 256×256, reflection padding is applied to extend each spatial dimension to 256×256. For evaluation, we maintain an 8-slice temporal dimension and allow flexible spatial dimensions of $64N$. If an input's spatial dimensions are not divisible by

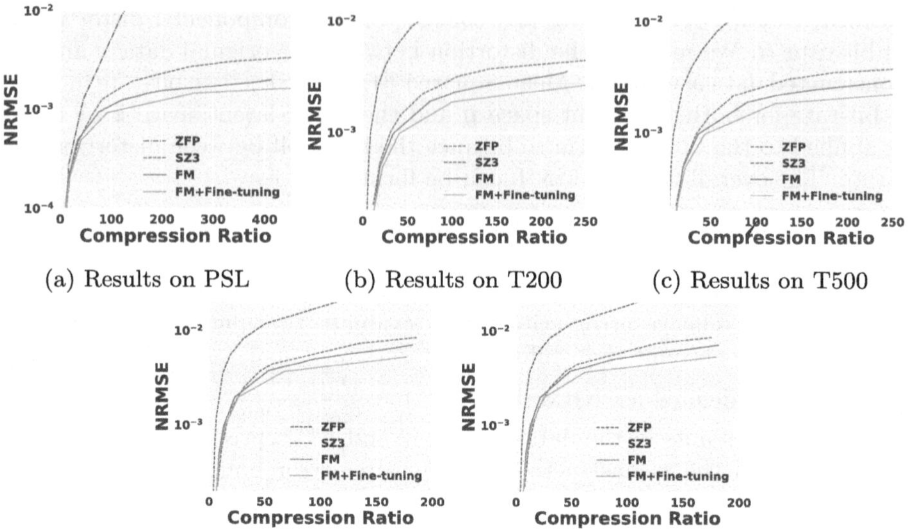

(a) Results on PSL (b) Results on T200 (c) Results on T500

(d) Results on VBOT (e) Results on UBOT

Fig. 3. FM evaluation on 5 variables of E3SM

64, reflection padding is applied to adjust each spatial dimension to the nearest multiple of 64. To prevent the model from being biased toward larger datasets, we balance dataset sizes by repeating samples from smaller datasets until each dataset has the same number of instances.

Generalization to Unseen Domains. To demonstrate the generalization ability of our FM to unseen domains, we evaluate it on the E3SM simulation dataset, which was not used for model training. We compare the performance of our method with two state-of-the-art error-bounded lossy compressors designed specifically for scientific data compression. We present results for our FM both with and without domain-specific fine-tuning, as shown in Fig. 3. Compared to SZ3 [15] and ZFP [7], our FM achieves comparable or higher compression ratios at the same reconstruction error level. Domain-specific fine-tuning further boosts compression ratios, providing significantly lower reconstruction error at high compression ratios and similar reconstruction error at low compression ratios. With domain-specific fine-tuning, Our method achieves up to a 4× higher compression ratio for the PSL variable and up to a 3× higher compression ratio for the two temperature and two wind speed variables, while maintaining the same NRMSE. We also visualize the reconstructed data for our method, SZ3, and ZFP at a compression ratio of 100 in Fig. 4. Our method shows lower reconstruction distortion compared to other compressors.

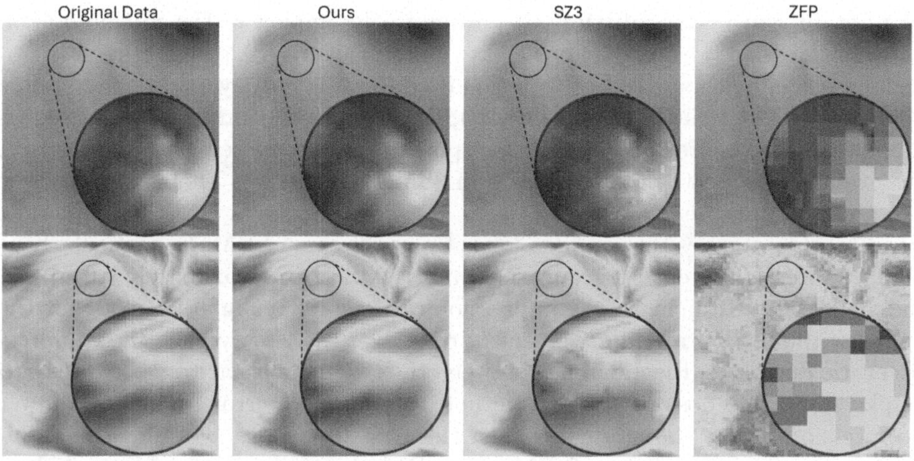

Fig. 4. Visualization of reconstructed data for our method, SZ3, and ZFP at the compression ratio of 100. The first row shows pressure data (PSL), and the second row shows temperature data (T200).

Adaptability to Data Dimensions. Although the model is trained with a spatial size of 256, it supports flexible input shapes and can accommodate spatial dimensions of $64N$, where $N \in \mathbb{Z}^+$. Scientific simulations often generate data with varying spatial dimensions. For using our FM model, data blocking should be applied as a preprocessing step and then the blocked data should be fed into the FM according to their specific requirements. In this experiment, we evaluate the adaptability of our FM model to different spatial resolutions using E3SM-PSL dataset as the testing dataset. We partition testing data into non-overlapping blocks and feeding the model with spatial resolutions of 64, 128, and 256, while keeping the temporal dimension fixed at 8. The results, presented in Fig. 5, demonstrate that our model can effectively handle different block sizes, showing promising compression performance. The result also shows that smaller block sizes generally result in lower compression ratios, as the model captures less structural information within each block. Larger blocks, on the other hand, allow the model to exploit more spatial correlations and patterns, enabling more efficient data representation.

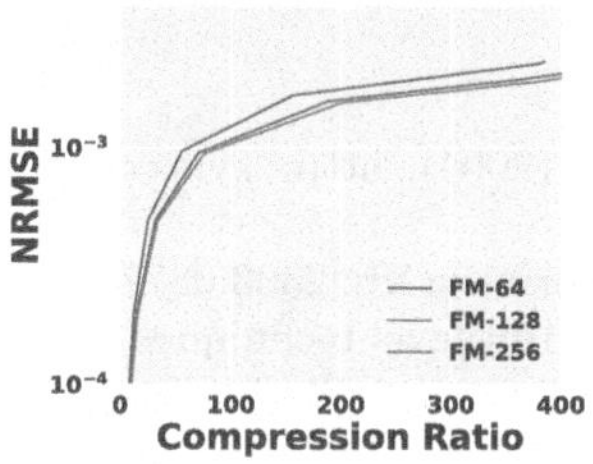

Fig. 5. Adaptability to data dimensions

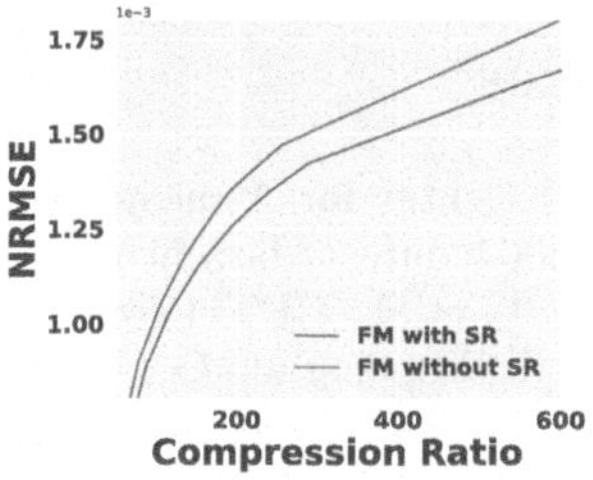

Fig. 6. Ablation study for SR module

Impact of Super-Resolution. To evaluate the impact of the SR module on data compression, we compare our FM to a baseline approach where the SR module is replaced with stacked convolutional and non-linear layers. In the baseline approach (refered as 'FM without SR'), 2D Transpose Convolution is used as the upsampling operator. The FM incorporating SR is refered as 'FM with SR'. We evaluate these two architectures on E3SM-PSL dataset. The results of these two architectures are presented in Fig. 6. As shown in the figure, the SR module significantly improves the compression ratio at the same NRMSE, particularly at high compression levels. For example, at an NRMSE of approximately 1.6×10^{-3}, the model with SR achieves a compression ratio of 498, compared to 384 for the model without SR—an improvement of roughly 30%. However, as the compression ratio decreases, the difference between the two models narrows because both employ the same post-processing method to guarantee error bounds. At lower compression levels, the storage cost of the post-processing becomes the dominant factor. The model size of 'FM with SR' is 1.8 million parameters, slightly larger than the 1.5 million parameters in 'FM without SR'. Despite this modest increase in size, the substantial improvement in performance demonstrates that the inclusion of the SR module is well justified. This improvement is achieved through more sophisticated hierarchical feature learning, rather than relying on simple upsampling or interpolation methods.

5 Conclusion

We present a robust and adaptable foundation model for lossy compression of scientific data, addressing challenges such as domain generalization, spatiotemporal correlations, and error control. By combining a variational autoencoder (VAE) with a hyper-prior structure and a super-resolution (SR) module, the model achieves up to 4× higher compression ratios and improved reconstruction quality compared to state-of-the-art methods. Experimental results demonstrate the model's ability to generalize across unseen scientific domains and data configurations, efficiently compressing diverse datasets. The integration of advanced architectures, such as SR networks, further enhances performance and meets the growing demand for efficient scientific data storage and processing. This foundation model represents a significant advancement in data compression, offering improved efficiency and better data integrity preservation.

References

1. National Center for Atmospheric Research (NCAR) (2004). http://vis.computer.org/vis2004contest/data.html
2. Achiam, J., et al.: GPT-4 technical report. arXiv preprint arXiv:2303.08774 (2023)
3. Ainsworth, M., Tugluk, O., Whitney, B., Klasky, S.: Multilevel techniques for compression and reduction of scientific data–the multivariate case. SIAM J. Sci. Comput. **41**(2), A1278–A1303 (2019)
4. Bai, H., Liang, X.: A very lightweight image super-resolution network. Sci. Rep. **14**(1), 13850 (2024)

5. Ballé, J., Minnen, D., Singh, S., Hwang, S.J., Johnston, N.: Variational image compression with a scale hyperprior. arXiv preprint arXiv:1802.01436 (2018)
6. Chen, X., Wang, X., Zhou, J., Qiao, Y., Dong, C.: Activating more pixels in image super-resolution transformer. In: Proceedings of the IEEE/CVF Conference on Computer Vision and Pattern Recognition, pp. 22367–22377 (2023)
7. Fox, A., Diffenderfer, J., Hittinger, J., Sanders, G., Lindstrom, P.: Stability analysis of inline ZFP compression for floating-point data in iterative methods. SIAM J. Sci. Comput. **42**(5), A2701–A2730 (2020)
8. Golaz, J.C., et al.: The DOE E3SM coupled model version 1: overview and evaluation at standard resolution. J. Adv. Model. Earth Syst. **11**(7), 2089–2129 (2019)
9. Haase, D., Amthor, M.: Rethinking depthwise separable convolutions: how intra-kernel correlations lead to improved MobileNets. In: Proceedings of the IEEE/CVF Conference on Computer Vision and Pattern Recognition, pp. 14600–14609 (2020)
10. Hersbach, H., et al.: The era5 global reanalysis. Q. J. R. Meteorol. Soc. **146**(730), 1999–2049 (2020)
11. Hui, Z., Gao, X., Yang, Y., Wang, X.: Lightweight image super-resolution with information multi-distillation network. In: Proceedings of the 27th ACM International Conference on Multimedia, pp. 2024–2032 (2019)
12. Kirillov, A., et al.: Segment anything. In: Proceedings of the IEEE/CVF International Conference on Computer Vision, pp. 4015–4026 (2023)
13. Li, X., Gong, Q., Lee, J., Klasky, S., Rangarajan, A., Ranka, S.: Machine learning techniques for data reduction of climate applications. arXiv preprint arXiv:2405.00879 (2024)
14. Li, X., Lee, J., Rangarajan, A., Ranka, S.: Attention based machine learning methods for data reduction with guaranteed error bounds. arXiv preprint arXiv:2409.05357 (2024)
15. Liang, X., et al.: SZ3: a modular framework for composing prediction-based error-bounded lossy compressors. IEEE Trans. Big Data **9**(2), 485–498 (2022)
16. Liu, J., et al.: SRN-SZ: deep leaning-based scientific error-bounded lossy compression with super-resolution neural networks. arXiv preprint arXiv:2309.04037 (2023)
17. Liu, J., et al.: Exploring autoencoder-based error-bounded compression for scientific data. In: 2021 IEEE International Conference on Cluster Computing (CLUSTER), pp. 294–306. IEEE (2021)
18. Liu, J., Di, S., Zhao, K., Liang, X., Chen, Z., Cappello, F.: FAZ: a flexible auto-tuned modular error-bounded compression framework for scientific data. In: Proceedings of the 37th International Conference on Supercomputing, pp. 1–13 (2023)
19. Liu, Z., Mao, H., Wu, C.Y., Feichtenhofer, C., Darrell, T., Xie, S.: A convnet for the 2020s. In: Proceedings of the IEEE/CVF Conference on Computer Vision and Pattern Recognition, pp. 11976–11986 (2022)
20. Mercea, V., Paraschiv, A.R., Lacatus, D.A., Marginean, A., Besliu-Ionescu, D.: A machine learning enhanced approach for automated sunquake detection in acoustic emission maps. Sol. Phys. **298**(1), 4 (2023)
21. Minnen, D., Ballé, J., Toderici, G.D.: Joint autoregressive and hierarchical priors for learned image compression. In: Advances in Neural Information Processing Systems, vol. 31 (2018)
22. Mun, S., Fowler, J.E.: DPCM for quantized block-based compressed sensing of images. In: 2012 Proceedings of the 20th European Signal Processing Conference (EUSIPCO), pp. 1424–1428. IEEE (2012)
23. Rissanen, J., Langdon, G.: Universal modeling and coding. IEEE Trans. Inf. Theory **27**(1), 12–23 (1981)

24. Wan, M., et al.: Johns Hopkins Turbulence Database (JHTDB) (2016)
25. Yoo, C.S., Lu, T., Chen, J.H., Law, C.K.: Direct numerical simulations of ignition of a lean n-heptane/air mixture with temperature inhomogeneities at constant volume: parametric study. Combust. Flame **158**(9), 1727–1741 (2011)

CANTER: A Novel Causal Model for Tourism Demand Forecasting

Xin Han[1], Haiyang Xia[2], Ye Zhu[1(✉)], Gang Li[1], and Rob Law[3]

[1] School of IT, Deakin University, Geelong, VIC 3125, Australia
{xinhan,ye.zhu}@ieee.org, gang.li@deakin.edu.au

[2] Research School of Management, Australian National University, Canberra 2600, Australia
haiyang.xia@anu.edu.au

[3] Asia-Pacific Academy of Economics and Management, Faculty of Business Administration, University of Macau, Macau, SAR, China
roblaw@um.edu.mo

Abstract. Tourism demand forecasting is fundamental to strategic planning and economic policy, particularly given the sector's substantial contribution to regional economies. While artificial intelligence (AI) models incorporating Search Engine Data (SED) have shown promise in tourism prediction, existing approaches often fail to capture the intricate causal dependencies among SED variables, thereby limiting their predictive capabilities. To address this limitation, we propose CANTER (Causal AI eNhanced Tourism dEmand foRecasting), a novel framework that leverages causal discovery and graph embedding techniques to integrate causal relationships among SED variables into AI-based forecasting models. The CANTER architecture comprises four key components: (1) temporal feature extraction from SED using sliding window analysis, (2) causal structure learning through the PCMCI algorithm, (3) causal graph representation learning via the Weave-Labelling model, and (4) integration of learned causal embeddings into predictive AI models. Empirical evaluation using Macau's visitor arrival data demonstrates that CANTER achieves superior forecasting accuracy across multiple time horizons compared to baseline models. Our findings establish causal AI's efficacy in forecasting tourism demand and provide methodological insights for advancing causality-aware data mining applications in tourism analytics.

Keywords: Tourism demand forecasting · AI-based forecasting · Causal discovery · Graph embedding · data-driven forecasting

1 Introduction

In the dynamic world of global travel, every decision–from staffing hotels to scheduling flights–hinges on a crucial element: the ability to accurately forecast tourism demand [17]. Traditionally reliant on static data such as exchange rates and income levels [13], forecasting now benefits from Search Engine Data

X. Wu et al. (Eds.): PAKDD 2025, LNAI 15875, pp. 381–393, 2025.
https://doi.org/10.1007/978-981-96-8295-9_28

(SED) [20]. Platforms like Google Trends provide real-time insights into traveler intent, offering a valuable resource for revolutionizing demand forecasting.

The appeal of SED lies in its ability to capture the pulse of potential tourists, offering daily or weekly snapshots of their interests [13]. Yet, despite its potential, existing forecasting models remain tethered to simplistic linear analyses, overlooking the rich tapestry of interconnected search terms that can reveal deeper insights. For instance, the common association between searches for "food" and "restaurant" signals a nuanced pattern of tourist behavior that traditional models fail to exploit.

The causal discovery field has emerged as a powerful tool for uncovering hidden relationships among variables [16]. The Granger causality test is widely used in the literature to examine direct causal links between SED and tourism arrivals [7]. However, its reliance on linearity and bivariate causality [11] limits its effectiveness in capturing the intricate interrelations among SED data. Thus, to truly harness the power of SED, two major challenges must be addressed. First, existing statistical methods often misinterpret the complex causal relationships among search terms [4], leading to inaccurate forecasts. Second, even when correlations are identified, translating these complex relationships into actionable insights for AI models is a formidable task.

To address these challenges, we introduce *CANTER* (Causal AI eNhanced Tourism dEmand foRecasting), a novel framework designed to uncover hidden causal relationships within SED and enhance the accuracy of AI-based forecasting models. The framework employs the *PCMCI* (Peter and Clark - Momentary Conditional Independence) model [16] to construct a causal graph that represents variables and their lagged causal effects as nodes and edges. This causal graph is then transformed into continuous representations using the *WL* (Weisfeiler-Lehman) embedding model [19]. These continuous representations serve as input features for AI-based models to predict tourism demand. A case study in Macau, a prominent tourism destination, demonstrates that CANTER not only surpasses traditional models but also establishes a new benchmark for precision in the field.

The main contributions of this study are as follows:

- We first identify the limitations of traditional AI-based models in capturing complex SED interrelations of tourism demand forecasting task.
- We propose a novel framework, CANTER, that integrates causal discovery and graph embedding techniques to enhance the accuracy of AI-based tourism demand forecasting models.
- We demonstrate the effectiveness of the CANTER framework in improving the accuracy of tourism demand forecasting through an empirical study.

The rest of this paper is organized as follows. Section 2 reviews related work on tourism demand forecasting and SED. Section 3 details the CANTER framework methodology. Section 4 evaluates its effectiveness through an empirical study, and Sect. 5 concludes with future research directions.

2 Related Work

2.1 Tourism Demand Forecasting Models

Tourism demand forecasting is crucial in tourism management, with methods generally classified into three frameworks: time series, economic, and AI-based models [17]. Time series models, such as ARIMA and SARIMA, utilize historical data trends and seasonality to predict demand [10]. Models like ARFRIMA, GARCH, and EGARCH have demonstrated superior predictive capabilities but are constrained by their linear assumptions. Economic models forecast demand based on variables such as income and prices, using econometric techniques like vector autoregressive and error correction models [3]. However, these models assume stable economic structures, limiting their responsiveness to dynamic shifts. AI-based models have become prominent due to their ability to manage nonlinear relationships and dynamic data [21]. These models include shallow approaches like Regression Trees and Support Vector Regression, as well as deep learning techniques such as LSTM networks [6].

Search engines serve as vital information sources for travelers, offering insights into traveler preferences and enhancing forecasting accuracy [7]. Tools such as Baidu Index and Google Trends have proven effective in predicting tourism demand in regions like China and South Korea [15]. Research highlights the dependencies between SED and tourism arrivals, emphasizing the significance of lag order for accurate forecasting [17]. However, semantic interdependencies among SED keywords are often overlooked, indicating a need to capture these relationships to enhance model performance. Moreover, the traditional methods, such as correlation and regression, do not account for common causes, potentially leading to spurious dependencies.

2.2 Causality Effects on Tourism Demand Forecasting

Recently, causal relationships have gained attention in tourism demand forecasting to understand how one variable affects another [3]. The Granger causality test is a common method used to capture causality, determining that X causes Y if X's past values improve the prediction of Y [1]. It has been applied to assess the predictive power of SED and its relation to tourist arrivals in Beijing [18]. However, Granger causality is limited to linear interdependencies and bivariate cases [2]. A more comprehensive approach would simultaneously identify all potential causal relations among SED.

Constraint-based causal learning algorithms infer causal relationships from time series data by constructing graphs based on conditional independence [14]. The Fast Causal Inference (FCI) and PC algorithms have been adapted to handle latent variables and temporal order in causal discovery [9]. These methods surpass Granger causality by accounting for latent variables and multivariable interactions. The PCMCI algorithm, a variation of the PC algorithm, excels in high-dimensional, highly correlated time series [16], and has been widely used for causal discovery in time series.

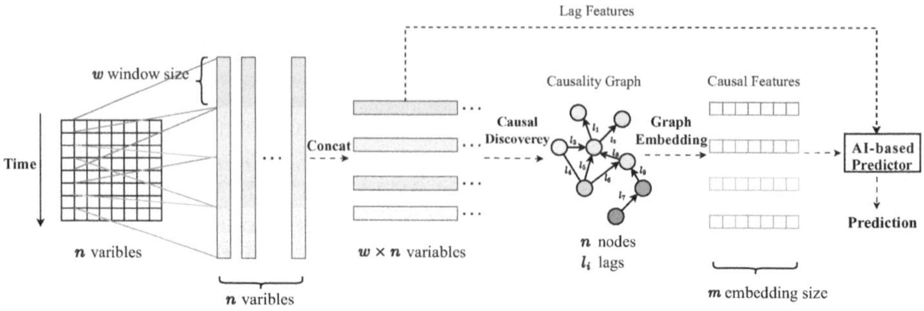

Fig. 1. The *CANTER* framework for tourism demand Forecasting

Research integrating AI models with causal discovery for forecasting is still developing [2]. There is a paucity of studies that incorporate causal relationships into AI methods for tourism demand forecasting. Discovering and encoding these relationships into AI models could provide additional information, potentially enhancing forecasting accuracy.

3 Methodology

This section first formulates the tourism demand forecasting problem then provides a detailed description of the proposed CANTER framework, which is divided into four key steps: lag feature extraction, tourism demand causal graph construction, causal graph embedding, and causal-enhanced tourism demand forecasting.

3.1 Formulation of the Tourism Demand Forecasting Problem

Let $Y \in \mathbb{R}^T$ be the historical series of tourism arrival volume over T time steps, where $y_t \in Y$ represents the volume of tourism arrival at time step t. The exogenous factors over T time steps are denoted by $X \in \mathbb{R}^{T \times n}$, where n indicates the number of exogenous factor, such as the n search intensity indicators from the search engine. For simplify, in subsequent analysis, we concatenate X and Y according to time t and denote it by $\tilde{X} = \{\tilde{x}_1, \tilde{x}_2, \cdots, \tilde{x}_T\}$, where $\tilde{x}_t = [x_t; y_t] \in \mathbb{R}^{n+1}$.

Definition 1. *The goal of demand forecasting is to construct a model $\mathcal{F}$ capable of predicting demand for the next q time steps as follows:*

$$\{\hat{y}\}_{T+1}^{T+q} = \mathcal{F}(\tilde{X}) \tag{1}$$

where $\{\hat{y}\}_{T+1}^{T+q}$ represents the predicted tourism arrival volume for the next q time steps.

3.2 Lag Feature Extraction

The lag feature extraction step uses a sliding window technique to capture the temporal relationship between each SED variable and the target value, a common method in tourism demand forecasting. This approach segments the dataset and incorporates lag characteristics of exogenous factors to improve prediction accuracy.

As shown in Fig. 1, the process involves two steps: constructing time series moving windows and combining lag features. A sliding window of fixed size captures time-varying lag features, which are then concatenated to form the lag feature representation.

3.3 Tourism Demand Causal Graph Construction

To capture the causal relationship between SED and tourism arrival, we employ the PCMCI method [16], a constraint-based approach that infers a causal graph among time series using independence tests and controls for false positives.

Definition 2 *(Causal Relationship [16]): A direct causal relationship between two variables $x^i \in \tilde{X}$ and $x^j \in \tilde{X}$ is inferred if, given a set of variables $\mathbb{C} \in \tilde{X} \setminus \{x^i, x^j\}$, changing the value of x^i affects the probability distribution of x^j while keeping $\mathbb{C}$ constant.*

Definition 3 *(Causal Graph): A causal graph is a directed graph $\mathcal{G} = \{\mathcal{V}, \mathcal{E}, \mathcal{X}\}$, where $\mathcal{V}$ is the set of nodes representing variables and $\mathcal{E}$ is the set of directed edges representing causal relationships between variables. Each edge $e_{ij} \in \mathcal{E}$ indicates a causal relationship from node i to node j. $\mathcal{X}$ is the feature matrix of the nodes.*

The PCMCI method comprises two steps: the conditional selection step (PC) and the momentary conditional independence test (MCI). The PC step identifies potential causal parents for each variable, while the MCI step eliminates spurious links by conditioning on relevant variables. Details of these steps and the test of independence are provided below.

PC step. In the PC step, all variables are initially considered interdependent and connected in an undirected graph. Conditional independence tests are performed for pairs of variables with increasing sizes of conditioning sets. The PC step iteratively adds a variable in $\mathbb{C}$ and tests the independence of two variables conditionally on $\mathbb{C}$. A significance threshold α is used to control the significance level of the conditional independence tests. If the two variables remain conditionally independent for all variables in $\mathbb{C}$, the edge between them is removed as they are considered independent. After this iteration, an undirected graph is constructed. Finally, the orientation of the undirected edges is then determined by applying a set of four recursive rules [16] to obtain a final partially directed graph $\mathcal{G}$.

MCI step. The MCI step takes this partially directed graph and evaluates the independence hypothesis for each edge, removing edges if the hypothesis is not rejected. Given any pair of variables $x^i, x^j \in \tilde{X}$, where $i \neq j$ and $\tilde{X}$ is the set of all variables, the MCI is defined as follows:

$$\text{MCI} : x^i_{(t-\tau)} \not\perp\!\!\!\perp x^j_t \mid \mathscr{P}\left(x^j_t\right) \backslash x^i_{(t-\tau)}, \mathscr{P}\left(x^i_{(t-\tau)}\right) \tag{2}$$

Where $\mathscr{P}(z)$ stands for parents of z in the graph, t runs through the samples up to the time series length τ that reflects the expected maximum causal time lag, and $\not\perp\!\!\!\perp$ indicates dependence. Here, the condition on the set of parents $\mathscr{P}\left(x^j_t\right)$, $\mathscr{P}\left(x^i_{t-\tau}\right)$ accounts for autocorrelation, allowing for the correct controlling of false-positive rates [16]. The MCI step outputs a directed graph $\mathcal{G}$, which represents the variables and their lagged causal relationships as nodes and edges.

3.4 Causal Graph Embedding

The causal graph represents pairwise causal relationships among variables but is incompatible with many AI algorithms requiring numeric vector representations. To address this, we use the WL embedding algorithm [19], which embeds the causal graph into a low-dimensional vector space. This scalable, unsupervised method leverages neighbor information to learn structural embeddings, offering a strong mathematical foundation and lower computational complexity compared to other graph embedding algorithms.

WL embedding is an iterative algorithm that refines the embedding of each node based on the embedding of its neighbors. Formally, given a causal graph $\mathcal{G} = \{\mathcal{V}, \mathcal{E}, \mathcal{X}\}$, the embedding process using WL embedding at iteration $h \geq 0$ is defined as follows:

$$\varphi^h(v) = \begin{cases} \mathcal{X}(v) & \text{if } h = 0 \\ \frac{1}{2}\left(\varphi^{h-1}(v) + \frac{1}{\deg(v)} \sum_{u \in \mathcal{N}(v)} \mathrm{w}(v,u) \cdot \varphi^{h-1}(u)\right) & \text{Otherwise} \end{cases} \tag{3}$$

where $\deg(v)$ is the degree of v, $\mathcal{N}(v)$ is the neighborhood of v, and the weight $\mathrm{w}(v, u)$ is set to 1 in our case.

Definition 4 *(Causal graph embedding). Given a causal graph $\mathcal{G} = \{\mathcal{V}, \mathcal{E}, \mathcal{X}\}$, the embedding of each node $v \in \mathcal{E}$ is defined as $\varphi^h(v) = [\varphi^0, \varphi^1, \cdots, \varphi^h]^\top$. Then the embedding matrix $\tilde{X}_{cf} \in \mathbb{R}^{T \times (n+1) \times (h+1)}$ is defined by stacking the embedding of all nodes in the graph.*

3.5 Caual Enhanced Tourism Demand Forecasting

Once the causal graph is embedded into a low-dimensional numerical vectors, the resulting casual feature $\tilde{X}_{cf}$ is incorporated into the forecasting model. This study incorporates $\tilde{X}_{cf}$ to various state-of-the-art AI models to improve their accuracy in tourism demand forecasting. Formally, the causal enhanced tourism demand forecasting model can be expressed as follows:

Definition 5 *(Causal Enhanced demand forecasting). The goal of causal demand forecasting is to construct a model $\mathcal{F}$ capable of predicting demand for the next q time steps as follows:*

$$\{\hat{y}\}_{T+1}^{T+q} = \mathcal{F}(\tilde{X}_{cf}; \tilde{X}) \tag{4}$$

where $\mathcal{F}$ is the AI based forecasting model such as SVR, ANN and LSTMs.

4 Experiments

4.1 Datasets

This study utilized monthly data on international tourism arrivals in Macau and Search Engine Data (SED) for selected keywords. To mitigate language and platform bias in SED [7], data from six major markets–Hong Kong, Malaysia, Philippines, Australia, the UK, and the USA–were included, collectively accounting for over 70% of Macau's international arrivals. Tourist arrival data from January 2011 to December 2019 (108 observations) were sourced from Macau's Statistics and Census Services (DSEC), excluding the COVID-19 period.

SED was gathered from Google Trends due to its widespread use and 66.7% market share, offering reliable search frequency data. Keywords were chosen from eight categories of Macau's tourist offerings: dining, lodging, transportation, tours, shopping, currency, clothing, and recreation.

4.2 Setup

We compared five commonly used forecasting models in tourism demand research [5,6,12]: linear regression (LASSO), artificial neural networks (ANN), random forest regression (RFR), support vector regression (SVR), and long short-term memory networks (LSTMs). We also evaluated their causal-enhanced versions: GS-LASSO, GS-ANN, GS-RFR, GS-SVR, and GS-LSTMs. Hyperparameters for each model are provided in the supplementary material. Both causal-enhanced model shared the same hyperparameters, with an additional parameter α set at 0.05 for causal graph generation in PCMCI.

Mean absolute percentage error (MAPE), mean absolute error (MAE), and root mean square error (RMSE) are used as valuation metrics. We further calculate the improvement rate (IR) of each valuation metric between the original and causal-enhanced models to demonstrate the extent of improvement. The IR of MAPE using the SVR model is given by Eq. 5.

$$\text{IRMAPE(SVR)} = \frac{\text{MAPE(SVR)} - \text{MAPE(GS-SVR)}}{\text{MAPE(SVR)}} \times 100\% \tag{5}$$

A positive IRMAPE indicates that the causal-enhanced SVR model outperforms the SVR model; a larger value signifies a greater improvement. Similar calculations were performed for MAE and RMSE.

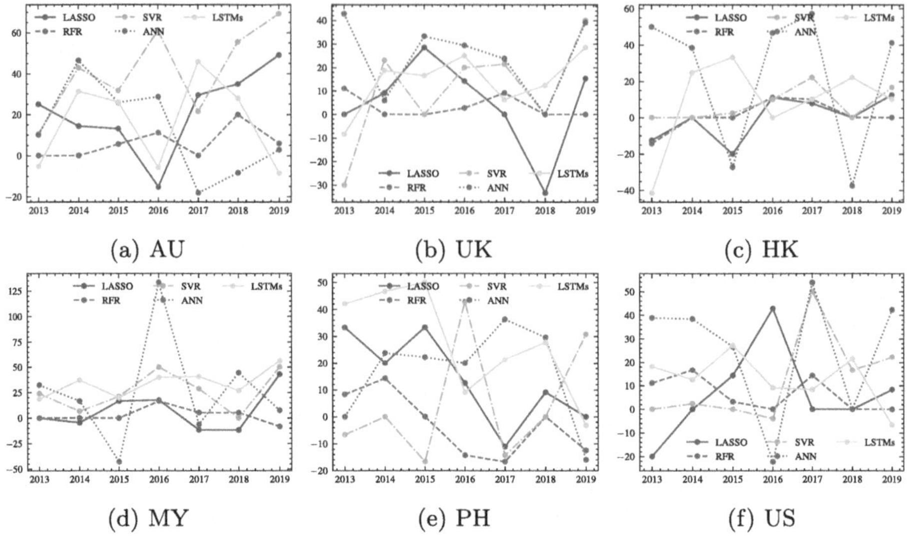

(a) AU (b) UK (c) HK

(d) MY (e) PH (f) US

Fig. 2. IRMAPE of one-step forecasting performance of benchmark models.

Table 1. Improvement of One-step Forecasting performance of benchmark models. The best results of each metric are highlighted in bold.

Alogorithms	Metrics	AU	UK	HK	MY	PH	MY	Average
GS-LASSO vs. LASSO	IRMAE	**21.13**	7.13	-1.84	12.99	13.11	14.39	11.15
	IRRMSE	**19.82**	1.11	-0.97	4.01	6.18	13.2	7.22
	IRMAPE	**21.50**	4.86	-0.17	7.02	13.88	6.5	8.93
GS-RFR vs. RFR	IRMAE	1.97	3.29	2.66	**5.90**	5.42	4.19	3.91
	IRRMSE	2.33	1.39	1.14	**6.25**	-0.10	1.92	2.16
	IRMAPE	6.08	3.28	0.97	2.69	-2.98	**6.47**	2.75
GS-SVR vs. SVR	IRMAE	**37.72**	8.39	7.73	23.48	9.14	9.66	16.02
	IRRMSE	**38.16**	16.90	6.47	21.16	7.21	9.01	16.48
	IRMAPE	**41.76**	10.64	7.34	25.76	5.14	12.47	17.18
GS-ANN vs. ANN	IRMAE	8.48	21.89	22.76	7.05	3.18	**22.06**	14.24
	IRRMSE	10.42	24.05	25.04	16.34	9.08	**25.55**	18.41
	IRMAPE	12.44	24.92	24.10	**26.43**	16.57	25.42	21.65
GS-LSTMs vs. LSTMs	IRMAE	10.24	12.89	6.74	**32.78**	25.72	10.96	16.55
	IRRMSE	9.77	11.86	1.16	**29.12**	21.80	14.26	14.66
	IRMAPE	15.92	14.20	8.45	**34.22**	27.69	12.88	18.89

4.3 Experimental Results

Performance on One-Step Forecasting. Figures 2 display the one-step forecasting metrics IRMAPE. The causal-enhanced model consistently outperforms

Table 2. Diebold-Mariano test results for comparison of predictive accuracy of GS-and single model. q is the number of steps of forecasting. P value < 0.05 is marked in bold.

Source Markets	Benchmark model	$q = 1$		$q = 3$		$q = 6$	
		DM Stat	P value	DM Stat	P value	DM Stat	P value
AU	GS-LASSO vs. LASSO	−2.5758	**0.01**	−2.1833	**0.02**	−2.11211	**0.02**
	GS-RFR vs. RFR	−2.6162	**0.01**	−2.5532	**0.01**	−2.11499	**0.02**
	GS-SVR vs. SVR	−2.4304	**0.01**	−2.3239	**0.01**	−2.5423	**0.01**
	GS-ANN vs. ANN	−3.7425	**0.00**	−5.0650	**0.00**	−5.13285	**0.00**
	GS-LSTMs vs. LSTMs	−3.6866	**0.00**	−5.4066	**0.00**	−5.14613	**0.00**
UK	GS-LASSO vs. LASSO	−0.6330	0.26	−2.8060	**0.00**	−0.1365	0.45
	GS-RFR vs. RFR	−2.9827	**0.00**	−2.6817	**0.00**	−3.27386	**0.00**
	GS-SVR vs. SVR	−3.2816	**0.00**	−2.7382	**0.00**	−3.44498	**0.00**
	GS-ANN vs. ANN	−1.4476	0.08	−2.2745	**0.01**	−1.39655	0.08
	GS-LSTMs vs. LSTMs	−3.7479	**0.00**	−4.7014	**0.00**	−4.75289	**0.00**
HK	GS-LASSO vs. LASSO	−1.9838	**0.03**	−5.4048	**0.00**	−5.52636	**0.00**
	GS-RFR vs. RFR	−0.1642	0.44	−0.7186	0.24	−0.74899	0.23
	GS-SVR vs. SVR	−1.3772	0.09	−1.9412	**0.03**	−1.72561	**0.04**
	GS-ANN vs. ANN	−1.6163	0.05	−1.6707	**0.05**	−1.16992	0.12
	GS-LSTMs vs. LSTMs	−2.2647	**0.01**	−2.0217	**0.02**	−1.79907	**0.04**
MY	GS-LASSO vs. LASSO	−0.5717	0.28	−1.1195	0.13	−0.87548	0.19
	GS-RFR vs. RFR	−1.3444	0.09	−0.7193	0.24	−1.15151	0.13
	GS-SVR vs. SVR	−0.3057	0.38	−1.0283	0.15	−0.45831	0.32
	GS-ANN vs. ANN	−0.4147	0.34	−4.3445	**0.00**	−0.41838	0.34
	GS-LSTMs vs. LSTMs	−0.6706	0.25	−4.9606	**0.00**	−4.2633	**0.00**
PH	GS-LASSO vs. LASSO	−5.3324	**0.00**	−1.0607	0.15	−4.91415	**0.00**
	GS-RFR vs. RFR	−4.6264	**0.00**	−4.9716	**0.00**	−4.61746	**0.00**
	GS-SVR vs. SVR	−6.1256	**0.00**	−5.6759	**0.00**	−5.56389	**0.00**
	GS-ANN vs. ANN	−3.3633	**0.00**	−2.8291	**0.00**	−2.5296	**0.01**
	GS-LSTMs vs. LSTMs	−0.2921	0.39	−0.3189	0.38	−2.88073	**0.00**
US	GS-LASSO vs. LASSO	−3.1780	**0.00**	−3.2097	**0.00**	−3.04461	**0.00**
	GS-RFR vs. RFR	−3.0287	**0.00**	−3.3546	**0.00**	−1.42347	0.08
	GS-SVR vs. SVR	−3.7672	**0.00**	−3.9299	**0.00**	−3.79304	**0.00**
	GS-ANN vs. ANN	−3.4979	**0.00**	−4.2981	**0.00**	−3.98129	**0.00**
	GS-LSTMs vs. LSTMs	−1.5182	0.07	−1.1180	0.13	−0.2572	0.40

LASSO, RFR, SVR, ANN, and LSTMs across various markets. Specifically, for Australian tourist arrivals, the causal model improves SVR performance by up to 41.76%. For UK visitors, it reduces ANN errors by up to 40%. In Hong Kong, it enhances LSTM performance by up to 8.45%. The model also decreases LSTM errors for Malaysian tourists by up to 34.22%, Philippine arrivals by up to 27.69%, and SVR errors for US arrivals by up to 21.65%.

Table 3. Improvement of Three-steps Forecasting performance of benchmark models. The best results of each metric are highlighted in bold.

Alogorithms	Metrics	AU	UK	HK	MY	PH	MY	Average
GS-LASSO vs. LASSO	RMAE	**23.47**	21.45	2.57	13.20	12.43	21.42	15.76
	IRRMSE	18.48	14.73	1.31	8.21	7.12	**20.61**	10.08
	IRMAPE	17.78	**21.92**	4.52	12.14	12.61	18.78	14.62
GS-RFR vs. RFR	RMAE	1.17	2.70	2.71	**5.37**	2.98	3.34	3.05
	IRRMSE	−1.95	2.27	2.91	**4.38**	3.72	3.02	2.39
	IRMAPE	−0.77	**5.75**	4.87	3.46	0.71	4.16	3.03
GS-SVR vs. SVR	RMAE	26.44	26.04	8.11	**36.63**	36.38	30.58	27.36
	IRRMSE	20.69	19.55	8.56	32.46	**44.42**	30.60	26.05
	IRMAPE	30.39	25.76	9.29	**37.72**	27.55	30.75	26.91
GS-ANN vs. ANN	RMAE	21.09	15.03	9.01	**27.89**	19.02	15.62	17.95
	IRRMSE	**21.22**	10.02	6.48	19.72	12.99	15.12	14.26
	IRMAPE	**24.98**	18.43	7.65	23.01	22.07	16.67	18.80
GS-LSTMs vs. LSTMs	RMAE	11.41	25.30	4.93	26.49	**37.76**	35.51	23.57
	IRRMSE	10.48	21.12	2.54	27.80	**40.31**	36.57	23.14
	IRMAPE	16.67	22.51	3.21	25.85	32.81	**35.09**	22.69

Table 1 summarizes the average improvement rates, demonstrating that causal-enhanced models outperform their single counterparts across all metrics. The model particularly excels with ANN, SVR, and LSTMs by leveraging causal relationships. LASSO achieves moderate improvements due to its linear nature, while RFR shows minimal gains, potentially due to data confounding and overfitting despite its effectiveness with nonlinear problems.

The Diebold-Mariano test [8] was used to assess performance improvements (see Table 2). The test's null hypothesis, which states there is no significant performance difference between models, is rejected in all cases. This indicates significant differences in forecasting performance, with negative Diebold-Mariano values showing that our proposed method outperforms all baseline methods.

Performance on Multi-step Forecasting. To assess the robustness of causal relationships in SED for tourism demand forecasting, we conducted experiments on three- and six-month forecasts. Tables 3 and 4 present the average improvement rates for these forecasts, showing that the causal-enhanced method generally outperforms baseline models, similar to one-step forecasting results. This is due to its ability to capture stable causal relationships, enhancing long-term forecasting performance. However, some baseline models, such as LASSO, perform comparably in specific cases like the three-month forecast for Hong Kong and the six-month forecast for the US, suggesting limited SED impact in these markets.

Table 4. Improvemnt of Six-steps Forecasting performance of benchmark models. The best results of each metric are highlighted in bold.

Alogorithms	Metrics	AU	UK	HK	MY	PH	MY	Average
GS-LASSO vs. LASSO	RMAE	18.14	**21.45**	2.30	12.71	12.43	3.24	11.71
	IRRMSE	**18.48**	10.22	0.78	10.08	7.12	5.02	8.62
	IRMAPE	17.78	**21.92**	4.52	10.55	12.61	1.63	11.50
GS-RFR vs. RFR	RMAE	3.73	**5.41**	1.11	5.37	2.98	3.34	3.66
	IRRMSE	2.91	**4.59**	2.64	4.38	3.72	3.49	3.62
	IRMAPE	2.59	**8.61**	1.30	3.46	0.71	4.16	3.47
GS-SVR vs. SVR	RMAE	18.26	35.00	6.40	36.63	**41.40**	30.58	28.04
	IRRMSE	12.46	28.47	3.68	32.46	**48.09**	30.60	25.96
	IRMAPE	16.68	34.33	6.12	**37.72**	33.27	30.75	26.48
GS-ANN vs. ANN	RMAE	21.09	19.95	16.97	21.03	20.46	**22.56**	20.34
	IRRMSE	21.22	15.84	13.4	13.53	15.76	**24.02**	17.30
	IRMAPE	**24.98**	20.81	11.22	16.71	22.07	23.02	19.80
GS-LSTMs vs. LSTMs	RMAE	16.90	15.55	8.49	**25.88**	18.51	16.39	16.95
	IRRMSE	20.61	17.03	6.91	**22.01**	21.58	17.02	17.52
	IRMAPE	17.76	13.86	5.81	**24.25**	14.23	18.13	15.67

Additionally, the Diebold-Mariano test results in Table 2 confirm the superiority of the causal-enhanced models over baselines for medium- and long-term forecasts. Overall, these findings highlight the enhanced generalization and robustness of our proposed method in forecasting performance.

5 Conclusions

Accurate tourism demand forecasting is vital for stakeholders' strategic decision-making, and search engine data (SED) has become a crucial tool due to its reflection on real-time tourist intentions. Despite the significant impact of SED, many models neglect the data dependencies that could enhance forecasting accuracy.

This study introduces CANTER, a causality-enhanced AI framework that integrates causal relationships among SED variables into tourism forecasting. CANTER constructs a comprehensive causality graph using the PCMCI algorithm, transforms it into numerical vectors with the WL embedding algorithm, and uses these vectors as features in the forecasting model. Validated through a case study of Macau's international tourism demand, CANTER outperforms baseline models, significantly improving accuracy for medium- and long-term forecasts using metrics like MAPE, MAE, and RMSE across several models.

Future research should account for dynamic causal changes over time and test the framework in other forecasting contexts to enhance its robustness and applicability.

References

1. Akinboade, O.A., Braimoh, L.A.: International tourism and economic development in South Africa: a granger causality test. Int. J. Tour. Res. **12**(2), 149–163 (2010). https://doi.org/10.1002/jtr.743
2. Assaad, C.K., Devijver, E., Gaussier, E.: Survey and evaluation of causal discovery methods for time series. J. Artif. Intell. Res. **73**, 767–819 (2022). https://doi.org/10.1613/jair.1.13428
3. Assaf, A.G., Scuderi, R.: Tourism demand analysis: directions for future research. Tour. Econ., 13548166221130466 (2022). https://doi.org/10.1177/13548166221130466
4. Bi, J.W., Li, C., Xu, H., Li, H.: Forecasting daily tourism demand for tourist attractions with big data: an ensemble deep learning method. J. Travel Res. **61**(8), 1719–1737 (2022). https://doi.org/10.1177/00472875211040569
5. Cao, Z., Li, G., Song, H.: Modelling the interdependence of tourism demand: the global vector autoregressive approach. Ann. Tour. Res. **67**, 1–13 (2017). https://doi.org/10.1016/j.annals.2017.07.019
6. Chen, J., Ying, Z., Zhang, C., Balezentis, T.: Forecasting tourism demand with search engine data: a hybrid CNN-BiLSTM model based on Boruta feature selection. Inf. Process. Manag. **61**(3), 103699 (2024). https://doi.org/10.1016/j.ipm.2024.103699
7. Dergiades, T., Mavragani, E., Pan, B.: Google trends and tourists' arrivals: emerging biases and proposed corrections. Tour. Manage. **66**, 108–120 (2018). https://doi.org/10.1016/j.tourman.2017.10.014
8. Diebold, F.X., Mariano, R.S.: Comparing predictive accuracy. J. Bus. Econ. Stat. **20**(1), 134–144 (2002). https://doi.org/10.1198/073500102753410444
9. Entner, D., Hoyer, P.O.: On causal discovery from time series data using FCI. Probab. Graph. Models, 121–128 (2010)
10. Goh, C., Law, R.: Modeling and forecasting tourism demand for arrivals with stochastic nonstationary seasonality and intervention. Tour. Manage. **23**(5), 499–510 (2002). https://doi.org/10.1016/S0261-5177(02)00009-2
11. Law, R., Li, G., Fong, D., Han, X.: Tourism demand forecasting: a deep learning approach. Ann. Tour. Res. **75**, 410–423 (2019). https://doi.org/10.1016/j.annals.2019.01.014
12. Li, C., Zheng, W., Ge, P.: Tourism demand forecasting with spatiotemporal features. Ann. Tour. Res. **94**, 103384 (2022). https://doi.org/10.1016/j.annals.2022.103384
13. Martins, L.F., Gan, Y., Ferreira-Lopes, A.: An empirical analysis of the influence of macroeconomic determinants on world tourism demand. Tour. Manage. **61**, 248–260 (2017). https://doi.org/10.1016/j.tourman.2017.01.008
14. Moraffah, R., et al.: Causal inference for time series analysis: problems, methods and evaluation. Knowl. Inf. Syst. **63**(12), 3041–3085 (2021). https://doi.org/10.1007/s10115-021-01621-0
15. Park, S., Lee, J., Song, W.: Short-term forecasting of Japanese tourist inflow to South Korea using google trends data. J. Travel Tourism Market. **34**(3), 357–368 (2017). https://doi.org/10.1080/10548408.2016.1170651
16. Runge, J., Nowack, P., Kretschmer, M., Flaxman, S., Sejdinovic, D.: Detecting and quantifying causal associations in large nonlinear time series datasets. Sci. Adv. **5**(11), eaau4996 (2019)

17. Song, H., Qiu, R., Park, J.: Progress in tourism demand research: theory and empirics. Tour. Manage. **94**, 104655 (2023). https://doi.org/10.1016/j.tourman.2022.104655
18. Sun, S., Wei, Y., Tsui, K.L., Wang, S.: Forecasting tourist arrivals with machine learning and internet search index. Tour. Manage. **70**, 1–10 (2019). https://doi.org/10.1016/j.tourman.2018.07.010
19. Togninalli, M., Ghisu, E., Llinares-López, F., Rieck, B., Borgwardt, K.: Wasserstein Weisfeiler-Lehman graph kernels. In: Advances in Neural Information Processing Systems, vol. 32 (2019)
20. Xue, G., Liu, S., Ren, L., Gong, D.: Forecasting hourly attraction tourist volume with search engine and social media data for decision support. Inf. Process. Manag. **60**(4), 103399 (2023). https://doi.org/10.1016/j.ipm.2023.103399
21. Zhang, Y., Li, G., Muskat, B., Law, R.: Tourism demand forecasting: a decomposed deep learning approach. J. Travel Res. **60**(5), 981–997 (2021). https://doi.org/10.1177/0047287520919522

Time-Aware Complex Attention Space for Temporal Knowledge Graph Completion

Rushan Geng[1] and Cuicui Luo[2(✉)]

[1] School of Computer Science and Technology, University of Chinese Academy of Sciences, Beijing, China
gengrushan23@mails.ucas.ac.cn
[2] International College, University of Chinese Academy of Sciences, Beijing, China
luocuicui@ucas.ac.cn

Abstract. Temporal knowledge graph embedding (TKGE) models play a crucial role in representing entities, relationships, and timestamps, enabling the inference of missing facts in dynamic environments. However, current models face a significant challenge in distinguishing representations of the same entity at different time points, limiting their ability to predict links accurately in temporal knowledge graphs (TKGs). In this paper, we introduce **Time-Aware Complex Attention Space (TACAS)**, a novel and straightforward model that directly incorporates temporal information into entity embeddings, allowing entities to evolve over time. This evolution enables TACAS to capture the temporal dynamics of relationships with greater precision. Moreover, TACAS leverages an attention mechanism within a complex space framework, which enhances the interaction between entities, relations, and time, improving the model's sensitivity to temporal information. Our experiments on four benchmark datasets demonstrate that TACAS outperforms majority of existing TKGE models.

Keywords: Temporal knowledge graph · Temporal knowledge graph embedding · Complex space · Attention mechanism · Time perspective

1 Introduction

Knowledge graphs (KGs) play a vital role in various semantic applications, such as question answering [1], recommendation systems [2], and information retrieval [3]. They organize and represent real-world knowledge in a machine-comprehensible format, enabling relational reasoning and structural representation learning. As real-world knowledge evolves over time, temporal knowledge graphs (TKGs) have garnered significant attention, as they can model the dynamic relationships between entities. Learning low-dimensional representations for entities and relations in knowledge graphs is an effective way to address this challenge.

X. Wu et al. (Eds.): PAKDD 2025, LNAI 15875, pp. 394–406, 2025.
https://doi.org/10.1007/978-981-96-8295-9_29

A temporal knowledge graph consists of quadruples (s, r, o, τ), where s and o represent the head and tail entities, r is the relation, and τ denotes the timestamp. For example, the fact "Barack Obama visited Malaysia on April 25, 2014" can be represented as (Barack Obama, Make a visit, Malaysia, 2014-04-25). Some facts are time-sensitive, such as Barack Obama's presidency, which lasted from 2009 to 2017. While some facts are stable over time, others change and require temporal precision in their representation. This dynamic nature highlights the importance of incorporating temporal information into entity representations to improve their relevance and accuracy.

Static knowledge graphs treat all facts as occurring at the same timestamp, which limits their ability to handle temporal information. Thus, temporal knowledge graph embedding (TKGE) models have been proposed to address these limitations. These models focus on incorporating temporal information into the representation of relations and can effectively model time. However, they often neglect the temporal evolution of entities, which is a critical aspect of many real-world scenarios.

There are two key challenges in current TKGE models. First, the evolution of entities over time is not adequately reflected, leading to reduced performance in link prediction tasks. We found performance degradation when representing inverse relations in TComplEx [4], indicating the need for temporal information in entity representations. Second, many TKGE models treat temporal information as a secondary feature, rather than integrating it deeply into the core representation of entities and relations. This limits their ability to capture temporal changes in relationships and entity attributes, reducing their effectiveness in tasks such as predicting future interactions or understanding past shifts.

In this paper, we propose a time-aware attention model to address these issues. Our model integrates time information into the complex space of entity representations, allowing relations to simultaneously capture both entity characteristics and temporal dynamics. We incorporate an attention mechanism that adjusts the influence of temporal aspects based on the context, improving the relationship prediction process. This approach enables a finer-grained analysis of how relationships evolve over time. Our model's time-aware attention mechanism captures subtle changes and patterns in temporal data that are often overlooked by traditional TKGE models.

The main contributions of this paper are as follows:

- We incorporate temporal information into the real and imaginary parts of entity representations, enabling entities to evolve over time for more precise temporal representations.
- We use an attention mechanism to enhance the interaction between the real and imaginary parts of entities, fostering synergy between these components.
- Extensive experiments on four public datasets demonstrate the superior performance of our model, validating its effectiveness and potential for real-world applications.

2 Related Work

Temporal knowledge graph embedding (TKGE) models often extend static knowledge graph methods. For example, TTransE [5] builds on TransE by adding temporal ordering constraints, modeling time-aware relationships. HyTE [6] projects entities and relations onto a time-specific hyperplane, linking time information with entities and relations. TeRo [7] represents the temporal evolution of entity embeddings as rotations in complex space, and RotateQVS [8] extends this by using quaternion rotations to capture temporal dynamics.

For temporal semantic matching, TA-DistMult [9] applies recurrent neural networks to learn time-aware relationship representations, using DistMult's standard scoring function. ChronoR [10], based on Rotate, represents relationships as rotations that include timestamps. DE-SimplE [11] combines historical entity embeddings with the SimplE framework for time-specific events. ATiSE [12] decomposes entity and relation evolution into trend, seasonal, and random components. TComplEx [4] enhances ComplEx by adding complex time embeddings for temporal graph completion. TeLM [13] uses fourth-order tensor decomposition with a time regularizer and multivector embeddings. BTDG [14] introduces block term decomposition for both discrete and continuous temporal facts.

HGE [15] embeds temporal facts into a product space consisting of complex, split-complex, and dual spaces, leveraging the geometric properties of each subspace to capture temporal patterns. Joint-MTComplEx [16] combines temporal and static modules, allowing joint learning to improve temporal and static information fusion for temporal knowledge graph completion.

3 Problem Formulation

A temporal knowledge graph $\mathcal{G}$ consists of quadruples represented as (s, r, o, τ), where $s, o \in \mathcal{E}$ denote subject (head) and object (tail) entities, respectively, and $r \in \mathcal{R}$ represents a relation connecting the subject to the object. The timestamp $\tau \in \mathcal{T}$ specifies the temporal context of the interaction between s and o under relation r. Here, $\mathcal{E}$, $\mathcal{R}$, and $\mathcal{T}$ represent the finite sets of entities, relation types, and timestamps, correspondingly. Temporal knowledge graph embedding techniques aim to map each entity, relation, and timestamp into a continuous vector space. These methods define a time-aware scoring function $f(s, r, o, \tau)$ to evaluate the likelihood of observing the quadruple (s, r, o, τ) in the graph. In response to queries like $(s, r, ?, \tau)$ or $(?, r, o, \tau)$, the goal of Temporal Knowledge Graph Completion is to infer the missing subject or object entity by leveraging known temporal facts. Ideally, the scoring function assigns higher scores to genuine facts over incorrect ones, reflecting their relative plausibility within the TKG framework.

4 Methodology

The TACAS model utilizes complex space to represent temporal knowledge graphs and incorporates an attention mechanism to analyze relationships and entities endowed with temporal attributes. We maintain that the informational construct of an entity is bifurcated into dual facets: firstly, the semantic segment, which is tasked with encapsulating the intrinsic semantic attributes inherent to the entity; secondly, the temporal segment, which is purposed to dynamically adjust and refine the semantics, thereby enhancing the precision and fidelity of the entity's information representation. We apply an attention mechanism to fuse the entities enriched with temporal information, aiming to achieve a more comprehensive representation.

4.1 Complex Embedding

Previous studies have shown that complex spaces are adept at capturing the inherent connections between entities more effectively. For the quadruples (s, r, o, t), the representation in complex space is defined as follows:

$$\begin{aligned} \mathbf{e}_s &= \mathbf{a}_s + i\mathbf{b}_s, \\ \mathbf{e}_r &= \mathbf{a}_r + i\mathbf{b}_r, \\ \mathbf{e}_o &= \mathbf{a}_o + i\mathbf{b}_o, \\ \mathbf{e}_\mathcal{T} &= \mathbf{a}_\mathcal{T} + i\mathbf{b}_\mathcal{T}, \end{aligned} \tag{1}$$

where $\mathbf{e}_s, \mathbf{e}_r, \mathbf{e}_o, \mathbf{e}_\mathcal{T} \in \mathbb{C}^k$, k is embedding rank. $\mathbf{a}_*$ and $\mathbf{b}_*$ denote the real and imaginary vector components, respectively. Unlike the TComplEx model, TACAS integrates temporal information directly into the entity embeddings, treating the entire embedding as a complex vector. Real and imaginary parts of the entity and time are concatenated as follows:

$$\begin{aligned} \mathbf{a}_s^t &= \mathbf{a}_s + \mathbf{a}_t, \\ \mathbf{b}_s^t &= \mathbf{b}_s + \mathbf{b}_t, \end{aligned} \tag{2}$$

where $\mathbf{a}_t$ and $\mathbf{b}_t$ are derived from the timestamp embedding $\mathbf{e}_t$. The time-aware entity embedding is defined as $\mathbf{e}_s^t = \mathbf{a}_s^t + i\mathbf{b}_s^t$, where $\mathbf{e}_s^t \in \mathbb{C}^k$. $\mathbf{e}_t$ is attuned to changes in time, which affords it greater variability and enables more accurate representations of entities across different time periods. Therefore, $\mathbf{e}_s$ and $\mathbf{e}_o$ incorporate temporal information and are updated to $\mathbf{e}_s^t$ and $\mathbf{e}_o^t$, respectively. The formula is as follow:

$$\begin{aligned} \mathbf{e}_s^t &= \mathbf{a}_s^t + i\mathbf{b}_s^t, \\ \mathbf{e}_o^t &= \mathbf{a}_o^t + i\mathbf{b}_o^t. \end{aligned} \tag{3}$$

4.2 Complex-Attention Module

In TKGs, relationships may exhibit varying frequencies of change, ranging from long-term static relationships to those that change frequently. Therefore, for each relationship, we use $\mathbf{a}_r$ to capture the temporal entity representations. We provide a complex attention mechanism to emphasize dynamic facts based on the characteristics of the relationship. The Complex-Attention module uses $\mathrm{Re}(s)_t$ and $\mathrm{Im}(s)_t$ as inputs to a Softmax function to compute attention scores:

$$(\mathbf{a}_{\mathrm{Re}}, \mathbf{b}_{\mathrm{Im}}) = \mathrm{Softmax}(\mathbf{c}^\top \mathbf{a}_s^t, \mathbf{c}^\top \mathbf{b}_s^t), \tag{4}$$

where $\mathbf{c} \in \mathbb{C}^{k/2}$ is an attention vector. $\mathbf{a}_{Re}$ and $\mathbf{a}_{Im}$ constitute a new relation $\mathbf{c}_r$, which we use to emphasize the most suitable temporal relationship in $\mathbf{e}_s^t$. As the changing frequencies of relations can be reflected through $\mathbf{a}_r$, TACAS is capable of simulating both static and dynamic logical and structural patterns in Temporal Knowledge Graphs. The formula is expressed as follows:

$$\begin{aligned} \mathbf{e}_{s,t} &= \mathrm{Att}(\mathbf{a}_s^t, \mathbf{b}_s^t; \mathbf{c}_r) \\ \mathbf{c}_r &= \mathbf{a}_{\mathrm{Re}} + i\mathbf{b}_{\mathrm{Im}}, \end{aligned} \tag{5}$$

where $\mathbf{c}_r \in \mathbb{C}^k$ denotes relation attention vector, $\mathbf{e}_{s,t}$ denotes entity with time information through complex-attention.

As TACAS operates in complex space, we employ the Hadamard product to map the relation

$$\begin{aligned} \boldsymbol{\xi}_{(\tau,r)} &= \mathbf{e}_\tau \circ \mathbf{e}_r \\ &= \mathbf{a}_\tau \mathbf{a}_r - \mathbf{b}_\tau \mathbf{b}_r + i(\mathbf{a}_\tau \mathbf{b}_r + \mathbf{b}_\tau \mathbf{a}_r) \end{aligned} \tag{6}$$

where $\circ$ denotes the Hadamard product. Following previous tensor factorization model [14], the score function of TACAS is denoted as:

$$\phi(s, r, o, t) = Re(\langle \mathbf{e}_{s,t}, \boldsymbol{\xi}_{(\tau,r)}, \overline{\mathbf{e}}_{o,t} \rangle). \tag{7}$$

where $Re(\cdot)$ is the real vector component.

4.3 Loss Function

Implementing a full multiclass log-softmax loss function combined with N3 regularization has demonstrated significant improvements in the performance of tensor decomposition-based temporal knowledge graph models. In our approach, we adopt this configuration and enhance it by incorporating reciprocal learning to streamline the training process.

For each relation r, we define an inverse relation r^{-1} and construct a quadruple (o, r^{-1}, s, t) for each training tuple (s, r, o, t). During the evaluation phase, a query $(?, r, o, t)$ is answered as $(o, r^{-1}, ?, t)$. The multiclass log-loss is mathematically expressed as:

$$\begin{aligned} \mathcal{L}_\omega = &- \log\left(\frac{\exp(f(s, r, o, t))}{\sum_{s' \in \mathcal{E}} \exp(f(s', r, o, t))}\right) \\ &- \log\left(\frac{\exp(f(o, r^{-1}, s, t))}{\sum_{o' \in \mathcal{E}} \exp(f(o', r^{-1}, s, t))}\right) \\ &+ \lambda_\mu \sum_{i=1}^{k} \left(\|\mathbf{e}_s\|_3^3 + \|\mathbf{e}_r\|_3^3 + \|\mathbf{e}_o\|_3^3\right), \end{aligned} \tag{8}$$

where λ_μ is the weight assigned to the N3 regularization. Assuming temporal proximity in embeddings, TComplex introduces a smoothing regularizer that enforces temporal continuity between embeddings at successive time intervals:

$$\mathcal{L}_{\mathcal{T}} = \sum_{i=1}^{n_{\mathcal{T}}-1} \|\mathbf{e}_{\mathcal{T}i+1} - \mathbf{e}_{\mathcal{T}_i}\|_p^p, \tag{9}$$

where $p = 3$ is specified for the N3 regularization. Moreover, we utilize the linear time regularizer from TeLM, which introduces a bias component between consecutive time embeddings:

$$\mathcal{L}_{\mathcal{T}} = \sum_{i=1}^{n_{\mathcal{T}}-1} \|\mathbf{e}_{\mathcal{T}_{i+1}} - \mathbf{e}_{\mathcal{T}_i} - \mathbf{e}_b\|_p^p, \tag{10}$$

where $\mathbf{e}_b$ denotes the bias embedding, initialized randomly and subsequently learned during training. This regularizer is designed to ensure that the differences between embeddings at adjacent time steps are minimized compared to those at greater temporal distances, thus promoting effective clustering and temporal embedding coherence.

The complete loss function is therefore formulated as:

$$\mathcal{L}_b = \frac{1}{b} \sum_{\omega \in \Omega_b} \mathcal{L}_\omega + \lambda_\tau \mathcal{L}_\tau, \tag{11}$$

where λ_τ represents the time regularization coefficient.

Table 1. Statistics of the experimental datasets. Time span indicates the range of years in which the events occur.

Datasets	Entities	Relations	Timastamps	Time Span	Training	Validation	Test
ICEWS14	6,869	230	365	A.D.2014	72,826	8,941	8,963
ICEWS05-15	10,094	251	4,017	A.D.2005-A.D.2015	368,962	46,275	46,092
GDELT	500	20	366	A.D.2015-A.D.2016	2,735,685	341,961	341,961
Wikidata12k	12,554	24	232	A.D.1709-A.D.2018	539,286	67,538	63,110

5 Experiment

5.1 Datasets

To evaluate the performance of the TACAS model, we selected four widely-used temporal knowledge graph (TKG) datasets: ICEWS14, ICEWS05-15, GDELT, and Wikidata12k. ICEWS14 and ICEWS05-15 are subsets of the Integrated Crisis Early Warning System (ICEWS) [17] and represent sparse temporal knowledge graphs. ICEWS14 contains events from 2014, while ICEWS05-15 spans events from 2005 to 2015. In contrast, GDELT [18] is a dense knowledge graph

Table 2. Performance comparison on temporal link prediction (metrics in %) on four TKG datasets ICEWS05-15, ICEWS14, GDELT, and Wikidata12k. All results are taken from the original papers.

Method	ICEWS05-15				ICEWS14				GDELT				Wikidata12k			
	MRR	Hits@1	Hits@3	Hits@10	MRR	Hits@1	Hits@3	Hits@10	MRR	Hits@1	Hits@3	Hits@10	MRR	Hits@1	Hits@3	Hits@10
TTansE	27.1	8.4	-	61.6	25.5	7.4	-	60.1	11.5	0.0	16.0	31.8	17.2	9.6	18.4	32.9
TComplEx	66.0	59.0	71.0	80.0	61.0	53.0	66.0	76.0	34.0	24.9	36.1	49.8	33.1	23.3	35.7	53.9
ATiSE	51.9	37.8	60.6	79.4	55.0	43.6	62.9	75.0	-	-	-	-	25.2	14.8	28.8	46.2
TimePlex	64.0	54.5	-	81.8	60.4	51.5	-	77.1	-	-	-	-	33.4	22.8	-	53.2
TeRo	58.6	46.9	66.8	79.5	56.2	46.8	62.1	73.2	24.5	15.4	26.4	42.0	29.9	19.8	32.9	50.7
ChronoR	67.5	59.6	72.3	82.0	62.5	54.7	66.9	77.3	36.6	29.2	37.9	53.8	-	-	-	-
BoxTE	66.7	58.2	71.9	82.0	61.3	52.8	66.4	76.3	-	-	-	-	-	-	-	-
TuckERTNT	63.8	55.9	68.6	78.3	60.4	52.1	65.5	75.3	-	-	-	-	-	-	-	-
TeAST	68.3	60.4	73.2	82.9	63.7	56.0	68.2	78.2	37.1	28.3	40.1	54.4	-	-	-	-
BDME	-	-	-	-	63.5	55.5	68.3	77.8	27.8	19.1	29.9	44.8	-	-	-	-
SANe	68.3	60.5	73.4	82.3	63.8	55.8	**68.8**	78.2	-	-	-	-	-	-	-	-
Joint-MTComplEx	68.3	60.1	73.6	83.2	63.6	55.6	68.1	78.6	-	-	-	-	**36.0**	27.0	**39.3**	54.1
TLT-KGE-HGE	**68.8**	**60.8**	74.0	**83.5**	63.4	55.0	68.5	**78.8**	37.1	27.7	40.2	55.6	27.4	-	-	-
TACAS	68.5	60.5	**73.4**	83.3	**64.0**	**56.1**	68.4	78.7	**41.3**	**33.0**	**44.3**	**57.4**	33.5	**23.7**	36.1	**54.6**

derived from the Global Database of Events, Language, and Tone, with over 3 million quadruples, 500 entities, and 20 relations. Wikidata12k is a subset of Wikidata, where each fact is annotated with a time interval. Following Dasgupta et al. [6], time intervals are discretized into multiple quadruples, each with a single timestamp, and month and day information is omitted, retaining only the year. A fixed month and day are added to each timestamp for these datasets. Table 1 summarizes the detailed statistics.

5.2 Baselines

We compare our proposed method with a set of TKGC methods include TTransE [5], TA-DistMult [9], HyTe [6], DE-SimplE [11], TComplEx and TNTComplEx [21], ATiSE [12], TimePlex [19], TeRo [7], ChronoR [10], TeLM [13], BoxTE [20], RotateQVS [8], TuckERTNT [22], BDME [23], TeAST [24], SANe [25], Joint-MTComplEx [16] HGE [15].

5.3 Evaluation Protocols

For each quadruple (s, r, o, t) from the test set, we generate two queries: $(s, r, ?, t)$ and $(?, r, o, t)$. Both queries are utilized concurrently for model optimization. It's important to emphasize that, conventionally, every quadruple (s, r, o, t) is supplemented with its reciprocal relation, denoted as (o, r^{-1}, s, t). Consequently, the query $(?, r, o, t)$ gets transformed to (o, r^{-1}, s, t). This alteration maintains the generality of the evaluation process. For performance evaluation, we resort to the Mean Reciprocal Rank (MRR) - which calculates the average of the inverse ranks across all cases - and Hits@N. The latter metric gauges the frequency with which the true entity candidate is positioned within the top N ranked candidates, specifically for N values of 1, 3, and 10.

Table 3. Ablation study on the ICEWS14, ICEWS05-15 Datasets. Other hyperparameters remained unchanged.

Method	ICEWS14				ICEWS05-15			
	MRR	Hits@1	Hits@3	Hits@10	MRR	Hits@1	Hits@3	Hits@10
TACAS	**64.0**	**56.1**	**68.4**	**78.7**	**68.5**	**60.5**	**73.4**	**83.3**
w/o $\mathbf{a}_s^t$	63.8	56.0	68.2	78.5	68.4	60.4	73.4	83.2
w/o $\mathbf{b}_s^t$	63.9	56.0	68.5	78.7	68.1	60.1	73.0	83.0
w/o $\mathbf{e}_t$	62.3	54.6	66.6	76.8	67.0	59.2	71.6	81.5
w/o Attention	63.2	55.2	67.3	77.9	67.5	59.5	72.5	82.6

5.4 Experiment Setup

We conducted experiments using the PyTorch deep learning framework. All experiments were performed on an NVIDIA Tesla A100 with 40 GB of memory. We utilized the Adagrad optimizer and performed grid search to find the optimal hyperparameters based on the performance on the validation set. Some parameter settings are as follows: We used a batch size of 1000 and selected an embedding rank of 2000. Our experiments were designed to run for up to 200 epochs, with a learning rate set at 0.1. We chose to overlook variance in our results since adjusting the random seed led to stable outcomes, indicating reliable performance across different experimental runs. The optimal hyperparameters for TACAS are as follows:

- ICEWS14: $\lambda_\mu = 0.0075, \lambda_\tau = 0.2$
- ICEWS05-15: $\lambda_\mu = 0.004, \lambda_\tau = 0.5$
- GDELT: $\lambda_\mu = 0.00005, \lambda_\tau = 0.05$
- Wikdata12k: $\lambda_\mu = 0.05, \lambda_\tau = 0.0025$

5.5 Experiment Result

The overall results of the entire experiment are presented in Table 2. The ICEWS05-15 dataset is similar to the ICEWS14 dataset, but with a larger time span for the timestamps. The TLT-KGE-HGE model integrates time and relation representations across multiple spaces, which allows it to achieve the best performance on the ICEWS05-15 dataset. On the ICEWS14 dataset, TACAS achieved the highest performance, demonstrating the effectiveness of incorporating temporal information into knowledge graph embedding models. Notably, on the GDELT dataset, TACAS showed substantial enhancement, with a 4% point increase in its MRR value. This suggests that TACAS has strong representational capabilities for dense datasets. This finding indicates that TACAS not only has robust capabilities for handling sparse datasets but also excels on dense datasets.

Table 4. The number of parameters of the models. $|\mathcal{E}|$, $|\mathcal{R}|$, $|\mathcal{T}|$, d, and m denote the number of entities, relations, timestamps, embedding dimension, and time windows, respectively.

Model	Number of Parameters
TComplEx	$(\|\mathcal{E}\| + \|\mathcal{R}\| \times 2 + \|\mathcal{T}\|) \times d \times 2$
TeRo	$(\|\mathcal{E}\| + \|\mathcal{R}\| \times 2) \times d \times 2 + \|\mathcal{T}\| \times d$
TeLM	$(\|\mathcal{E}\| + \|\mathcal{R}\| \times 2 + \|\mathcal{T}\|) \times d \times 2 + d$
TeAST	$(\|\mathcal{E}\| + \|\mathcal{R}\| \times 2 + \|\mathcal{T}\| \times 2) \times d \times 2$
RotateQVS	$(\|\mathcal{E}\| + \|\mathcal{R}\| \times 2 + \|\mathcal{T}\|) \times d \times 4$
TACAS	$(\|\mathcal{E}\| + \|\mathcal{R}\| \times 3 + \|\mathcal{T}\| \times 2) \times d \times 2$

Regarding the temporal span, events in the ICEWS dataset typically cover just a few years, whereas events in the Wikidata12k dataset can span nearly 3,000 years. TACAS also achieved the second-highest performance on this dataset. Joint-MTComplEx uses multiple models for joint training, combining static and dynamic approaches. This enhances its ability to represent static information, leading to better performance; however, it comes with increased computational costs during training.

5.6 Ablation Study

The results of our ablation studies are presented in Table 3. We primarily investigated the impact of removing various components, including the real and imaginary parts of the temporal information, all temporal information, and the Complex-Attention mechanism. For each dataset, all other settings within the TACAS model remained unchanged.

In these experiments, we systematically removed the real and imaginary components of the temporal information appended to the entities. This alteration led to an imbalance in the entity information, resulting in an asymmetrical structure of the model and a substantial decline in performance. Furthermore, completely removing all temporal information resulted in further performance degradation, bringing the model's effectiveness closer to that of the TComplEx model. The removal of the Attention module also led to a performance decline. These results underscore the efficacy of the Complex-Attention mechanism in seamlessly integrating temporal and entity information, highlighting its critical role in enhancing model performance.

5.7 Complex Analysis

To assess the computational efficiency and scalability of the TACAS model, we compared the number of parameters used during training across several models, as presented in Table 4. The TACAS model shows a slight increase in the number of parameters compared to TComplEx, which can be attributed to the additional parameters required for incorporating temporal information and attention

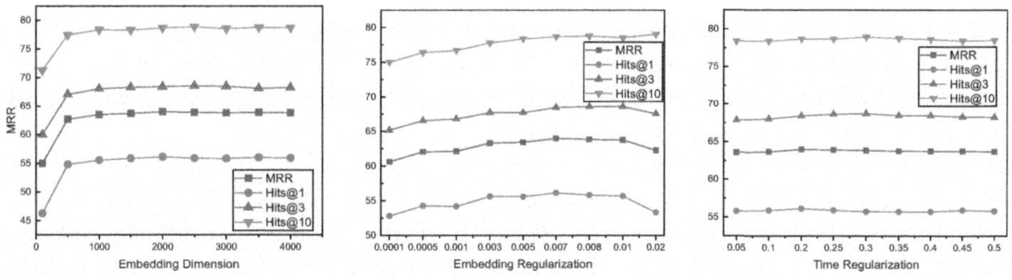

(a) (b) (c)

Fig. 1. Subfigure (a) illustrates the impact of embedding dimensions on performance. Subfigure (b) explores how the weights of embedding regularization affect performance. Subfigure (c) analyzes the specific effects of time regularization weights on performance.

weights. This increase, however, is a necessary trade-off for the enhanced performance TACAS delivers, enabling more precise modeling of temporal dynamics and improved link prediction in temporal knowledge graphs.

5.8 Hyperparameter Analysis

We conducted experiments to investigate the impact of various hyperparameters on the performance of TACAS. Initially, we experimented with different embedding dimensions, selecting from the set 100, 500, 1000, 2000, 2500, 3000, 3500, 4000, and conducted tests on the ICEWS14 dataset while keeping other parameters constant. As illustrated in Fig. 1(a), TACAS performance improves with increasing dimensions and stabilizes around a dimension of 2500.

Further analysis on the impact of λ_μ and λ_τ on performance is shown in Fig. 1(b) and Fig. 1(c). We varied λ_μ within the set $\{0.0001, 0.0005, 0.001, 0.003, 0.005, 0.007, 0.008, 0.01, 0.02\}$ and λ_τ within $\{0.05, 0.1, 0.2, 0.25, 0.3, 0.35, 0.4, 0.45, 0.5\}$. The results indicate that $\lambda_\mu = 0.07$ yielded the highest performance, and the model's performance remained stable with subsequent variations, demonstrating the robustness of the TACAS model. The effect of λ_τ on model performance was minimal, with optimal performance achieved at $\lambda_\tau = 0.2$, further confirming the robustness of our proposed model.

6 Conclusion

In this paper, we propose TACAS, a novel Temporal Knowledge Graph Completion method that improves the modeling of temporal knowledge graphs. TACAS integrates timestamps into entity representations, allowing entities to evolve and capture dynamic interactions over time with greater accuracy. Our experiments on benchmark datasets show that TACAS outperforms existing methods in modeling dynamic entity behaviors. The attention mechanism used in TACAS effectively captures the complex interactions between entities, relations, and time, improving the accuracy of temporal relationship representations. Extensive testing confirms that TACAS enhances the quality of temporal knowledge graph representation. We hope that TACAS will inspire further advancements in modeling temporal data in knowledge graphs.

Acknowledgements. This work was supported by the National Natural Science Foundation of China under Grant No. 72210107001, the Beijing Natural Science Foundation under Grant No. IS23128, the Fundamental Research Funds for the Central Universities, and the CAS PIFI International Outstanding Team Project (Grant No. 2024PG0013).

References

1. Saxena, A., Tripathi, A., Talukdar, P.: Improving multi-hop question answering over knowledge graphs using knowledge base embeddings. In: Proceedings of the 58th Annual Meeting of the Association for Computational Linguistics, pp. 4498–4507 (2020)
2. He, Z., Hui, B., Zhang, S., Xiao, C., Zhong, T., Zhou, F.: Exploring indirect entity relations for knowledge graph enhanced recommender system. Expert Syst. Appl. **213**, 118984 (2023)
3. Xiong, C., Callan, J., Liu, T.-Y.: Word-entity duet representations for document ranking. In: Proceedings of the 40th International ACM SIGIR Conference on Research and Development in Information Retrieval, pp. 763–772 (2017)
4. Lacroix, T., Obozinski, G., Usunier, N.: Tensor decompositions for temporal knowledge base completion. In: International Conference on Learning Representations, pp. 4816–4821. Association for Computational Linguistics, Brussels, Belgium (2019)
5. Leblay, J., Chekol, M.W.: Deriving validity time in knowledge graph. In: Companion Proceedings of the Web Conference 2018, pp. 1771–1776 (2018)
6. Dasgupta, S.S., Ray, S.N., Talukdar, P.: HyTE: hyperplane-based temporally aware knowledge graph embedding. In: Proceedings of the 2018 Conference on Empirical Methods in Natural Language Processing, pp. 2001–2011 (2018)
7. Xu, C., Nayyeri, M., Alkhoury, F., Yazdi, H.S., Lehmann, J.: TeRo: a time-aware knowledge graph embedding via temporal rotation. In: Proceedings of the 28th International Conference on Computational Linguistics, p. 1583–1593 (2020)
8. Chen, K., Wang, Y., Li, Y., Li, A.: RotateQVS: representing temporal information as rotations in quaternion vector space for temporal knowledge graph completion. In: Proceedings of the 60th Annual Meeting of the Association for Computational Linguistics (Volume 1: Long Papers), pp. 5843–5857 (2022)
9. Garcia-Duran, A., Dumančić, S., Niepert, M.: Learning sequence encoders for temporal knowledge graph completion. In: Proceedings of the 2018 Conference on Empirical Methods in Natural Language Processing, pp. 4816–4821 (2018)
10. Sadeghian, A., Armandpour, M., Colas, A., Wang, D.Z.: ChronoR: rotation based temporal knowledge graph embedding. In: Proceedings of the AAAI Conference on Artificial Intelligence, vol. 35, pp. 6471–6479 (2021)
11. Goel, R., Kazemi, S.M., Brubaker, M., Poupart, P.: Diachronic embedding for temporal knowledge graph completion. In: Proceedings of the AAAI Conference on Artificial Intelligence, vol. 34, pp. 3988–3995 (2020)

12. Xu, C., Nayyeri, M., Alkhoury, F., Yazdi, H., Lehmann, J.: Temporal knowledge graph completion based on time series Gaussian embedding. In: Pan, J.Z., et al. (eds.) ISWC 2020. LNCS, vol. 12506, pp. 654–671. Springer, Cham (2020). https://doi.org/10.1007/978-3-030-62419-4_37
13. Xu, C., Chen, Y.-Y., Nayyeri, M., Lehmann, J.: Temporal knowledge graph completion using a linear temporal regularizer and multivector embeddings. In: Proceedings of the 2021 Conference of the North American Chapter of the Association for Computational Linguistics: Human Language Technologies, pp. 2569–2578 (2021)
14. Lai, Y., Chen, C., Zheng, Z., Zhang, Y.: Block term decomposition with distinct time granularities for temporal knowledge graph completion. Expert Syst. Appl. **201**, 117036 (2022)
15. Pan, J., Nayyeri, M., Li, Y., Staab, S.: HGE: embedding temporal knowledge graphs in a product space of heterogeneous geometric subspaces. In: Proceedings of the AAAI Conference on Artificial Intelligence, vol. 38, pp. 8913–8920 (2024)
16. Zhang, F., Chen, H., Shi, Y., Cheng, J., Lin, J.: Joint framework for tensor decomposition-based temporal knowledge graph completion. Inf. Sci. **654**, 119853 (2024)
17. Lautenschlager, J., Shellman, S., Ward, M.: ICEWS event aggregations. Harvard Dataverse **3**, 28 (2015)
18. Leetaru, K., Schrodt, P.A.: GDELT: global data on events, location, and tone, 1979–2012. In: ISA Annual Convention, vol. 2, pp. 1–49. Citeseer (2013)
19. Jain, P., Rathi, S., Chakrabarti, S., et al.: Temporal knowledge base completion: new algorithms and evaluation protocols. In: Proceedings of the 2020 Conference on Empirical Methods in Natural Language Processing (EMNLP), pp. 3733–3747 (2020)
20. Messner, J., Abboud, R., Ceylan, I.I.: Temporal knowledge graph completion using box embeddings. In: Proceedings of the AAAI Conference on Artificial Intelligence, vol. 36, pp. 7779–7787 (2022)
21. Lacroix, T., Usunier, N., Obozinski, G.: Canonical tensor decomposition for knowledge base completion. In: International Conference on Machine Learning, pp. 2863–2872. PMLR (2018)
22. Shao, P., Zhang, D., Yang, G., Tao, J., Che, F., Liu, T.: Tucker decomposition-based temporal knowledge graph completion. Knowl.-Based Syst. **238**, 107841 (2022)
23. Yue, L., Ren, Y., Zeng, Y., Zhang, J., Zeng, K., Wan, J.: Block decomposition with multi-granularity embedding for temporal knowledge graph completion. In: International Conference on Database Systems for Advanced Applications, pp. 706–715. Springer (2023)
24. Li, J., Su, X., Gao, G.: TeAST: temporal knowledge graph embedding via Archimedean spiral timeline. In: Proceedings of the 61st Annual Meeting of the Association for Computational Linguistics (Volume 1: Long Papers), pp. 15 460–15 474 (2023)
25. Li, Y., Zhang, X., Zhang, B., Ren, H.: Each snapshot to each space: space adaptation for temporal knowledge graph completion. In: International Semantic Web Conference, pp. 248–266. Springer (2022)
26. Zhang, F., Zhang, Z., Ao, X., Zhuang, F., Xu, Y., He, Q.: Along the time: timeline-traced embedding for temporal knowledge graph completion. In: Proceedings of the 31st ACM International Conference on Information & Knowledge Management, pp. 2529–2538 (2022)

27. Yang, J., Shi, Y., Tong, X., Wang, R., Chen, T., Ying, X.: Improving knowledge graph embedding using affine transformations of entities corresponding to each relation. In: Findings of the Association for Computational Linguistics: EMNLP 2021, pp. 508–517 (2021)
28. Nickel, M., Tresp, V., Kriegel, H.-P.: A three-way model for collective learning on multi-relational data. In: Proceedings of the 28th International Conference on International Conference on Machine Learning, pp. 809–816 (2011)
29. Yang, J., et al.: Learning hierarchy-aware quaternion knowledge graph embeddings with representing relations as 3D rotations. In: Proceedings of the 29th International Conference on Computational Linguistics, pp. 2011–2023 (2022)
30. Dettmers, T., Minervini, P., Stenetorp, P., Riedel, S.: Convolutional 2D knowledge graph embeddings. In: Proceedings of the AAAI Conference on Artificial Intelligence, vol. 32 (2018)
31. Le, T., Le, N., Le, B.: Knowledge graph embedding by relational rotation and complex convolution for link prediction. Expert Syst. Appl. **214**, 119122 (2023)

Adaptive Extraction of Variable-Length Subsequence Patterns in Noisy Time Series

Ke Zhang[1,2], Jiangyong Duan[1(✉)], Tiantian Yang[1,2], Hongfei Wang[1], and Congmin Lv[1]

[1] The Key Laboratory of Space Utilization, Technology and Engineering Center for Space Utilization, Chinese Academy of Sciences, Beijing 100094, China
{zhangke19,duanjy,yangtiantian23,Whf,lvcongmin}@csu.ac.cn
[2] University of Chinese Academy of Sciences, Beijing 100190, China

Abstract. Time series contains subsequence patterns of unknown lengths and expressions. Extracting these patterns accurately is essential for uncovering the information within the time series. To this end, we propose a clustering framework to determine the subsequence lengths and segmentation positions adaptively by minimizing intra-class error under the time series cover constraints. Besides, we introduce a virtual noise cluster in our framework to differentiate noise data from normal subsequence patterns, reducing the impact of noise data on the extraction of time series subsequence patterns. We also employ a nested optimization algorithm with a dynamic optimization routine and three optimization operations to refine the subsequence patterns: combining patterns that tend to appear adjacent to each other, splitting patterns with large intra-class errors, and removing clusters when the algorithm gets stuck in local optima. We compare our method with state-of-the-art methods on synthetic and real datasets. Both quantitative and qualitative results demonstrate the effectiveness of our model.

Keywords: Time series · Patterns extraction · Variable-length subsequence · Subsequence clustering

1 Introduction

Time series is widely used in various fields such as industrial fault diagnosis, human activity recognition, healthcare, finance, engineering, etc. [23] Extracting the patterns in time series facilitates the completion of data mining downstream tasks including rule discovery, classification, prediction, and anomaly detection [13]. Subsequence clustering is a common pattern extraction method which divides the time series into several subsequence segments and clusters them according to similarity, which is still a quite challenging problem [12].

The challenge lies in the fact that it need to simultaneously estimate the variable-length and accurate split positions of subsequence patterns, which are unknown in advance. Additionally, noise subsequences may widely exist in time

X. Wu et al. (Eds.): PAKDD 2025, LNAI 15875, pp. 407–419, 2025.
https://doi.org/10.1007/978-981-96-8295-9_30

series, which complicate the accurate patterns extraction. Figure 1 displays a synthetic dataset designed to illustrate our motivation. The input time series in Fig. 1a includes two subsequence patterns with different lengths and several noise patterns as shown in Fig. 1b. To extract subsequence patterns, it is necessary to obtain the variable-length of different subsequence patterns and their corresponding segmentation positions adaptively. Besides, the noise data need to be excluded simultaneously to enhance the accuracy of the clustering results.

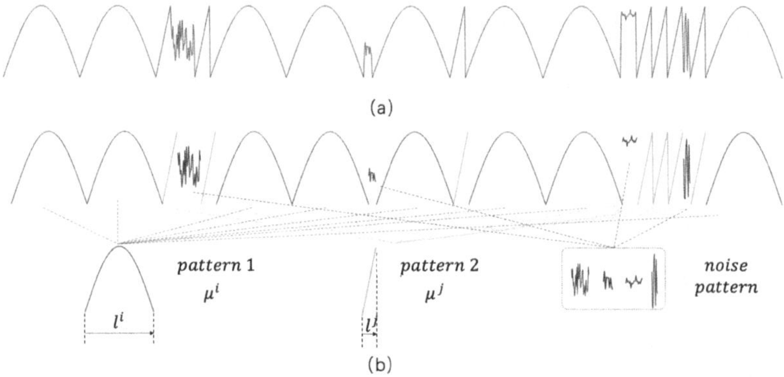

Fig. 1. Time series with variable-length patterns and noise data.

Considerable research on time series subsequence clustering has been published in recent years. Keogh et al. [17] found that traditional clustering methods using fixed-length sliding windows to segment subsequences such as K-means [10], fuzzy C-means [5], hierarchical methods [3], and self-organizing maps [8] are meaningless because of the trivial matches produced by the overlap of adjacent sliding windows. To avoid meaningless subsequence clustering results, motif discovery methods have been proposed to identify frequently occurring similar subsequences within time series. Michele et al. [18] proposed a scalable motif discovery algorithm which can effectively find motifs of all lengths within a given range. To enhance the precision of motif discovery, Sara et al. [4] integrated DTW method into the distance measure to achieve warping patterns detection. Cartwright et al. [7] used symbolic aggregate approximation and the minimum description length to identify locally similar motifs with boundary misalignment. However, motif discovery methods aim to extract the most frequently occurring patterns in time series. As a result, extracted frequent patterns may be some part of intrinsic pattern or some combination of intrinsic patterns. Other common subsequence clustering methods include segmented methods such as Toeplitz inverse covariance-based clustering method [14] and its improved method Context-aware segmentation and clustering method [15] require data confirming to specific distributions.

To address the limitations mentioned above, we propose a new subsequence clustering method. Our model adaptively determines the lengths and splitting positions of subsequence patterns with unknown lengths in time series. Besides, we naturally distinguish noise patterns from normal subsequence patterns in a unified clustering framework, thus eliminating the impact of noise data on subsequence clustering results. Furthermore, we propose a dynamic optimization algorithm to refine our results. We compare our method with state-of-the-art methods both on synthetic and real time series datasets, demonstrating the effectiveness of our model.

2 Our Model

In this section, firstly we illustrate our motivation. Then we describe how we exclude the noise data naturally and extract subsequence patterns of unknown lengths with our framework. Finally, we describe the optimization algorithm and estimate the computational complexity.

2.1 Motivation

We define the problem of extracting variable-length subsequence patterns. For a given time series $X_{1:T}$ of length T, the goal is to identify k clusters $C = \{C^i, i = 1, \ldots, k\}$ representing k different subsequence patterns.

Some issues exist in practical applications, making subsequence extraction quite challenging. As shown in Fig. 1, *Pattern 1* and *Pattern 2* have different lengths. Using a fixed length to segment the series can lead to inaccuracies. If the chosen length exceeds the actual length of the pattern, the extracted patterns may include partial information from adjacent patterns. Conversely, if the chosen length is shorter than the pattern's actual length, parts of the pattern that should be included are missed. Inappropriate length selection leads to inaccurate pattern extraction results. Additionally, the position of the subsequence segmentation is also crucial to the results.

Moreover, real time series usually contains noise data which do not indicate meaningful patterns. These noise make accurate pattern extraction more difficult. We need to identify irregular noise patterns from other regular patterns.

To deal with these challenges, we propose a new clustering framework for variable-length subsequence pattern extraction and introduce a virtual noise subsequence cluster in our model to isolate noise data into a separate cluster by minimizing intra-class errors.

2.2 Adaptive Subsequence Clustering Model

Our model extract variable-length subsequence patterns by simultaneously optimizing the unknown lengths subsequence cluster centers μ and the corresponding subsequence segmentation Z by minimizing the following subsequence intra-cluster loss under certain time series constraints:

$$\min_{C,Z,\mu} f(C, Z, \mu) = \sum_{i=1}^{k} (\sum_{j=1}^{n_i} \|x_{i,j} - \mu^i\|^2)/|\mu^i| + \sum_{x \in N} loss_{noise}(x, \tau), \tag{1}$$
$$s.t. \quad x_{i,j} \cup N = X, l_{min} \leq |x_{i,j}| \leq l_{max}$$

In Eq. 1, C represents the clusters set, which indicates different normal subsequence patterns and noise patterns. N represents the set of noise subsequence patterns. Z is the set of subsequence segmentations, containing the start and end positions of each subsequence. $x_{i,j}$ denotes the j-th subsequence in the i-th cluster set. μ represents the cluster centers while μ^i represents the cluster center of the i-th cluster. $|\mu^i|$ indicates the length of μ^i which means the length of subsequences in i-th cluster. τ is the noise cluster center while $loss_{noise}$ is the loss value when subsequence x classified as noise. The constraint $x_{i,j} \cup N = X$ ensures all data points be allocated to the corresponding clusters. l_{min} and l_{max} are the minimum and maximum lengths respectively for subsequence division. Within this range, we search the optimal division lengths and positions.

In our objective function 1, the cluster of subsequence x is determined by the distance d between x and each cluster centers μ, thus the i-th cluster set $C^i = \{x|d(x,\mu^i) \leq d(x,\mu^j)\}$. As for non-noise subsequence patterns, $d = \|x - \mu\|^2/|\mu|$. As for subsequences that deviate from the normal patterns and should not belong to any specific normal cluster as the blue patterns shown in Fig. 1, we specify a fixed noise loss $loss_{noise}$ to represent d. When the distance between a subsequence and other cluster centers exceeds this fixed noise loss value, the subsequence is considered to be noise and assigned to cluster N.

For the noise loss, it should be related to the noise subsequence pattern length, i.e., the longer the noise subsequence length, the larger the loss value. Besides, noise loss is assumed to be related to the variance of the entire time series, i.e., when time series varies rapidly, the noise loss will be somewhat large. These two assumptions are intuitive. In our model, we empirically set the noise loss to be $loss_{noise} = len_{noise} \times var(X) \times \omega$. len_{noise} represents the length of the noise subsequence, which is usually chosen to be relatively short because we suppose that longer noise subsequences can be represented by multiple short noise segments. $var(X)$ denotes the variance of the entire sequence X while ω is a coefficient.

Our model solves the subsequence extraction problems from several aspects. Firstly, we optimize both the variable-length cluster centers μ and the corresponding segments Z simultaneously. Secondly, we introduce virtual noise cluster to isolate irregular noise subsequences from normal patterns into a separate cluster, which enhances the accuracy of subsequence clustering. Thirdly, we deal with trivial matches generated by the sliding window methods resulting in meaningless results, our model's constraint ensures that each data point only needs to exist in any one cluster, thus more meaningful non-trivial matched subsequence patterns can be obtained.

2.3 Optimization

Our algorithm is optimized by two nested processes: the inner process collaboratively optimizes the subsequence segments Z and the cluster centers μ under fixed subsequence lengths. The outer process performs subsequence cluster combination, splitting, and removing operations.

For inner optimization, we use a dynamic optimization method to minimize the objective function. Assuming we have already obtained the subsequence segmentation Z_{p-1} at position $p-1$ and its corresponding model loss D_{p-1}, the process of calculating the subsequence segments Z_p and the loss D_p within time series X at position p is as follows:

$$Z_p = \{Z_J, X_{p-|\mu_i|+1:p}\} \tag{2}$$

$$D_p = D_J + \begin{cases} \|X_{p-|\mu^i|+1:p} - \mu^I\|^2/|\mu^I|, & X_{p-|\mu_i|+1:p} \notin N \\ loss_{noise}, & X_{p-|\mu_i|+1:p} \in N \end{cases} \tag{3}$$

where

$$I, J = \underset{i=1:k, j=p-|\mu^i|:p-1}{\arg\min} \quad D_j + \begin{cases} \|X_{p-|\mu^i|+1:p} - \mu^i\|^2/|\mu^i|, & X_{p-|\mu_i|+1:p} \notin N \\ loss_{noise}, & X_{p-|\mu_i|+1:p} \in N \end{cases} \tag{4}$$

after we obtain the optimal subsequence segments Z, the cluster center μ^i will be updated as follows:

$$\mu^i = \begin{cases} \frac{1}{|C^i|} \sum_{d(x,\mu^i) \leq d(x,\mu^j)} x, & x \notin N \\ \tau, & x \in N \end{cases} \tag{5}$$

In the above process, given an initial subsequence pattern cluster, we globally optimize Z and μ alternatively, so the value of objective function 1 reduces at every iteration and ultimately converges. For initialization, to extract subsequence patterns of variable-length, we segment subsequences into different lengths within the given length range as the initial subsequence set and obtain subsequence patterns through inner optimization process.

In the outer process, we conduct combination, splitting and removing operations to ensure we obtain appropriate subsequence patterns. For two patterns that frequently appear adjacent to each other in Fig. 2a, we assume they belong to the same cluster, thus we merge these two clusters into one. For clusters with high intra-cluster errors in Fig. 2b, we suppose they contain patterns that do not belong to the same cluster, so we split them into two clusters. When the model gets stuck in a local optimum and cannot converge, or when the number of clusters exceeds the specified k after the splitting operation, we perform the removing operation. The outer process algorithm is described in Algorithm 1. The loss value of the objective function is decreasing during the optimization process, which ensures the convergence eventually.

Fig. 2. Cluster optimization. (a) Clusters combination, (b) cluster splitting.

Algorithm 1. Our Optimization Algorithm

```
Input: time series X_{1:T}, cluster number k, minimal subsequence length l_{min}, maximal subsequence length l_{max}, noise loss loss_{noise}
Output: subsequence clusters set C, subsequence segmentation set Z
C = Initialization(X, k, l_{min}, l_{max}, loss_{noise})
K = ClusterNumber(C)
Z = SubsequencePartion(C)
Loss = ComputeLoss(X, C, Z)
while K >= k do
    C = ClusterSplit(X, C_{x∉N}) + C_{x∈N}
    C = ClusterCombine(X, C_{x∉N}) + C_{x∈N}
    Z = SubsequencePartion(C)
    Loss = ComputeLoss(X, C, Z)
    K = ClusterNumber(C)
    if K > preK then
        C = RemoveOneCluster(X, C_{x∉N}) + C_{x∈N}
        Z = SubsequencePartion(C)
        tmpLoss = ComputeLoss(X, C, Z)
        if tmpLoss > Loss then
            C = RemoveOneCluster(X, C_{x∉N}) + C_{x∈N}
        end if
    else if K = preK and curLoss > preLoss then
        tmpC = RemoveOneCluster(X, C_{x∉N}) + C_{x∈N}
        tmpK = ClusterNumber(tmpC)
        if tmpK >= K then
            C = tmpC
        end if
    end if
end while
return C, Z
```

2.4 Time Complexity

The optimization process of the model includes two nested routines. For the inner optimization, the time complexity of optimizing Z and μ all equal to $O(Tkl_{\max})$. For the outer optimization, the complexity is $O(Tkl_{\max}^3)$ for splitting operation since at most $(l_{\max} - l_{\min})k$ clusters with up to $l_{\max}$ positions in each need to be split. The complexity is $O(Tkl_{\max}^3)$ for combination since at most $(l_{\max}-l_{\min})k$ clusters with up to $2l_{max}-1$ positions in each need to be evaluated. For cluster removing the time complexity is $O(Tk^2l_{\max}^3)$. In summary, the overall time complexity is linearly proportional to the length T of the time series.

3 Experiments

In this section, we conduct comparisons on both synthetic and real datasets with state-of-the-art methods to demonstrate the effectiveness of our model.

3.1 Experimental Settings

Baselines. The baselines can be divided into three categories. The first category is subsequence clustering methods based on fixed length sliding window: **K-means** [10], **DTWGAK** [9], **NeuralGas** [11], **K-medoids** [16], **K-shape** [20]. The second category are based on segmentation: **TICC** [14], **CASC** [15]. The third category are motif discovery methods: **K-motif** [17], **MDLKmotif** [21], **VLmotif** [19], **EXTRACT** [6], **SCRIMP++** [24], **SWAMP** [4], **SLIM** [7].

Evaluation Metrics. We analyze the results in quantitative and qualitative evaluations. For quantitative evaluation, as for the input time series $X_{1:T}$, we define the accuracy rate $acc = \frac{\sum_{i=1}^{T} 1_{M_i=A_i}}{T}$ where M_i and A_i represent the calculated cluster label and the ground truth label respectively. For qualitative analysis, we display the estimated cluster centers, extracted subsequence samples and segmentation results within time series. For sliding window methods, we display the cluster centers and segmentation results. For segmentation methods, we only show the segmentation results because they do not produce cluster centers. For motif discovery methods, we display the extracted cluster centers and the subsequence samples because they obtain similar subsequences without segment the whole time series.

3.2 Experiments on Synthetic Data

In this section, we construct a synthetic dataset containing two types of subsequence patterns with lengths of 20 and 50 which shown in Fig. 3a. Besides, we introduce noise data of unknown lengths into the dataset. The noise differs significantly in expression from the two normal subsequence patterns. We compare the results obtained by our method with and without introducing noise cluster. We also compare with baselines on this dataset.

For experiment settings, we set subsequence lengths ranging from 3 to 60 with cluster number $k = 3$ in our model. The len_{noise} is 5 and ω was empirically set to 0.5. We set window lengths of 5, 20, and 50 for fixed sliding windows methods. For motif discovery methods, we set subsequence lengths and cluster numbers consistent with us if required, other parameters followed the default

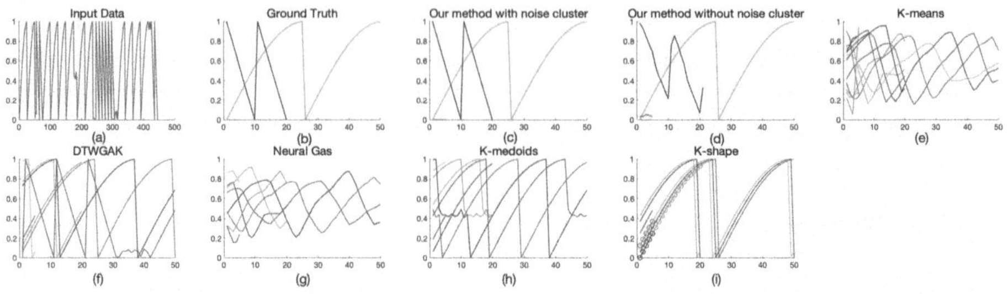

Fig. 3. Extracted clustering centers on synthetic data for different methods.

settings. The visualization results are displayed in Fig. 3, 4 and 5. We also give the quantitative analysis results for the methods generating segmentation results.

Figure 3b shows the cluster centers ground truth of two types of patterns. The results of our model with and without the introduction of synthetic noise cluster are in Fig. 3c and Fig. 3d. In Fig. 3c, the blue and green patterns represent the normal subsequence patterns we extracted, and the red pattern represents the noise cluster center. We obtain ideal results compared to the ground-truth. Figure 3d shows the result of our model without introducing noise cluster and the result are influenced by the noise subsequence patterns. The results of the baselines based on sliding window are shown from Fig. 3e to 3i. For methods that update the cluster centers based on the mean value of the subsequences, such as K-means and Neural Gas, extracting meaningless cluster centers which similar to sine waves. Although DTWGAK, K-medoids and K-shape are capable to extract part of the meaningful patterns, they are unable to extract variable-length subsequence patterns because of the fixed window length.

Figure 4 displays subsequence pattern samples extracted by our method and motif discovery methods. Figure 4a shows the ground truth of the subsequence patterns. After introducing a noise cluster, our model can extract results as shown in Fig. 4b that are ideal compared to the ground truth in Fig. 4a for non-noise subsequences. We exclude the noise impact on pattern extraction. The result of our model without noise cluster is shown in Fig. 4c. Without consideration of noise, the presence of noise data has a certain impact on the results of pattern extraction. For the methods based on motif discovery in Fig. 4d - 4j, they are all capable of extracting frequently occurring patterns to some extent. However, methods such as Kmotif, VLmotif, EXTRACT, SWAMP and SCRIMP++, extracting part of the subsequence patterns or combining multiple patterns because they cannot estimate proper segmentation lengths adaptively. MDLKmotif and SLIM can estimate variable-length patterns but they also extract part of the patterns or combine two patterns.

Figure 5 shows the results of subsequence segmentation by sliding window methods and segmentation methods, where the same color indicates the same pattern. Our method in Fig. 5b demonstrates identical result to the ground truth

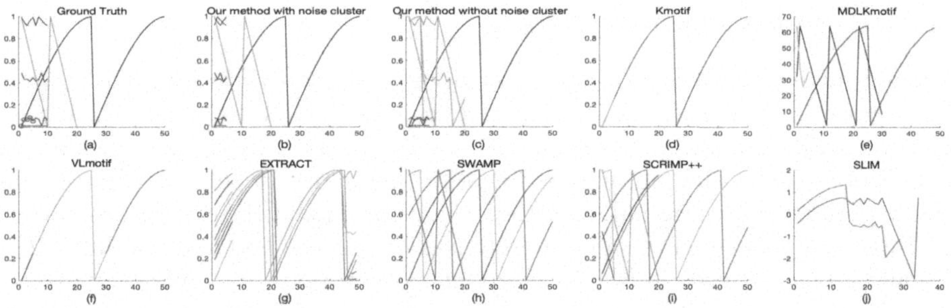

Fig. 4. Extracted subsequence samples on synthetic data for different methods.

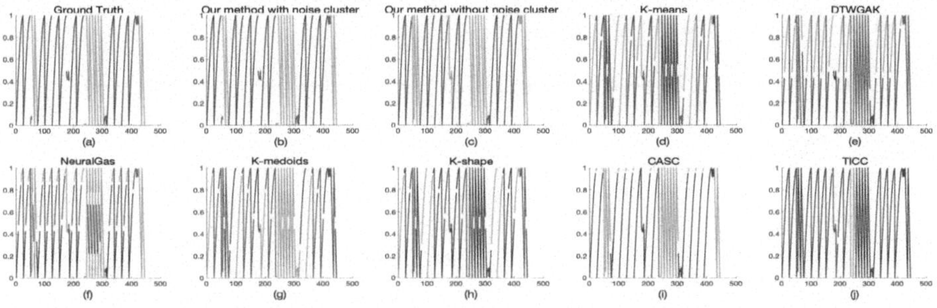

Fig. 5. Segmentation results on synthetic data for different methods.

in Fig. 5a. Without introducing noise cluster, our results are shown in Fig. 5c. The results of methods based on sliding window as shown in Fig. 5d–5h are not ideal because they fail to extract subsequence patterns of variable-length. For the segmentation-based methods in Fig. 5i and 5j, satisfactory results are not be obtained when the data does not match the model's distribution. The accuracy rate *acc* of the segmentation results is listed in Table 1. Our methods produce satisfied results compared with other methods.

Table 1. Segmentation Accuracy on Synthetic Data (%)

Method	Accuracy	Method	Accuracy	Method	Accuracy
Our method with noise cluster	100	DTWGAK	65.39	K-shape	38.88
Our method without noise cluster	93.48	NeuralGas	44.49	CASC	89.66
K-means	40	K-medoids	44.27	TICC	74.16

3.3 Experiments on Real Data

For real time series, we test various algorithms on a temperature time series from NAB dataset [22]. We set the subsequence lengths from 20 to 80 and the cluster number k is 3. len_{noise} is 5 while ω is 0.5. In addition to the length and the number of clusters, the parameters for other methods are set according to the default settings of the original algorithm. Figure 6 displays the cluster centers and Fig. 7 displays subsequence samples obtained by our method and baselines. The blue and green patterns in Fig. 7a represent normal patterns while red patterns are noise. We estimate two proper variable-length similar subsequence patterns, with high overlap within each cluster but distinct differences between the clusters. Other methods did not extract subsequence patterns with clear differences between classes.

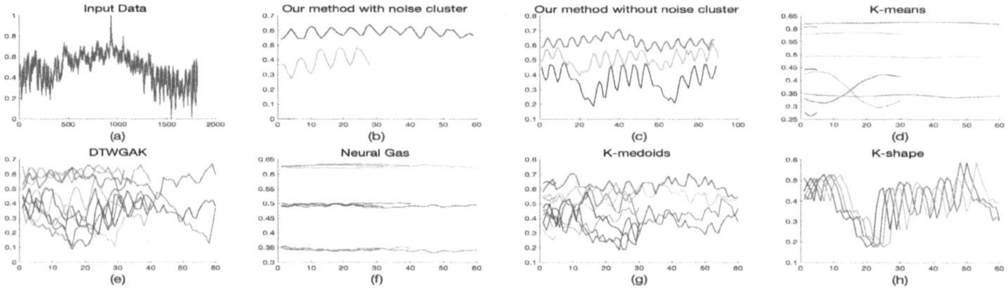

Fig. 6. Extracted clustering centers on real data for different methods.

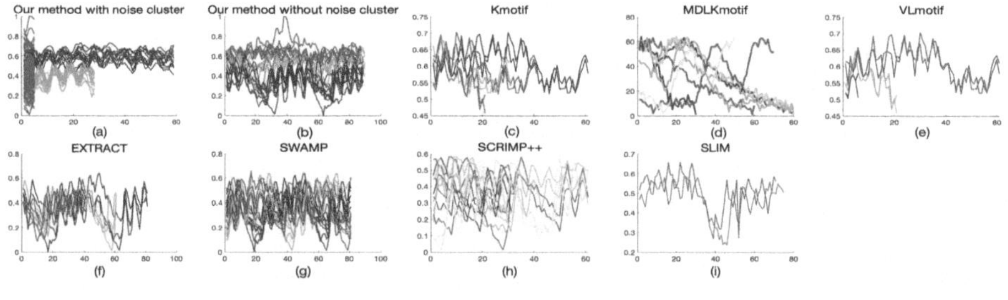

Fig. 7. Extracted subsequence samples on real data for different methods.

Figure 8 shows the comparing results of subsequence segmentation by sliding window methods and segmentation methods. By introducing noise cluster, we extracted two potentially noteworthy temperature change trends from the quite irregular time series and identified some noise data that differ from the normal patterns in Fig. 7a. In Fig. 8a, we segmented the time series according to the

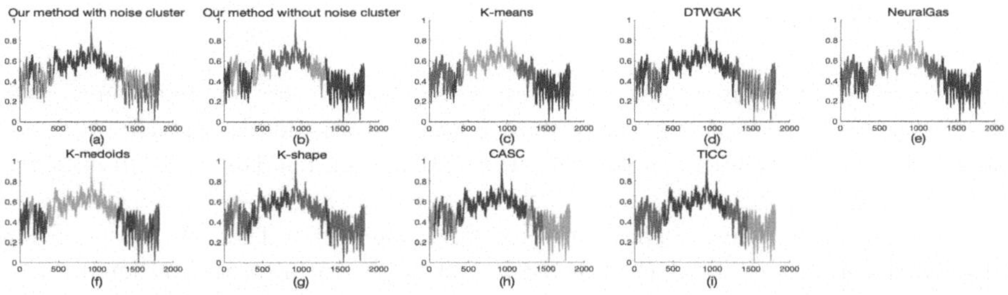

Fig. 8. Segmentation results on real data for different methods.

extracted patterns, and the results seems reasonable. For other methods, abrupt noise data affects the segmentation results.

In addition, we also validate the effectiveness of our model on the temperature dataset [2] (the first two rows) and ECG dataset [1] (the last row). For the 3 datasets, we set l_{min}, l_{max}, k, len_{noise} and ω as (30,60,3,5,0.5), (30,60,4,5,0.5), (40,120,3,5,0.5). The results are shown in Fig. 9. Figure 9b is the extracted variable subsequence cluster centers, Fig. 9c is the segmentation results and Fig. 9d is the extracted subsequences and corresponding clustering results. The red patterns are noise while other colors represent normal patterns. Our model identifies meaningful variable-length patterns of temperature or heart rate variations in time series and distinguishes noise patterns that differ from the normal ones simultaneously.

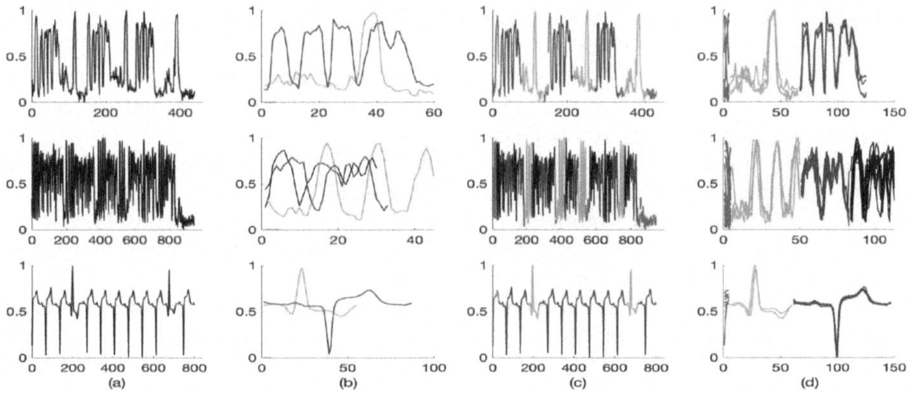

Fig. 9. Subsequence extraction results on real time series. (a) Input data, (b) cluster centers, (c) segmentation results, (d) subsequences in each clusters.

4 Conclusions

This paper presents an algorithmic framework to extract variable-length subsequence patterns adaptively. Our model can automatically identify the segmentations and expressions of subsequences of varying lengths while minimizing intra-class error. Additionally, for real time series that may contain noise data, we effectively eliminate the influence of noise subsequences on the extraction of normal patterns in our framework. For future research, we plan to further improve our model by incorporating appropriate similarity measures learned from data, making it more suitable for real application scenarios.

Acknowledgments. This work was supported by the Development and Construction of Ground Station Network and Measurement, Transportation and Control System for Ground Application System under Grant E5GZ15010507 and the Prospective Study on the Development and Utilization of Earth-Moon Space under Grant E4GZ150305.

References

1. Physiobank atm. https://archive.physionet.org/cgi-bin/atm/ATM
2. Temperature data. http://alumni.cs.ucr.edu/~rakthant/TSEpenthesis/
3. Aghabozorgi, S., Shirkhorshidi, A.S., Wah, T.Y.: Time-series clustering-a decade review. Inf. Syst. **53**, 16–38 (2015)
4. Alaee, S., Kamgar, K., Keogh, E.: Matrix profile XXII: exact discovery of time series motifs under DTW. In: ICDM, pp. 900–905 (2020)
5. Bezdek, J.C.: Pattern Recognition with Fuzzy Objective Function Algorithms. Springer (2013)
6. Blalock, D.W., Guttag, J.V.: Extract: strong examples from weakly-labeled sensor data. In: ICDM, pp. 799–804 (2016)
7. Cartwright, E., Crane, M., Ruskin, H.J.: Side-length-independent motif (SLIM): Motif discovery and volatility analysis in time series-SAX, MDL and the matrix profile. Forcasting **4**(1), 219–237 (2022)
8. Cherif, A., Cardot, H., Boné, R.: SOM time series clustering and prediction with recurrent neural networks. Neurocomputing **74**(11), 1936–1944 (2011)
9. Cuturi, M.: Fast global alignment kernels. In: ICML, pp. 929–936 (2011)
10. Das, G., Lin, K.I., Mannila, H., Renganathan, G., Smyth, P.: Rule discovery from time series. In: KDD, vol. 98, pp. 16–22 (1998)
11. Dimitriadou, E., Hornik, K.: Cclust: convex clustering methods and clustering indexes (2023). https://cran.r-project.org/web/packages/cclust/
12. Duan, J., Guo, L.: Variable-length subsequence clustering in time series. TKDE **34**(2), 983–995 (2022)
13. Ezugwu, A.E., et al.: A comprehensive survey of clustering algorithms: state-of-the-art machine learning applications, taxonomy, challenges, and future research prospects. Eng. Appl. Artif. Intell. **110**(104743) (2022)
14. Hallac, D., Vare, S., Boyd, S., Leskovec, J.: Toeplitz inverse covariance-based clustering of multivariate time series data. In: IJCAI, pp. 5254–5258 (2018)
15. Jain, S., Hallac, D., Sosic, R., Leskovec, J.: CASC: context-aware segmentation and clustering for motif discovery in noisy time series data. arXiv preprint arXiv:1809.01819 **124**, 1–8 (2018)

16. Kaufman, L., Rousseeuw, P.J.: Finding Groups in Data: An Introduction to Cluster Analysis. Wiley (2009)
17. Keogh, E., Lin, J.: Clustering of time-series subsequences is meaningless: implications for previous and future research. Knowl. Inf. Syst. **8**(2), 154–177 (2005)
18. Linardi, M., et al.: Matrix profile X: VALMOD-scalable discovery of variable-length motifs in data series. In: COMAD, pp. 1053–1066 (2018)
19. Nunthanid, P., Niennattrakul, V., Ratanamahatana, C.A.: Discovery of variable length time series motif. In: ECIT, ThaiLand, pp. 472–475 (2011)
20. Paparrizos, J., Gravano, L.: k-shape: efficient and accurate clustering of time series. In: SIGMOD, pp. 1855–1870 (2015)
21. Rakthanmanon, T., Keogh, E.J., Lonardi, S., Evans, S.: Time series epenthesis: clustering time series streams requires ignoring some data. In: ICDM, pp. 547–556 (2011)
22. Williams, L., O'Kane, C.O., Franks, A.L.: Numenta anomaly benchmark: a benchmark for anomaly detection algorithms (2016). https://github.com/numenta/NAB
23. Zhang, K., et al.: Self-supervised learning for time series analysis: taxonomy, progress, and prospects. IEEE Trans. Pattern Anal. Mach. Intell. (2024)
24. Zhu, Y., Yeh, C.C.M., Zimmerman, Z., Kamgar, K., Keogh, E.: Matrix profile XI: SCRIMP++ : time series motif discovery at interactive speeds. In: ICDM, pp. 837–846 (2018)

Hunting Inside N-Quantiles of Outliers (Hino)

Jessy Colonval(✉) and Fabrice Bouquet

Université de Franche-Comté, Institut Femto-ST, CNRS, 16 Route de Gray, Besançon, France
{jessy.colonval,fabrice.bouquet}@femto-st.fr

Abstract. This paper presents an approach (called **Hino**) to detect outliers present in a data set, also called aberrations or anomalies. These data may reduce the quality of data analysis and lead to erroneous results. In the case of learning algorithms, they can deviate their behavior, i.e. reduce their efficiency. Thus, outlier detection is crucial where it improves performance by providing better data quality and reduces the influence of outliers. Inter Quartiles Range (**IQR**) is a popular statistical detection method, which has the advantage of being simple and fast in calculation time. It is based on the distribution quartiles of a data set and considers the most extreme values as outliers. This means that this method only searches for **point outliers**, which is a restrictive and naive approach. Indeed, nothing prevents an element from having an extreme value while remaining consistent with the rest of the elements. The proposed method is also a statistical detection method based on quantiles, but it looks for **contextual outliers** instead of **point outliers** and consider the context of a point to determine whether it is an outlier or not. The effectiveness of **Hino** is compared with the original **IQR** method and other approaches, including **Isolation Forest**, **SVM**, and **LOF**, using 16 real and 278 synthetic data sets.

Keywords: Outliers detection · Data filtering · Interquartile range · Machine learning

1 Introduction

An important principle in the use of artificial intelligence algorithms (AIs) is the quality of the learning data. So, the implementation of tools to help filter the data is a critical point [10]. However, data sets may contain outliers, which can be defined as "an observation (or subset of observations) which appears to be inconsistent with the remainder of that set of data" [3]. Outliers can reduce the quality of data analysis and lead to erroneous results. And in the case of AIs, they can deviate behavior of AIs and reduce their efficiency with two effects:

1. Weaken the predictive power of the model obtained after the learning phase.
2. Weaken the score obtained from the model during its validation phase.

X. Wu et al. (Eds.): PAKDD 2025, LNAI 15875, pp. 420–431, 2025.
https://doi.org/10.1007/978-981-96-8295-9_31

Therefore, it is essential to identify and eliminate them directly. Thus, the quality of the data is improving and outlier's influence is reducing [9].

Three categories of outliers are identified [8]:

1. **Points outliers** are isolated individuals, they are far from other data.
2. **Contextual outliers** are isolated individuals in a specific context, i.e. according to their relations between their attributes and their behavioral values.
3. **Collective outliers** are a set of individuals consistent with each other but isolated from the rest of the data.

The **IQR** approach considers the distribution of a point to determine if it is isolated which makes it a statistical detection method. And since the context is not taken into consideration, then this method detects **point outliers**. Moreover, this method considers that a point with at least one of these values isolated is sufficient to consider it as an outlier. However, we will see that this method is ineffective for detecting aberrations. The objective is to extend the **IQR** approach so that it is able to detect **contextual outliers** and tolerate that the values of a point can be isolated a minimum number of times before considering the point fully isolated. This new approach is called **Hunting inside n-quantiles of outliers (Hino)**.

Since **Hino** is looking for contextual outliers and the data sets used in the experimentation are spatial (see Sect. 5), then the terminology defines two types of attributes [1, Chapter 11]:

1. **Behavioral attributes** are attributes of interest measured for each point, commonly called *class*. For example: the type of glass, the presence of a heart abnormality, or the description of an image. A behavioral value is a value of this attribute. For example, a behavioral attribute that describes 7 different glass categories, here **1** would be one of the existing behavioral values.
2. **Contextual attributes** is an attribute expressing the characteristics of a point, commonly called *feature*. For example: the composition of a glass pane, the number of heartbeats per minute, or the color of a pixel.

The Sect. 2 present the **IQR** method of detection by quantiles, its functioning and the problems it raises. Then, the Sect. 3 present the **Hino** approach, its functioning and its meta-parameters. Follow-up of the Sect. 4 which proposes a methodology to generate the synthetic data sets used (see Sect. 4.1) and equations to calculate the **Hino** meta-parameters (see Sect. 4.2). Finally, the performance detection of **Hino** is compared with three methods (**IQR**, **Isolation Forest**, **SVM** and **LOF**) on a panel of 278 synthetic data sets (see Sect. 5.1) and on a panel of 16 real data sets (see Sect. 5.2) to conclude with a discussion of some biases present in the evaluation of real data sets and on the values used for the meta-parameters (see Sect. 5.3).

2 Detection Method Based on the Interquartile Range

In general, the median divides a data set into two more or less equal parts (depending on if the number of elements is even or odd), so approximately half of the elements are below the median and the rest are above it. On the same principle, it is possible to make this division into two unequal parts such that a percentage p of the data is less than a certain number and a percentage $1 - p$ is greater than that number. This number is called the $100p$ *empirical percentile* or the *pth empirical quantile* and is denoted by $q_n(p)$ [5].

To calculate the interquartile range, the data set is divide into four parts. Using the previous definition we define: the 25th empirical percentile $q_n(0.25)$ is called the *lower quartile* (Q1) and the 75th empirical percentile $q_n(0.75)$ is called the *upper quartile* (Q3). Together with the median, they allow the division of a data set into four parts more or less equal in number of elements. The distance between these two quartiles gives an indication of the skewness of the data set. This distance is called **IQR** (**I**nter **Q**uartile **R**ange), and it equals to:

$$IQR = Q3 - Q1 = q_n(0.75) - q_n(0.25) \tag{1}$$

The interquartile range is used to describe the spread of a distribution. Thus, it is possible to detect outliers by considering the most scattered elements as outliers. For this, it is necessary to define thresholds that will cover the majority of the points so that only the most scattered ones are considered as outliers. The lower fence is equal to $Q1 - 1.5 * IQR$ and the upper fence is equal to $Q3 + 1.5 * IQR$. The constant 1.5 allows controlling the sensitivity of the interval and thus the rule of detection. A larger one would cause outliers to be considered healthy, while a smaller one would cause some points to be detected as outliers. Thus, with a Gaussian distribution, then a constant equal to 1.5 gives fences covering -2.698σ (see Eq. 2) and 2.698σ or 99.3% of the points.

$$\begin{aligned} Q1 - 1.5 * IQR &= Q1 - 1.5 * (Q3 - Q1) \\ &= -0.6745\sigma - 1.5 * (0.6745\sigma - (-0.6745\sigma)) \\ &= -0.6745\sigma - 1.5 * 1,349\sigma = -2.698\sigma \end{aligned} \tag{2}$$

Figure ??, gives a visualization of quartiles and fences as a box plot and relates them to a Gaussian distribution. The majority of the points are included between these fences and only a small part of the points, the most extreme, will be excluded. Vinutha et al. [11] use the interquartile range as an outlier detection method. For each contextual attribute, this range is calculated and all points within are considered as outliers.

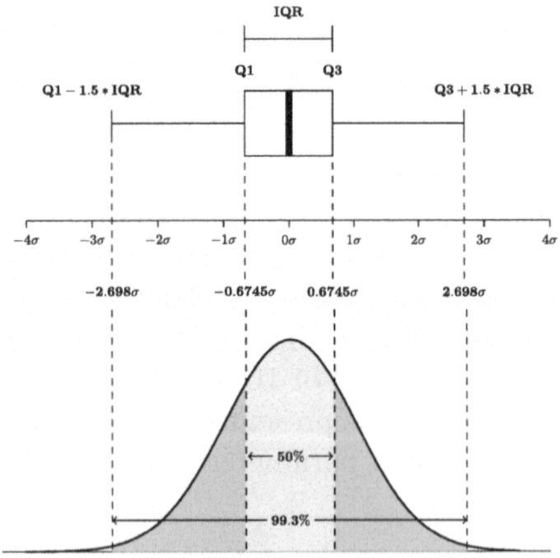

Fig. 1. Box plot and the Gaussian probability density of interquartile range

This method have three problems that make hazardous the use of the interquartile range as an outlier detection method. First, it removes the points that have extreme values who aren't always outliers. Indeed, nothing prevents an inlier to be extreme. Second, it doesn't consider the individual distribution of a behavioral value. One of them can be essentially present on the extreme values which would be suppressed with this approach. Third, there is no **tolerance limit** to determine if a point is an outlier. With only one extreme values among the contextual attribute, the large data sets are disadvantaged and risk to have a high false positivity rate because a point will have more chance to have an extreme value.

3 Extension of the IQR Method

As a reminder, an outlier can be defined as an abrupt change. **IQR** detects only **point outliers** and considers as outliers the points with at least one extreme values. As explained previously, this approach is problematic. Thus, our objective is to extend **IQR** to be able to detect **contextual outliers** and tolerate the isolation of a point for few attributes before being an outlier.

For this purpose, we propose to separate the data set into several quantiles for each contextual attribute, instead of only 4. For each quantile, we check that all the behavioral values that constitute it are also present in adjacent quantiles. Otherwise, all representatives of absent values are considered isolated for this attribute. Except in the case where all the representatives of a value are present in a single quantile. A point isolated for few attributes is not necessarily an outlier. Indeed, depending on the total number of attributes, an isolation for just one of them does not have the same impact. For example, an isolation for 1 attribute out of 10 does not have the same significance for 100. Our approach tolerates a certain level of isolation before identifying a point as an outlier. To summarize, an outlier is a point "too often" isolated.

Our extension is summarized in the Algorithm 1. It takes as parameters the data set (P), the number of quantiles used (n_qtils), the tolerance limit ($limit$) and the maximum percentage of points that can be considered as outliers (p_{max}) to preserve the consistency of data set. And it returns the mapping between the points in P and a Boolean to indicate if they are outliers. Except P, the other arguments strongly influence the detection result. However, it is difficult to determine the best values for each of them to obtain an optimal outlier detection.

- p_{max} is set at 20%, greater, we assume the data set will be distorted.
- n_qtils is calculated according to the number of points they must contain. They must be small enough so that the points they contain share similar characteristics and make it easier to detect abrupt changes.
- $limit$ is calculated according to the number of contextual attributes and behavioral values. When the number of contextual attributes and/or behavioral values are high, there's a greater chance of obtaining many isolated values.

This algorithm has been implemented in Python and its time complexity is equal in the worst case to $O(p(ac + a + 1))$, while in the best case to $O(p(ac + \frac{a}{a+1} + 1))$, where p is the number of points, a is the number of attributes and

Algorithm 1: Hino.

```
input : Let P be a set of n points
        Let n_qtils ≥ 2 be the number of quantiles
        Let limit > 0 be the maximum of breaking rule
        Let p_max be the maximum percentage of outliers
output: Let result be an mapping indicating if a point is an outlier
A_c be the set of contextual attributes labels in P
/* Counters that track the frequency of isolated points.                */
n_cdt ← {(p, 0)|p ∈ P}
for a ∈ A_c do
    qtils ← A list of list that groups the points divided by n_qtils quantiles on
      the attribute a
    for i ∈ [0, card(qtils)[ do
        /* Gets the behavioral values of the previous, next and
           current quantile, respectively.                              */
        prev_cls ← ∅
        if i > 0 then
            prev_cls ← Behavioral values in qtils[i − 1]
        next_cls ← ∅
        if i < card(qtils) − 2 then
            next_cls ← Behavioral values in qtils[i + 1]
        cur_cls ← Behavioral values in qtils[i]
        /* Missing behavioral values are those in the current
           quantile, but absent from the two adjacent ones.             */
        missing_cls ← cur_cls \ (prev_cls ∪ next_cls)
        /* If one is missing, then the counters of their points in the
           current quantile are increased.                              */
        if card(missing_cls) > 0 then
            for pq ∈ qtils[i] do
                pq_cls ← Behavioral value of pq
                all_here ← True if all of pq_cls are in qtils[i]
                /* In that case their counters are not incremented.     */
                if pq_cls ∈ missing_cls and not all_here then
                    n_cdt[pq] + +
/* Repeat with a limit incremented if a behavioral value is fully
   identified as an outlier or if more than p_max points are
   outliers.                                                            */
repeat
    result ← {(p, False)|p ∈ P}
    for (p, count) ∈ n_cdt do
        result[p] = count > limit
    limit + +
until (countTrue(result) ≤ p_max and not viableCls(result))
```

c is the number of behavioral values. Plus, its memory complexity is equal to $O(p(a + \frac{a}{a+1}))$.

4 Determination of Meta-parameters

The first meta-parameters is the number of quantiles and it equals to $\frac{p}{c+1}$ to ensure that each quantile is large enough so that each behavioral value is represented at least once. The second meta-parameter is the maximum percentage of outliers set at 20%, we assume that over this value a data set may become inconsistency. The third meta-parameter is the **tolerance limit**, which is determined by the number of attributes and behavioral values. However, building an equation from these values is difficult, so it's estimated empirically using a regression study based on synthetic data sets (see Sect. 4.2).

4.1 Generation of Synthetics Data Sets and These Outliers

Synthetic data sets are created with the same algorithm designed to generate *Madelon* [6]. They all have only one behavioral attribute whose values are homogeneously distributed. And all contextual attributes are useful for establishing the behavioral value. The original generation does not contain any outliers, but they are added after by using different approaches of *machine learning* algorithms (**ML**) from the Python library *scikit-learn*[1]: *SVC*[2], *K Neighbors Classifier*(See footnote 2), *Random Forest Classifier*[3], *Extra Trees Classifier*(See footnote 3), *Gradient Boosting Classifier*(See footnote 2) and *Logistic Regression*[4].

The goal is to detect the most useful points for predictions in order to change their behavioral value. Thus, the chance of having strong outliers which are harmful to the predictions is increased. While a purely random method wouldn't prevent the selection of points that are not very useful for the prediction. For this purpose, the data sets are randomly divided 10 times into a pair of training **T** (70%) and test **V** (30%) sets. For each pair, the **ML** are trained with **T** and predict the points in **V**. The number of times each point is wrongly predicted are counted. Thus, each point will have an associated counter with a value between 0 and $60 = 10 * 6$ (**ML** algorithms). Those with the lowest values will be the points that have been most frequently correctly predicted. Finally, the n points that will become outliers are the first n with the lowest scores. Where there are more than 2 behavioral values, the new one is chosen randomly.

4.2 Tolerance Limit Estimation

The tolerance limit is determined by a regression study based on 346 data sets which contain 5% of outliers. Plus, they have distinct characteristics in order to cover a large set of attribute and behavioral value number (see Table 1). The number of points is small in order to perform a large amount of computation in a reasonable time. Since **Hino** works with quantiles of fixed sizes, so the number of points has no influence on the outlier detection.

[1] https://scikit-learn.org/stable/index.html.
[2] All meta-parameters have the default values.
[3] All meta-parameters have the default values, except `n_estimator` set to 200.
[4] All meta-parameters have the default values, except `max_iter` set to 1 000.

This study consists of determining what is the optimal **tolerance limit** for each data set. It is based on the assumption that there is a relationship between this limit, the numbers of contextual attributes and behavioral values. More concretely, all possible **tolerance limits** are used in order to determine which one is the best. A detection is performed for each limit and a **sensitivity** and **specificity** score is calculated. These scores are determined with the numbers of true positive (**TP**), true negative (**TN**), false positive (**FP**) and false negative (**FN**).

Table 1. data sets used in regression study with a step of 10 for a and 1 for c

Points (p)	Attributes (a)	Behavioral values (c)
250	10 ... 50, 25	2 ... 5
500	10 ... 70, 25, 75	2 ... 5
1 000	10 ... 100, 25, 75	2 ... 7
2 500	10 ... 100, 25, 75	2 ... 8
5 000	10 ... 100, 25, 75, 125	2 ... 11

$$sensitivity = \frac{TP}{TP + FN} \quad (3) \qquad specificity = \frac{TN}{TN + FP} \quad (4)$$

The **sensitivity** measures the ability of a test to give a positive result when a hypothesis is verified, and it is calculated using Eq. 3. It corresponds to the percentage of correctly detected outliers. While the **specificity** measures the ability of a test to give a negative result when the hypothesis is not verified, and it is calculated using Eq. 4. It corresponds to the percentage of undetected points that are actually healthy. These metrics are interpreted together, a good **tolerance limit** must give these scores as close to 100% as possible.

Plus, a good **tolerance limit** must not remove all points of a behavioral value, and it must not remove too many points, e.g. 20%. The optimal limit is selected from the remaining ones, and is the one with the highest sum of **sensitivity** and **specificity**. Thus, those found for each data set are visualized on a 3D graph according to the number of contextual attributes and behavioral values in Fig. 2a. This limit is shown to increase with the number of attributes and behavioral values, which confirms the relationship between them.

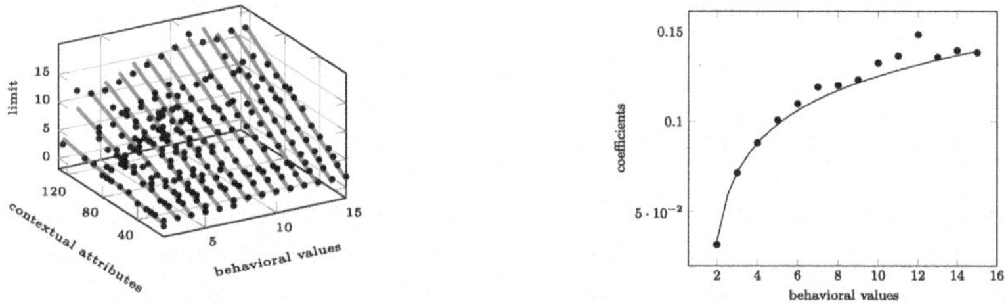

(a) Graph of linear regression computed on optimal limits obtained

(b) Logarithmic regression of linear's coefficients

[c = 2]0.03183a [c = 3]0.07163a [c = 4]0.08832a [c = 5]0.1009a [c = 6]0.1099a [c = 7]0.1193a [c = 8]0.1202a
[c = 9]0.1234a [c = 10]0.1327a [c = 11]0.1367a [c = 12]0.1486a [c = 13]0.136a [c = 14]0.1397a [c = 15]0.1386a

(c) Linear function for each behavioral value

Fig. 2. Regression study

From this relation, a second regression is performed in order to determine the slope of each linear line in Fig. 2a according to the number behavioral values. Note that the equations of these lines are shown in Fig. 2b. As shown in Fig. 2c, these slopes do increase with the characteristics of the data set, but logarithmically. Thus, this regression gives the following equation:

$$0.0205 * log_2(-1.623730 + c) + 0.062579 \quad (5)$$

Finally, it is possible to establish an equation that will approach the optimal **tolerance limit** with the number of attributes and behavioral values:

$$\lfloor (0.0205 * log_2(-1.623730 + c) + 0.062579) * a \rceil \quad (6)$$

The result is only an estimation and is not guaranteed to be the optimal **tolerance limit**. This is due to the choices made on the characteristics of the synthetic data sets and the approximations made during the regressions. However, we can be confident that this estimation gives a viable solution.

5 Experimentation

The efficiency of the **IQR** (see Sect. 2), **Isolation Forest** [7], **SVM** [2], **LOF** [4] and **Hino** (see Sect. 3) methods are compared to their ability to correctly detect outliers on two kinds of data sets: synthetics (different from those in Table 1) and real. Computations have been performed on the supercomputer facilities of the *Mésocentre de calcul de Franche-ComtÃl'*. The results obtains are present in a GitHub directory[5].

5.1 On Synthetic Data Sets

An additional 278 data sets with their outliers are created with the same way as those used in the Sect. 4.1, which has the characteristics present in Table ??. **LOF**, **Isolation Forest** and **SVM** come from the Python library *scikit-learn* and performed with the default settings. Meta-parameters used for **Hino** are already established in Sect. 4.2.

Table 2. Data sets used in evaluation with a step of 10 for a and 1 for c

Points (p)	Attributes (a)	Behavioral values (c)
375	10 . . . 50	2 . . . 5
1 250	10 . . . 100	2 . . . 7
3 000	10 . . . 100	2 . . . 8
6 000	10 . . . 100	2 . . . 9
10 000	10, 50, 75, 100	2, 3, 5, 7, 9, 11
25 000	10, 50, 75, 100	2, 3, 5, 7, 9, 11

Outliers detected are compared with the real ones in order to determine a **sensitivity** and **specificity**. Results are given in Fig. 3 and the best method is the one that has these two scores closest to 100%. The results conclude:

- **IQR** have a linear relationship between these scores, it means that the number of false positives is proportional to the number of outliers detected.

[5] https://github.com/JessyColonval/Hino.

- The whole **sensitivity** of **Isolation forest** remains <20% (the majority is <10% or = 0%) while its **specificity** is between 80% and 100%. It means few or no outliers are removed, but it's always removing inliers.
- **SVM** and **LOF** preserve correctly the inliers with a score of **specificity** >96%. However, they remove a fewer percent of outlier, i.e. <5%.
- The **sensitivity** of **Hino** are < 60% (the majority are between 10 and 40%), while the **specificity** are >70% (the majority are > 80%). It means that generally, it removes more outlier than inliers unlike **IQR** but also a larger number than the **Isolation Forest**.

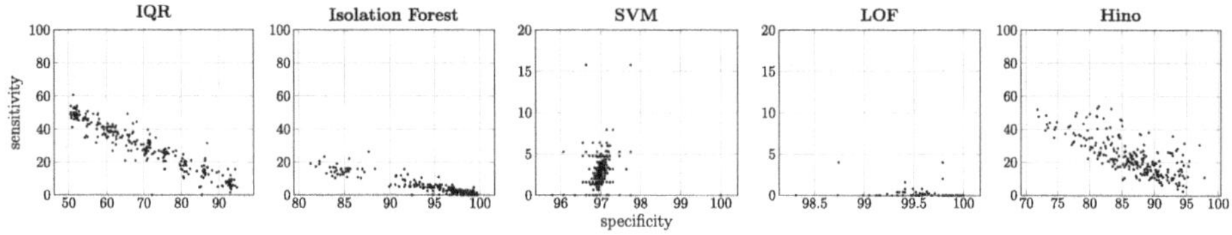

Fig. 3. Specificity and sensitivity on the 5 methods

5.2 On Real Data Sets

In order to confirm the effectiveness of **Hino** compared to state-of-the-art methods, the outlier detection is computed on 16 real data sets from *UCI Machine Learning Repository*[6]. Few are used, due to the difficulty of finding ones with expected characteristics. Their contextual attributes are numerical values and they represents a panel of different characteristics (see Table 3) in order to observe the behavior of these methods according to these different characteristics.

Table 3. Characteristics of real data sets.

Data sets	Points	Attributes	Behavioral values	Data sets	Points	Attributes	Behavioral values
annthyroid	7 200	21	3	multiple features	2 000	619	10
breastW	683	9	2	musk	6 598	166	2
cardio	2 126	21	10	parkinson	756	752	2
glass	214	9	6	pendigits	10 002	16	10
ionosphere	351	33	2	satimage2	6 435	36	6
isolet	7 797	617	26	shuttle	58 000	9	7
letter recognition	20 000	16	26	wine	178	13	3
mammo-graphy	11 183	6	2	wine-quality	1 599	11	6

Two detections are analyzed with **Hino**: one with the meta-parameters calculation described in Sect. 4.2, and other one with the best solution among all possible **tolerance limits**. The goal is to validate the **tolerance limit** provide by equation 6 and observe the variation with the optimal one. Once detected, these outliers are removed from the data set, and the relevance of this removal is measured using two metrics by using 6 **ML** algorithms described in Sect. 4.1:

- **Cross-validation** measures the prediction performance of these algorithms. We assume that the presence of outliers reduces the performance of these algorithms, then it determines which removal was the most beneficial.

[6] https://archive.ics.uci.edu/ml/index.php.

- **False positive** measure the performance of these **ML** algorithms to correctly predict the outliers detected. We assume that the outliers removed from the training set cannot be correctly predicted by these algorithms..

Table 4 summarizes and presents these results in 2 separate columns:

- **Cross-validation** cells contain the percentage of cross-validation (1st line); the standard deviation (2nd line) and **tolerance limit** (only for **Hino**) | percentage of outliers | ✓ if all behavioral values are kept, ✗ otherwise (3rd line). **Tolerance limits** annotated with a '*' are those that have been modified to obtain a percentage of outliers <20%.
- **False positive** cells contain the percentage of points falsely detected as outliers, and the standard deviation of this percentage.

Cases where no outliers have been detected are indicated by the value **None** with the **cross validation** of the full data set. Cells with **Error** imply that the computations of the metrics could not be done because of an overly distorted data set. For example, when a behavioral value has only one representative.

A method is considered better when: it doesn't suppress any behavioral value; the number of points removed isn't too high; the cross-validation is higher and the false positive is lower. Sometimes the choice is not obvious, and the method that is considered better depends on the weight given to these different metrics. These choices are symbolized by two color: **gray** when one solution is better; **light gray** when multiple solutions are closed and better than the others.

According to Table 4, **Hino** gives equivalent or better results for almost all data sets with the estimation of the meta-parameters proposed in Sect. 4 (column **Hino**). Except for *wine*, where the best solutions are given by **SVM** and **LOF**. But by manually changing the **tolerance limits**, it is possible to get better results (column **Best of Hino**). Thus, out of 16 data sets, the results are clearly improved for 8, remain equivalent for 5 and do not change for the last 3.

5.3 Discussion

The use of **ML** algorithms to compute **false positives** introduces a bias. Indeed, a point detected as an outlier, but correctly predicted, does not necessarily mean that the detection has failed. In a case where too many inliers have been removed from the training set, then it is possible that the **ML** algorithms no longer have enough information to correctly predict the outliers. Thus, these outliers would be wrongly predicted because the training set is too distorted. This situation is less likely when few points are removed, but it should be read carefully and in addition to the other metrics.

The way the **tolerance limit** equation is established in Sect. 4.2 is strongly related to the maximum percentage of inlier that is agreed to be removed (i.e. 20%). A different hypothesis would change this equation and the resulting detection. Plus, for the preservation of data sets, the algorithm shifts these limit when the maximum number of outliers is reached or when a behavioral value is completely removed. In practical, in Table 4, the limits of 9 detections has been

Table 4. Detection methods comparison on real data sets

<table>
<tr><th rowspan="2">Data sets</th><th colspan="2">IQR</th><th colspan="2">Isolate Forest</th><th colspan="2">SVM</th><th colspan="2">LOF</th><th colspan="2">Hino</th><th colspan="2">Best of Hino</th></tr>
<tr><th>Cross validation</th><th>False positive</th><th>Cross validation</th><th>False positive</th><th>Cross validation</th><th>False positive</th><th>Cross validation</th><th>False positive</th><th>Cross validation</th><th>False positive</th><th>Cross validation</th><th>False positive</th></tr>
<tr><td>annthyroid</td><td colspan="2">Error
53.6%</td><td>97.68%
±0.19
5.54% | ✓</td><td>94.22%
±0.07</td><td>97.12%
±0.22
3.01% | ✓</td><td>97.98%
±0.06</td><td>98.12%
±0.22
11.22% | ✓</td><td>94.72%
±0.12</td><td>97.65%
±0.19
2 | 10.4% | ✓</td><td>94.43%
±0.06</td><td colspan="2">No change</td></tr>
<tr><td>breastW</td><td>97.13%
±1.12
28.1% | ✓</td><td>91.66%
±0.36</td><td>98.61%
±0.29
37.63% | ✓</td><td>94.22%
±0.90</td><td>97.09%
±0.91
7.32% | ✓</td><td>99.67%
±0.0</td><td>96.73%
±1.19
24.60% | ✓</td><td>95.08%
±0.25</td><td>98.24%
±0.86
2* | 12.7% | ✓</td><td>89.66%
±0.17</td><td>97.61%
±1.18
3 | 3.95% | ✓</td><td>81.48%
±0.00</td></tr>
<tr><td>cardio</td><td>81.4%
±1.94
56.3% | ✗</td><td>60.35%
±0.19</td><td>81.23%
±1.26
16.04% | ✓</td><td>72.71%
±0.35</td><td>82.06%
±1.5
3.15% | ✓</td><td>71.67%
±0.3</td><td>82.45%
±1.69
1.69% | ✓</td><td>71.67%
±3.15</td><td>83.16%
±1.30
3 | 5.9% | ✓</td><td>39.72%
±0.59</td><td>84.76%
±1.3
2 | 12.61% | ✓</td><td>43.77%
±0.57</td></tr>
<tr><td>glass</td><td>73.46%
±4.87
36.4% | ✗</td><td>28.12%
±1.09</td><td>74.75%
±5.05
16.82% | ✓</td><td>48.7%
±1.71</td><td>72.28%
±4.66
1.87% | ✓</td><td>49.17%
±1.76</td><td>73.49%
±5.15
15.89% | ✓</td><td>40.98%
±1.25</td><td>73.95%
±4.21
2* | 9.35% | ✓</td><td>29.25%
±1.29</td><td>72.16%
±5.55
4 | 0.93% | ✓</td><td>0.00%
±0.00</td></tr>
<tr><td>ionosphere</td><td>95.12%
±2.58
48.1% | ✓</td><td>59.47%
±0.43</td><td>90.81%
±1.97
32.48% | ✓</td><td>86.54%
±0.52</td><td>91.21%
±1.88
2.85% | ✓</td><td>91.67%
±0.00</td><td>94.23%
±2.00
34.76% | ✓</td><td>55.61%
±0.72</td><td>92.46%
±2.19
6* | 16.8% | ✓</td><td>78.02%
±0.47</td><td>92.57%
±2.26
9 | 4.56% | ✓</td><td>63.54%
±0.00</td></tr>
<tr><td>isolet</td><td colspan="2">Error
96.6%</td><td>93.1%
±0.49
14.21% | ✓</td><td>93.4%
±0.14</td><td>93.53%
±0.44
2.98% | ✓</td><td>90.55%
±0.25</td><td>93.60%
±0.50
0.05% | ✓</td><td>75.00%
±0.00</td><td>95.08%
±0.40
97 | 13.7% | ✓</td><td>76.61%
±0.23</td><td colspan="2">No change</td></tr>
<tr><td>letter recognition</td><td>90.04%
±0.38
47.4% | ✓</td><td>68.37%
±0.23</td><td>91.48%
±0.27
13.57% | ✓</td><td>88.15%
±0.09</td><td>91.64%
±0.32
3.01% | ✓</td><td>92.75%
±0.12</td><td>91.48%
±0.25
0.19% | ✓</td><td>94.52%
±0.23</td><td colspan="2">None - 91.72%
3 | 0.0%</td><td>91.93%
±0.24
0 | 8.16% | ✓</td><td>64.47%
±0.15</td></tr>
<tr><td>mammo-graphy</td><td>99.41%
±0.04
37.6% | ✓</td><td>94.33%
±0.03</td><td>99.26%
±0.03
19.88% | ✓</td><td>91.06%
±0.2</td><td>98.69%
±0.12
3.08% | ✓</td><td>91.23%
±0.16</td><td>98.66%
±0.00
1.78% | ✓</td><td>90.98%
±0.19</td><td>99.85%
±0.03
1* | 10.7% | ✓</td><td>83.75%
±0.05</td><td>99.24%
±0.15
3 | 1.11% | ✓</td><td>27.28%
±0.47</td></tr>
<tr><td>multiple features</td><td>Error
97.65% | ✗</td><td>35.3%
±0.74</td><td>99.0%
±0.45
35.45% | ✓</td><td>91.15%
±0.2</td><td>98.29%
±0.45
3.00% | ✓</td><td>90.14%
±0.54</td><td>98.33%
±0.47
0.05% | ✓</td><td>3.33%
±10.54</td><td>99.43%
±0.25
78 | 17.9% | ✓</td><td>85.2%
±0.37</td><td>98.7%
±0.34
119 | 0.7% | ✓</td><td>25.6%
±2.18</td></tr>
<tr><td>musk</td><td>91.38%
±4.61
97.4% | ✓</td><td>59.22%
±0.37</td><td>96.39%
±0.34
6.68% | ✓</td><td>98.78%
±0.06</td><td>96.45%
±0.27
2.99% | ✓</td><td>99.31%
±0.25</td><td>96.54%
±0.35
0.32% | ✓</td><td>93.17%
±0.41</td><td>98.31%
±0.23
6 | 4.5% | ✓</td><td>43.46%
±0.31</td><td>97.47%
±0.36
11 | 1.55% | ✓</td><td>22.91%
±0.72</td></tr>
<tr><td>parkinson</td><td colspan="2">Error
100%</td><td>87.04%
±1.78
2.78% | ✓</td><td>86.19%
±1.14</td><td>86.45%
±1.9
3.04% | ✓</td><td>87.03%
±1.08</td><td>86.67%
±1.67
3.04% | ✓</td><td>83.26%
±0.79</td><td>96.93%
±0.80
95* | 19.8% | ✓</td><td>9.48%
±0.29</td><td>96.94%
±1.06
96 | 19.71% | ✓</td><td>9.47%
±0.29</td></tr>
<tr><td>pendigits</td><td>98.58%
±0.20
4.7% | ✓</td><td>85.18%
±0.17</td><td>98.87%
±0.23
51.91% | ✗</td><td>50.45%
±0.17</td><td>98.52%
±0.16
2.98% | ✓</td><td>94.99%
±0.16</td><td>98.67%
±0.20
1.56% | ✓</td><td>76.34%
±0.45</td><td>98.63%
±0.20
2 | 16.7% | ✓</td><td>97.27%
±0.04</td><td>98.58%
±0.12
9 | 0.01% | ✓</td><td>0.00%
±0.00</td></tr>
<tr><td>satimage2</td><td>89.19%
±0.46
8.4% | ✓</td><td>96.65%
±0.16</td><td>88.48%
±0.75
19.40% | ✓</td><td>58.24%
±4.04</td><td>89.28%
±0.52
2.97% | ✓</td><td>97.56%
±0.06</td><td>89.89%
±0.52
1.54% | ✓</td><td>76.35%
±0.49</td><td>90.04%
±0.48
9* | 18.9% | ✓</td><td>88.20%
±0.10</td><td>90.10%
±0.48
18 | 0.05% | ✓</td><td>93.89%
±1.76</td></tr>
<tr><td>shuttle</td><td>99.83%
±0.05
79.6% | ✗</td><td>81.91%
±0.0</td><td>99.78%
±0.03
15.62% | ✗</td><td>60.8%
±0.04</td><td colspan="2">Error
2.79%</td><td colspan="2">Error
19.00%</td><td>99.36%
±0.03
6* | 0.02% | ✓</td><td>56.97%
±1.28</td><td>99.36%
±0.03
7 | 0.01% | ✓</td><td>51.67%
±4.56</td></tr>
<tr><td>wine</td><td>98.03%
±1.66
9.6% | ✓</td><td>90.2%
±0.0</td><td>95.96%
±2.81
13.48% | ✓</td><td>95.83%
±0.00</td><td>98.3%
±1.45
4.49% | ✓</td><td>88.12%
±1.01</td><td>98.11%
±1.54
2.81% | ✓</td><td>70.67%
±1.41</td><td>96.94%
±2.37
2* | 8.4% | ✓</td><td>95.56%
±0.0</td><td>98.14%
±1.35
3 | 1.12% | ✓</td><td>91.67%
±0.00</td></tr>
<tr><td>winequality</td><td>64.42%
±2.17
25.2% | ✓</td><td>51.85%
±0.61</td><td>63.99%
±1.95
16.45% | ✓</td><td>46.05%
±0.57</td><td>63.48%
±1.85
2.88% | ✓</td><td>35.72%
±1.16</td><td>63.66%
±1.69
2.94% | ✓</td><td>41.31%
±1.02</td><td>63.08%
±2.15
7* | 1.06% | ✓</td><td>0.00%
±0.00</td><td colspan="2">No change</td></tr>
</table>

shifted: 7 cause the maximum number of outliers and 2 cause the integrity of behavioral values. Despite this precaution, the **tolerance limit** doesn't always provide the optimal solution. As Table 4 shows, there is a much better solution in 7 over 16 cases. Thus, the **tolerance limit** suit better as a meta-parameter.

6 Conclusion

This paper presented a parameterized method of outlier detection based on the quantile principle and a calculation to determine the parameters to be used. This method was compared with a state-of-the-art method (**IQR**, **Isolation Forest**, **SVM** and **LOF**) on 278 synthetic data sets and 16 real data sets. It was clearly shown that the **Hino** approach was the more efficient. The computational efficiency of meta-parameters for **Hino** has been discussed and can be studied to adapt it to take into account the context of data sets. Finally, this approach can be easily integrated into standard machine learning library (as scikit-Learn or Tensor Flow) to help preparation of data sets.

References

1. Aggarwal, C.C.: Outlier Analysis. Springer International Publishing, Cham, 2 edn. (2017). https://doi.org/10.1007/978-3-319-47578-3
2. Amer, M., Goldstein, M., Abdennadher, S.: Enhancing one-class support vector machines for unsupervised anomaly detection. In: Proceedings of the ACM SIGKDD Workshop on Outlier Detection and Description, pp. 8–15. ODD '13, Association for Computing Machinery, New York, NY, USA (Aug 2013). https://doi.org/10.1145/2500853.2500857
3. Barnett, V., Lewis, T., others: Outliers in statistical data, vol. 3. Wiley New York (1994). https://doi.org/10.1057/jors.1995.142
4. Breunig, M., Kriegel, H.P., Ng, R., Sander, J.: LOF: Identifying Density-Based Local Outliers. vol. 29, pp. 93–104 (Jun 2000). https://doi.org/10.1145/342009.335388
5. Dekking, F.M., Kraaikamp, C., Lopuhaä, H.P., Meester, L.E.: Exploratory data analysis: numerical summaries. In: A Modern Introduction to Probability and Statistics. STS, pp. 231–243. Springer, London (2005). https://doi.org/10.1007/1-84628-168-7_16
6. Guyon, I.: Design of experiments for the NIPS 2003 variable selection benchmark. In: undefined (2003). https://doi.org/10.1007/978-3-540-35488-8_10, https://www.semanticscholar.org/paper/Design-of-experiments-for-the-NIPS-2003-variable-Guyon/b979fa88ca448fb08633f961131f45214b1cf109
7. Liu, F.T., Ting, K.M., Zhou, Z.H.: Isolation-based anomaly detection. ACM Trans. Knowl. Disc. Data **6**(1), 3:1–3:39 (Mar 2012). https://doi.org/10.1145/2133360.2133363
8. Sadik, S., Gruenwald, L.: Research issues in outlier detection for data streams. ACM SIGKDD Explor. Newsl **15**(1), 33–40 (2014). https://doi.org/10.1145/2594473.2594479
9. Souiden, I., Omri, M.N., Brahmi, Z.: A survey of outlier detection in high dimensional data streams. Comput. Sci. Rev. **44**, 100463 (May2022). https://doi.org/10.1016/j.cosrev.2022.100463, https://www.sciencedirect.com/science/article/pii/S1574013722000107
10. Thung, F., Wang, S., Lo, D., Jiang, L.: An empirical study of bugs in machine learning systems. In: 2012 IEEE 23rd International Symposium on Software Reliability Engineering, pp. 271–280 (Nov 2012). https://doi.org/10.1109/ISSRE.2012.22, iSSN: 2332-6549
11. Vinutha, H.P., Poornima, B., Sagar, B.M.: Detection of outliers using interquartile range technique from intrusion dataset. In: Satapathy, S.C., Tavares, J., Bhateja, V., Mohanty, J.R. (eds.) Information and Decision Sciences. AISC, vol. 701, pp. 511–518. Springer, Singapore (2018). https://doi.org/10.1007/978-981-10-7563-6_53

Fast Approximation Algorithm for Euclidean Minimum Spanning Tree Building in High Dimensions

Keito Kido, Daichi Amagata(✉), and Takahiro Hara

The University of Osaka, Osaka, Japan
{kido.keito,amagata.daichi,hara}@ist.osaka-u.ac.jp

Abstract. This paper addresses the problem of building the Euclidean minimum spanning tree (EMST) of a high-dimensional dataset. The EMST problem has a number of applications, including visualization, cosmology, and chemical sciences, to name a few. Recently, due to machine learning-based embedding techniques, many objects are represented as high-dimensional vectors. The above applications, hence, require an efficient EMST algorithm in high dimensions. Borůvka's algorithm is a representative one, but it incurs $O(n^2 \log n)$ time for n points. The state-of-the-art EMST algorithms improve practical time costs, but they assume low-dimensional points. In high dimensions, such techniques are generally reduced to exhaustive searches, suggesting that computing the exact EMST incurs a high computational cost. Fortunately, many applications require fast response time while allowing approximate results. We therefore propose a new approximate EMST algorithm for high-dimensional points. We conduct experiments on real datasets, and the experimental results demonstrate that our algorithm yields an almost correct answer in a much faster time than that of state-of-the-art algorithms.

Keywords: EMST · High-dimensional vectors · Algorithm

1 Introduction

This paper addresses the EMST (Euclidean Minimum Spanning Tree) problem, a classic but still important problem with about 100 years of research history [8]. Given a set P of n points in $\mathbb{R}^d$, the EMST of P is the spanning tree with the minimum weight in the complete graph of P, where the weight of a spanning tree is the sum of the distances between connected points. The EMST problem has a number of applications, such as clustering [29], chemical science [26], cosmology [21], visualization [3], among others. Recently, many objects are represented as high-dimensional vectors, due to the proliferation of machine learning-based embedding techniques. The above applications also have this observation, so efficient computing of EMST in high dimensions is required.

Challenge. Efficiently computing EMST has been extensively studied, and both theoretical algorithms (e.g., [7]) and practical ones (e.g., [31]) were devised. Most

X. Wu et al. (Eds.): PAKDD 2025, LNAI 15875, pp. 432–443, 2025.
https://doi.org/10.1007/978-981-96-8295-9_32

of the state-of-the-art practical algorithms [19,27,31] assume low-dimensional spaces. Some theoretical algorithms also have this assumption [7,11], whereas others only estimate the weight of the EMST [10] or assume a distributed computing environment, which is different from our assumption.

The state-of-the-art practical algorithms, which assume low dimensions, employ a tree-based index (e.g., kd-tree). Some of them improve Borůvka's algorithm [8], which incurs $O(n^2 \log n)$ time, with the tree index, whereas the others employ well-separated pair decomposition (WSPD) [9], and the tree index is used to run WSPD efficiently. In high dimensions, because of the curse of dimensionality, tree indices do not function well, and it is hard to find well-separated pairs. Therefore, their approaches are reduced to simple iterations of the closest-pair finding with $O(n^2 \log n)$ time exhaustive searches.

As with this paper, the state-of-the-art approximation algorithm [23] assumes high dimensions. This algorithm, however, considers only dimensionality reduction of P and simply uses Borůvka's algorithm on the dataset with reduced dimensionality. Hence, it still incurs $O(n^2 \log n)$ time.

Contribution. Motivated by the importance of EMST in high dimensions and the efficiency issue in existing works, we propose a new approximate EMST algorithm. The high computational cost of the state-of-the-art algorithms is derived from their dependence on closest-pair computations (in Borůvka's algorithm). We therefore design our algorithm not to have this limitation, and our algorithm does not rely on Borůvka's algorithm.

Conceptually, our algorithm builds a *sparse* graph $G = (P, E)$, where P is a set of vertices (i.e., vertices are points) and E is a set of edges, such that $e \in E$ connects only close points. To obtain such G, our algorithm employs an approximate k nearest neighbor graph and approximate k nearest neighbor searches. After G is obtained, we can obtain the minimum spanning tree of G with Kruskal's algorithm [16]. This approach does not run any closest pair findings but still makes the necessary edges. Our algorithm therefore yields a spanning tree of P similar to its exact EMST.

We summarize the main contributions of this paper as follows:

- We propose a new approximation algorithm for the EMST problem. This algorithm returns an approximate EMST in $\tilde{O}(n)$ time in practice. ($\tilde{O}(\cdot)$ hides any polylog factors.)
- We conduct extensive experiments on real-world datasets. The experimental results demonstrate that our algorithm returns an almost correct answer in a much faster time than state-of-the-art algorithms. We also confirm how similar the result of our algorithm is to the exact EMST by visualization.

2 Preliminary

2.1 Problem Definition

Let P be a set of high-dimensional points $\in \mathbb{R}^d$, where $d = O(1)$. We use $dist(p, p')$ to denote the Euclidean distance between points p and $p' \in P$. The Euclidean minimum spanning tree of P is defined below.

Definition 1 (Euclidean minimum spanning tree). *Given a set P of d-dimensional points, we can have an undirected graph $G = (P, E)$ where E is a set of edges between two points in P. The weight of an edge (p, p') is $dist(p, p')$. Let $W(G)$ be the weight of G and*

$$W(G) = \sum_{(p,p') \in E} dist(p, p'). \tag{1}$$

Consider the complete graph G_{comp} of P, i.e., there is an edge between every pair of points in P. The Euclidean minimum spanning tree (EMST) T^* of P is the spanning sub-graph of G_{comp} with the minimum weight.

Because computing the exact EMST involves many closest-pair computations, it incurs a significant computational cost in high dimensions, which can be seen in Sect. 2.2. We therefore consider an approximation algorithm for the EMST problem. That is, we return a spanning tree T of P which may have $T \neq T^*$. It is clear that $W(T) \geq W(T^*)$. When $W(T) \approx W(T^*)$, T is highly accurate. The objective of this paper is to devise an approximate EMST algorithm that yields such a T quickly.

2.2 Borůvka's Algorithm

The most straightforward approach to computing the EMST of P is first to compute G_{comp} and then run Kruskal's algorithm [16]. This approach needs $O(n^2)$ space, so it is not feasible when n is large.

Borůvka's algorithm does not have this high space complexity, and this is a building block of the state-of-the-art exact and approximate EMST algorithms [19,23]. In this algorithm, given P, we first consider each $p \in P$ as a graph of a single vertex. For each $p \in P$, Borůvka's algorithm computes its nearest neighbor point and makes an edge between p and its nearest neighbor. In the next iteration, for each disjoint graph, Borůvka's algorithm computes the nearest graph (i.e., the closest pair of points belonging to different graphs) and makes an edge between them. This is repeated until there is no disjoint graph.

Although this algorithm requires only $O(n)$ space, it incurs $O(n^2 \log n)$ time, which is exactly the same cost as that of the most straightforward approach. The state-of-the-art algorithms [19,23] hence suffer from this high cost.

2.3 Relevant Techniques

Our algorithm is based on an approximate k nearest neighbor graph (AkNNG) and approximate nearest neighbor search (ANNS). Because our algorithm employs the state-of-the-art AkNNG construction algorithm and ANNS one, we introduce them briefly.

AkNNG Construction. Because computing the exact kNNG is computationally expensive and incurs $O(n^2)$, we use an AkNNG. To construct this, we employ NNDescent [14]. For each point in P, NNDescent first adds k random edges (i.e., neighbors). Next, for each point $p \in P$, it traverses p's neighbors and their neighbors, and then it updates p's k neighbors. This is repeated until the updates are converged or a given iteration number is reached.

For fixed parameters, NNDESCENT needs only $O(n)$ space and $O(n)$ time [4]. In addition, it returns an AkNNG with high accuracy empirically.

ANNS. The ANNS problem has been extensively studied, and the current state-of-the-art approach is based on proximity graph [17]. Among existing proximity graph algorithms, we use HNSW [18] because it is the de-facto standard algorithm. HNSW also consumes only $O(n)$ space and empirically returns ANNS results with high recall. Note that HNSW is not an AkNNG-based graph. Thus, we cannot obtain an AkNNG from the HNSW structure.

3 Related Work

MST on Graphs. The MST problem is one of the oldest studies in computational geometry. The input of this problem is a graph $G = (V, E)$, where V is a set of vertices. This problem has three famous algorithms: Prim's [25], Kruskal's [16], and Borůvka's. These algorithms all need $O(|E| \log |V|)$ time to obtain the MST of G. The algorithm introduced in Sect. 2.2 is actually an extended version of the original Borůvka algorithm. (The original Borůvka algorithm iteratively selects the edge with the minimum weight for each sub-graph, thus the extended version needs to compute the nearest neighbor for each point in each iteration.) Prim's and Kruskal's algorithms can deal with EMST only when a graph of P is obtained.

EMST. Different from the above problem, EMST requires a set of d-dimensional points as its input. Section 2.2 introduced the state-of-the-art algorithms based on Borůvka's algorithm. Its GPU implementation was devised in [27], but this is a different setting from ours. We hence review the other EMST works.

Literature [28] shows that the EMST can be obtained in $O(n \log n)$ time in 2-dimensional spaces. However, its proposed algorithm needs $O(n^2 \log n)$ time when $d > 3$. Yao devised an $O((n \log n)^{1.8})$ time algorithm for three-dimensional spaces [32].

In [2], a closest-pair-based algorithm is proposed, but it incurs $O(n^2 \log^d)$ time in high-dimensional spaces. To improve the closest pair findings practically, a WSPD-based algorithm was considered in [22]. A theoretical algorithm using WSPD was analyzed in [7]. The running time of this WSPD algorithm grows exponentially w.r.t. d, suggesting that the WSPD-based approach is not appropriate in high dimensions. [26] proposed a similar technique to ours. However, it does not guarantee to return a spanning tree, and our algorithm does not have this issue.

Some theoretical studies, such as [10], do *not* compute the EMST of P but estimate the weight of the EMST. In addition, some works, e.g., [15], address the EMST problem in distributed computing environments. Different from this setting, we assume a single machine.

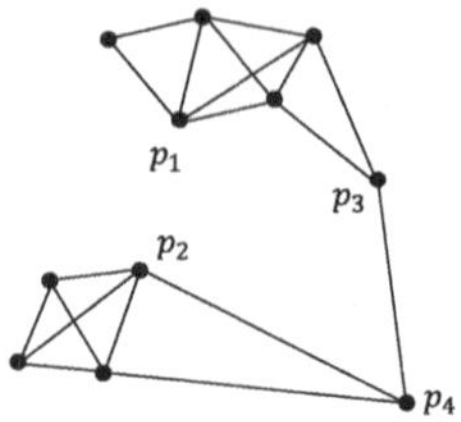

Fig. 1. Approximate k nearest neighbor graph with an edge (p_3, p_4) that prevents from obtaining the EMST (if we import the minimum spanning tree from this AkNNG). The edge (p_1, p_2) is necessary to obtain a spanning tree with less weight.

4 Proposed Algorithm

Main Idea. The straightforward algorithm builds the complete graph G_{comp} of P, which is time- and space-consuming. To alleviate this space inefficiency, Borůvka's algorithm incrementally computes the edges of the minimum spanning tree of G_{comp}. However, Borůvka's algorithm does not solve the time inefficiency issue. Our new idea for overcoming this challenge is to *build a sparse graph of P and then add only promising edges to this sparse graph in a batch.*

For each $p \in P$, the sparse graph has much fewer edges than the complete graph (e.g., $5 \ll n$), so it is space-efficient. Moreover, to obtain an approximate yet highly accurate spanning tree, this sparse graph needs to have the edges in the EMST. Because each point in the EMST is connected to its close points, we implement this idea by exploiting an approximate k nearest neighbor graph (AkNNG).

It is important to notice that the AkNNG can have disjoint sub-graphs and detour paths (even if the AkNNG contains spanning trees). Figure 1 depicts a detour example in an AkNNG. Although we can obtain a spanning tree from this AkNNG, we cannot obtain the EMST. Any spanning trees of this AkNNG definitely have the edge (p_3, p_4), but the EMST does not have this edge, and the upper and bottom parts of this AkNNG should be connected by (p_1, p_2). To avoid such disjoint sub-graphs and detour paths, we add promising edges to the AkNNG by exploiting approximate nearest neighbor searches.

Our algorithm is orthogonal to any AkNNG building and ANNS algorithms (so our algorithm can be considered as a framework). Therefore, by carefully selecting efficient AkNNG building and ANNS techniques, our algorithm can quickly return a spanning tree with high accuracy. To obtain a practically and theoretically fast algorithm (by putting a reasonable assumption), we employ NNDescent and HNSW as AkNNG building and ANNS algorithms, respectively. Although our algorithm may be simple, its novel idea yields high efficiency (and accuracy) in practice.

The Algorithm. Algorithm 1 details our proposed algorithm. Note that $\mathcal{H}$, the HNSW of P, is pre-built offline.

Algorithm 1: APPROXIMATE EMST BUILDING

Input: P, $\mathcal{H}$ (HNSW of P), and k
1 $G \leftarrow$ NNDESCENT(P, k)
2 $P' \leftarrow$ KRUSKAL(G) ▷ obtain a set of points (vertices) with a single degree in the MST of each sub-graph
3 $k \leftarrow k \times 2$
4 **foreach** $p_i \in P'$ **do**
5     $S \leftarrow$ ANNS($k, \mathcal{H}$)
6   **foreach** $p_j \in S$ **do**
7   Add edge (p_i, p_j) into G if (p_i, p_j) is not in E
8 **while** 1 **do**
9     $f \leftarrow$ ISDISJOINT(G) ▷ checked by depth-first-search
10   **if** f *is true* **then**
11   Run lines 2–7
12   **else**
13   **break**
14 $T \leftarrow$ KRUSKAL(G) ▷ return the MST in G
15 **return** T

1. We first run NNDESCENT with a very small k to obtain a sparse graph G of P, i.e., an AkNNG. Note that we use G as an undirected graph, like [6,30].
2. Next, we run Kruskal's algorithm on G and obtain the MST of G. Notice that this MST may contain disjoint sub-graphs, and our objective here is not to obtain the MST itself but to obtain a set P' of points (vertices of G) with a single degree in this MST. Our idea here is that such points exist in *boundaries* between different connected components and between different dense spaces (as seen in Fig. 1).
3. To remove such disjoint sub-graphs and detours, we run ANNS by using a larger k and each of these points. Specifically, for each $p_i \in P'$, we run AkNNS on $\mathcal{H}$ by using p_i as a query point and obtain the AkNNS result S. We add the edge (p_i, p_j) to G for each $p_j \in S$. It is important to notice that $|P'| \ll n$ in practice. That is, different from the existing techniques, our algorithm does not run (A)NNS for "every" point in P, thanks to the idea in step (2). This approach thus saves computational costs better than the existing state-of-the-art algorithms.
4. After that, we confirm whether G is a single connected component (i.e., whether it has some disjoint sub-graphs) by using depth-first-search. If so, we go to step (5). Otherwise, to make G a single connected component, we run steps (2) and (3) by increasing k. This step (4) is repeated until no disjoint sub-graphs appear.
5. We run Kruskal's algorithm on G and return T, the MST of G.

Example 1. Figure 2 illustrates an example of our algorithm. Given a set P of points (Fig. 2(a)), we build G_1, an AkNNG of P ($k = 2$ in Fig. 2(b)). We compute

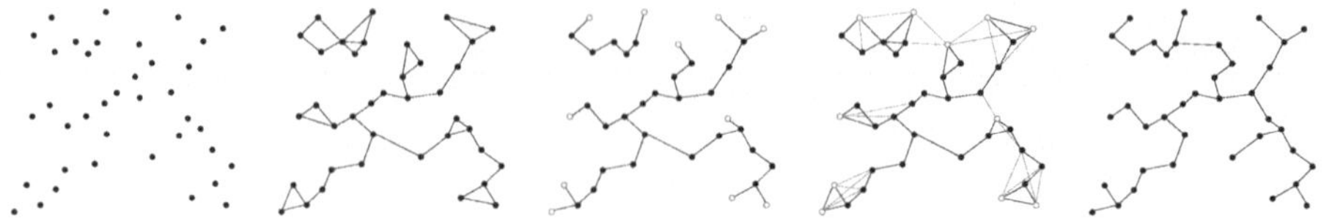

(a) A point set P (b) G_1: AkNNG of P ($k = 2$) (c) MST of G_1 (edges are not removed in our algorithm) (d) G_2: new edges (dashed red lines) are added to G_1 (e) MST of G_2

Fig. 2. Example of our algorithm.

a set P' of points with a single degree in the MST of G_1, i.e., red circles in Fig. 2(c). We set $k = 4$ and run AkNNS for each point in P' and add edges to G_1 as shown in Fig. 2(d). As the resulting graph G_2 does not have disjoint sub-graphs, we run Kruskal's algorithm to obtain the MST of G_2. We finally return it (Fig. 2(e)).

This example clarifies the following observations: (i) Our algorithm adds edges that avoid disjoint sub-graphs and detours (e.g., compare Figs. 2(c) and 2(e)) while minimizing redundant edges. (ii) This graph building is achieved by only a single AkNNG construction and some additional ANNS (i.e., the number of query points is much less than n).

Remark 1 (PARALLELIZABILITY). Our algorithm is easy to parallelize: NNDESCENT is parallelizable [14], and the for loop in line 4 is also parallelizable because each ANNS is done independently. Therefore, multi-threading improves the practical efficiency of our algorithm.

Time Analysis. For fixed parameters, NNDESCENT needs $O(n)$ time [4,5]. The time of Kruskal's algorithm on the AkNNG is $O(nk \log n) = O(n \log n)$ time [16]. Let $\mathcal{T}(n)$ be the search time of HNSW on a set of n points[1]. Because the number of points (vertices) with a single degree can be $O(n)$, lines 4–7 need $O(n) \times \mathcal{T}(n)$ time. Line 9 requires $O(n)$ time, so each iteration of the while loop needs at most $O(n) \times (\mathcal{T}(n) + \log n)$ time. As we double k in each iteration, the number of iterations is $O(\log n)$. Given the above facts, our algorithm needs $O(n \log n) \times \mathcal{T}(n) + O(n \log^2 n)$ time.

In practice, the original HNSW paper [18] claims that $\mathcal{T}(n) = \text{polylog}(n)$. Furthermore, we observe that the number of iterations of the while loop is a small constant (e.g., less than three). Our algorithm, therefore, runs in $\tilde{O}(n)$ time in practice, which is much faster than the existing algorithms and is confirmed in our experiments.

[1] The theoretical upper-bound of $\mathcal{T}(n)$ has not been given.

5 Experiment

We report our experimental results in this section. All experiments were conducted on a Ubuntu 22.04 LTS machine with 28 core Intel Xeon Platinum 8280L CPU@2.7 GHz and 742 GB RAM.

5.1 Setting

Dataset. We used FastText [20], GloVe [24], MNIST [13], and SIFT [1]. FastText (GloVe) is a set of 300-dimensional (200-dimensional) 1 (1.2) million word embedding vectors, and MNIST is a set of 784-dimensional 60,000 digit images. SIFT is a set of 128-dimensional 1 million sift image feature vectors.

Evaluated Algorithm. We evaluated the following EMST algorithms.

- Ours[2]: our algorithm proposed in Sect. 4. We set $k = 5$.
- Borůvka: a state-of-the-art exact EMST algorithm.
- Borůvka + RP [23]: a state-of-the-art approximate EMST algorithm. The reduced dimensionality was 20, as with the original paper.
- DualTree [19]: a state-of-the-art exact EMST algorithm for low-dimensional points. We used MLPACK implementation [12].

All algorithms were implemented in C++ and compiled by g++ 11.4.0 with -O3 optimization.

Ours and Borůvka (+ RP) enable parallel processing, and we used OpenMP for multi-threading. By default, we used 28 threads. DualTree is a sequential algorithm and cannot be parallelized.

Criteria. We measured the time to build the exact or approximate EMST. Note that we set two hours as the time limit. For approximation algorithms, we measured the accuracy, which is $\frac{W(T)}{W(T^*)}$, where T is the spanning tree of P returned by an approximation algorithm.

5.2 Efficiency Result

We first compare our algorithm with the state-of-the-art ones while varying the data size by random sampling. Figure 3 shows the experimental results. Except for the case where a given dataset is very small (i.e., 20% sampling rate in MNIST), our algorithm outperforms the comparison algorithms. DualTree could not terminate within the time limit on the large datasets (FastText, GloVe, and SIFT) even when the sampling rate is 20%. Borůvka has a similar result, and it terminated on the large datasets only when the sampling rate is 20%. Borůvka + RP a bit improves this result, but it still could not terminate on the large datasets when the sampling rate is more than 60%.

On the other hand, our algorithm does not have the above critical issues and returns its result within a reasonable time (i.e., less than 3 min). We observe

[2] https://github.com/KAKu99/approx_emst.

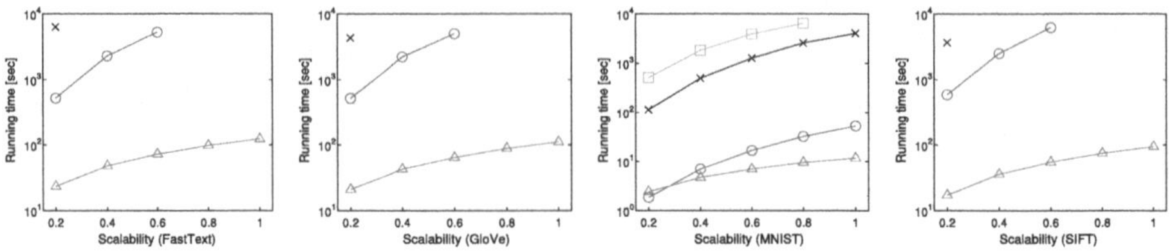

(a) FastText (b) GloVe (c) MNIST (d) SIFT

Fig. 3. Running time vs. data size. "□" shows DualTree, "×" shows Borůvka, "○" shows Borůvka + RP, and "△" shows Ours.

Table 1. Decomposed time [sec]

Dataset	Ours ($k = 5$)				Variant of ours ($k = 20$)			
	FastText	Glove	MNIST	SIFT	FastText	Glove	MNIST	SIFT
NNDESCENT	73.91	58.48	10.10	63.20	351.89	249.57	35.20	224.85
ANNS	23.23	22.69	0.80	7.28	–	–	–	–
KRUSKAL	9.04	9.83	0.26	8.18	8.34	9.83	0.32	7.96

that our algorithm is at least 55 times faster than Borůvka + RP on the three large datasets. The comparison algorithms have a quadratic scale of n, but our algorithm scales linearly to n. (Plots are log-scale.) This comparison result clarifies that our new approach functions well in high dimensions and is much better than the existing techniques.

Decomposed Time. To show the efficiency of our algorithm, we use a variant of our algorithm as a comparison. This variant runs NNDESCENT with a reasonably large k that provides an AkNNG with no detours and disjoint sub-graphs (so it does not run ANNS). Then, this algorithm returns the MST of the AkNNG via Kruskal's algorithm.

Table 1 shows the time of each component in the algorithms. It can be observed that the variant used a bit larger k than our algorithm, but its total time is at least twice than that of our algorithm. NNDESCENT with a larger k is an intuitive and easy approach to obtaining an AkNNG whose MST is similar to the exact EMST. However, this approach makes many redundant edges, incurring in vain computations. On the other hand, our algorithm uses a small k to alleviate the cost of NNDESCENT and employs ANNS to add only necessary edges. The comparison result demonstrates that this idea minimizes the total time needed to compute an approximate EMST.

5.3 Accuracy Result

We show the accuracy of each approximation algorithm in Table 2. The exact algorithm did not terminate within the time limit when all points in a given

Table 2. Accuracy

Dataset	FastText	Glove	MNIST	SIFT
Ours	1.01	1.02	1.00	1.01
Borůvka + RP	1.36	1.32	1.28	1.31

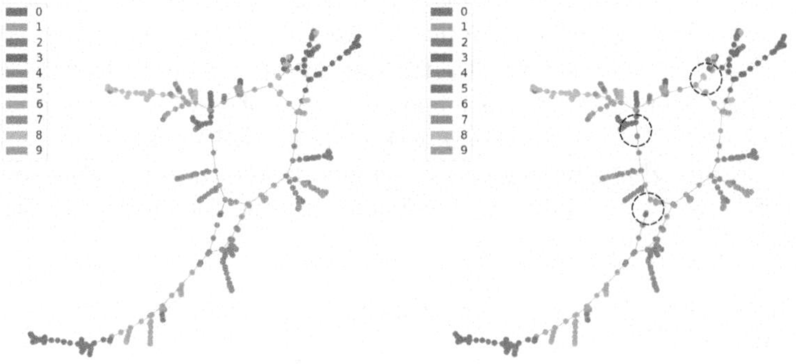

(a) T^*: EMST

(b) T: the spanning tree returned by our algorithm

Fig. 4. Visualization of spanning trees of 300 random samples of MNIST (best viewed in color). (Color figure online)

dataset were used (except MNIST). We therefore used 20% points to measure the accuracy.

It can be seen that our algorithm returns almost perfect results, and its accuracy is better than that of the state-of-the-art approximation algorithm Borůvka + RP. This result again justifies that our algorithm is reasonable. Even when using the exact algorithm (i.e., Borůvka), the dimensionality reduction by random projection incurs a large accuracy loss. In addition, recall that our algorithm is much faster than Borůvka + RP. That is, the performance of our algorithm dominates that of Borůvka + RP.

Visualization. Last, to investigate the effectiveness of our algorithm, we visually compare T (the spanning tree returned by our algorithm) with T^* (the exact EMST). We used 30 random samples of each class in MNIST (i.e., 300 samples in total) to easily confirm the accuracy of T. For visualization, we used the spring layout model.

Figure 4 depicts the visualization result. Note that each color represents the corresponding digit. Consistent with the result in Table 2, T is almost perfect, and $\frac{W(T)}{W(T^*)} = 1.002$. Figure 4(b) shows that T has only slight differences from T^* (specified by dashed circles). These minor differences do not lose the global structure of a given dataset, thus it does not bother subsequent applications that want to use EMST.

6 Conclusion

The Euclidean minimum spanning tree problem has many important applications, and objects in these applications are usually represented as high-dimensional vectors nowadays. Existing EMST algorithms are optimized for low-dimensional points, so they are not efficient for high-dimensional points. Motivated by these observations, this paper addressed the EMST problem in high dimensions. Because it is practically hard to obtain the exact EMST of many points quickly, we considered returning an approximate EMST. We proposed a fast algorithm that returns a spanning tree of P whose weight is similar to that of its EMST. We conducted experiments on real-world datasets, and the results demonstrate that our algorithm is much faster than the state-of-the-art EMST algorithms. In addition, we confirmed its high accuracy quantitatively and qualitatively.

Acknowledgements. This work was partially supported by AIP Acceleration Research JPMJCR23U2, JST.

References

1. http://corpus-texmex.irisa.fr/
2. Agarwal, P.K., Edelsbrunner, H., Schwarzkopf, O., Welzl, E.: Euclidean minimum spanning trees and bichromatic closest pairs. In: SoCG, pp. 203–210 (1990)
3. Alcaide, D., Aerts, J.: Spanning trees as approximation of data structures. IEEE Trans. Visual Comput. Graphics **27**(10), 3994–4008 (2020)
4. Amagata, D., Onizuka, M., Hara, T.: Fast and exact outlier detection in metric spaces: a proximity graph-based approach. In: SIGMOD, pp. 36–48 (2021)
5. Amagata, D., Onizuka, M., Hara, T.: Fast, exact, and parallel-friendly outlier detection algorithms with proximity graph in metric spaces. VLDB J. **31**(4), 797–821 (2022)
6. Arai, Y., Amagata, D., Fujita, S., Hara, T.: LGTM: a fast and accurate KNN search algorithm in high-dimensional spaces. In: DEXA, pp. 220–231 (2021)
7. Arya, S., Mount, D.M.: A fast and simple algorithm for computing approximate euclidean minimum spanning trees. In: SODA, pp. 1220–1233 (2016)
8. Borůvka, O.: O jistém problému minimálním. Práce Moravské přírodovědecké společnosti **3**(3), 37–58 (1926)
9. Callahan, P.B., Kosaraju, S.R.: Faster algorithms for some geometric graph problems in higher dimensions. In: SODA, vol. 93, pp. 291–300 (1993)
10. Chen, X., Cohen-Addad, V., Jayaram, R., Levi, A., Waingarten, E.: Streaming euclidean MST to a constant factor. In: STOC, pp. 156–169 (2023)
11. Chen, X., Jayaram, R., Levi, A., Waingarten, E.: New streaming algorithms for high dimensional EMD and MST. In: StOC, pp. 222–233 (2022)
12. Curtin, R.R., et al.: MLPACK: a scalable C++ machine learning library. J. Mach. Learn. Res. **14**(1), 801–805 (2013)
13. Deng, L.: The MNIST database of handwritten digit images for machine learning research. IEEE Signal Process. Mag. **29**(6), 141–142 (2012)

14. Dong, W., Moses, C., Li, K.: Efficient k-nearest neighbor graph construction for generic similarity measures. In: World Wide Web, pp. 577–586 (2011)
15. Jayaram, R., Mirrokni, V., Narayanan, S., Zhong, P.: Massively parallel algorithms for high-dimensional euclidean minimum spanning tree. In: SODA, pp. 3960–3996 (2024)
16. Kruskal, J.B.: On the shortest spanning subtree of a graph and the traveling salesman problem. Proc. Amer. Math. Soc. **7**(1), 48–50 (1956)
17. Li, W., et al.: Approximate nearest neighbor search on high dimensional data-experiments, analyses, and improvement. IEEE Trans. Knowl. Data Eng. **32**(8), 1475–1488 (2019)
18. Malkov, Y.A., Yashunin, D.A.: Efficient and robust approximate nearest neighbor search using hierarchical navigable small world graphs. IEEE Trans. Pattern Anal. Mach. Intell. **42**(4), 824–836 (2018)
19. March, W.B., Ram, P., Gray, A.G.: Fast euclidean minimum spanning tree: algorithm, analysis, and applications. In: SIGKDD, pp. 603–612 (2010)
20. Mikolov, T., Grave, É., Bojanowski, P., Puhrsch, C., Joulin, A.: Advances in pre-training distributed word representations. In: LREC (2018)
21. Naidoo, K., et al.: Beyond two-point statistics: using the minimum spanning tree as a tool for cosmology. Mon. Not. R. Astron. Soc. **491**(2), 1709–1726 (2020)
22. Narasimhan, G., Zhu, J., Zachariasen, M.: Experiments with computing geometric minimum spanning trees. In: ALENEX, pp. 183–196 (2000)
23. Narayanan, S., Silwal, S., Indyk, P., Zamir, O.: Randomized dimensionality reduction for facility location and single-linkage clustering. In: ICML, pp. 7948–7957 (2021)
24. Pennington, J., Socher, R., Manning, C.D.: Glove: global vectors for word representation. In: EMNLP, pp. 1532–1543 (2014)
25. Prim, R.C.: Shortest connection networks and some generalizations. Bell Syst. Tech. J. **36**(6), 1389–1401 (1957)
26. Probst, D., Reymond, J.-L.: Visualization of very large high-dimensional data sets as minimum spanning trees. J. Cheminform. **12**(1), 1–13 (2020). https://doi.org/10.1186/s13321-020-0416-x
27. Prokopenko, A., Sao, P., Lebrun-Grandié, D.: A single-tree algorithm to compute the euclidean minimum spanning tree on GPUs. In: ICPP, pp. 1–10 (2022)
28. Shamos, M.I., Hoey, D.: Closest-point problems. In: SFCS, pp. 151–162 (1975)
29. Sibson, R.: Slink: an optimally efficient algorithm for the single-link cluster method. Comput. J. **16**(1), 30–34 (1973)
30. Uemura, R., Amagata, D., Hara, T.: An efficient framework for approximate nearest neighbor search on high-dimensional multi-metric data. In: SISAP, pp. 3–17 (2024)
31. Wang, Y., Yu, S., Gu, Y., Shun, J.: Fast parallel algorithms for euclidean minimum spanning tree and hierarchical spatial clustering. In: SIGMOD, pp. 1982–1995 (2021)
32. Yao, A.: On constructing minimum spanning trees in k-dimensional spaces and related problems. SIAM J. Comput. **11**(4), 721–736 (1982)

ShuttleSHAP: A Turn-Based Feature Attribution Approach for Analyzing Forecasting Models in Badminton

Wei-Yao Wang[1(✉)], Wen-Chih Peng[1], and Wei Wang[2]

[1] National Yang Ming Chiao Tung University, Hsinchu, Taiwan
sf1638.cs05@nctu.edu.tw, wcpengcs@nycu.edu.tw
[2] University of California, Los Angeles, USA
weiwang@cs.ucla.edu

Abstract. Agent forecasting systems have been explored to investigate agent patterns and improve decision-making in various domains, e.g., pedestrian predictions and sports analytics. Badminton represents a fascinating example of a multifaceted turn-based sport, requiring both sophisticated tactic developments and alternate-dependent decision-making. Recent deep learning approaches for player tactic forecasting in badminton show promising performance partially attributed to effective reasoning about rally-player interactions. However, a critical hurdle lies in the unclear functionality of which features are learned for simulating players' behaviors by black-box models, where existing explainers are not equipped with *turn-based* and *multi-output* attributions. To support these functionalities, we propose a turn-based feature attribution approach, ShuttleSHAP, for analyzing badminton forecasting models based on variants of Shapley values. ShuttleSHAP is a model-agnostic explainer aiming to quantify contribution by not only temporal but also player aspects in terms of fine-grained cues. Incorporating ShuttleSHAP into the state-of-the-art turn-based model on the benchmark dataset reveals that it is, in fact, insignificant to reason about the initial few strokes, while conventional sequential models have greater impacts.

Keywords: Sports Analytics · Attributions · Badminton Forecasting

1 Introduction

The research of applying deep neural networks to simulate patterns has broad applicability across various domains. Whether it involves investigating previous and future behaviors for sports analytics [3] or financial markets [10], these scenarios can be effectively framed as agent forecasting systems characterized by intricate interactions and essential decision-making processes. Automatically predicting agents' patterns is crucial for improving the effectiveness of decision-making and goal planning.

X. Wu et al. (Eds.): PAKDD 2025, LNAI 15875, pp. 444–456, 2025.
https://doi.org/10.1007/978-981-96-8295-9_33

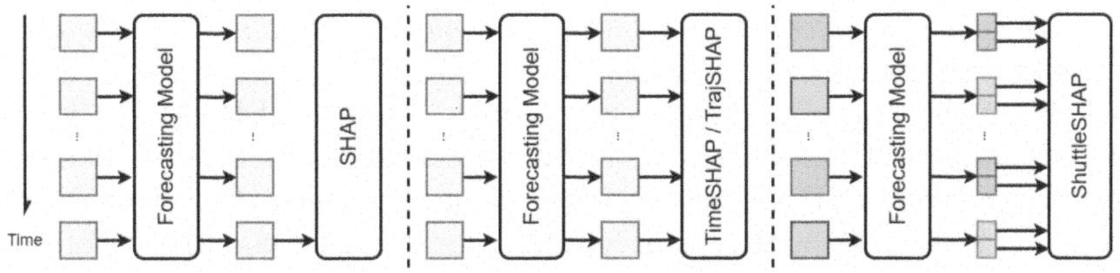

Fig. 1. Comparisons of SHAP [11], TimeSHAP [2] and TrajSHAP [13], and our proposed ShuttleSHAP. Each square denotes a component. Red and blue are different players in a turn-based sequence, where existing works (left and middle) view them as the same entities. Two components in a timestamp in the red part represent the shot type and area, where existing explainers fail to incorporate multiple components.

Over the past few years, turn-based sports analytics for action valuing [18,19], action detection [9], and strategy development [7,17] have attracted researchers due to the sophisticated characteristics that are composed of two players returning strokes alternatively to form a rally, and rapid changes in players' tactics within a match or even a rally. Therefore, we focus on turn-based sports and adopt badminton as an illustration example since there are no public stroke-level datasets in other turn-based sports [21]. Recently, [21] proposed the stroke forecasting task for badminton matches with a promising performance by the turn-based designs of not only rally but also player encoders. Despite the impressive results, it is challenging to understand how such turn-based forecasting networks make predictions, and the potential for performance improvements [20], and therefore they are often viewed as black-box models. Understanding the behaviors of black-box models is crucial in high-stakes decisions, especially in sports, given their multifaceted nature. These concerns notoriously prevent coaches and practitioners from the development of forecasting models to enhance decision-making in physical training and pre-match tactical investigation [15].

To mitigate the aforementioned shortcomings of existing black-box models and potential directions for further progress, we focus on model-agnostic and post-hoc explainers since our goal is to investigate the behaviors from the existing methods for turn-based analytics. A potential solution is to apply a game theory-based approach, Shapley values (SHAP) [11]. However, SHAP only measures instance-wise feature contributions, which is not applicable to sequential decision-making tasks. [2] proposed TimeSHAP to explain the predictions on fraud detection tasks, and [13] extended SHAP to the trajectory prediction task (named TrajSHAP for short). Despite the progress, the temporal-wise explainer is insufficient for turn-based sports analytics due to the compositions of forming a sequence (i.e., rally) alternatively by two players instead of typical sequences with the same targets. That is, TimeSHAP and TrajSHAP fail to capture players' attributions within a rally as illustrated in Fig. 1. Furthermore, there are multiple outputs of each stroke (i.e., shot types and area coordinates); adopting existing explainers suffers from the single-performance quantification of each prediction of the model. The question, therefore, naturally arises: *Is there a model-*

agnostic explainer that can further attribute players' contributions in terms of multifaceted performance?

In light of the above challenges, we propose ShuttleSHAP, a turn-based feature attribution framework, to analyze forecasting models in badminton. Specifically, ShuttleSHAP incorporates Shapley values [11,14] to develop a model-agnostic feature attribution method for turn-based stroke forecasting models. Moreover, the player influence attribution is proposed by aggregating the contributions of players in an alternative-playing sequence to quantify how well the model leverages player information for forecasting future behaviors. In this manner, ShuttleSHAP is able to measure the contributions not only from the temporal perspective (i.e., *how do the initial few strokes, specifically the ones from the past, influence the predictions*) but also from the player perspective (i.e., *how do players influence the predictions*) in terms of multi-faceted performance (i.e., *the performance of shot types and area coordinates*).

By leveraging our analysis of the representative state-of-the-art turn-based forecasting model ShuttleNet [21] and the best-performing sequential model TF [5] observed in [21], we unearthed that 1) Turn-based behaviors do play a critical role in both models in terms of the performance of shot types and area coordinates; 2) Past strokes contribute insignificantly to the area performance and shot type performance for ShuttleNet, while TF without turn-based designs, which refers not explicitly consider player styles in the model [21], predicts future behaviors based on past strokes.

In brief, our contributions are three-fold:

- We address, for the first time, feature attribution for the turn-based forecasting problem to break new ground for researchers and domain experts to investigate the actual cues about the model behavior. ShuttleSHAP is a model-agnostic explainer that can be applied to different forecasting models and can be extended to other turn-based sports (e.g., tennis).
- ShuttleSHAP reinforces the capability to quantify the contributions of not only the temporal aspect but also the player aspect. Moreover, it supports multi-output performance attribution for sequential prediction tasks.
- We uncovered that the turn-based model does not significantly utilize past (i.e., the first few) strokes for area predictions. In addition, player interactions have important cues for reasoning future strokes for both approaches. We also study local analysis from real-world match scenarios using our plugin deployed in the systems.

2 Preliminaries

2.1 The Stroke Forecasting Task

Problem Definition. A turn-based rally in badminton is denoted as $R = [r_1, r_2, \cdots, r_{|R|}]$ consisting of multiple strokes, where $r_i = (p_i, s_i, a_i)$ is the i-th stroke composed of one of two players returning alternatively (A denotes the served player and B otherwise, e.g., $[p_1^A, p_2^B, \cdots, p_{|R|-1}^A, p_{|R|}^B]; p_i \in P)$[1],

[1] i is omitted to denote players for simplicity in the following.

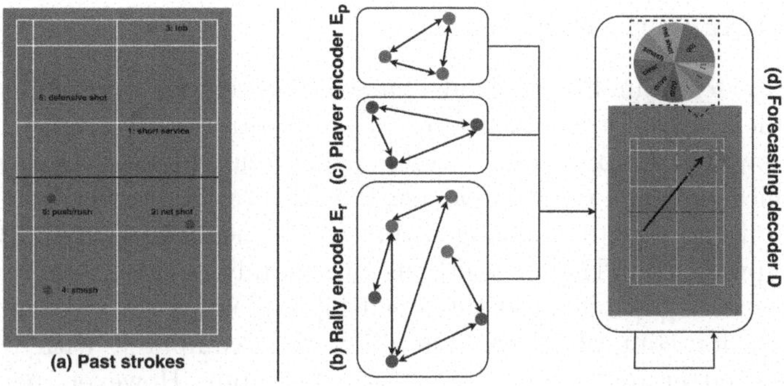

Fig. 2. General framework of existing turn-based forecasting approaches: Given the past τ strokes (a), the rally encoder (b) captures the rally situation as the rally embeddings, and the player encoder (c) captures the overall styles of both players. Afterwards, the forecasting decoder (d) predicts future strokes autoregressively. Note that conventional sequence prediction models (e.g., TF [5]) do not have the player encoder to explicitly consider players' styles.

the shot type $s_i \in S$, and the 2D area coordinates representing the destination locations $a_i \in \mathbb{R}^2$. The stroke forecasting problem is defined as forecasting the future strokes consisting of shot types and area coordinates $R_{\tau+1:|R|} = [(s_{\tau+1}, a_{\tau+1}), \cdots, (s_{|R|}, a_{|R|})]$ by giving the previous $\tau \in \{2, 4, 8\}$ strokes $R_{1:\tau} = [(s_1, a_1), \cdots, (s_\tau, a_\tau)]$ and the two players for each rally. Generally, simulating players' future behaviors requires models to learn past situations in a rally and the styles of both players.

Existing Turn-Based Forecasting Models. Figure 2 (b)-(d) summarizes the general framework that unifies state-of-the-art turn-based forecasting and most sequence forecasting methods. Given past strokes of the rally (a): (b) A rally encoder E_r that captures previous strokes in a rally $H_r = E_r(R_{1:\tau})$. (c) A player encoder E_p that learns the overall styles of both players separately as $H_p^A, H_p^B = E_p(R_{1:\tau}^A, R_{2:\tau}^B)$, where $R_{1:\tau}^A = [(p^A, s_1, a_1), (p^A, s_3, a_3), \cdots]$ and $R_{2:\tau}^B = [(p^B, s_2, a_2), (p^B, s_4, a_4), \cdots]$. (d)A forecasting decoder D that predicts future strokes based on the outputs of (b) and (c), and continues the process autoregressively $R_{\tau+1:|R|} = D(H_r, H_p^A, H_p^B)$. It is noted that each shot type, the area coordinates, and player information are first transformed to d-dimensional embeddings, respectively. The existing turn-based forecasting approach [21] adopts Transformer-based extractors for E_r and E_p, respectively, and then fuses them with a position-aware technique. As the stroke forecasting task remains unexplored, a best-performing sequential prediction model [5] is also employed following the original paper to investigate the behaviors compared with turn-based models, where the model does not contain a player encoder and the fusion technique; that is, $R_{\tau+1:|R|} = D(H_r)$.

2.2 Model-Agnostic Explainers

Model-agnostic explainable methods have pioneered interpretability in deep learning models due to their flexibility to plug into already trained models. A feature is viewed as important if shuffling its value changes the performance, since it implies that the model relies on the feature for predictions, while it is considered unimportant if unchanged due to the negligence of the feature for making the prediction. Earlier methods are mainly divided into two groups: perturbation-based [4] and gradient-based [6] explainers. The former perturbs input features for computing relevant differences compared with the original ones, and the latter introduces noises into the input. However, perturbation-based explainers violate the *sensitivity axiom*, while gradient-based explainers violate the *implementation invariance axiom*, i.e., the attributions are always identical for two functionally equivalent networks [16].

[11] introduced SHAP, which is the only method that fulfills these desirable axioms based on Shapley values [14], that maps an input $N = \{r_1, \cdots, r_n\}$ to a game, where players are the individual features r_i and the payout is the model behavior (e.g., prediction or performance). Formally, the contributions of the individual feature r_i can be quantified by its Shapley values $\phi(r_i)$:

$$\phi(r_i) = \sum_{S \subseteq N \setminus \{r_i\}} \frac{|S|!(|N| - 1 - |S|)!}{|N|!} (f(S \cup \{r_i\}) - f(S)), \tag{1}$$

where S is a subset of features excluding the i-th feature and f represents the model behavior. Our ShuttleSHAP not only preserves these axioms on top of SHAP, but also extends it to support multi-output multifaceted attributions.

3 The Proposed ShuttleSHAP

3.1 Prerequisite

What Model Behavior f is to be Attributed? We are interested in the improvement of understanding which shot types and area coordinates are used by models to perform well; that is, quantifying the contribution of each input stroke to the *performance* of a given model, which is also a common choice of observing model behaviors, e.g., [13]. Therefore, the behavior of f is defined as the relevant error of the output prediction: $f := \mathcal{L}(\hat{R}_{\tau+1:|R|}, R_{\tau+1:|R|})$, where $\mathcal{L}$ can be auxiliary losses. Following the setting of the original task, $\mathcal{L}$ includes cross-entropy loss (CE) $\mathcal{L}_{type}$ for shot type prediction and negative log-likelihood loss (NLL) $\mathcal{L}_{area}$ for area coordinate prediction:

$$\mathcal{L} = \mathcal{L}_{type} + \mathcal{L}_{area}; \mathcal{L}_{type} = CE(s_i, \hat{s}_i), \mathcal{L}_{area} = NLL(x_i, y_i, \hat{x}_i, \hat{y}_i). \tag{2}$$

In this manner, it is able to quantify the microscopic attributions with respect to multifaceted predictions and the macroscopic attribution of a stroke. To provide the macroscopic attribution, the scores from shot type and area coordinate predictions are averaged to view both outputs as having the same importance.

How to Design Dropped Features for S? Since the exact computation of Shapley values requires exponential complexity, a common solution for efficiency is to replace the dropped features with the reference value, e.g., set as zero [1]. However, setting the reference value for the dropped features to either random/mean values from the dataset or zero padding values introduces some unexpected signals to represent the process of a badminton rally (e.g., always returning the stroke to the top-left corner with a padded shot type).

To design a *reasonable* reference to represent the dropped features, the reference stroke is introduced to be a *defensive stroke returning to the middle of the half-court*, which offers a consistent single-reaction strategy similar to the shuttle serving machine and can defensively return most strokes back to the opponent for simulating their tactics: $(s_i, a_i) \leftarrow (s', a')$, where the i-th shot type is replaced with a defensive shot s', and the i-th area coordinates are replaced with the coordinates in the middle of a half-court a'. This modified version of Shapley values enables us to quantify the strokes that we aim to investigate relative to the consistent single-reaction strategy, and to apply this technique to different existing models. The proposed design of reference values is generic to be adapted to other turn-based sports (e.g., tennis); that is, setting the reference values as the defensive shot that is more likely to be returned.

It is noted that the justification for quantifying its contribution through the extent of predictive enhancement is not always valid, contrary to prevailing perceptions – [8] (Sect. 3.3 and Example 7) argued that evaluating causal potency by omitting the feature for prediction is fundamentally flawed and demonstrated that it should be randomized instead. Therefore, the ablative experiments of randomizing dropped features are excluded in this paper since it is inconclusive based on flawed modifications.

3.2 Past Strokes Attributions

The real-time decision-making of players based on the immediate situation is considered by their instant reactions for returning to the next stroke instead of the strokes far from now [12], which piques our curiosity about how the models take advantage of past information for future simulations. We define *past* strokes as *the first τ strokes*. With the foundation of ShuttleSHAP, quantifying the contribution of past strokes in terms of multi-performance is possible, while existing explainers fail to attribute multiple outputs in a timestamp.

To delve into how a model considers past strokes for future predictions, the shot type is replaced with the defensive shot, and the area coordinates are changed to the middle of a half-court from the second stroke:

$$R_{2:\tau} \leftarrow [(p_2, s', a'), \cdots, (p_\tau, s', a')]. \quad (3)$$

The reason to retain the first stroke is that the first stroke is the service shot, which has limited choices: a short service to the half-middle court or a long service to the backcourt, while it is impossible to be a defensive shot to the middle court due to the serving area rule. Therefore, the first stroke is contained

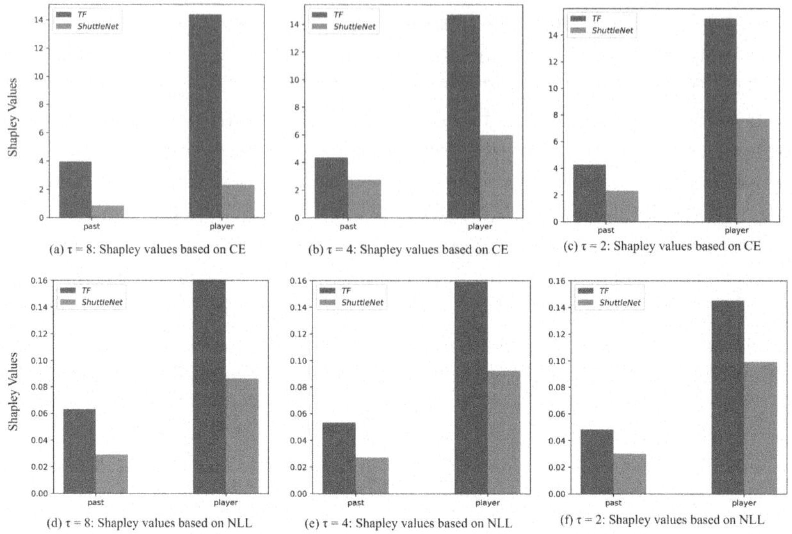

(a) τ = 8: Shapley values based on CE (b) τ = 4: Shapley values based on CE (c) τ = 2: Shapley values based on CE
(d) τ = 8: Shapley values based on NLL (e) τ = 4: Shapley values based on NLL (f) τ = 2: Shapley values based on NLL

Fig. 3. Shapley values in three different given strokes (τ) of ShuttleNet and TF on the stroke forecasting benchmark, where the x-axis denotes the attribution targets.

in a rally to not only reflect the proper starting stroke but also to remove noise from the simulation.

After obtaining the contribution of each stroke, aggregating these Shapley values over several strokes is essential to draw conclusions about the model behavior on the turn-based dataset. Formally, given the Shapley values of a stroke in a rally $\phi(r_i)$, the aggregated value of a rally ϕ_R^P is $\phi_R^P = \rho(\phi_{r_i \in R_{\tau+1:|R|}}(r_i))$, where ρ denotes the aggregation function.

The potential selection of ρ is to adopt the max operation to choose the most important timestamp for each sequence, similar to [13]; this is problematic when measuring multiple strokes with true causal influences throughout different scenarios. For instance, some transition strokes are returned to organize their tactics, which are still useful for providing explanations for simulation. Therefore, we opt for the average operation for ρ to represent the Shapley values of a rally in terms of the past strokes attributions. For the global aggregation (i.e., the entire dataset), the average operator is also incorporated. A higher non-negative value suggests that the previous strokes positively contribute to predicting future strokes, while a zero score is that none of the past strokes are used for predictions.

3.3 Player Influence Attributions

In addition to temporal attributions, one of the characteristics of badminton is multiple alter-returning agents in a sequence, which is different from the conventional sequential tasks composed of a single agent in an entire sequence. This is considered in the state-of-the-art model, ShuttleNet [21], which motivates us

Table 1. Retrained performance of TF and ShuttleNet on modified player information.

	$\tau = 8$			$\tau = 4$			$\tau = 2$		
Model	CE	MSE	MAE	CE	MSE	MAE	CE	MSE	MAE
TF	2.4398	1.7125	1.4247	2.4392	1.6792	1.3974	2.4925	1.7102	1.4116
w/o player	1.9951	2.8248	1.8980	1.9833	2.7121	1.8643	1.9869	2.8332	1.9008
Difference	0.4447	-1.1123	-0.4733	0.4559	-1.0329	-0.4669	0.5056	-1.1230	-0.4892
ShuttleNet	1.9854	1.5950	1.3993	2.0568	1.6101	1.3954	2.1107	1.6140	1.3960
w/o player	2.0001	2.7777	1.8840	1.9818	2.7549	1.8733	1.9856	2.4514	1.7931
Difference	-0.0147	-1.1123	-0.4733	0.0750	-1.1448	-0.4779	0.1251	-0.8374	-0.3971

to propose the player influence attribution technique aiming to examine to what extent both players contribute to the predictive performance.

For each target player, the same dropping strategy of the past stroke attributions is adopted to replace the shot type and the area coordinates with the defensive shot and the middle of the half-court, respectively. Furthermore, the target player of each stroke is changed to the other player, e.g., p^A is modified as p^B if the target player is A and vice versa. Formally, the replacements are:

$$R^A_{3:|R|} \leftarrow [(p^B, s'_3, a'_3), (p^B, s'_5, a'_5), \cdots]; R^B_{2:|R|} \leftarrow [(p^A, s'_2, a'_2), (p^A, s'_4, a'_4), \cdots]. \tag{4}$$

This imputation makes the ever-changing player act as a static serving machine against himself/herself, where the relevance error compared with the performance of the original one can then be used to represent the influence of the players. Note that Eq. (4) is executed independently to obtain the attributions of both players, and the first stroke retains the same due to the serving stroke.

To perform contribution aggregation, the attributions of multiple Shapley values are first averaged for each player of each stroke, and the attributions of two players of a stroke are then aggregated since we hypothesize that players consider that their own styles and those of their opponents have equal importance when returning the next stroke. Afterwards, the same procedure used for past stroke attributions is adopted to obtain the player influence attributions, considering the same importance over multiple strokes.

4 Experiments and Analysis

4.1 Experimental Setup

Benchmarks. As there is no existing forecasting work tackling other turn-based sports as discussed in §1, ShuttleSHAP is employed on a state-of-the-art turn-based model, ShuttleNet [21], and a best-performing conventional sequential model, TF [5], as introduced in §2.1 to incorporate the badminton dataset with stroke-level records from [21]. The dataset contains 31 professional men's singles

Table 2. Retrained performance of TF and ShuttleNet on different past strokes.

	$\tau = 8$			$\tau = 4$			$\tau = 2$		
Model	CE	MSE	MAE	CE	MSE	MAE	CE	MSE	MAE
TF	2.4398	1.7125	1.4247	2.4392	1.6792	1.3974	2.4925	1.7102	1.4116
w/o past	2.4813	1.6870	1.4143	2.4496	1.6358	1.3846	2.4274	1.7015	1.4132
Difference	-0.0415	0.0255	0.0104	-0.0104	0.0434	0.0128	0.0651	0.0087	-0.0016
ShuttleNet	1.9854	1.5950	1.3993	2.0568	1.6101	1.3954	2.1107	1.6140	1.3960
w/o past	1.9924	1.5949	1.3953	2.0697	1.5830	1.3939	2.1249	1.6216	1.3939
Difference	-0.0007	0.0001	0.0040	-0.0129	0.0271	0.0015	-0.0142	-0.0076	0.0021

and women's singles players (P) playing 75 matches, which includes 180 sets, 4,325 rallies, and 43,191 strokes. The set of shot types S consists of *clear, net shot, smash, push/rush, drop, drive, lob, defensive shot, short service,* and *long service*, which are defined by [18]. We follow the same splitting method to split the dataset for training and testing. The hyper-parameters of both models are implemented following the original paper.

Evaluation Metrics. Following [21], we adopt cross-entropy (CE) for the shot type measurement, and mean square error (MSE) and mean absolute error (MAE) are used for the area coordinates measurements. As discussed in §3.1, CE and NLL are used for reporting the contributions of models. Three different given strokes $\tau = 8, 4$, and 2 following the stroke forecasting task are conducted to investigate different behaviors in terms of model predictions.

4.2 Can ShuttleSHAP Identify Attributions?

Figure 3 summarizes the contributions of shot type performance (CE) and area performance (NLL) for the player influences (player) and for the removal of the first τ strokes (past). It is worth noting that the performance of all models (i.e., the first row of each model in Tables 1 and 2) is similar to the results in the original paper.

A1: Analysis on Player Influences. From the results of *player* influences, it is observed that the attributed values of players have significant impacts in terms of the shot type (CE) and area predictions (NLL) for TF and ShuttleNet compared with the *past* stroke attributions. In addition, ShuttleNet highly depends on the player styles when the number of given past strokes reduces (orange), while TF decreases the contributions for predicting future landing locations (blue). These behaviors suggest that turn-based behaviors indeed play a critical role in simulating future strokes including shot types and landing locations.

To further evaluate our findings, we retrain these models from scratch without considering the styles of players. This can be achieved by imputing one player and the corresponding strokes with Eq. (4) in the training set respectively, and then

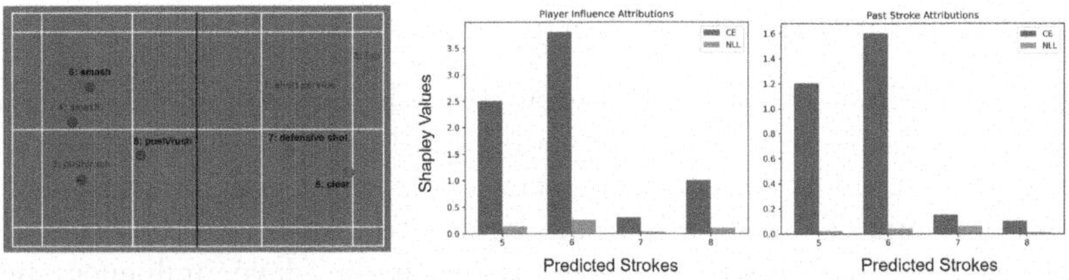

Fig. 4. Case study of applying ShuttleSHAP on ShuttleNet for a scenario where the styles of players are important for returning the next strokes. The first four strokes in grey are the given strokes. CE/NLL stands for the Shapley values based on the corresponding losses (Eq. 2). In the left figure, the landing locations in the left court are denoted in blue dots and those in the right court are denoted in red dots. In the middle and right figures, the Shapley values from five to eight predicted strokes are illustrated in terms of shot types and landing locations. (Color figure online)

average the performance results. Table 1 reports the original and the retrained results of two models with different past strokes (denoted as w/o player). Similar to the analysis, the results indicate that both models show substantial differences with and without player influences, suggesting that players' styles influence the predictions. Notably, the performance of shot type predictions for both models can be improved when neglecting the considerations of players' styles, which leaves a potential direction to tackle the multi-output turn-based sequences.

A2: Analysis on Past Stokes. The comparison of *player* and *past* stroke attribution in Fig. 3 reveals that for both models, shot types and landing locations make a relatively minor contribution compared with player influences. Moreover, the attributions of past strokes expose that ShuttleNet insignificantly takes advantage of past strokes for predicting future strokes compared with TF (a-c). This implies that sequential models without turn-based designs rely on implicit player influences as well as previous situations in the rally, while ShuttleNet is able to simulate future behaviors by primarily focusing on player styles.

This result draws our attention to whether the performance of ShuttleNet remains similar if the context of the past strokes is dropped as the contribution of each stroke is quantified to the performance of a given model (§3.1). Therefore, the same process is conducted to observe the relative difference between considering and removing (w/o past) past strokes presented in Table 2, where w/o past is implemented by injecting Eq. (3) into the training set. When discarding the information from past strokes, the performance of ShuttleNet remains stable in terms of all metrics and different given strokes, while the difference of TF impacts more substantially, although not as much, when neglecting player information. This demonstrates that past strokes are not considered significant in either model, whereas player influences play a vital role in simulating future strokes from encoding past strokes either implicitly or explicitly. Moreover, it also highlights that ShuttleSHAP is able to identify the contributions of each feature in terms of multifaceted performance and the performance yields consistent with our findings.

4.3 Usage of ShuttleSHAP: Local Analysis

The analysis of the contributions of each stroke in terms of shot type and area predictions can be viewed as understanding the behaviors of the model to help domain experts reduce the concerns when deploying the simulation model for pre-match training. To that end, a paradigm is to investigate the practicability by applying ShuttleSHAP to ShuttleNet with the player influence and past stroke attributions. Figure 4 demonstrates a real-world scenario from a high-ranking men's singles match where the first four strokes in grey text are given for the model to forecast the future strokes in black text in the left figure. The model predicts the fifth stroke with a clear to the backcourt after receiving the aggressive smash, and then the player on the right court launches another attack by returning another smash. Afterwards, the player on the left court has limited option to return and is forced to return the defensive shot to the middle court, and then loses the rally from failing to return back to the push/rush stroke. ShuttleNet equipped with our method attributes the fifth and sixth strokes significantly to player influences as well as past strokes in terms of shot types and landing locations, which shows that both players return strokes mainly based on their styles. This can be seen in the third and fourth strokes, where the former is a passive stroke with a lob and the latter is an aggressive stroke with a smash. Therefore, we can conclude that the left player plays more passively and the right player plays more aggressively, which is also confirmed by domain experts. With the help of ShuttleSHAP providing quantitative attributions, badminton experts and researchers are able to analyze and interpret forecasting results from the model in terms of microscopic cues for the specific rally.

5 Conclusions

This paper presents ShuttleSHAP, the first feature attribution method for the turn-based forecasting problem. Distinct from existing model-agnostic explainers, ShuttleSHAP analyzes the actual cues that are used by models to improve forecasting performance not only from the temporal but also from the player aspect. Meanwhile, it enables multi-output contributions that can quantify shot type as well as area attributions. Our proposed method quantifies causal influence in the context of the stroke forecasting task for benchmarking how well existing models make predictions based on such influence. Furthermore, we reveal that the turn-based model does not rely on past strokes for future simulation. As our proposed explainer was flexibly designed in a model-agnostic manner, we expect ShuttleSHAP to be useful to other applied scientists and domain experts in sports analytics, especially those whose goal is to perform turn-based and multi-output attributions. Since this paper identifies insights into existing turn-based models, future work could investigate more advanced methods to improve forecasting applications.

References

1. Ancona, M., Öztireli, C., Gross, M.H.: Explaining deep neural networks with a polynomial time algorithm for shapley value approximation. In: ICML. Proceedings of Machine Learning Research, vol. 97, pp. 272–281. PMLR (2019)
2. Bento, J., Saleiro, P., Cruz, A.F., Figueiredo, M.A.T., Bizarro, P.: Timeshap: explaining recurrent models through sequence perturbations. In: KDD, pp. 2565–2573. ACM (2021)
3. Chang, K., Wang, W., Peng, W.: Where will players move next? dynamic graphs and hierarchical fusion for movement forecasting in badminton. In: AAAI, pp. 6998–7005. AAAI Press (2023)
4. Feng, S., Wallace, E., Grissom II, A., Iyyer, M., Rodriguez, P., Boyd-Graber, J.: Pathologies of neural models make interpretations difficult. In: Proceedings of the 2018 Conference on Empirical Methods in Natural Language Processing, pp. 3719–3728. Association for Computational Linguistics (Oct-Nov 2018). https://doi.org/10.18653/v1/D18-1407
5. Giuliari, F., Hasan, I., Cristani, M., Galasso, F.: Transformer networks for trajectory forecasting. In: 25th International Conference on Pattern Recognition, pp. 10335–10342. IEEE (2020)
6. Han, X., Wallace, B.C., Tsvetkov, Y.: Explaining black box predictions and unveiling data artifacts through influence functions. In: ACL, pp. 5553–5563. Association for Computational Linguistics (2020)
7. Huang, L., Hsueh, N., Chien, Y., Wang, W., Wang, K., Peng, W.: A reinforcement learning badminton environment for simulating player tactics (student abstract). In: AAAI, pp. 16232–16233. AAAI Press (2023)
8. Janzing, D., Balduzzi, D., Grosse-Wentrup, M., Schölkopf, B.: Quantifying causal influences (2013)
9. Li, W., Liu, X., An, K., Qin, C., Cheng, Y.: Table tennis track detection based on temporal feature multiplexing network. Sensors **23**(3), 1726 (2023)
10. Lien, Y., Lin, Y., Wang, Y.: Uncertainty awareness for predicting noisy stock price movements. In: ECML/PKDD (6). Lecture Notes in Computer Science, vol. 13718, pp. 154–169. Springer (2022)
11. Lundberg, S.M., Lee, S.: A unified approach to interpreting model predictions. In: NIPS, pp. 4765–4774 (2017)
12. Macquet, A.C., Fleurance, P.: Naturalistic decision-making in expert badminton players. Ergonomics **50**(9), 1433–1450 (2007)
13. Makansi, O., et al.: You mostly walk alone: Analyzing feature attribution in trajectory prediction. In: ICLR, OpenReview.net (2022)
14. Shapley, L.S.: A value for n-person games. Classics in game theory **69** (1997)
15. Sun, X., Davis, J., Schulte, O., Liu, G.: Cracking the black box: distilling deep sports analytics. In: KDD '20: The 26th ACM SIGKDD Conference on Knowledge Discovery and Data Mining, pp. 3154–3162 (2020)
16. Sundararajan, M., Taly, A., Yan, Q.: Axiomatic attribution for deep networks. In: ICML, Proceedings of Machine Learning Research, vol. 70, pp. 3319–3328. PMLR (2017)
17. Wang, W.: Modeling turn-based sequences for player tactic applications in badminton matches. In: CIKM, pp. 5128–5131. ACM (2022)
18. Wang, W., Chan, T., Peng, W., Yang, H., Wang, C., Fan, Y.: How is the stroke? inferring shot influence in badminton matches via long short-term dependencies. ACM Trans. Intell. Syst. Technol. **14**(1), 7:1–7:22 (2023)

19. Wang, W., Chan, T., Yang, H., Wang, C., Fan, Y., Peng, W.: Exploring the long short-term dependencies to infer shot influence in badminton matches. In: ICDM, pp. 1397–1402. IEEE (2021)
20. Wang, W.Y., Du, W.W., Peng, W.C., Ik, T.U.: Benchmarking stroke forecasting with stroke-level badminton dataset (2024)
21. Wang, W., Shuai, H., Chang, K., Peng, W.: Shuttlenet: position-aware fusion of rally progress and player styles for stroke forecasting in badminton. In: AAAI, pp. 4219–4227. AAAI Press (2022)

Offline Map Matching Based on Localization Error Distribution Modeling

Ruilin Xu[1], Yuchen Song[2], Kaijie Li[2], Xitong Gao[2], Kejiang Ye[2], Fan Zhang[2], and Juanjuan Zhao[2(✉)]

[1] Southern University of Science and Technology, Shenzhen, China
xurl2024@mail.sustech.edu.cn

[2] Shenzhen Institute of Advanced Technology, Chinese Academy of Sciences, Shenzhen, China
{yc.song,kj.li,xt.gao,kj.ye,jj.zhao}@siat.ac.cn

Abstract. Offline map matching involves aligning historical trajectories of mobile objects, which may have positional errors, with digital maps. This is essential for applications in intelligent transportation systems (ITS), such as route analysis and traffic pattern mining. Existing methods have two main limitations: (i) they assume a uniform Localization Error Distribution (LED) across urban areas, neglecting environmental factors that lead to suboptimal path search ranges, and (ii) they struggle to efficiently handle local non-shortest paths and detours. To address these issues, we propose a novel offline map matching method for sparse trajectories, called LNSP, which integrates LED modeling and non-shortest path detection. Key innovations include: (i) leveraging public transit trajectories with fixed routes to model LED in finer detail across different city regions, optimizing path search ranges, and (ii) scoring paths using sub-region dependency LED and a sliding window, which reduces global map matching errors. Experimental results using real-world bus and taxi trajectory datasets demonstrate that the LNSP algorithm significantly outperforms existing methods in both efficiency and matching accuracy.

Keywords: GPS error distribution · Map matching · Hybrid trajectories

1 Introduction

Offline map matching (MM) aligns historical trajectory data with digital maps, correcting GPS errors from signal reflection and multipath interference. It is crucial for ITS, aiding in route planning and traffic pattern analysis. Trajectory sampling frequency is often reduced to save storage and communication costs. However, lower sampling rates increase ambiguity in route identification between GPS points. Current methods use dynamic programming algorithms based on local shortest paths for MM [1], but they face limitations due to poor LED modeling and inaccurate path scoring.

X. Wu et al. (Eds.): PAKDD 2025, LNAI 15875, pp. 457–470, 2025.
https://doi.org/10.1007/978-981-96-8295-9_34

1) LED modeling: LED affects how paths are searched, but current MM methods often assume LED is the same across all urban areas, ignoring differences caused by factors like mountains and buildings. To avoid missing the correct road, the search radius is often set wide. However, as the number of trajectory points increases (e.g., to hundreds), the search space becomes too large, significantly reducing matching efficiency.
2) Path scoring: Current methods primarily rely on the geographical distance between trajectory points and candidate path to get the matching score. However a closer between the location point and the candidate road does not necessarily indicate a higher match accuracy since the different LED. This leads to inaccuracies in path scoring, especially when trajectory points are sparse, or the vehicle does not follow the shortest path.

To address these limitations, this paper integrates bus trajectories, which have fixed routes, and proposes an offline MM method that accurately models LED and detects non-shortest paths. The main contributions include:

1) It uses bus trajectories with fixed routes to capture localization error patterns in fine-grained urban areas. To address the uneven coverage of bus trajectories, hierarchical spectral clustering is applied to group city grids, identifying sub-regions composed of geographically adjacent grids with similar LED.
2) It improves the robustness and efficiency of map matching by adjusting the search radius of candidate routes, scoring each route based on region-dependent LEDs, and connecting local sub-trajectories using overlapping sliding windows.
3) Experiments conducted on real trajectory dataset of buses and taxis in Shenzhen, China for one month show that our method significantly outperforms existing mainstream methods in both map matching efficiency and accuracy.

2 Related Work

Offline Map Matching is designed to precisely match historical and complete trajectory data with electronic maps. It is crucial for vehicle trajectory analysis, traffic pattern mining, and trajectory compression [2].

Some offline matching algorithms assume global point correlation of a trajectory based on the shortest path. Srivatsa et al. [3] identify candidate paths using the shortest distances from all trajectory points to eahc candidate path as the weight. Rahmani et al. [1] consider both path length and match with travel time in calculating path weight. [4–7] also simultaneously consider positions and time factors between adjacent GPS trajectory points, transforming the matching process into an optimization problem where the optimal solution corresponds to the matched roads. For example, in [4], candidate road segments are organized into a matching graph based on adjacent-time trajectory points, and the best matching path is derived from this graph, representing the matching result. Similarly, in [6], a voting mechanism, Hidden Markov Model, and A* algorithm are employed to solve the optimization model. Quddus et al. [8]

include distance weights and heading differences. Since the shortest path is not always followed in real scenarios [9], Zhu et al. [10] proposed a trajectory segmentation optimization model using dynamic programming to align trajectory and path curves. Chambers et al. [11] break down the global path into k local shortest paths, using the Fréchet distance between trajectory and path curves as the weight.

Existing methods use a larger search radius to tolerate positioning errors, which can include irrelevant roads and increase the matching burden [12].

3 Overview

3.1 Terms and Problem Definition

Table 1. Definitions of Terms

Term	Definition
GPS Point	A GPS point $z = (lon, lat, t)$ is the position of a vehicle, consisting of longitude, latitude, and timestamp.
Trajectory	A trajectory is an ordered sequence of GPS points $Z = (z_1, z_2, \ldots, z_{\lvert Z \rvert})$ collected by a moving vehicle, where z_1 is the start, $z_{\lvert Z \rvert}$ is the end, and $z_i.t \leq z_{i+1}.t$ for all i.
Road Network	A directed graph $\mathcal{G} = (\mathcal{V}, \mathcal{E})$, where $\mathcal{V}$ and $\mathcal{E}$ is the set of intersections and directed road segments. A segment $e \in \mathcal{E}$ is associated with an identifier id, speed limit v, length l, start point $start$, end point end, and intermediate points.
Candidate Road	A candidate road is a road segment within a specified radius of a GPS point, where the point may potentially match.
Candidate Matched Point	It is the closest position on a candidate road to a target GPS point, denoted as s. The road and the error is $s.c$ and $s.err$.
Path	A path between two GPS points p_{start} and p_{end} is a sequence of road segments $P_{p_{\text{start}} \rightarrow p_{\text{end}}} = e_1 \rightarrow e_2 \rightarrow \cdots \rightarrow e_n$, where e_1 and e_n correspond to the segments containing p_{start} and p_{end}, and $e_k.end = e_{k+1}.start$.

Problem Definition: Given a trajectory Z and a road network G, the goal is to estimate the actual travel path P that matches Z (Table 1).

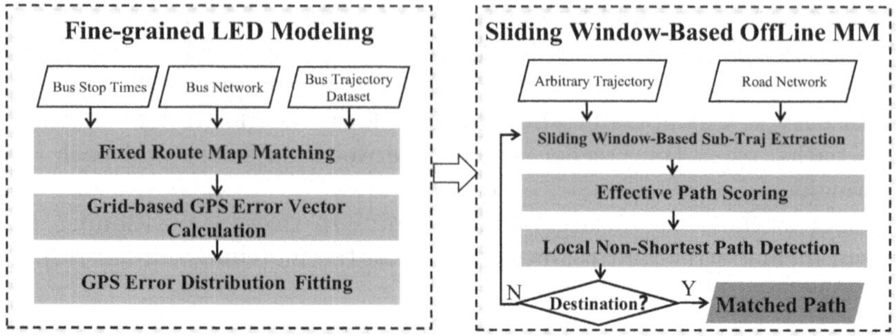

Fig. 1. LNSP Framework.

3.2 Framework

The LNSP framework, shown in Fig. 1, consists of two main components: Fine-grained LED Modeling and Sliding Window-Based Offline MM. Fine-grained LED Modeling integrates bus travel paths, maps historical trajectories to roads, and constructs a detailed LED function for urban areas. The Sliding Window-Based Offline MM enhances offline map matching efficiency and accuracy by adjusting the search radius for candidate roads, scoring paths using LED, and linking sub-trajectories with overlapping windows.

4 Methodology

4.1 Fine-Grained LED Modeling

This section leverages bus trajectories with fixed routes to capture the fine-grained LED. To improve efficiency and accuracy, we segment bus trajectories using the actual arrival and departure times of the bus at each station, narrowing the search scope.

The city area is divided into equal-sized grids, denoted as $\{g_1, g_2, \cdots, g_M\}$, M is the total number of grids. Bus trajectory points are mapped into these grids, and a GPS error histogram vector $\boldsymbol{V}_j$ is created for each grid g_j, where each element represents a probability distribution for a specific error interval. To address uneven GPS point distribution and sparse data in some grids, we use the spatial similarity of nearby grids. Graph message passing is applied to fill in missing data by averaging error histograms from adjacent grids. Then, spectral clustering is used to group neighboring grids with similar GPS error patterns into same sub-region.

The similarity of localization error distributions between two grids g_i and g_j is calculated as followings: For adjacent grids, we calculate similarity using a Gaussian kernel: $W_{ij} = \exp\left(-\frac{\|\boldsymbol{V}_i - \boldsymbol{V}_j\|^2}{2\sigma^2}\right)$, where $\boldsymbol{V}_i$ and $\boldsymbol{V}_j$ are the error histogram vectors for grids g_i and g_j, and σ is the Gaussian kernel's bandwidth parameter.

For non-adjacent grids, the similarity is set to 0 to keep the cells of same subregion connected. Therefore, We create K_g subregions $\{\Gamma_1, \Gamma_2, \ldots, \Gamma_{K_g}\}$. Each subregion consists of connected grid cells, and each grid is assigned a subregion identifier, stored in an array that maps grid indices to their respective subregions.

Finally, a LED function is fitted for each subregion g_j. Based on the vector $\boldsymbol{V}_j$, we select an appropriate probability distribution U_i (e.g., Gaussian, mixture of Gaussians, exponential, or log-normal) and fit it iteratively. The parameters are determined using Maximum Likelihood Estimation (MLE), and the fit's effectiveness is evaluated using the Akaike Information Criterion (AIC [13]). The cumulative distribution function as A_i. For these subregions that don't fit any common distributions, we use the error histogram vector to represent them.

4.2 Sliding Window-Based OffLine MM

The Basic Processing Flow. Given a trajectory $Z = (z_1, z_2, \ldots, z_n)$ and a fixed-length window of size W_{len}, the window starts at the first point (w_{left}) and ends at the W_{len}-th point (w_{right}), as shown in Fig. 2. For the local trajectory within this window, we first identify the grids where the starting point $z_{w_{\text{left}}}$ and ending point $z_{w_{\text{right}}}$ are located, along with their corresponding LED functions. GPS errors that exceed a certain threshold (e.g., 99% cumulative probability) define the search radius for candidate points. This gives us two sets of candidate points: - $S_{w_{\text{left}}} = (s^1_{w_{\text{left}}}, s^2_{w_{\text{left}}}, \ldots, s^m_{w_{\text{left}}})$ for the start points, and - $S_{w_{\text{right}}} = (s^1_{w_{\text{right}}}, s^2_{w_{\text{right}}}, \ldots, s^n_{w_{\text{right}}})$ for the end points. Next, Dijkstra's algorithm is used to find the shortest path $P_{s^i_{w_{\text{left}}} \to s^j_{w_{\text{right}}}}$ between each pair of start and end candidates, resulting in up to $m \times n$ possible path segments. For each shortest path, a fixed path matching method is applied, and the path score $P_{s^i_{w_{\text{left}}} \to s^j_{w_{\text{right}}}}.f$ is computed, as described in the next section. Finally, we select the top k paths with the highest scores.

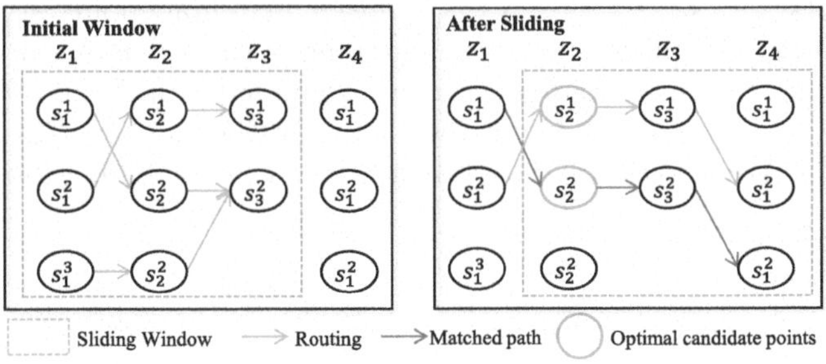

Fig. 2. Sliding Window Matching Process.

The window is then slid backward by a fixed length W_{len}, ensuring the overlap between windows is at least W_{olen} (as shown by the green window in Fig. 2). The values of w_{left} and w_{right} are updated accordingly. The starting point in the new window is now $z_{w_{left}}$, and the candidate points are defined as the points $P.s_{w_{left}}$ corresponding to $z_{w_{left}}$ in the retained paths from the previous window (illustrated by the blue-marked points in Fig. 2, with $W_{olen} = 2$). These candidate points are considered more optimal for positioning $z_{w_{left}}$ based on the previous window. Each candidate point $s^j_{w_{left}}$ retains its associated path $s^j_{w_{left}}.P$ from the previous window, as well as its predecessor point $s^j_{w_{left}}.pres$ and predecessor route $s^j_{w_{left}}.prer$.

Using the same method as the initial window, candidate points for the endpoint are identified, and the shortest paths are computed. From the new window onward, each path gets a score and an accumulated score $P.af$, which combines the current score with the previous window's score at the starting point (in the initial window, $P.af = P.f$). The accumulated score af for each path $P_{s^i_{w_{left}} \to s^j_{w_{right}}}$ in the new window is calculated as:

$$P_{s^i_{w_{left}} \to s^j_{w_{right}}}.af = \begin{cases} P_{s^i_{w_{left}} \to s^j_{w_{right}}}.f & w_{left} = 1 \\ P_{s^i_{w_{left}} \to s^j_{w_{right}}}.f + s^j_{w_{left}}.P.af & w_{left} > 1 \end{cases} \quad (1)$$

Finally, the top k paths with the highest wf values are retained in the window. The window continues to slide backward, performing the same calculations.

Once the window slides to the end of the trajectory and the calculations are finished (Fig. 2), each retained path has a cumulative score wf, representing the total weight of the path and its preceding routes. By backtracking from the endpoint of the path with the largest wf, using the preserved predecessor routes, the final matched path is identified.

Path Scoring Based on Sub-region LED. Existing methods for calculating path weights rely on the spatial similarity between the trajectory and the path, assuming a uniform error distribution across all regions. This paper refines the scoring method. After matching the shortest path P to the trajectory $Z = (z_{w_{left}}, z_{w_{left}+1}, \ldots, z_{w_{right}})$ and obtaining the set of matching points $S = (s_{w_{left}}, s_{w_{left}+1}, \ldots, s_{w_{right}})$, each matching point s_i corresponding to a positioning point z_i has a matching error value $s_i.err$. The probability $F_{sr_{s_i}}(s_i.err)$ of s_i matching z_i is calculated based on the sub-region sr_{s_i} where s_i is located and its corresponding LED function $U_{sr_{s_i}}$. The path weight score P is the sum of the probabilities of all matching points in S by $P.f = \sum_{i=w_{left}}^{w_{right}} F_{sr_{s_i}}(s_i.err)$.

Detection of Local Non-Shortest Paths. The sliding window method effectively handles global non-shortest path (NSP) driving, ensuring continuity in local shortest path matching and the correlation between consecutive trajectories. However, finding suitable window parameters for all trajectories is challenging, and shorter NSP segments can still occur in rare cases. For example, as shown in Fig. 3, a detour of less than 100 m due to route regulations can cause the shortest path matching algorithm to fail. This paper analyzes the data and explores the geometric and topological characteristics of urban roads. It

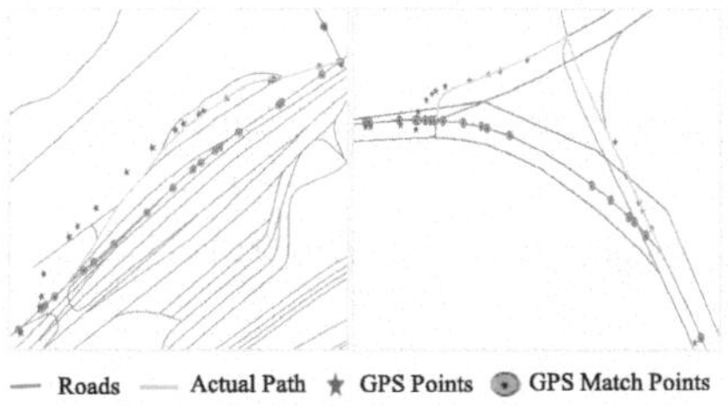

Fig. 3. Local Non-Shortest Path.

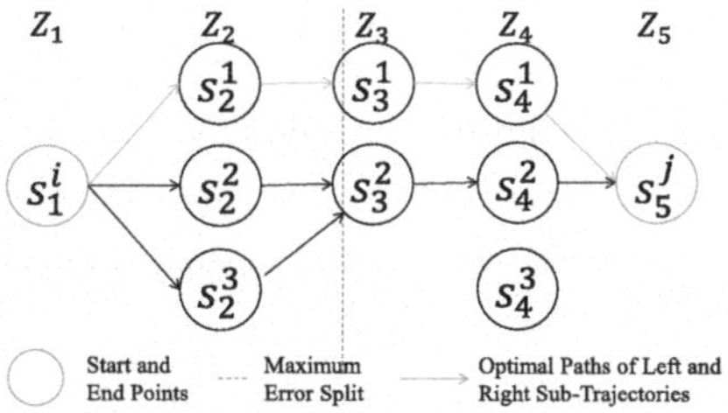

Fig. 4. Local NP Detection.

finds that when a vehicle deviates from the shortest path, it usually first moves away from the path and then returns. This incorrect matching is marked by a sequence of points with large errors, where the positioning points drift away and then gradually return, a common feature of detour trajectories.

Based on this characteristic, the paper proposes a method to detect local NSP segments. If a sequence of matching points contains a segment of $NSP.n$ consecutive points with error distances greater than $NSP.err$, where the maximum error exceeds $NSP.\max\_err$, the segment length exceeds $NSP.len$, and the error distribution significantly deviates, the segment is flagged as potentially containing a local NSP. The trajectory is then split at the point where the error exceeds $NSP.\max\_err$ (see Fig. 4). The left half of the trajectory is recursively analyzed for shortest path matching, producing a list of possible path sequences, $leftSequenceList$. The endpoints of each path in $leftSequenceList$ are used as candidates for the right half of the trajectory, which is recursively matched with the fixed endpoint of the original sequence, resulting in $rightSequenceList$. The path with the highest score is selected and combined with the best path from $leftSequenceList$ to form a new matching path, $nseq$. If $nseq$ has a higher score than the original sequence, seq is replaced with $nseq$.

This recursive method enhances performance by splitting the trajectory at the maximum error point into two shortest-path segments. It handles multiple local non-shortest path segments and terminates when the split segment length is below a threshold $NSP.len$ or if the optimal match matches the original sequence, avoiding excessive splits.

5 Experiment

5.1 Experiment Settings

Dataset. The used datasets include four types in Shenzhen, China: road networks, public transit routes, bus and taxi trajectories. In addition to analyzing GPS error distribution, the bus trajectory data is employed for testing, which assumes the paths are unknown.

Road Network Data: It is sourced from OpenStreetMap, focusing on the high-density road network in Futian District. The specific attributes of the road network are listed in Table 2.

Table 2. Basic Attributes of Target Road Network

Attribute	Value	Attribute	Value	Attribute	Value
Node Number	2813	Shortest Road	4.5 m	Road Number	6892
Road Degree	2.45	Road Length	6.9 m	Longest Road Length	2910 m

Bus Lines: Each bus line is made up of a series of stops and the road segments it travels. These segments were collected by Shenzhen Transportation Department using high-precision GPS devices.

Taxi Trajectories: It spans from February 1 to February 28, 2023. The taxis' onboard devices record location every 5 to 30 s, including latitude, longitude, time, speed, direction, and passenger status. The location sequence is divided into individual trips (trajectories) by detecting transitions between passenger-carrying and non-carrying states. To obtain the actual path as the map matching ground truth for each trajectory, we selected only high-sampling-rate trajectories (with intervals of less than 3 s per point), ensuring good matching results even with conventional methods. The results from dynamic programming algorithm-based map matching are used as the ground truth. For validation, we downsampled these trajectories to create low-sampling-rate trajectories (ranging from 5 s to 2 min) by randomly discarding some points. The closer the matching results under low sampling rates are to those under high sampling rates, the better the matching performance.

Bus Trajectories: We use bus trajectories for the same period as the taxi data. Bus GPS devices record location every 5 s and provide additional details, such as bus line and direction (inbound or outbound).

Metrics. Matching accuracy and efficiency are critical performance metrics for map matching, and they are widely used in existing researches [1,3,8]. We evaluate the results using three metrics: Accuracy of Point Matching (Acc), Path Matching Precision (Prc), Average Matching Time (Mt).

$$\text{Acc} = \frac{\text{Correct matches}}{\text{Total points}}, \quad \text{Prc} = \frac{\text{Length of correct paths}}{\text{Total path length}}, \quad \text{Mt} = \frac{\text{Matched trajectories}}{\text{Total matching time}}$$

Baselines. We use three widely adopted Offline map matching algorithms as benchmarks: SPT [1], stMM [8], and k-spMM [11].

SPT [1]: is a global shortest-path based matching algorithm that considers path-trajectory length similarity, the Fréchet distance between trajectory and path curves, and the matching of global path length with travel time.

stMM [8]: it adds two weights to the traditional curve distance and azimuth difference: the distance between the shortest path and vehicle trajectory, and the heading difference between the path and trajectory.

k-spMM [11]: is a local shortest-path matching algorithm that splits the global path into K local shortest paths using dynamic programming, with the Fréchet distance as the weight, aiming to minimize the overall matching cost.

Parameter Settings. We set all parameters of sliding Window based on test result. The best setting is: Minimum trajectory length within the window $W_{len} = 600\,\text{m}$, minimum overlap length between adjacent sliding windows $W_{olen} = 300\,\text{m}$, and the number of paths retained within a single window $k = 4$.

5.2 Analysis of Experimental Results

Region Division and Localization Error Distribution. We divided the Futian District into 100-meter square grids. Spectral clustering is used to group the cells based on LED, resulting in 179 sub-regions. The average errors are shown in Table 3. Most sub-regions have average errors between 0 and 10, with only a few having higher errors. This demonstrates that dynamically adjusting the candidate road search radius based on error distributions is effective, avoiding the need for a large fixed global search radius and validating our approach.

For the 179 divided regions, Table 4 shows the LED functions and Fig. 5 provides some examples. Among these, 54.1% of the sub-regions' LED fit Gaussian or mixture of Gaussians distributions, while a small number fit log-normal or exponential distributions. For 24.5% of the sub-regions that cannot be effectively fitted, GPS histogram vectors are used instead.

Table 3. Sub-Regions with Different Error Ranges

Average Error(m)	[0, 5)	[5, 10)	[10, 15)	[15, 20)	[20, 25)	[25, 100)
Number of Sub-Regions	42	93	24	12	10	8

Table 4. Sub-Regions with Different LED Functions

DFF	Number	DFF	Number
Gaussian Distribution	64	Log-Normal Distribution	14
Gaussian Mixture Distribution	33	Exponential Distribution	24
Not-Fitting	44		

Road Network Matching Accuracy. All methods are evaluated on both bus and taxi trajectories under various GPS sampling rates ranging from 10 to 120 s, as illustrated in Fig. 6 and Fig. 7.

The LNSP significantly outperforms the SPT [1] and stMM [8] algorithms. For example, the accuracy (ACC) for bus trajectories improves by 12.28% and

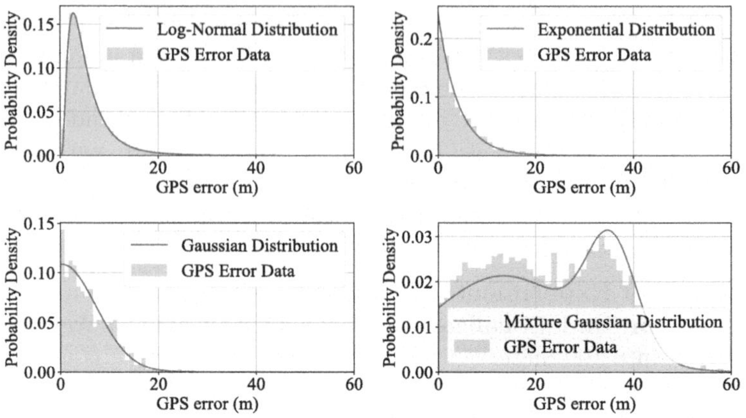

Fig. 5. Examples of Sub-region GPS Error Distribution Fitting.

16.9% compared to SPT and stMM, with precision metrics increasing by 13.33% and 17.65% at a 5-s sampling rate. This advantage is due to the fact that vehicle trajectories do not always follow the global shortest paths. Compared to k-spMM [11], LNSP also performs better at high sampling rates. At a 5-s average sampling rate, accuracy and precision improve by 2.09% and 1.86%, respectively. k-spMM struggles in areas with significant GPS errors, where the candidate paths may not be unique, and trajectory curves can deviate greatly from the actual path, leading to incorrect matches.

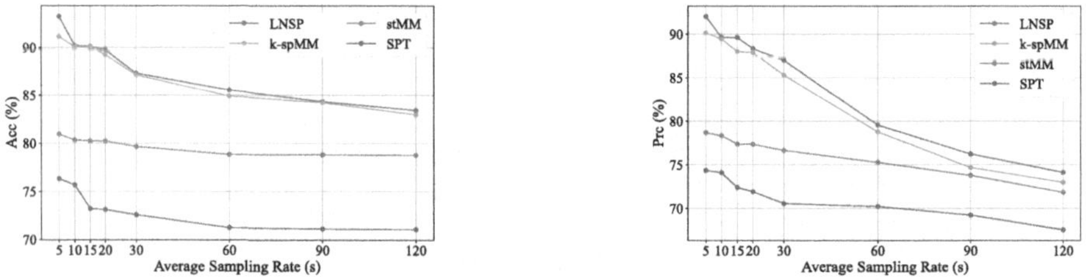

(a) (b)

Fig. 6. Overall Performance for Bus Trajectories.

As shown in Fig. 8, the trajectory in the right half is located in an area with significant errors, resulting in a smaller Fréchet distance between the incorrectly matched path and the actual trajectory. Unlike Fréchet distance, our method LNSP more accurately describes the likelihood of errors between the trajectory and the matching path by calculating the total weight of the LED probabilities for each matching point's subregion along the path. In the case shown in Fig. 8, the LNSP model improves the matching by recognizing the error distribution

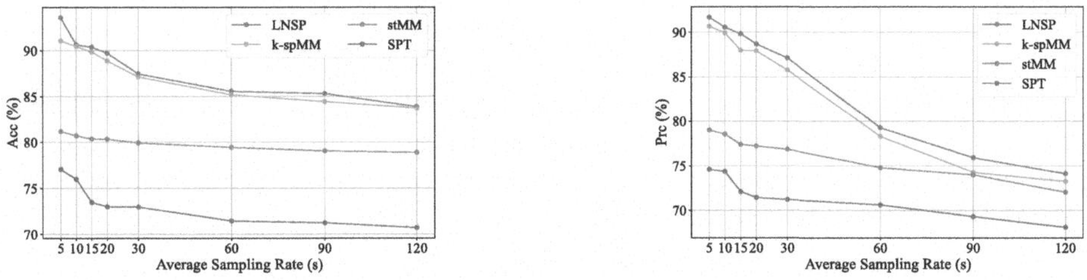

(a) (b)

Fig. 7. Overall Performance for Taxi Trajectories.

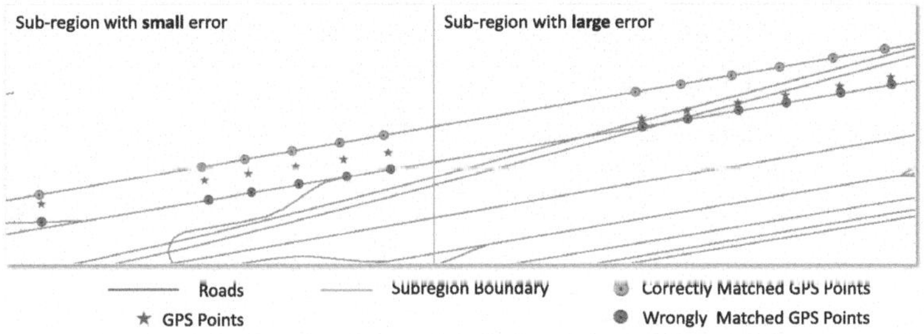

Fig. 8. Example of Incorrect Matching with the k-spMM Model When Encountering Consecutive Large Errors.

of the region, increasing the weights of the correct candidate points in both the left and right halves, thereby successfully matching the correct trajectory path. Additionally, the algorithm smooths the connections between the matching paths of each local trajectory segment using a sliding window, strengthening the relationship between local and global trajectories. As a result, the LNSP model demonstrates higher matching accuracy.

However, in the taxi trajectory matching results presented in Fig. 7, the matching accuracy of LNSP shows slight degradation at medium to low sampling rates. This is due to the more complex road environments for taxis compared to buses, and the GPS error distribution for some taxi data, especially those operating in non-bus routes, is estimated through interpolation. Despite this, LNSP still maintains an advantage at high sampling rates.

Matching Efficiency. We conducted experiments to measure the matching time of different methods on the same computer configuration. The results in Table 5 show that the LNSP model has slightly lower matching efficiency than SPT [1] and stMM [8] due to its more complex local non-shortest path detection and correction, which require additional computational resources (further explained in the next subsection). However, LNSP outperforms the k-spMM model, which uses time-consuming dynamic programming to search for optimal shortest path combinations in complex road networks [14]. In contrast, LNSP first matches the shortest path and then detects errors and segments potential non-shortest paths based on maximum error points, leveraging the geometric and topological features of the road network. This results in faster and more efficient trajectory segmentation, reducing matching time. Additionally, LNSP dynamically adjusts the candidate search range based on error distribution, reducing the number of candidate points compared to k-spMM's fixed search radius.

Table 5. Comparison of Model Matching Time Metric (Mt)

Model	Mt on Bus (ms)	Mt on Taxi (ms)
StMM	116	137
SPT	128	125
LNSP	135	168
k-spMM	262	272

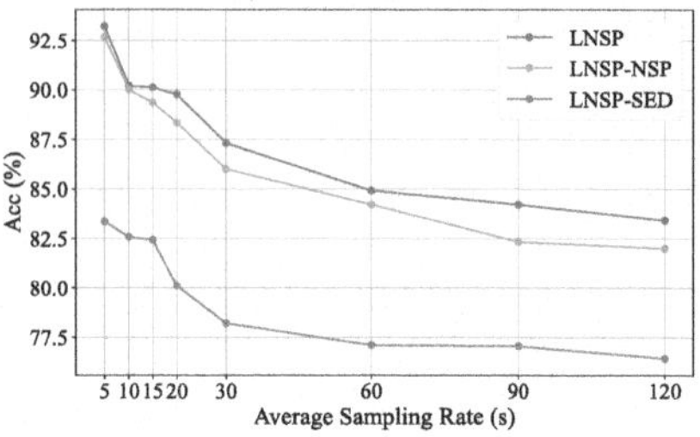

(a) ACC of bus trajectories

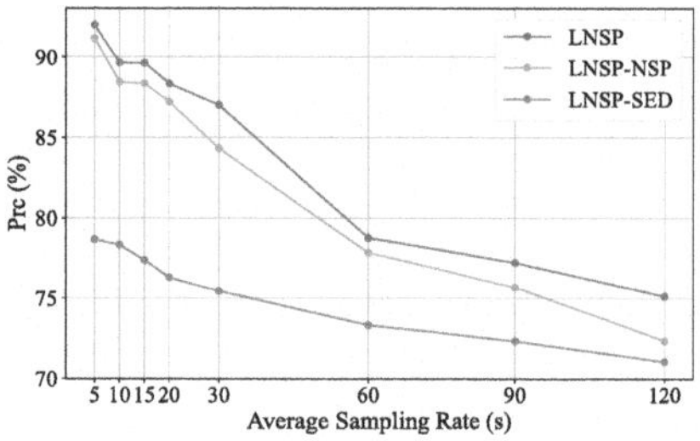

(b) Prc of bus trajectories

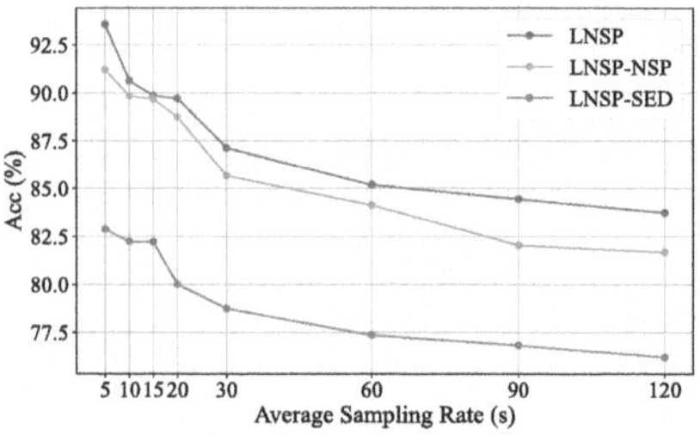

(c) ACC of taxi trajectories

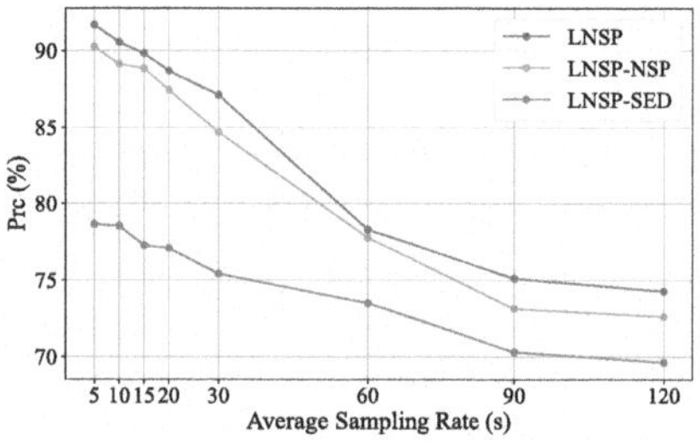

(d) Prc of taxi trajectories

Fig. 9. Acc & Prc of Ablation Models for Taxi/Bus Trajectories.

Ablation Study. To further analyze the contributions of the sub-region LED-based candidate point search radius setting and path weight scoring (SED) module, as well as the local non-shortest path detection and correction (NSP) module, we conducted experiments by separately removing these modules. The two models LNSP−NSP (with the SED module removed) and LNSP−SED (with the NSP module removed) were compared. The matching accuracy metrics are illustrated in Fig. 9a and Fig. 9c. The removal of the local non-shortest path detection leads to a significant decrease in Acc and Prc metrics, indicating that, in the GPS trajectory dataset used in this experiment, deviations from the shortest path are common in local paths. The NSP module's ability to detect and correct these deviations confirms the validity of the approach.

6 Conclusion

This paper proposes LNSP, a novel offline map matching method that effectively addresses key limitations in handling localization errors and path scoring for sparse trajectory data. By leveraging bus trajectories with fixed routes, LNSP provides a more accurate, region-specific model for localization error distribution (LED), which enhances the efficiency of the path search process. Additionally, the use of a sliding window technique and region-dependent LED scoring significantly improves matching accuracy, particularly in urban environments with varying geographical factors. Experimental results using real-world bus and taxi datasets from Shenzhen show that LNSP outperforms existing methods in both efficiency and accuracy, offering a robust solution for intelligent transportation systems.

Acknowledgments. This work is supported by Guangdong Basic and Applied Basic Research Foundation (No. 2023B1515130002), Shenzhen Industrial Application Projects of undertaking the National key R & D Program of China (No. CJGJZD20210408091600002), Consultation Research Projects of Guangdong Institute of Engineering Science and Technology Development Strategy (No. 2024-GD-5-02).

References

1. Rahmani, M., Koutsopoulos, H.N.: Path inference from sparse floating car data for urban networks. Transp. Res. Part C Emerg. Technol. **30**, 41–54 (2013). https://doi.org/10.1016/j.trc.2013.02.002
2. Zuo, Y.M., Lin, X.L., Ma, S., Jiang, J.H.: Road network aware online trajectory compression. Ruan Jian Xue Bao/J. Softw. **29**, 734–755 (2018). https://doi.org/10.13328/j.cnki.jos.005438
3. Srivatsa, M., Ganti, R.K., Wang, J., Kolar, V.: Map matching: facts and myths. In: Proceedings of the 21st ACM SIGSPATIAL (2013). https://api.semanticscholar.org/CorpusID:16066643
4. Lou, Y., Zhang, C., Zheng, Y., Xie, X., Wang, W., Huang, Y.: Map-matching for low-sampling-rate GPS trajectories. In: Proceedings of the 17th ACM SIGSPATIAL International Conference on Advances in Geographic Information Systems, pp. 352–361 (2009)

5. Newson, P., Krumm, J.: Hidden Markov map matching through noise and sparseness. In: Proceedings of the 17th ACM SIGSPATIAL International Conference on Advances in Geographic Information Systems, pp. 336–343 (2009)
6. Giovannini, L.: A novel map-matching procedure for low-sampling GPS data with applications to traffic flow analysis (2011)
7. Zheng, K., Zheng, Y., Xie, X., Zhou, X.: Reducing uncertainty of low-sampling-rate trajectories. In: 2012 IEEE 28th International Conference on Data Engineering, pp. 1144–1155. IEEE (2012)
8. Quddus, M., Washington, S.: Shortest path and vehicle trajectory aided map-matching for low frequency GPS data. Transp. Res. Part C Emerg. Technol. **55**, 328–339 (2015). https://doi.org/10.1016/j.trc.2015.02.017, engineering and Applied Sciences Optimization (OPT-i) - Professor Matthew G. Karlaftis Memorial Issue
9. Zhu, S., Levinson, D.: Do people use the shortest path? An empirical test of Wardrop's first principle. PLOS ONE **10**(8), 1–18 (2015). https://doi.org/10.1371/journal.pone.0134322
10. Zhu, L., Gonder, J., Holden, J.: Trajectory segmentation map-matching approach for large-scale, high-resolution GPS data. Transp. Res. Rec. **2645**(1), 67–75 (2017). https://doi.org/10.3141/2645-08. Publisher Copyright: © 2017, SAGE Publications Ltd. All rights reserved
11. Chambers, E., Fasy, B.T., Wang, Y., Wenk, C.: Map-matching using shortest paths. In: Proceedings of the 3rd International Workshop on Interactive and Spatial Computing, IWISC 2018, pp. 44–51. Association for Computing Machinery, New York (2018). https://doi.org/10.1145/3191801.3191812
12. Kubicka, M., Cela, A., Mounier, H., Niculescu, S.I.: Comparative study and application-oriented classification of vehicular map-matching methods. IEEE Intell. Transp. Syst. Mag. **10**(2), 150–166 (2018). https://doi.org/10.1109/MITS.2018.2806630
13. Akaike, H.: A new look at the statistical model identification. IEEE Trans. Autom. Control **19**(6), 716–723 (1974). https://doi.org/10.1109/TAC.1974.1100705
14. Chao, P., Xu, Y., Hua, W., Zhou, X.: A survey on map-matching algorithms. ArXiv **abs/1910.13065** (2019). https://api.semanticscholar.org/CorpusID:204949888

Author Index

X. Wu et al. (Eds.): PAKDD 2025, LNAI 15875, pp. 471–473, 2025.
https://doi.org/10.1007/978-981-96-8295-9

The manufacturer's authorised representative in the EU is Springer Nature Customer Service Centre GmbH, Europaplatz 3, 69115 Heidelberg, Germany. If you have any concerns regarding our products, please contact ProductSafety@springernature.com

Printed and bound by CPI Group (UK) Ltd, Croydon, CR0 4YY
01/10/2026
02225908-0018